Evolution of Sanitation and Wastewater Technologies through the Centuries

Evolution of Sanitation and Wastewater Technologies through the Centuries

Edited by

A. N. Angelakis and J. B. Rose

IWA PUBLISHING

Published by **IWA Publishing**
Alliance House
12 Caxton Street
London SW1H 0QS, UK
Telephone: +44 (0)20 7654 5500
Fax: +44 (0)20 7654 5555
Email: publications@iwap.co.uk
Web: www.iwapublishing.com

First published 2014

British Library Cataloguing in Publication Data
A CIP catalogue record for this book is available from the British Library

ISBN 9781780404844 (Paperback)
ISBN 9781780404851 (eBook)

Contents

List of reviewers

Dr. Mohammad Ebrahim Azari Najaf Abad, KU Leuven, Hydraulics Division, Kasteelpark Arenberg 40, Leuven, 3001, Belgium, mohammad.azari@student.kuleuven.be

Prof. (Em.) Stylianos Alexiou, Univ. of Crete, Iraklion, Hellas.

Dr. Andreas N. Angelakis, Institute of Iraklion, National Foundation for Agricultural Research (N.AG. RE.F.), 71307 Iraklion, Hellas, info@a-angelakis.gr

Dr. Akissa Bahri, Coordinator, African Water Facility, African Development Bank, 13, rue du Ghana, B.P. 323–1002 Tunis Belvédère, Tunisia, A.BAHRI@AFDB.ORG.

Prof. Constantin Canavas, Hamburg University of Applied Sciences, Faculty of Life Sciences, D-21033 Hamburg, GERMANY, costas.canavas@phl.uoc.gr

Dr. Manos Chalkiadakis, University of Oxford, St Antony's College, Oxford, UK, manoschalkiadakis@yahoo.gr

Prof. Giovanni De Feo, University of Salerno, Department of Civil Engineering, via Ponte don Melillo, 1–84084 Fisciano (SA), Italy, g.defeo@unisa.it

Prof. Walter Dragoni, Dipartimento di Scienze della Terra, Università di Perugia, Piazza Università 1-06123 – Perugia, Italy, wd2698@yahoo.com

Prof. Pengfei Du, School of Environment, Tsinghua University, Beijing 100084, P.R. China, dupf@tsinghua.edu.cn

Prof. Angela F Danil de Namor, Department of Chemistry, University of Surrey, Guildford, Surrey GU2 7XH, UK, a.danil-de-namor@surrey.ac.uk

Dr. Giorgos Facorellis, Department of Antiquities and Works of Art Conservation, Technical Educational Institute of Athens, Aghiou Spyridonos, 12210 Athens, Hellas, yfacorellis@yahoo.com

Dr. Maria Galanaki, Iraklion, Hellas, galanakimaria@hotmail.com

Prof. Hemda Garelick, Department of Natural Sciences, Middlesex University, The Burroughs, London W4 4BT, UK, H.Garelick@mdx.ac.uk

Prof. Fatma El Gohari, National Research Center, Water Pollution Research Department El behooth Str., Dokki, Cairo, Egypt, fgohary@hotmail.com

Prof. Yuri Gorokhovich, Lehman College, City University of New York, Department of Environmental, Geographical and Geological Sciences Gillet Hall 315, 250 Bedford Park Blvd. West, Bronx, NY 10468, USA, yuri.gorokhovich@lehman.cuny.edu

Prof. Mooyoung Han, Dept. of Civil and Environmental Eng., #35–302 Seoul National University, Shinrimdong Kwanakgu, Seoul, Korea, myhan@snu.ac.kr

Prof. Benoit Haut, Université Libre de Bruxelles, C.P. 165/67, 50 Av. F.D. Roosevelt, 1050 Bruxelles, Belgium, bhaut@ulb.ac.be

Prof. Francesc Hernandez-Sancho, Department of Applied Economics II. University of Valencia, Spain, Francesc.Hernandez@uv.es

Prof. Blanca Jiménez, Division of Water Sciences, Secretary of the International Hydrological Programme (IHP) UNESCO, Paris, France, bjimenezc@iingen.unam.mx

Prof. Ioannis Kalavrouziotis, Chair, Department of Environment and Natural Resources Management, University of Western Hellas, Agrinio, Hellas, ikalavru@cc.uoi.gr and ikalabro@yahoo.com

Dr. Eleni I. Kanetaki, 67 Dimocharous str., 11521 Athens, Hellas, eleni.kanetaki@gmail.com

Prof. George P. Karatzas, School of Environmental Engineering, Technical University of Crete, Chania, Hellas, karatzas@mred.tuc.gr

Prof. Nicholas Kathijotes, Department of Civil Engineering & Geomatics Cyprus University of Technology P. O. Box 50329, Limassol, 3603 Cyprus, nicholas.kathijotes@cut.ac.cy

Prof. Tapio Katko, Tampere University of Technology, Institute of Environmental Engineering and Biotechnology, Finland, tapio.katko@tut.fi

Prof. Albert Koenig, Department of Civil Engineering, The University of Hong Kong, Hong Kong SAR, P.R. China, kalbert@hkucc.hku.hk.

Prof. Kostas Komnitsas, School of Mineral Resources Engineering, Technical University of Crete, Polytechnioupolis, Akrotiri, 73100 Chania, Hellas, komni@mred.tuc.gr

Prof. Demetris Koutsoyiannis, Department of Water Resources, School of Civil Engineering, National Technical University of Athens, Zografos Campus, 1570 Athens, Hellas, dk@itia.ntua.gr

Prof. Jens A. Krasilnikoff, Department of History and Area Studies, Aarhus University, Aarhus, Demark, hisjk@hum.au.dk

Dr. Arun Kumar, Indian Institute of Technology Delhi Civil Engineering (Block 5, Room 204) New Delhi 110016, India, arunku@civil.iitd.ac.in

Dr. Pietro Laureano, IPOGEA,Vico Conservatorio s.n., 75100 Matera, Italy, ipogea@ipogea.org

Dr. Giusy Lofrano, Sanitary Environmental Engineering Division Department of Civil Engineering, University of Salerno, Fisciano, Salerno-Italy, giusylofrano@gmail.com

Prof. Yiannis Lolos, Department of history and Archaeology, University of Thessaly, Volos, Hellas, ylolos@otenet.gr

Prof. Gerasimos Lyberatos, School of Chemical Engineering, National Technical University of Athens, Zografos Campus, 1570 Athens, Hellas, lyberatos@chemeng.ntua.gr

Dr. Olfa Mahjoub, Institut National de Recherche en Genie Rural, Eaux et Forets (INRGREF), Tunis, Tunisia, olfama@gmail.com.

Prof. Nikos Manassis, Department of Water Resources, School of Civil Engineering, National Technical University of Athens, Zografos Campus, 1570 Athens, Hellas, nikos@itia.ntua.gr.

Prof. Laila Mandi, Environmental Sciences, University Cadi Ayyad, Marrakech, Morocco, mandi@uca.ma

Dr. Andrea Mangano, Water Resources Consultant, former Director of international Operation of Acea Spa, Rome, Via Ticino 7, 00198 Roma, Italy, andreamangano@yahoo.it.

Prof. Jure Margeta, University of Split, Faculty of Civil Engineering Architecture and Geodesy 21000 Split, Matice Hrvatske 15, Croatia, margeta@gradst.hr

Prof. Aradhana Mehra, Dept. of Environmental Geochemistry and Health Faculty of Education, Health and Sciences, University of Derby, Kedleston Road, Derby DE22 1GB, UK, a.mehra@derby.ac.uk

Prof. Gideon Oron, Environment Water Resources, The Institutes for Desert Research, Ben-Gurion University of The Negev, Israel, gidi@bgu.ac.il

Prof. Nikolaos Paranychianakis, Technical University of Crete, Chania, Hellas, niko.paranychianakis@enveng.tuc.gr

Prof. Paolo Roccaro, Department of Civil and Environmental Engineering, University of Catania, Viale Andrea Doria, 6, 95125 Catania, Italy, proccaro@dica.unict.it.

Dr. Apostolos Roubelakis, Dept of Cardiothoracic Surgery, Southampton University Hospitals, Southampton SO16 6YD, UK, roube@hotmail.com

Prof. Kalliopi A. Roubelakis Angelakis, Dept. of Biology, University of Crete, Iraklion, Hellas, poproube@biology.uoc.gr

Prof. Joan B. Rose, Dept. of Fisheries and Wildlife, Michigan State University, East Lansing, MI 48824, USA, rosejo@msu.edu

Prof. Miquel Salgot, Soil Science Laboratory, Facultat de Farmàcia, Universitat de Barcelona, Barcelona, Spain, salgot@ub.edu

Prof. Anna I. Strataridaki, Department of History and Education, University of Crete, Rethymnon, Hellas, astrat@edc.uoc.gr

Prof. Marco Tallini, Dept of Structural, Water and Soil Engineering, L'Aquila University, Italy, marco.tallini@univaq.it

Prof. Aldo Tamburrino, Universidad de Chile, Santiago, Chile, atamburr@ing.uchile.cl

Dr. Maria Tsibidou Avloniti, ΙΣΤ' ΕΠΚΑ, Thessaloniki, Hellas, mtsimbidou@yahoo.gr

Prof. Vassilios A. Tsihrintzis, Laboratory of Reclamation Works and Water Resources Management, National Technical University of Athens, Zografos Campus, 1570 Athens, Hellas, tsihrin@otenet.gr

Prof. Kostas Voudouris, Lab. of Engineering Geology & Hydrogeology, School of Geology, Aristotle University, 54124 Thessaloniki, Hellas, kvoudour@geo.auth.gr

Prof. Jiří Wanner, Department of Water Technology and Environmental Engineering, Institute of Chemical Technology, CZ-166 28, Technická 5, Prague 6, Czech Republic, jiri.wanner@vscht.cz

Dr. -Ing. Peter Wilderer, EASA Institute of Advanced Studies on Sustainability, Munich, Germany, peter.wilderer@mytum.de

Dr. Nikos Zarkadoulas, Department of Water Resources, School of Civil Engineering, National Technical University of Athens, Zografos Campus, 1570 Athens, Hellas, ps03014@yahoo.gr

Prof. Nina Zarrineh, Vrije Universiteit Brussel (VUB), Dept. Hydrology and Hydrological Engineering, Pleinlaan 2, B-1050 Brussels, Belgium, Nina.Zarrineh@vub.ac.be

Dr. Xiao Yun Zheng, Yunnan Academy of Social Sciences, Kunming, China, zhengxy68@163.com

Prof. Anastasios Zoumpoulis, Dept. of Chemistry, Aristotle University, 54124 Thessaloniki, Hellas, zoubouli@chem.auth.gr

List of authors

Dr. Seyed Ali Mahmoudian, Shahid Abdollah Zadeh St. Keshavarz Bulvd, National Water and Wastewater Engineering Company, Tehran, Iran, s.a.mahmoudian@gmail.com or seyedali@nww.co.ir

Dr. Andreas N. Angelakis, Institute of Iraklion, National Foundation for Agricultural Research (N.AG. RE.F.), 71307 Iraklion, Hellas, info@a-angelakis.gr

Mr. Georgios Antoniou, Deinokratous 73, 11521Athens, Hellas, antonioug@tee.gr

Dr. Nick Apostolidis, Global Development Leader, 201 Charlotte St, Brisbane QLD, 4000 Australia, napost13@gmail.com

Ms. Panagiota Azina, Sewerage Board of Paphos, POB 60474, Paphos, 8103 Cyprus, pk.azina@edu.cut.ac.cy

Ing. Ondrej Benes, VEOLIA VODA ČESKÁ REPUBLIKA, a.s., Prague, Czech Republic, ondrej.benes@veoliavoda.cz

Dr. Hanane Benqlilou, National Office for Electricity and Potable Water ONEE, International Institute for Water and Sanitation IEA, Bouregreg Treatment Plant, BP Rabat-Chellah, 10002 Rabat, Morocco, hbenqlilou@iea.ma

Prof. Semir Bensaid, National Office for Electricity and Potable Water ONEE, International Institute for Water and Sanitation IEA, Bouregreg Treatment Plant, BP Rabat-Chellah, 10002 Rabat, Morocco, sbensaid@iea.ma

Dr. Lamprini E. Chioti, Depart. of Archaeology and History, University of Athens, Athens, Chelidoreon 14, 14564 Athens, Hellas, chioti_lambrini@hotmail.com

Dr. Efstathios D. Chiotis, Chelidoreon 14, Athens 14564, Hellas, former Director at the Institute of Geology and Mineral Exploration, Athens, Hellas, echiotis@otenet.gr

Dr. Manolis G. Dialynas, DIALYNAS SA, Environ. Techn., 71304, Iraklion, Hellas, md@dialynas.com

Dr. Renato Drusiani, FederUtility, Rome, Italy, acqua@federutility.it

Prof. Giovanni De Feo, University of Salerno, Department of Industrial Engineering, via Ponte don Melillo, 1–84084 Fisciano (SA), Italy, g.defeo@unisa.it.

Dr. Sabino De Gisi, ENEA, Water Resource Management Laboratory (UTVALAMB-IDR), via Martiri di Monte Sole 4, 40129 (BO), Italy, sabino.degisi@enea.it

Dr. Santino Di Berardino, Laboratório Nacional de Energia e Geologia, I.P., Portugal, santino.diberardino@lneg.pt

Prof. Mohammed El Faïz, Cadi Ayad University Marrakech, Morocco, menaraagdal@gmail.com.

Prof. Fatma El Gohari, National Research Center, Water Pollution Research Department El behooth Str., Dokki, Cairo, Egypt, fgohary@hotmail.com

Dr. Hilal Franz Fardin, Paris 8 University, Ladyss (UFR 7533 – CNRS), French Institute of Pondicherry (Umifre 21 – CNRS/MAEE) Social Sciences Dept., France, franz.fardin@ifpindia.org.

Dr. Maria Galanaki, Iraklion, Hellas, galanakimaria@hotmail.com

Prof. Emmanuele Gautier-Costard, Geography Department, University Paris 8 and Laboratory of Physical Geography (CNRS/UMR 8591), 1 place Aristide Briand, 92195, Meudon Cedex, France, Emmanuele.Gautier@cnrs-bellevue.fr

Prof. Marwan Haddad, Civil Engineering Department, Water and Environmental Studies Institute (WESI), An-Najah National University, Nablus, Palestine, haddadm@najah.edu

Prof. Mooyoung Han, Dept. of Civil and Environmental Eng., #35–302 Seoul National University, Shinrimdong Kwanakgu, Seoul, Korea, myhan@snu.ac.kr

Prof. Jacques Haury, Department of Ecology, Agrocampus Ouest, 65 rue de Saint-Brieuc, CS 84215, 35042 Rennes, France, jacques.haury@agrocampus-ouest.fr

Prof. Francesc Hernandez-Sancho, Department of Applied Economics II. University of Valencia, Spain, Francesc.Hernandez@uv.es

Dr. Annick Hollé, Laboratory of Social Dynamics and Spatial Recomposition (CNRS/UMR 7533), Geography Department, University Paris 8, 2 rue de la Liberté, 93526, Saint-Denis Cedex, France, annick.holle@free.fr

Dr. Meisha Hunter, Li/Saltzman Architects, PC, 62 White Street, New York, NY 10013, USA, mhunter@lisaltzman.com

Dr. Jaroslav Jásek, Pražské vodovody a kanalizace, a.s., Prague, Czech Republic, jaroslav.jasek@pvk.cz

Mr. Jalal Jooshesh, No.14, Payam St, Ershad Bulvd,Toossab Consulting Engineer, Mashhad, Iran, j.jooshesh@toossab.net

Dr. Asimina Kaiafa, School of Architecture, Aristotle Univ., 54124 Thessaloniki, Hellas, minakasar@gmail.com

Dr. Eleni I. Kanetaki, 67 Dimocharous str., 11521 Athens, Hellas, eleni.kanetaki@gmail.com

Prof. Nicholas Kathijotes, Department of Civil Engineering & Geomatics Cyprus University of Technology P. O. Box 50329, Limassol, 3603 Cyprus, nicholas.kathijotes@cut.ac.cy.

Dr. Eliza Kavoulaki, Archeological Museum of Iraklion, 71202 Iraklion, Hellas, panexam@yahoo.gr

Dr. Saifullah Khan, Institute of Geography, Urban and Regional Planning, University of Peshawar, KPK, Pakistan, Saifullahkhan33@gmail.com

Dr. Mikyoung Kim, Korea Institute of Construction Technology 283, Goyangdae-Ro, Ilsanseo-Gu, Goyang-Si Gyeonggi-Do, 411-712 Republic of Korea, mkkim@kict.re.kr

Prof. Kostas Komnitsas, School of Mineral Resources Engineering, Technical University of Crete, Polytechnioupolis, Akrotiri, 73100 Chania, Hellas, komni@mred.tuc.gr

Dr. Sasha Koo-Oshima, The Office of Water, EPA, 1200 Pennsylvania Ave., NW, Mail Code 4101 M, Washington, D.C. 20460-0001, USA, Koo-Oshima.Sasha@epamail.epa.gov

Mr. Roman Llagostera, City Council of Barcelona, Torrent de l'Olla, 218–220, 08012 Barcelona, Spain, rllagostera@bcn.cat

Prof. Gerasimos Lyberatos, School of Chemical Engineering, National Technical University of Athens, Zografos Campus, 1570 Athens, Hellas, lyberatos@chemeng.ntua.gr

Dr. Olfa Mahjoub, Institut National de Recherche en Genie Rural, Eaux et Forets (INRGREF), Tunis, Tunisia, olfama@gmail.com.

Dr. Katja Marasovič, University of Split, Faculty of Civil Engineering Architecture and Geodesy 21000 Split, Matice Hrvatske 15, Croatia; katja.marasovic2@gmail.com

Prof. Jure Margeta, University of Split, Faculty of Civil Engineering Architecture and Geodesy 21000 Split, Matice Hrvatske 15, Croatia; margeta@gradst.hr

Mr. Pere Malgrat, CLABSA, Acer, 16. 08038 Barcelona, Spain, pere@clabsa.es

Prof. Gianmarco Margaritora, Roma Sapienza University Rome, Italy, gianmarco.margaritora@uniroma1.it

Prof. Jure Margeta, University of Split, Faculty of Civil Engineering Architecture and Geodesy 21000 Split, Matice Hrvatske 15, Croatia, margeta@gradst.hr

Mrs. Montse Martinez, CLABSA, Acer, 16. 08038 Barcelona, Spain, montsem@clabsa.es

Prof. Pierluigi Martini, Associazione Idrotecnica Italiana, Rome, Italy, segreteria@idrotecnicaitaliana.it

Prof. Aradhana Mehra, Faculty of Education, Health and Sciences, School of Science, Univ. of Derby, Derby DE22 1GB, UK, A.Mehra@derby.ac.uk

Dr. Vasilios Melfos, School of Geology, Aristotle Univ., 54124 Thessaloniki, Hellas, melfosv@geo.auth.gr

Dr. Maria Molinos-Senante, Department of Mathematics for Economics. University of Valencia, Spain. Maria.Molinos@uv.es

Dr. Saeid Neirizi, No.14, Payam St, Ershad Bulvd, Toossab Consulting Engineer, Mashhad, Iran, s.neirizi@toossab.net

Ing. Jan Palas, Národní Technické Muzeum, Prague, Czech Republic, jan.palas@ntm.cz

Dr. Eleni Papanikolaou, School of Geology, Aristotle Univ., 54124 Thessaloniki, Hellas, papelen@geo.auth.gr

Dr. Chrysoula Papacharalampou, School of Geology, Aristotle Univ., 54124 Thessaloniki, Hellas, x.papaxaralampou@gmail.com

Prof. Nymphodora Papassiopi, School of Mining and Metallurgical Engineering, National Technical University of Athens, Zografos Campus, 15780 Athens, Hellas, papasiop@metal.ntua.gr

Dr. Snježana Perojevič University of Zagreb, Faculty Architecture, Mediterranean centre for built heritage 21000 Split, Bosanska 4, Croatia, snjezana.perojevic@gmail.com

Dipl.-Ing. Alexander Reyes-Knoche, PhD candidate, International Independent Consultant, Management Consulting and Development Cooperation, Allmendweg 5, D-68526 Ladenburg, Germany, alexander.reyes-knoche@web.de

Prof. Paolo Roccaro, Department of Civil and Environmental Engineering, University of Catania, Viale Andrea Doria, 6, 95125 Catania, Italy, proccaro@dica.unict.it.

Prof. Joan B. Rose, Dept. of Fisheries and Wildlife, Michigan State Univ., East Lansing, MI 48824, USA, rosejo@msu.edu

Prof. Miquel Salgot, Soil Science Laboratory, Facultat de Farmàcia, Universitat de Barcelona, Joan XXIII, s/n. 08028 Barcelona, Spain, salgot@ub.edu

Dr. Antonella E. Santamaria, Department of Civil and Environmental Engineering, University of Catania, Viale Andrea Doria, 6, 95125 Catania, Italy, asantama@dica.unict.it

Dr. Mohamed Thameur Chaibi, National Research Institute for Rural Engineering, Water and Forestry P.O. Box 10, 2080, Ariana, Tunisia, olfama@gmail.com

Dr. Vasileios Aim. Tzanakakis, Agricultural Services, Region of Crete, Iraklion, Hellas, vetzanakakis@cyta.gr

Prof. Federico G. A. Vagliasindi, Department of Civil and Environmental Engineering, University of Catania, Viale Andrea Doria, 6, 95125 Catania, Italy, fvaglias@dica.unict.it.

Mrs. Cristina Vila, City Council of Barcelona, c. Tarragona, 183, 08014 Barcelona, Spain, cvilaru@bcn.cat

Prof. Kostas Voudouris, Lab. of Engineering Geology & Hydrogeology, School of Geology, Aristotle Univ., 54124 Thessaloniki, Hellas, kvoudour@geo.auth.gr

Dr. Heikki S. Vuorinen, Hjelt Institute, Department of Public Health, FI-00014 University of Helsinki, Finland, heikki.vuorinen@helsinki.fi

Prof. Anthimos Xenidis, National Technical University of Athens, School of Mining and Metallurgical Engineering, Zografou Campus, 15780 Athens, Hellas, axen@metal.ntua.gr

Dr. Alessandro Zanobini, ACEA, Roma ,Italy, a.zanobini@aceaspa.it.

Dr. Xiao Yun Zheng Xiao Yun, Yunnan Academy of Social Sciences, Kunming, China, zhengxy68@163.com

Dr. Yao Yu Zheng, PAYAP University, Mae-Kow Chiang Mai, 50000 Thailand, yypayap@163.com

Prolegomena: *Probing the past and facing the future*

'Ομοια γάρ ως επί το πολύ τα μέλλοντα τοις γεγονόσι.
Most future facts are based on those in the past.
Euripides, 480–406 BC, Ancient Greek Tragic

It is well documented that most of the technological developments relevant to water supply and wastewater are not achievements of present-day engineers but date back to more than five thousand years ago. Already during the Bronze Age such. developments were driven by the necessities to make efficient use of natural resources, to make civilizations more resistant to destructive natural elements, and to improve the standards of life, both at the public and private level. With respect to the latter, Minoans in the island of Crete (*ca.* 3200–1100 BC) and an unknown civilization (*ca.* 2600–1900 BC) in Indus valley at Mohenjo-Daro, Harappa, and Lothal developed advanced, comfortable, and hygienic lifestyles, as manifested by efficient sewerage systems, bathrooms and flushing toilets, which can be compared to the modern wastewater system of today, re-established in Europe and North America only a century and half ago. The amazing evolution and development of structures for bathing, sanitary and other purgatory installations can be traced from the Minoan palaces and houses at Mohenjo-Daro, Harappa, and Lothal up to the cities of Ancient Egypt, of the Hellenistic period, of the Chinese Dynasties and Empires to the facilities built during the Roman period. It is interesting that very unsanitary conditions and overcrowding were widespread throughout Europe and Asia during the Middle Ages, resulting periodically in cataclysmic pandemics such as the Plague of Justinian (541–542AD) and the Black Death (1347–1351AD), which killed tens of millions of people and radically altered societies.

The book 'Evolution of Sanitation and Wastewater Technologies Through the Centurie' presents the major achievements in the scientific fields of sanitation throughout the millennia from a wide geographic perspective. It provides valuable insights into ancient wastewater technologies and management with their apparent characteristics of durability, adaptability to the environment, and sustainability. A comparison of the water technological developments in several civilizations is undertaken. These technologies are the underpinning of modern achievements in sanitary engineering and wastewater management practices. It is the best proof that 'the past is the key for the future.'

A timeline of historical developments associated with sanitation and wastewater management worldwide through the last 5500 years of humankind's history is considered. A chronological order is followed with emphasis to the major periods and the corresponding worldwide civilizations.

Rapid technological progress in the twentieth century created a disregard for past water technologies that were considered to be far behind the present state of knowledge and engineering. There is a great deal of unresolved problems, still, related to the wastewater management principles, such as decentralization of treatment processes, durability of the water projects, cost effectiveness, and sustainability issues such as protection from floods and droughts. In the developing world, such problems were intensified to an unprecedented degree.

Moreover, new problems have arisen such as the contamination of surface and groundwater. Naturally, intensification of unresolved problems led societies to revisit the past and to reinvestigate the successful past achievements. To their surprise, those who attempted this retrospect, based on archaeological, historical, and technical evidence were impressed by two things: the similarity of principles with present ones and the advanced level of wastewater engineering and management practices in the past.

Most likely urbanization will continue to increase. The proportion of people living urban areas might rise to 80–90% of the global population. This means that innovative sanitation technologies must be developed to regulate and manage municipal wastewater and stormwater. Many of these technologies will be based on decentralized principles, will serve different sizes of buildings from single-family homes to high-rise buildings, public or commercial buildings. Treated water might be readily reused locally for various purposes such as toilet flushing, watering gardens or car washing. Sludge from decentralized plants can be used as fertilizer as demonstrated in the past at various locations in the world. Also, measures and technologies for harvesting rainwater in order to reduce the flood risk and increase water availability need to be further developed.

There is much we can learn from past technologies and practices that were implemented. This includes for example design philosophy, adaptation to the environment, and decentralization management of water and wastewater projects, architectural tied to operation aspects, and sustainability as a design principle. It is time to think about managing the complete water cycle, to do so with impunity we look towards *probing the past, forging the future.*

70 authors and or co-authors from several disciplines and regions of the planet developed the chapters in this book. The disciplines include Archaeology, History, Engineering, Life sciences, Health sciences, History of Medicine, Environmental sciences, Biology and Geosciences. The geographical coverage is very wide, with prominence in the Mediterranean world. However, several other civilizations from other parts of the world, such as Asia (Iran, India, China, and Korea), Central Europe, and South America are also covered. The themes of the Chapters included are from prehistoric to medieval and even modern times. All Chapters submitted were peer-reviewed by at least two reviewers and the Editors.

The book is organized in six parts. The first three chapters in the first Part are introductory mainly referring to pre historical civilizations. The eleven chapters in the second Part refer to historical civilizations (including Archaic, Classical, Hellenistic and Roman) over the globe. The six chapters in the third Part, should be considered as case studies; major cities with long histories are included. The following two Chapters in the fourth Part address the history of water borne diseases. The final four chapters in the fifth Part mainly deal with cases in the modern times. The last chapter (sixth Part), summarizes and synthesizes the conclusions, comparison, and lessons learned with some commentary on the future.

We appreciate the efforts and contributions of the authors who have written a compilation of the labours of humankind to bring sanitation to the people and cities. We are particularly grateful to Dr. Peter A.

Wilderer, Soc. Acad. European Academy of Sciences and Arts, who was extremely helpful in reviewing chapters and providing advice on the overall content of the book.

Andreas N. Angelakis
National Agricultural Research Foundation, Institute of Iraklion,
P.O. Box 2229, 71307 Iraklion, Hellas
info@a-angelakis.gr

Joan B. Rose
13 Natural Resources Building, Michigan State University,
East Lansing, MI 48824, USA
rosejo@msu.edu

Chapter 1

Sanitation and wastewater technologies in Minoan Era

A. N. Angelakis, E. Kavoulaki, and E. G. Dialynas

1.1 PROLEGOMENA

> '*. . . all the sewers were still working! It was very interesting for me to see the water in the drainages and sewers so big that a man could enter. I doubt if there are other examples of ancient sewerages working after 4 thousand years. . .*' Angelo Mosso, during his visit at the palace of Phaistos in the beginning of 20th century (Mosso, 1907).

The history of water supply and wastewater engineering on Crete dates back more than *ca.* 4500 years. From the Early Minoan period (*ca.* 3000–2000 BC) issues related to water supply were considered of great importance and were accordingly developed. Archaeological and other evidence indicate that during the Bronze Age advanced water management and sanitary techniques were practiced in several settlements on Crete (Angelakis & Koutsoyiannis, 2003). The emergence of the palaces revealed a remarkable development of water management in the urban context. Moreover, during the Middle Minoan period (*ca.* 2000–1700 BC) and the beginning of the Late Minoan periods (*ca.* 1700–1100 BC) a 'cultural explosion' occurred on the island. A striking indication of this is manifested, *inter alia*, in the advanced wastewater and stormwater management techniques practicede at that time. These included various scientific fields of water resources, such as wells and ground-water hydrology, aqueducts, and domestic water supply, according to local condition in terms of climate and geomorphology (De Feo *et al.*, 2012). Additionally, the construction and use of sanitary facilities, even the recreational uses of water, signify attitudes of life and taste (Lyrintzis & Angelakis, 2006). Furthermore, numerous very advanced and wonderful water wastewater systems, including aqueducts, cisterns, filtering systems, rainfall-harvesting systems, terracotta pipes for water supply, fountains, baths, sewers, and lavatories were practiced in several Minoan palaces and other settlements (Angelakis *et al.*, 2005; Antoniou & Angelakis, 2013).

The Minoans came up with many innovations, such as those related to water and wastewater management. In their palaces, they had running water and flushing lavatories. Thus, all wastes were transferred outside the building, and away from the living places. In their houses, the bathtubs were moveable, allowing to move them to different rooms. The palaces of the Minoans were very open without lockers, served as community centers, and people could walk into any room they wanted To serve their gods, the Minoans had ritual baths in their palaces (Angelakis & Spyridakis, 1996a; De Feo *et al.*, 2012).

One of the salient characteristics of Minoan Era was the hydraulic function of the storm water and wastewater in the palaces and towns. In the entire structure of the Minoan palace nothing is more remarkable than the elaborate sewerage system that runs throughout its domestic quarter and adjoining halls. Evans (1921–1935) and MacDonald and Driessen (1988), described the course of these sewers and drew plans of what they considered to be their original form.

At Knossos, the Minoans took advantage of the steep grade of the land to devise a drainage system with lavatories, sinks and manholes. Archaeologists have found pipes laid in depths from just under the surface in one area to more than 3 m deep in others. Such a plan of the entire sewerage system of Knossos was previously reported (Angelakis & Spyridakis, 1996a; Antoniou & Angelakis, 2013; MacDonald & Driessen, 1988 and others). This plan could provide the visitor with a basic orientation of the site and help him to view the entire network in an integrated manner. A stone-by-stone description of the sewerage system with reference to the architecture above has been given as well by MacDonald and Driessen (1988). The total sewerage system, including outlets and tributaries, exceeds one hundred and fifty meters. The diminutive size of the channels and other obstacles prevented a more thorough examination.

The Minoan palaces of Knossos, Phaistos, Malia, Zakros, and Galatas were the seats of a powerful centralized administration that controlled trade and carried out large-scale public works, ranging from the construction of roads to water supply and drainage and sewerage systems. It is evident that during the Minoan Era extensive drainage systems and elaborate structures were planned, designed and built to protect the growing population centers and the agricultural land. In several Minoan palaces and other settlements, one of the most important elements was the provision and distribution of water and the transfer of stormwater and sewerage in drains by means of hydraulic systems. Rainwater from the flat roofs of the palace at Knossos was carried off by vertical pipes; one of these, located in the eastern wing, was emptied into a stone sewer-head from which a stone channel carried the flow of storm water (Angelakis & Spyridakis, 2013).

Many remains of sanitary structures have been found in the greater region of the ancient Hellenic world. Some of them are dated even in the Minoan Era (e.g., Knossos, Phaistos, Malia, Gournia, and Tylissos sites). The installation of hygiene can be classified as a characteristic factor of living's standard and economic prosperity, both in domestic and public uses. Because of these reasons lavatories have become, and often still, are considered as luxury elements (Antoniou, 2007). The bathing facilities remained mainly public until the establishment of Etruscan civilization (*ca.* 800–100 BC) in the central Italy, which appears to develop a rational planning of urban areas with a clear distinction between public and private spaces with specific water systems supplying workshops (Angelakis *et al.*, 2012a). Nevertheless, this is a tradition still maintained in several communities in the Middle East. On the other hand, the sanitary facilities had been widely applied in domestic-residential buildings (Antoniou & Angelakis, 2013).

In this Chapter significant developments relevant to the hygienic lifestyle in Crete, Hellas during the Bronze Age are considered. Special references to sanitary and relevant structures, such as sewers, drains, bathrooms, and lavatory are made, following a short review of Crete physical setting.

1.2 PHYSICAL SETTING OF THE ISLAND CRETE

1.2.1 Location

In modern geographical terms, Hellas is located in southern Europe, between Albania and Turkey, surrounded by the Aegean, Ionian and Mediterranean seas. Crete is a Hellenic mountainous island located at the eastern Mediterranean, in the southern part of the Aegean sea. The island of Crete measures approx. 260 km east to west and 60 km north to south at its widest point; its landmass covers an area of 8336 km^2 with a coastline of 1046 km (Angelakis *et al.*, 2012a). Due to its position between Asia, Africa and Europe,

Crete holds a strategic location, as it forms a natural and vital bridge between the three continents. This unique geographical position determines its historical course throughout both antiquity and modern times. During the ancient period the island was wracked by earthquakes, volcanic eruptions, and winter storms (Gorokhovich *et al.*, 2011). A map of Crete with the major Minoan archaeological sites considered in this Chapter is shown in Figure 1.1.

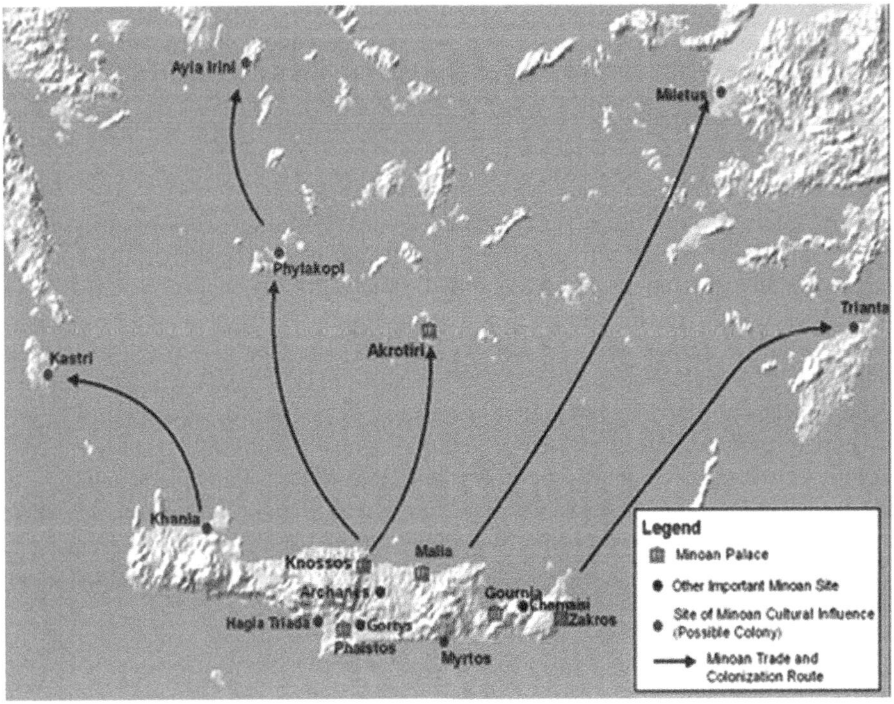

Figure 1.1 Map of crete with major minoan archaeological sites with emphasis in Minoan waste water, and stormwater systems (adapted from http://www.anthrogenica.com/showthread.php?715-The-Minoan-World – (1900–1425-BC)).

1.2.2 Climatic conditions

Today, Crete straddles two major climatic zones, the Mediterranean and the North African, mainly falling within the former. As such, the climate in Crete is primarily temperate. The atmosphere can be quite humid, depending on the proximity to the sea, whilst winter is fairly mild. Snowfall is common on the mountains between November and May, but rare in the low lying areas, especially near the coast. However, deviations from this regime are not uncommon, such as the heavy cold snap that swept the island in February 2004, when the whole island was blanketed with snow. In the summer, average temperatures reach the high 30's and the low 15°C, with maxima touch the upper 40's°C (Angelakis *et al.*, 2012a). The south coast, including the Messara valley and Asterousia mountains,belongs to the North African climatic zone, thus enjoying significantly more sunny days and high temperatures throughout the year. In southern Crete, date palms bear fruit, and swallows remain year–round rather than migrate to Africa.

It is known from several studies on climate variations in the Mediterranean region during the Holocene period, that a number of different climatic periods have occurred during the past 5000 years (e.g., cold period, *ca.* 4500–3000; cold and humid period, *ca.* 3000–2200; and a warm period, *ca.* 2200–1400 BC) (Angelakis & Spyridakis, 1996). Despite varying climatic conditions over the past 5000 years, it may be conjectured that a clear abundance of water resources cannot be documented in the sites of significant cultural development in Crete, including centres such as Knossos, Zakros and Phaistos. Given these climatic and hydrological considerations, early Cretan urban societies had to develop innovative technological means to capture, store, and convey water even from long distances; moreover, legislation and institutions to manage water more effectively had to be established (Angelakis & Koutsoyiannis, 2003; Angelakis *et al.*, 2012a).

1.2.3 Hydrology

Cretan hydrology varies greatly from west to east and from north to south. Geologically, numerous water basins exist, yet the island is officially considered as an independent river basin district (RBD). The atmospheric precipitation is mainly frontal, linked to 'the interaction of contrasting air masses in eastward moving depressions', and orographic due to the existence of three main mountainous formations (Grove & Contario, 1995). Therefore, it exhibits intense spatial and temporal variation; it decreases from west to east and from north to south (Voudouries *et al.*, 2006), whilst also increases with altitude. In particular, the average precipitation ranges from 440 mm/yr on the plain of Ierapetra (southestern Crete) to 2000 mm/yr in the Askifou uplands (northwestern Crete). The mean annual precipitation in eastern Crete measures 816 mm/yr while in western Crete it measures 927 mm/yr. Moreover, annual precipitation is divided into a wet and a dry season; the first one lasting from October to March, and the second one from April to September. The atmospheric precipitation in Crete indicates intense spatial and temporal variation. Generally the precipitation decreases from west to east and from north to south and also increases with altitude.

In Crete the air temperature increases from the west (16.96°C Alikianos station) to the east (18.33°C Siteia station). The mean annual temperature ranges from 17°C to 20°C. The prevailing wind direction is north and north-westerly. High speed winds can occur any time during the year. The driest months of the year are June and July (with a mean relative humidity ranging from 48.9% recorded at Souda station to 59.9% in Iraklion station). The most humid month is December (Angelakis *et al.*, 2012a).

Potential evapotranspiration (ET), as estimated using the Penman-Monteith method, a system which provides the most accurate estimates, varies from 1240 mm/yr to 1570 mm/yr. Within the annual circle, the monthly ET rate changes from about 25 mm in winter to 225 mm in summer. The mean annual actual ET has been estimated to represent 75% to 85% of the mean annual precipitation in low elevation areas (less than 300 m asl) while dropping to 50% to 70% in high elevation areas (Angelakis *et al.*, 2012a).

1.3 MAJOR SANITARY TECHNOLOGIES IN MINOAN ERA

The island of Crete was the center of Europe's first advanced civilization, the Minoan (Mays *et al.*, 2007). The earliest human settlements on the island date back to the Neolithic period (*ca.* 5700–3000 BC). Ancient Knossos was one of these major Neolithic settlements. Later on, Knossos became the capital of the Minoan Crete which reached its peak during the Bronze Age (*ca.* 3000–1100 BC). Soon, its cultural influence and trade relationships extended beyond the borders of the Cretan island reaching destinations as far as Cyprus, Egypt and Anatolia. Contacts of Minoans with other prehistoric civilizations (e.g., European, Asian, and even North American) has been reported (Mariolakos, 2012).

The Cretans were well-known for their navy which dominated the Aegean sea, their artistic pottery, and their luxurious palaces and villas.

Cultural advancements can be observed throughout the third and second millennia BC, but great progress was made in Crete, especially during and after the Middle Minoan period (*ca.* 2000–1700 BC), when the population in its central and southern regions increased, towns were developed, the first palaces were built, and Crete achieved a prosperous and uniform culture. In the early phases of the Late Minoan period (*ca.* 1700–1100 BC), Crete appears to have prospered even more, as evidenced by the larger houses and more luxurious palaces of this period (Koutsoyiannis *et al.*, 2008). At this time, the flourishing arts, improvements in metalwork along with the construction of better-equipped palaces, and an excellent road system, reveal a wealthy, highly cultured, well-organized society and government. However, one of prominent characteristics of the Minoan Era was the architectural and the hydraulic function of the sanitary structures, such as sewers, drains, bathrooms, and lavatory (Angelakis & Spyridakis, 1996a).

1.3.1 Use of harvested water in minoan crete

A systematic evolution of water management in ancient Hellas began on Crete during the early Bronze Age, for example, the Early Minoan period (*ca.* 3000–2000 BC). Wells, cisterns, water distribution, fountains, and even recreational functions existed. In prehistoric Crete rivers and springs provided people with water. In the Early Minoan period II, a variety of technologies, such as wells, cisterns, and aqueducts were used. Also the Minoan architecture, including flat rooftops, light wells, and open courts played an important role in water management. The rooftops and open courts acted as catch basins to collect rainwater, which was sent to storage areas or cisterns, like those found in Knossos, Phaistos, Archanes, Myrtos Pyrgos, and Chamaizi (Angelakis & Spyridakis, 2013).

Anthropological studies in developing countries show that in communities without direct access to potable water sources, the domestic water use was in the range of 10–20 L/inh.d (Mays *et al.*, 2012). In such villages all water had to be stored in water cisterns and/or carried in pots and buckets some distance from the water source to the home. For a family of 6, 60–120 L of water each day had to be transferred. Based on this it is assumed that water consumption during the Minoan Era was not exceeding 15 L/inh.d at any time of its long period.

Historically, drinking water has been considered as the clear water. Considering the scientific knowledge of that era, this simplification was totally justified, even without the tools of chemistry and microbiology. Even today, clarity (and probably taste) is the main criterion for classifying water as suitable for human consumption (Mays *et al.*, 2012). One of the salient characteristics of the Minoan Era (*ca.* 3200–1100 BC) was the treatment devices used for water supply in palaces, towns, and villages from the beginning of the Bronze Age. It is truly amazing that the most common water quality modification technique for providing suitable domestic water supplies was known to Minoan engineers. Thus, according to Dafner (1921), a strange, oblong device with an opening in one of its ends, was used to treat domestic water (Figure 1.2a). The device was constructed in a similar manner and with the same material as the terracotta water pipes. Spanakis (1981) theorized this device as a hydraulic filter which was probably connected to a water supply reservoir by a rope passing through its outside holds. Its operation relied on local, high speed, turbulent conditions in order to continuously clean the porous surface thus allowing the continuous flow of filtered water to the jar. For cleaning purposes, after extensive solids accumulation, it was possible to release it from the pipe end by loosening the rope in the holes. A similar clarifying device was used by Egyptians in the tomb of Amenophis II and later in the tomb of Rameses II. This device allowed impurities to settle out of the water and then the clarified water was siphoned off and stored for later use (Mays *et al.*, 2012; Antoniou & Angelakis, 2013).

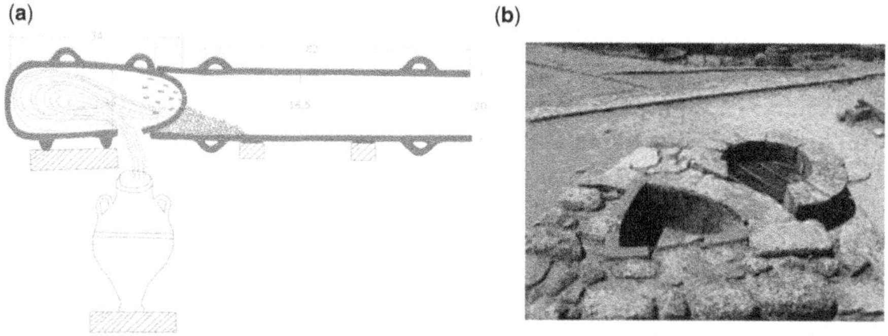

Figure 1.2 Minoan water filters: (a) Ceramic filter and (b) special cistern with sandy filter in the Phaistos palace (with permission of A. N. Angelakis).

In addition to the terracotta filters, in some cases, cisterns were associated with small canals collecting water from rainfall and from mountain streams (Viollet, 2003). It is therefore possible that cisterns were connected to the sandy filters, such as that shown in Figure 1.2b.

Other terracotta infiltration devices were discovered in Agios Mamas, location of the spring from where water was transported to the Tylissos village (Figure 1.3). In the Tylissos houses, water was transported through an aqueduct to the three Minoan houses from the Agios Mama spring to a distance of about 3 km. These devices were filled with charcoal, thus playing the role of activated carbon treatment processes for removing both organic and inorganic constituents. In addition to these devices, a terracotta pipeline of 42 m length, similar to those used in Knossos, was discovered in the north-western site of House B (Mays *et al.*, 2012).

Figure 1.3 Two terracotta conical tubes used probably as refinery devises from Agios Mamas spring in Tylissos (Archeaological Museum of Iraklion), Iraklion (with permission of A. N. Angelakis).

1.3.2 Sewerage and drainage systems

It is evident that during the Minoan Era extensive systems and elaborate structures for water supply, irrigation and drainage were planned, designed and built to supply the growing population centres and agriculture with water (Angelakis & Koutsoyiannis, 2003). In several Minoan palaces discovered by archaeologists in the 20th century, one of the most important elements was the provision and distribution of water and the sewerage and drainage systems by means of sophisticated hydraulic works. It is worth to mention that an impressive and common feature of the Minoan palaces and towns is their famous sewage and drainage systems. These are mainly stone structures. Stone conduits formed drains which led rainwater from the courts outside the palaces, to eliminate the risk of flooding.

The major water supply and drainage works in the Minoan palaces and towns were executed in the Early Minoan period. Some of these were repaired and cleaned out afterwards, while others were abandoned (Driessen & Schoep, 1994). Only very few sewers and drains of an urban character were constructed later on. Sakellarakis and Sapouna–Sakellaraki (1991) suggested that drains at Archanes, excavated by Evans, actually date to the early Minoan period, and were repaired afterwards. In the Late Minoan there is little evidence for urban sewerage and drainage system constructions (Driessen & Schoep, 1994).

In several Minoan palaces and towns well established drainage and sewerage systems existed, which are even today in good functional condition. As the hill in the palace of Knossos was periodically drenched by torrential rains, a runoff system was a necessity. Stormwater from the flat roofs of the palace at Knossos was carried off by vertical pipes; one of these, located in the eastern wing, emptied into a stone sewer head from which a stone channel carried the flow of stormwater (Evans, 1921–1935). It began with channels in the flat surfaces, which were zigzag and contained catchment basins to control the water velocity. In addition to the management of the rainwater from the roofs, rainwater from courts and the overflows from the cisterns was carried down into buried central sewer and drain stone-made conduits (Figure 1.4a). Stone channels was used to drain water accumulated during strorm rains from within courts, from flat or sloping roofs and from light-wells in the interiors of buildings (Shaw, 1973). While conduits of large cross section were stone made, those with smaller cross section were made from terracotta in the form of U-shaped tiles. Such terracotta conduits were discovered as parts of sewer and drain networks at the Knossos and Phaistos palaces and other Minoan settlements (Figure 1.4b; see also Angelakis *et al.*, 2012b). It is noted that U-shaped terracotta drains covered with stone slabs, running below the streets, were also found in Habuba Kabira, a small Sumerian city of the Bronze Age (Angelakis *et al.*, 2012b).

Figure 1.4 Minoan drains: (a) stone-made in the palace of Knossos and (b) U-shaped terracotta at Phaistos palace (modified from Angelakis *et al.*, 2012b).

In the palace of Knossos, runoff from a part of the central court was handled by a very capacious underground channel built of stone and lined with cement; it ran beneath the passage leading from the north entrance and received several flows from various quarters. The most fully explored pan of the palace sewerage and drainage system is the portion which ran beneath the floors of the Residential Quarter. This formed a great loop with its high point located under the light-well, next to the Grand Staircase, and emptied *via* a combined channel down the slope to the east of the palace (Evans, 1921–1935). In the area of the Hall of the Double-Axes and the Queen's Megaron with its associated chambers, it received the wastewater and rainwater of no less than five light-wells; it also served a lavatory on the lowest floor, and was connected with three vertical shafts. The latter, evidently, received storm water from the roof and was probably connected with lavatories on the upper floors.

The sewerage and drainage system was built of stone blocks lined with cement and measured about 79 by 38 cm per section. Probably the upper system was open. The sewers and drains, then, were large enough to permit men to enter them for cleaning or maintenance; in fact, manholes were provided for that purpose in the parts that were covered. Airshafts at intervals also helped to ventilate sewers (Graham, 1987). Finally, the drainage and sewerage system of the palace, running underneath most of the palace, with baffles and overflows was slowing down the runoff and prevented flooding. A plan of the entire sewerage and drainage system of Knossos is illustrated in Figure 1.5.

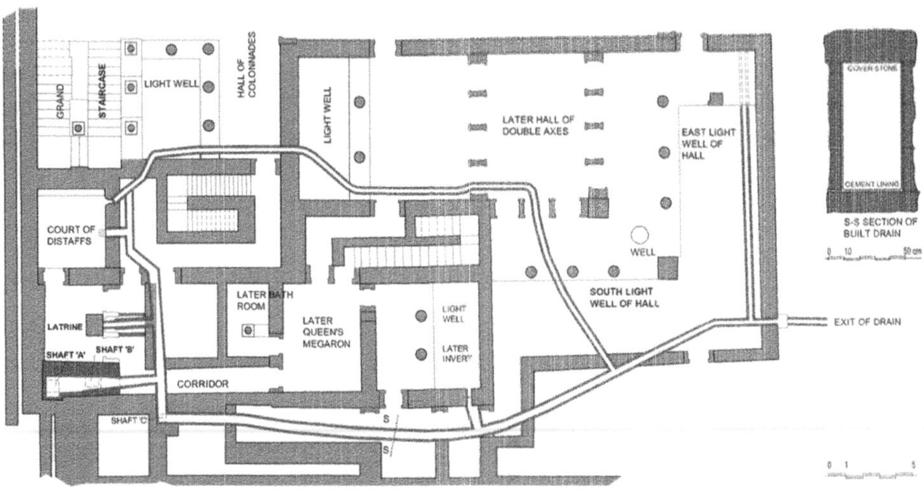

Figure 1.5 The sewerage and drainage system in the palace of Knossos beneath the Domestic Quarter (modified from Antoniou & Angelakis, 2013).

Certainly, the plumbing arrangements and especially the sewers in the Minoan towns were carefully planned. Covered stone, slab-built sewerage and drainage systems in many towns to carry away sewage and rainwater including storm waters are evident. The remains at the palace of Knossos palace show clearly how rainwater was drained from the roof by way of light-wells and used to flush out sewage from bathrooms and lavatories.Minoan palaces and towns were equipped with elaborate storm drainage and sewerage systems (Figure 1.6). In fact; all palaces had applied strategies to dispose wastewater (MacDonald & Driessen, 1990). Open terracotta and stone conduits were used to convey and remove stormwater and limited quantities of wastewater. Pipes, however, were scarcely used for this purpose.

Larger sewers, sometimes large enough for a man to enter and clean them, were used in Minoan palaces of Knossos and Phaistos (Angelakis & Spyridakis, 2013).

Figure 1.6 Minoan water sewerage and drainage systems: (a) Part of central system at Knossos palace and (b) part of the Minoan drainage system at the 'Little palace' at Knossos (with permission of A. N. Angelakis).

The drainage and sewerage system of Zakros was quite dense and of high water-engineering standards as at the palace of Knossos and other Minoan setlements (Platon, 1974). Zakros provides us with well-preserved remains of sophisticated networks in which descending shafts and well-constructed stone sewers, large enough to permit the passage of a man, play a part. Yet, there is evidence that the entire system was not effective in times of intense or extended storms. However, due to the privileged location of the site in a natural slope, the final disposal of wastewater and storm waters at sea was easily attained. Platon (1974) found three basic types of conduits in the sewerage system of Zakros: a clay-conduit in reversed n-shape, another built up with stones and a third, narrow type, constructed with stones but open at the top. A small section of the Zakros sewerage system, as it appears today, is shown in Figure 1.7a.

One of the most advanced Minoan sanitary and storm sewer systems was discovered in Hagia Triadha (close to the south coast of Crete, few km west of Phaistos). The Italian writer Angelo Mosso, who visited the villa of Hagia Triadha at the beginning of the 20th century and inspected the storm sewer system (Figure 1.7b) noticed that all the sewers of the villa functioned perfectly and was amazed to see stormwater come out of sewers, 4000 years after their construction (Mosso, 1907). Gray (1940) who related this story and quoted Mosso (see the quotation in the beginning of the paper) added the following statement:

> *Perhaps we also may be permitted to doubt whether our modern sewerage systems will still be functioning after even one thousand years.*

Figure 1.7 Parts of the sanitary and storm sewerage systems: (a) in Zakros palace and (b) Hagia Triada villa (modified from Angelakis *et al.*, 2005).

It may be concluded that in the entire history of mankind there exists no other example of a sewerage and drainage system still functional more than 4000 years after its original construction. Thus, the existence today of several Minoan archaeological sites could be due mainly to their very advanced sewerage and drainage systems.

1.3.3 Bathrooms and lustral basins

It should be noted that at that time, in addition to sewers, bathrooms were not considered necessary, merely convenient, and most palaces did not have them. Although the function of Minoan rooms is difficult to define, Evans (1921–1935), who discovered the palace of Knossos, identified three rooms as bathrooms. The main type, which resembles the bathrooms discovered at Phaistos and Malia, is that found near the Queen's Megaron in the palace. However, there is a difference in the level of the floor and the consequent absence of steps. Evans (1921–1935 referred to these rooms as *'lustral chambers or basins'*. Also, Graham (1984) assumed that a room which started out as a lustral chamber or basin later became an ordinary bathroom. In fact, investigations by Platon (1990) led to assume that this also happened twice in the houses at Tylissos; and a careful study could show the same to be true for the bathroom of the Queen's Hall in the palace of Minos (Graham, 1984). Thus, it could be suported that Minoans in their latter days began putting cleanliness before godliness (Mays & Angelakis, 2012). Lustral basins at palaces in Kato Zakros and in Malia are shown in Figure 1.8.

It must be noticed that, in spite of the view of scholars (Alexiou, 1964), the absence of bath facilities in some sanitary cisterns, and their existence in other places, should not be considered coincidental. Graham (1987) and Platon (1974) have reported that cisterns were used for the cleansing of both body and soul. Also, most Minoan baths were connected to independent septic systems in the outside, a practice indicative of the advanced water resources management and environmental techniques of that period (Angelakis *et al.*, 2005).

In Phaistos, a luxurious bathroom is located in southwest corner of King's megaron decorated with wall paintings and with usual steps descending into it (Figure 1.9a). Also, its walls and floor are faced with alabaster slabs, with an attic on the east side. In the western side, an alabaster slab forming a step with a hole has been interpreted as a lavatory. In addition, a bath of similar in function to that of the so called Queen's Hall found in the Phaistos, was found in Da house in Malia (Figure 1.9b).

Figure 1.8 Lustral basins: (a) at the palace in Kato Zakros and (b) at the palace in Malia (with permission of A. N. Angelakis).

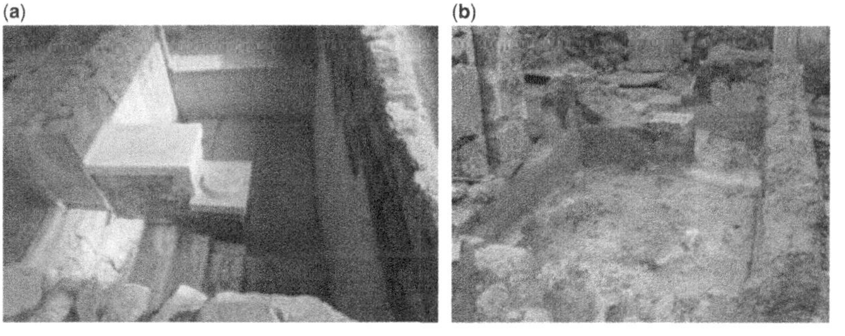

Figure 1.9 Bathrooms: (a) In the King's Megaron at Phaistos and (b) in the Da House at Malia (with permission of A. N. Angelakis).

Similar bathrooms have been reported in other Minoan sites. One of them, discovered in the palace of Zakros by Platon (1974), appears in Figure 1.10. Platon (1990) has provided us with some preliminary statistical data on Minoan cisterns, bathrooms and other sanitary facilities. She concluded that, in terms of chronology, most of them should be placed in the middle Minoan period; in regards to location, 16 are found next to domestic rooms, seven near holy altars and two in palace entrances. In only two instances various facilities for baths were found, seven were filled up with earth and two had been rebuilt and converted into bathrooms. Also, in 14 of these sites various holy objects were found, while in cement coats were indicated (Platon, 1990). In addition, frescos related to various holy subjects, were found only in two sanitary cistern-bathrooms located in Knossos and Zakros palaces. It should be noticed that in spite of the view of some scholars (Alexiou, 1964), the absence of bath facilities in some sanitary cisterns and their existence in others should not be considered coincidental. Graham (1987) and Platon (1974) have reported that cisterns were used for the cleaning of both the body and soul. Also the most Minoan baths were connected to independent septic systems in outside, practice indicative of the advanced wastewater management and environmental techniques of that Era (Angelakis & Spyridakis, 1996).

In several bathrooms clay tubs were used. A variety of such tubs have been discovered in Minoan sites. The clay tubs in the Minoan bathrooms were filled and emptied more likely by hand rather than by direct connection to the sewers.

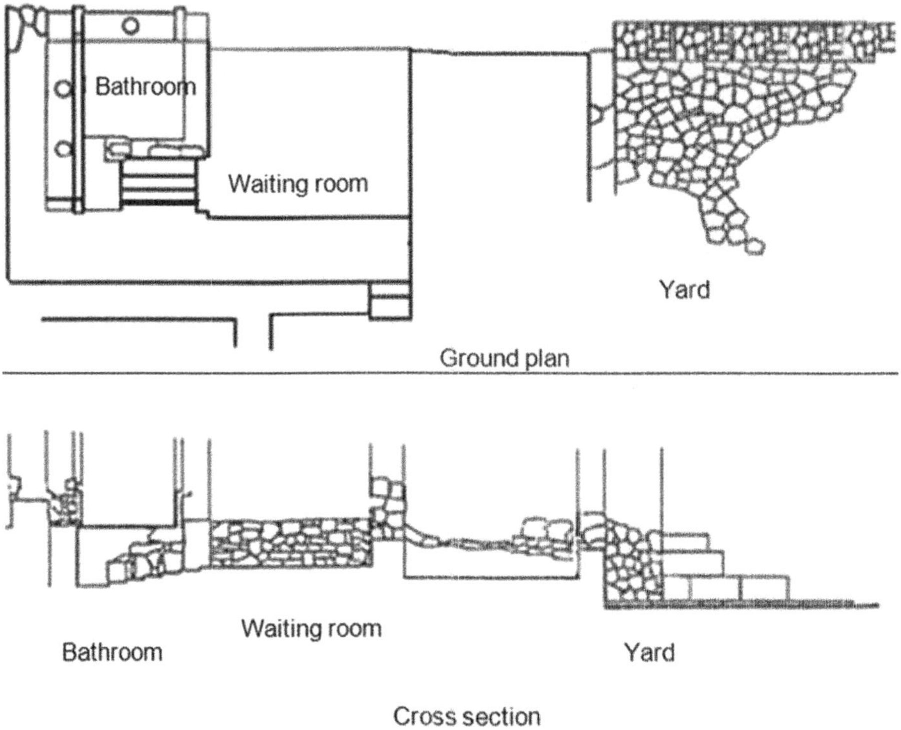

Figure 1.10 Bathroom in the palace of Zakros: Ground plan (up) and cross section (down) (modified from Platon, 1974).

The *Caravanserai*, as named by A. Evans (1921–1935), a rest house just south of the palace, is one of the most important regional monuments of the palace of Knossos, which according to the view of the excavator, was manufactured during the MMIII-LM IA period. In the eastern part of the complex accommodation facilities were found, and also warehouses and other facilities, as evidenced by the plant seeds in few jarsm, while in the western part fragments of clay bathtubs came to light, which in the opinion of Evans, were parts of hot baths. West of the wider area is located the 'Room of the Spring' (Evans, 1921–1935).

The most important part of the complex was restored by Evans (designed by the architect – partner Piet de Jong) with the same logic and technique applied to the palace. The western part of the renovated room of the complex was reserved for public baths (Figure 1.11). The entrance was from the north site with three tiers and from the south side *via* a scale with five levels from the inner courtyard. The space of the waiting room (length 5.75 m and width 2.10 m) occupied a tank measuring 1.52 × 1.38 m^2. The walls were made of hewn limestone. Around the tank, the Minoans could seat and rest in the opinion of Evans (Figure 1.12a). He also believed that a footbath for the weary travellers was supplied by a direct pipe, and the overflow discharged by another conduit (Figure 1.12b); a branch of the water channel also served as a drinking trough (Angelakis & Spyridakis, 1996a).

The filling and emptying of the tank was very pretentious system: A stone conduit leading water from the south under the steps of the tank, where it connects with a clay pipe, in which a parabola controlled flow. On the western side of the tank from a small hole below the water slides immediately find a way out of the rightmost entry, which concentrated on a small stone basin, from which they could drink water animals. In the northwest corner of the tank was the overflow channel, which ran through the wall diagonally. This should be associated with a north-western direction conduit, who communicated with the overflow channel of the neighbouring fountain. In order to empty and clean the tank was a circular hole in the north-eastern corner, leading the water to drain below. Thus, it appears that six different conduits were consisted the complex system of the water cistern thereof.

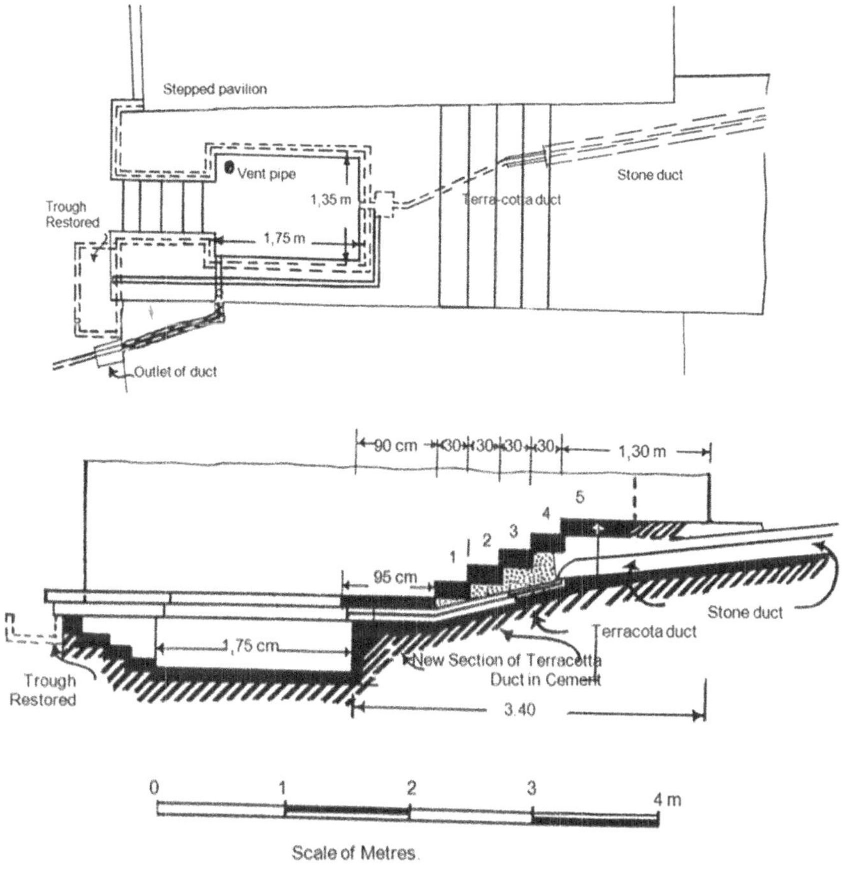

Figure 1.11 The complex in *Caravanserai* reserved for public baths: Plan (up) and section (down) (Modified from Evans, 1921–1935).

In the western part of the band's *Caravanserai* is the restored 'Room of Spring' or 'Minoan Fountain.' It is a rectangular underground structure with a central cistern-basin in tiled floor, which was accessible from the courtyard of the adjacent building with three descending stairs at the end of which the right was a circular slot door. The walls of the 'Room of the Fountain' was constructed of large limestone blocks and slabs of internally coated gypsum.

The water depth in the cistern had reached the 45 cm, as a result of an artesian spring naturally from the earth, among the pebbles forming the bottom. The plate of the entrance was very worn, perhaps because of the filling process of the vessel with water. An open channel overflow is visible below this plate to B. Then, under the yard. Subsequently, easternmost it was twisted and met another line from the overflow drainage system of the footbath. From there, they both joined and continued down the ancient road.

Figure 1.12 (a) Tank in *Caravanserai* which in the opinion of Evans was used for footbaths and (b) the drain of the wastewater (with permission of E. Kavoulaki and A. N. Angelakis, respectively).

1.3.4 Roads

Paved roads connected the palaces and towns of Crete through the interior of the island. The first paved road network was constructed in the middle Minoan period (*ca.* 1700–1450) and reached impressive density during the late Minoan period (after *ca.* 1450). While ancient paved roads are difficult to identify today in the countryside, several well preserved examples appear in the Minoan ruins of Gournia and other Minoan palaces (Driessen & Schoep, 1994).

Most streets in Minoan settlements date to Middle Minoan period; later ruins follow earlier alignments, beneath the pavings, and some middle Minoan houses were actually constructed on top of the earlier streets. This is the case at Gournia, Malia, and Palekastro, where the street system probably dates to Middle Minoan period but was subsequently repaired. Likewise, at Kommos in the southern area of Iraklion, a fine paved road preceded the construction of Building T. A good paradigm is the ancient road in Knossos known as Royal Road, which is considered the 'oldest road in Europe' (Figure 1.13a). The Royal Road appears to be a Late Minoan reconstruction of an earlier pavement (Driessen & Schoep, 1994). Several grandstands were made in middle Minoan times along the Royal Road, with only part being rebuilt at the very end of late Minoan times. It connects the palace (Theatre area), the House of Frescos, and the Little palace. It is very well paved with stone slabs and on either side, there are cement wings and drains (Pendlebury, 1964).

Figure 1.13 Drains in central roads: (a) The 'Royal Road' known as the 'oldest road in Europe' in the palace of Knossos and (b) at Palekastro town (with permission of A. N. Angelakis).

In addition a similar situation is observed in the town of Palekastro in the eastern Crete where the central road of the town is well designed and constructed by considering the drainage of the rainwater shown in the left side of the road (Figure 1.13b). All the central sewers were beneath the streets (e.g., Knossos and Phaistos), were drainage systems were on both sides of the streets and stairs.

Among the advanced rainwater drainage techniques practiced in Minoan Crete is the drainage system found in the outer stairway of the northeast wing of the palace at Knossos near the Kairatos river. On one site of stairway there is a small open drain that follows a parabolic flow instead of the line of the stairs (Figure 1.14). In this manner the rainwater, not only flows down smoothly without causing any erosion, but it does not become a nuisance to the people using the stair (Angelakis & Spyridakis, 1996a).

By the Late Brinze Age period a well-planned drainage system and paved roads in Hagia Irini town in Kea island is referred (Weisman, 2008). By this time, Hagia Irini had become a strongly Minoanized settlement. This Minoanization can be seen in all aspects of life, from the types of ceramic vessels used in everyday activities; to the proliferation of Minoan iconography in local art and religion; to the appearance of the dwelling-place of the local ruler, who was probably Minoan or whose authority was derived from Crete at the very least. Weisman (2008) is describe the creation of a town-wide drainage system consisting of a network of stone and terracotta drains was put in place beneath the paving stones of the town streets. Some of these pipes were linked to upstairs rooms and at least two rooms in House A had sluices leading out into covered drains in the alleys. The drainage system was used not only for indoor plumbing but also to carry off rainwater that would otherwise flood cellars and spoil goods.

Figure 1.14 Part of restored Minoan stairway with parabolic runnels in the north eastern entrance (with permission of A. N. Angelakis).

On the other hand in the Pompeii, 1000 years later, the streets were a sort of open channel conveying water coming from public fountains, rainwater and segregate sewage. Therefore, as shown in Figure 1.15, there were raised sidewalks (50–60 cm high) in the streets with stepping stones (pondera) at the street corners to enable pedestrians to cross from one side to the other without stepping down (De Feo *et al.*, 2010).

Flood protection of 'Minoan viaduct'. The 'Minoan viaduct' is one of the most impressive structures in the archaeological site of Knossos palace and the bulkier technical work of Minoan Crete, discovered so far. According to the excavator A. Evans, about half of its height was saved. It consists of four columns (width 3.2–4.60 m), made of carved limestone alternating with stepped openings, probably for the free passage of stormwater flowing from the steep hillside. The stepped openings (width 2.30–3.10 m) are from three to four tiers. In the opinion of Evans, it was arched and was made with ekforic system in order to be very well protected (Figure 1.16).

The Minoan palace means of entry so passed over the 'Viaduct', a height of about 10 m from ground level, uniting the two banks of the stream Vlichia. This was the main road linking the south to the north coast of Crete. The people coming from the interior of the island were rested in the *Caravanserai* and then continue north through the 'Viaduct' and the 'Stepped Portio', a majestic building with steps and colonnade entering the palace.

1.3.5 Toilets or lavatories

In several Minoan houses, the lavatories were usually located in the private living rooms and well positioned. In most cases the evidence for the identification of a lavatory is little more than the existence of a sewer at the floor level passing through the exterior wall and connecting with the outside central sewerage and drainage system. In some cases there are traces of some sort of provision for a stone or wooden seat.

Figure 1.15 Stepping stones (pondera) in Pompeii (with permission of G. De Feo).

Figure 1.16 Part of the original wall of the 'Minoan Viaduct' with the stepped openings in the *Caravanserai* at Knossos palace (with permission of A. N. Angelakis).

One of the most interesting rooms in the ground-floor in the residential quarter of the Knossos palace was identified as a lavatory. The queen's Megaron contained an example of the first water-flushing toilet system adjoining the bathroom. This toilet was a seat over a drain that was flushed by pouring water from a jug. The bathtub located in the adjoining bathroom similarly had to be filled by someone heating, carrying, and pouring water, and must have been drained by overturning into a floor drain or by bailing. Remains of a clay tube were found just outside the door of the room. Apparently, water was poured through a hole in the floor immediately outside the lavatory door; an under-floor channel linked the hole with the vertical clay pipe under the lavatory seat (Castleden, 1993). The lavatory could thus be flushed even during a rainless summer, either by an attendant outside the lavatory or by the user. At certain times of the year the drains in the palace of Knossos may have adequately flushed out by the rainfall that fell into the light-wells; but, in general, Evans (1921–1935) supposed that water was poured into the lavatories to flush them. He also observed that there was sufficient at the end of the seat at Queen's Megaron lavatory in Knossos for a large pitcher. This toilet and bathtub were exceptional structures within the 1300-room complex. Evans (1921–1935) concludes with evidence satisfaction that: *'As an anticipation of scientific methods of sanitation, the system of which we have here the record has been attained by few nations even at the present day* (of about a century ago).

The lavatory is similar in function to that of the so called Queen's Megaron and those found in the Phaistos and Malia palaces and in some of Minoans towns and houses as well. Fortunately, one of the houses near the palace at Malia, known as Da (Figure 1.17a), contains a lavatory seat in nearly perfect condition, since it was made not of wood, like the seat of the palace of Knossos in Knossos, but of solid stone (Antoniou & Angelakis, 2013). This stone seat is 68.50 cm long by 45.50 cm wide front to back and its surface is 34–38 cm above the floor (Figure 1.17b). It is built directly against an outside wall through which a large sewer passes. Like in Knossos, the structure was evidently intended to be used as a seat rather than as a stand; thus, it resembles the 'Egyptian' toilet more closely than the so-called 'Turkish toilet' type found in the palaces at Mari and Alalakh in Syria (Angelakis *et al.*, 2005). However, there is a substantial difference of those lavatories compared to the Minoans, which is due to their flushing processes and their connections to the sewers. The lavatory illustrated in Figure 1.16b is probably the earliest flush lavatory in history. A similar lavatory has been discovered in the west side of the so called Queen's apartment at Phaistos. It was connected to a closed sewer, part of which still exists. Another lavatory sewer was discovered in House C at Tylissos (Angelakis & Spyridakis, 1996a). In addition, most Minoan lavatories were located near or next to the bathrooms (e.g., that in Queen's Megaron at Knossos, Queen's Apartment at Phaistos, and that in Da house in Malia).

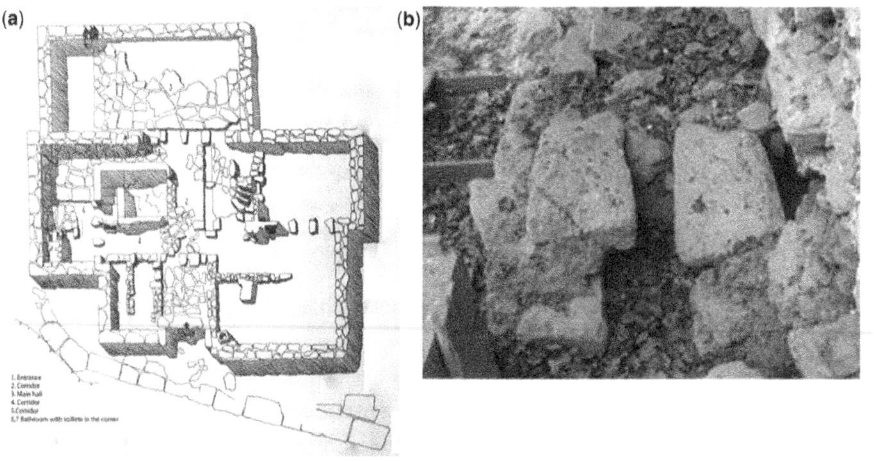

Figure 1.17 Toilet in the House Da, Malia: (a) Layout of the house (adapted from Graham, 1984) and (b) a recent photo of toilet (no 6 in the layout plan) (with permission of A. N. Angelakis).

Some palaces had lavatories with flushing systems operated by pouring water in a conduit (Angelakis *et al.*, 2005). However, the best example of such an installation was found in the island of Thera (Santorini) in the Cyclades (Palyvou, 2005). This is the most eloquent and best-preserved example belonging to *ca.* 1550 BC, in the Bronze Age settlement of Akrotiri, which shares the same cultural context with that of Crete.

At certain times of the year, the sewers in the palace of Knossos may have been flushed out adequately by the rain that fell into the light-wells; however, it appears that water was poured into the lavatories to flush them. In fact, Evans noted the existence of sufficient space for placing a large pitcher at one end of the seat at Knossos and concluded with evident delight (Angelakis *et al.*, 2005).

Also a remarkably sophisticated plumbing and drainage system in the House A is described in Hagia Irini in Cycladic Kea island in the late Bronge Age (Weisman, 2008). The toilet in Room 24 of that House had a drain encrusted with lime deposits that conducted water into the drain beneath the paving slabs of the street below. Cummer and Schofield (1984) suggest that this toilet had a flush mechanism similar to that of Knossos palace toilet. They stated: 'A down spout from the roof probably emptied into the head of the drain, producing a strong, if seasonal, flow of water across the outlet from Room 24, perhaps rivaling the flush toilets of Knossos.' In the next room no 23, the adjacent lightwell, had drains which were connected to those of Room 24, to ensure the removal of rainwater.

On the other hand in Roman period, 1000 years later, lavatories did not have advanced plumbing for flushing. They used usually pots that were emptied outside. For this purpose, extensive streets washing program was applied in most Roman cities. Thus, most of the water used for washing the streets eventually ended up in the sewers. At that period the hygienic conditions in both public and private toilets must have been very poor, and consequently intestinal diseases were diffused (De Feo *et al.*, 2010). Dysentery, typhoid fever and different kinds of diarrhoeas are likely candidates for diagnoses.

1.3.6 Outlets and disposal and reuse sites

The sewerage and drainage systems of Minoan palaces and towns have a lot of similarities in the principles of their design and construction. The sanitation drainage in the palace of Knossos was through a closed system leading to a sewer apart from the hill. The end section of the main part of the sewerage system of the palace is shown in Figure 1.18a. The outlet of the Phaistos palace system appears to be similar (Figure 1.18b). Note that Evans (1921–1935) and MacDonald and Driessen (1988) considered that the main part of the system had been planned and constructed originally in Middle Minoan time. The main disposal sites at the Knossos and Zakros palaces were directed to the Kairatos river and to the sea, respectively. However, there are indications that in the palace of Phaistos and in the villa of Hagia Triadha cisterns were also used as disposal sites of surface water, along with appropriate landforms.

Figure 1.18 Outlet of the central Minoan sewerage and drainage systems: (a) palace of Knossos and (b) palace of Phaistos (with permission of A. N. Angelakis).

As indicated in the Introduction, several periods during the Minoan Era were under severe water shortages. Thus, reuse of water was a necessity. In order to maximize and utilize the meagre water resources available, it was necessary to use and reuse water. Certain human activities produce light water. Today, used water is called 'gray water'. This includes bath water which may be reused to wash floors. Bath or cooking water might be given to animals, or used to irrigate house plants, since only soap but not harmful detergents had yet been invented in ancient times (Crouch, 1996).

Non-potable water included mainly storm runoff and sewage. Storm water could be utilized as irrigation water, especially valuable when enriched by sewage placed up along the way which converted it into fertilizer. Houses and public buildings customarily had gutters and drains that connected with the sewers under the streets. These, like their modern counter pans existed under the gates and were emptied into the dry bed of the nearest river, which served as a drainage channel. Today, it is generally recognized that wastewater was first used for irrigation in Minoan times (Angelakis & Spyridakis, 1996b).

Any type of used water could serve to flush the latrine into the sewer. Various examples of this practice are known from ancient Corinth, where the overflow water from fountains was utilized in latrines and baths. Similarly, in Roman times water from swimming pools and plunge baths was channelled through latrines for Hushing purposes. This, like so many Roman wastewater management practices, was derived from Minoan sources.

As mentioned above, the drains at one site of the northeaster stairway-entrance of the palace at Knossos follows a parabolic flow instead of the line of the stairs. In this manner, the storm water not only flows down smoothly without causing any erosion, but does not become a nuisance to people entering the palace (Angelakis & Spyridakis, 1996a). These runnels with their parabolic curves following the turns of outer stairways, and the further arrangement of their channels, secure the confinement of sediment in intermediary little tanks, placed at proper intervals, to allow it to settle to the bottom (Michailidou, 1992). This elaborate method of collecting the maximum amount of rain-water, freed from impurities, in a downstream tank for washing or other purposes, is only one indication of the highly skilled hydraulic knowledge attained by the Minoans. The special suitability of rain-water for washing linen, encourages the assumption that the tank was used for this purpose and that Minoan maidens may have come here from the palace halls above to do the royal laundry. Similarly, in the villa of Hagia Triadha, surface water from the stormwater system passed through a rectangular cistern (about $1.6 \times 2.0 \times 6.0$ m). It may be speculated that water from this cistern was probably used for washing or other reuses.

It must be concluded, therefore, that Minoan hydrologists and engineers were aware, to some degree, of the principles of siphon, communicating vessels, Pascal's law and Archimedes' principle of what we call today water and environmental sciences.

1.4 CONCLUSIONS

In general, the use of traditional knowledge does not directly apply techniques of the past but instead, try 'to understand the logic of this model of knowledge' (Laureano, 2007). Traditional knowledge allowed ancient societies to keep ecosystems in balance, carry out outstanding technical, artistic, and architectural work that has been universally admired. The use of traditional knowledge has been able to renew and adapt itself. Traditional knowledge incorporates innovation in a dynamic fashion, subject to the test of a long term, achieving local and environmental sustainability.

Minoans originally and Hellenes and Romans thereafter are considered pioneers in the development of the basic water and wastewater technologies, and especially on sanitation in urban environment, in a sustainable way. Minoans lived in harmony with nature and environment. Until Classical times, Minoan plumbing and drainage were the most developed in what, was then, the western world. Complex open-topped sewerage and drainage systems carried storm water and sewage. Crete may be the home of the first urban sewers and drains, a 'flush' toilet, a ground-floor bath, and an overhead water cistern.

The evolution of urban wastewater and stormwater management in ancient Hellas began in Crete during the Middle Bronze and the beginning of the Late Bronze Age. Numerous remarkable developments occurred at various stages of Minoan Era which flourished during these particular periods. One of its salient characteristics was the architectural and hydraulic function of its sewerage systems in the Minoan palaces and several other settlements. It might be implied, therefore, that Minoan master craftsmen in Bronze Age Crete were aware of some of the basic principles of what we call today sanitation, wastewater, and environmental technologies. Archaeological and other evidence indicate that Minoan technicians originally, during the Bronze Age and by Hellenic philosophers, mathematicians, and engineers thereafter developed the basic principles of hydraulics. With respect to the wastewater management, Minoans developed an advanced, comfortable and hygienic lifestyle, as manifested from flushing lavatories, public and private baths and very effective sewers and drains. The hydraulic and architectural function of sewer systems in palaces and towns are regarded as one of the salient characteristics of the Minoan civilization. In this Chapter the most characteristic examples of extant hydraulic works and technologies relevant to wastewater and stormwater developed during the Bronze Age in Crete, Hellas are presented and discussed.

The Minoans were forerunners of modern sanitation systems. Archaeological studies have established unequivocally that, the origin of modern technologies of water management dates back to ancient Crete (Lofrano & Brown, 2010). The status of urban sewage and stormwater drainage systems in Minoan Crete is well documented in this Chapter. Thousands of years ago, the Minoans implemented several outstanding projects to cope with the watewater managemet and violent floods. Some of these projects are still in use today. These projects evolved from the experience and knowledge accumulated through the long coexistence of people with nature. The concepts behind these ancient wastewater and stormwater management practices, such as low-impact development and sustainable drainage systems, are similar to the technology applied in modern stormwater management. The Minoan knowledge behind these achievements is seen in the design concepts and the features of these projects. These features help us to understand better their applications in the contemporary environment. In today's more complex environment, integrating traditional and advanced philosophy with modern technologies is extremely useful in building rural wastewater and stormwater management systems.

It was concluded that during the ancient Hellenic civilizations extensive drainage and sewerage systems and elaborate sanitary structures were implemented. Such technologies were practiced as early as in the Minoan and Mycenaean civilizations and continued at a similar level in other places during the prehistoric Hellenic civilizations. In light of these historical and archaeological evidences, it turns out that the progress of present day in urban water and wastewater technologies as well as in comfortable and hygienic living is not as significant in terms of evolution as we tend to believe (Koutsoyiannis & Angelakis, 2003). Many civilizations, which were great centers of power and culture, were built on locations that could not support the populations that developed. Now we find ourselves in similar situations in many places around the world. How do we balance the mega water projects with the methods of traditional knowledge? De Feo *et al.* (2011) explored the legacies and lessons on urban water and wastewater management learned from the ancient Hellenic civilizations. They summarized the lessons learned as follows: (a) The meaning of sustainability in modern times should be re-evaluated in light of ancient public works and management practices. Technological developments based on sound engineering principles can have extended life span. And (b) Security, with respect to water and wastewater, is of critical importance in the sustainability of a population.

Finally, Gray (1940) gives clearly the great interest of the Minoans in sanitation as follows: '*We often hear to talk about the 'modern' hygiene as it was something that developed recently and there seems to be a prevailing idea that urban sanitation is something very modern established somewhere in the middle of the last 19th century. Perhaps these ideas are trying to strengthen a somewhat problematic idea in the modern culture [.....], But when examined in the light of history that is anything but new or recent. Indeed, in the light of history, it is surprising, if not bitterness, that man has gone so poorly, if at all, about four thousand years [.....]. Archaeological evidences shows that people [Minoans] had to proceed very comfortable and healthy living, with a considerable degree of beauty and luxury [.....]. This was done about four thousand years ago.'*

Acknowledgements

The authors are grateful for Prof. (Em.) S. Alexiou for his critical comments.

1.5 REFERENCES

Alexiou S. (1964). Minoan Civilization. Sons of S. Alexiou, Iraklion, Greece, (in Hellenic).

Angelakis A. N. and Koutsoyiannis D. (2003). Urban water resources management in ancient Greek times. In: The Encycl. of Water Sciences, B. A. Stewart and T. Howell (eds), Markel Dekker Inc., Madison Ave. New York, N.Y., USA, pp. 999–1008.

Angelakis A. N. and Spyridakis S. V. (1996a). The status of water resources in Minoan times: a preliminary study. In: Diachronic Climatic Impacts on Water Resources with Emphasis on Mediterranean Region, A. N. Angelakis and A. S. Issar (eds), Springer-Verlag, Heidelberg, Germany, pp. 161–191.

Angelakis A. N. and Spyridakis S. V. (1996b). Wastewater Management in Minoan Times. In: Proc. of the Meeting on Protection and Restoration of Environment, August 28–30, Chania, Hellas, pp. 549–558.

Angelakis A. N. and Spyridakis S. V. (2013). Major urban water and wastewater systems in Minoan Crete, Greece. *Water Science and Technology, Water Supply*, **13**(3), 564–573.

Angelakis A. N., Koutsoyiannis D. and Tchobanoglous G. (2005). Urban wastewater and stormwater technologies in the ancient Greece. *Water Research*, **39**(1), 210–220.

Angelakis A. N., Dialynas M. G. and Despotakis V. (2012a). Evolution of water supply technologies in Crete, Greece through the centuries. In: Evolution of Water Supply Throughout Millennia, A. N. Angelakis, L. W. Mays, D. Koutsoyiannis and N. Mamassis (eds), IWA Publishing, London, UK, pp. 227–258.

Angelakis A. N., Koutsoyiannis D. and Papanikolaou P. (2012b). On the geometry of the Minoan water conduits. In: Proc. of the 3rd IWA International Symposium on Water and Wastewater Technologies in Ancient Civilizations, May 28–30, 2009, Istanbul, Turkey, 2009, pp. 172–177.

Antoniou G. P. (2007). Lavatories in Ancient Greece. *Water Science and Technology, Water Supply*, **7**(1), 155–164.

Antoniou A. and Angelakis A. N. (2014). Sanitation Technologies in Ancient Greece. *Journal of Early Medicine Issue on Latrines, Hygiene and Health* (in press).

Castleden R. (1993). Minoans: Life in Bronze Age Crete. Routledge, London, UK.

Chamonard J. (1924). *Le Quartier du Theatre.* Contient vol. 1: Etude sur l'habitation délienne à l'époque hellénistique. vol. 2: Construction et technique. vol. 3: Planches. Paris, France: 1922–1924. *Collection Exploration archéologique de Délos*, **8**, 1-2-3.

Crouch D. C. (1996). Avoiding water shortages: some Ancient Greek solutions. In: Diachronic Climatic Impacts on Water Resources with Emphasis on Mediterranean Region, A. N. Angelakis and A. S. Issar (eds), Springer-Verlag, Heidelberg, Germany, pp. 129–159.

Cummer W. W. and Schofield E. (1984). Keos III. Ayia Irini: House A. Philipp von Zabern, Mainz, Germany.

De Feo G., Laureano P., Drusiani R. and Angelakis A. N. (2010). Water and wastewater management technologies through the centuries. *Water Science and Technology: Water Supply*, **10**(3), 337–349.

De Feo G., Mays L. W. and Angelakis A. N. (2011). Water and wastewater management technologies in Ancient Greek and Roman Civilizations. In: Treatise on Water Science, P. Wilderer (ed.), Academic Press, Oxford, UK, Vol. **1**, pp. 3–22.

De Feo G., Antoniou G. P., Mays L., Dragoni W., Fardin H. F., El-Gohary F., Laureano P., Kanetaki E. I., Zheng X. Y. and Angelakis A. N. (2013). Historical development of waste water management. In: Handbook of Engineering Hydrology, S. Eslamian, (ed.), Taylor & Francis Group, 6000 Broken Sound Parkway NW, Boca Raton, Florida 33487, USA, Vol. 2, Ch. 9 (in press).

Defner, M. (1921). Διυλιστηριο *Υστορομινωικης Εποχος*. Αρχαιολογικη Εφημεριδα (Water purification filter of Late Minoan Era. Archaeological Journal), Iraklion, Hellas, Vol. **78** (in Hellenic).

Delorme J. (1960). Gymnasion: étude sur les monuments consacrés à l'éducation en Grèce, des origines à l'empire Romain. De Boccard Paris, France.

Driessen J. and Schoep I. (1994). The architect and the scribe: Political implications of architectural and administrative changes on MM 11-LM IIIA Crete. In: Proceedings of the 5th International Aegean Conference/5e Rencontre égéenne internationale, April 10–13, 1994, Robert Laffineur and Wolf-Dietrich Niemeier, University of Heidelberg, Archäologisches Institut, Vol. 2, pp. 649–664.

Evans S. A. (1921–1935). The Palace of Minos at Knossos: A Comparative Account of the Successive Stages of the Early Cretan Civilization as Illustrated by the Discoveries. Macmillan and Co., London, UK, Vols. I–IV (reprinted by Biblo and Tannen, New York, 1964).

Graham J. W. (1987). The Palaces of Crete. Revised Ed. Princeton University Press, Princeton, New Jersey, USA.

Gray H. F. (1940). Sewerage in Ancient and medieval times. *Sewage Works J.*, **12**(5), 939–946.

Koutsoyiannis D. and Angelakis A. N. (2003). Hydrologic and hydraulic sciences and technologies in Ancient Greek times. In: The Encycl. of Water Sciences, B. A. Stewart and T. Howell (eds), Markel Dekker Inc., Madison Ave. New York, N.Y., USA, pp. 415–417.

Koutsoyiannis D., Zarkadoulas N., Angelakis A. N. and Tchobanoglous G. (2008). Urban water management in Ancient Greece: legacies and lessons. *ASCE, Journal of Water Resources Planning & Management*, **134**(1), 45–54.

Laureano P. (2007). Ancient water techniques for proper management of Mediterranean ecosystems. *Water Science and Technology: Water Supply*, **7**(1), 237–244.

Lawrence A. W. (1957). Greek Architecture. (The Pelican History of Art.). West Drayton: Penguin Books, Harmondworth, London, UK, pp. xxxiv + 327.

Lofrano G. and Brown J. (2010). Wastewater management through the ages: a history of mankind. *Science Total Environment*, **408**(22), 5254–5264.

Lyrintzis A. and Angelakis A. N. (2006). Is the 'Labyrinth' a water catchment technology? A preliminary approach. In: IWA Specialty Conference: 1st International Symposium on Water and Wastewater Technologies in Ancient Civilizations, A. N. Angelakis and D. Koutsoyannis (eds), National Foundation for Agr. Research, Athens, Iraklion, Hellas, pp. 163–174.

MacDonald C. F. and Driessen J. M. (1988). The drainage system of the domestic quarter in the palace at Knossos. *British School of Athens*, **83**, 235–358.

MacDonald C. F. and Driessen J. M. (1990). The Storm Drains of the East Wing at Knossos. In: *L'habitat égéen préhistorique* (The prehistoric habitation of Aegean), P. Darcque and R. Treuil (eds), *Bulletin de Correspondance Hellénique*, Athens, Hellas, Supplément 19, pp. 141–146.

Mariolakos I. 2012. Οι ξεχασμενες γεωγραφικες και φυσικο-ωκεανογραφικες γνωσεις των πρωνστiρικνν Ελλιηνων (in Hellenic). http://www.ekke.gr/estia/Cooper/Mariolakos/New_Mariolakos_greek.pdf (accessed 9 Febraury 2013).

Mays L. W. and Angelakis A. N. (2012). Gods and goddesses of water. In: Evolution of Water Supply Throughout Millennia, A. N. Angelakis, L. W. Mays, D. Koutsoyiannis and N. Mamassis (eds), IWA Publishing, London, UK, pp. 1–18.

Mays L. W., Sklivaniotis M. and Angelakis A. N. (2012). Water for human consumption through the history. In: Evolution of Water Supply Throughout Millennia, A. N. Angelakis, L. W. Mays, D. Koutsoyiannis and N. Mamassis (eds), IWA Publishing, London, UK, pp. 19–42.

Michailidou A. (1992). Knossos. A Completed Guide to the Palace of Minos. Ekdotiki Athenon S.A., Athens, Hellas.

Mosso A. (1907). Escursioni nel Mediterraneo e gli scavi di Creta. Treves, Milano, Italy.

Neudecker R. (1994). Die Pracht der Latrine: zum Wandel öffentlicher Bedürfnisanstalten in der kaiserzeitlichen Stadt München. Pfeil, Munich, Germany.

Orlandos A. (1940). The pole of the roman building located northern of horologe of Andronikos (Kiristou). In: Proceeding of Athens Academy, Athens, Hellas (in Hellenic).

Orlandos A. and Travlos L. (1986). Dictionary Ancient Architectural Terms. The Athens Archaeological Association, Athens, Hellas (in Hellenic).

Palyvou C. (2005). Akrotiri Thera: An Architecture of Affluence 3500 Years Old. INSTAP Philadelphia, USA, pp. 41–42, 51–53.

Pendlebury J. D. S. (1964). The Archaeology of Crete. Methuen's Handbooks of Archaeology, London 1939. Reprinted 1964. London, UK.

Platon M. (1990). New indications for the problems of purgatory cisterns and bathrooms in Minoan World. In: Proceedings of the Sixth International Cretologic Congress. Literary Association Chrysostomos, Chania, Hellas, **A2**, 141–155 (in Hellenic).

Platon N. (1974). Zakros, The New Minoan Palace. The Athens Archaeological Society, Athens, Hellas (in Hellenic).

Robinson D. M. (1934). Excavations at Olynthus pt. 2. Architecture and Sculpture. John Hopkins Press, Baltimore, USA.

Sakellarakis J. A. and Sapouna-Sakellaraki E. (1991). Crete Archanes. Ekdotiki of Athenon Athens, Hellas, ISBN 960-213-234-5 (guidebook in Hellenic).

Scatzmann P. (1932). Kos I, Asklepieion. A.I.d.D R, Berlin, Germany.

Shaw J. W. (1973). Minoan architecture: materials and techniques. In: Annurario della scuola archeologica di Atene et delle missioni italiani in Oriente. Instituto Poligrafico dello Stato, Rome, Italy, 49, pp. 1–257.

Spanakis S. (1981). The Water Supply of Iraklion, 828–1939. The Technical Chamber of Greece, Iraklion, Hellas (in Hellenic).

Viollet P.-L. (2003). The predecessors of European Hydraulic Engineers: Minoans of Crete and Mycenaeans of Greece (2100–1200 BC). In: XXX IAHR Congress, Theme E: Linkage Between Education Research and Professional Development in Water Engineering, August 2003, Thessaloniki, Hellas.

Voudouris K., Mavrommatis Th., Daskalaki P. and Soulios G. (2006). Rainfall variations in Crete island (Greece) and their impacts on water resources. *Publicaciones del Instituto Geologico y Minero de España.* Serie: Hidrogeologia y aguas subterráneas No 18, Karst, climate change and groundwater, Madrid, pp. 453–463.

Weisman S. (2008). An Analysis of the Late Bronze Age Site Of Ayia Irini, Keos. http://askthevaledictorian.files.wordpress.com/2011/11/keos1.pdf (accessed 9 May 2014).

Chapter 2

Sanitation and wastewater technologies in Harappa/Indus valley civilization (*ca.* 2600–1900 BC)

Saifullah Khan

2.1 INTRODUCTION

Historians know less about the civilization in the Indus Valley than about those to the west. They have not yet deciphered the Indus system of writing. Evidence comes largely from archaeological digs, although many sites remain unexplored, and floods probably washed away others long ago. At its height, however, the civilization of the Indus Valley influenced an area much larger than did either Mesopotamia or Egypt.

No one is sure how human settlement began in the Indo-Pakistan subcontinent. Perhaps people who were arrived by sea from Africa settled in the south of the Indus valley. Northern migrants may have made their way through the Khyber Pass in the Hindu Kush Mountains. Archaeologists have found evidence in the highlands of agriculture and domesticated sheep and goats dating to about 7000 BC. By *ca.* 3200 BC people were farming in villages along the Indus River (Smith, 2006).

Around 2500 BC, while Egyptians were building pyramids, people in the Indus Valley were laying the bricks for India's first cities. They built strong levees, or earthen walls, to keep water out of their cities. When these were not enough, they constructed human-made islands to raise the cities above possible floodwaters. Archaeologists have found the ruins of more than 100 settlements along the Indus and its tributaries mostly in modern-day Pakistan. The largest cities were Kalibangan, Moen-Jo-Daro, and Harappa. Indus Valley Civilization is sometimes called Indus Valley Civilization , because of the many archaeological discoveries made at that site (Carry, 2012).

One of the most remarkable achievements of the Indus Valley people was their sophisticated city planning. The cities of the early Mesopotamians were a jumble of buildings connected by a maze of winding streets. In contrast, the people of the Indus laid out their cities on a precise grid system. Cities featured a fortified area called a citadel, which contained the major buildings of the city. Buildings were constructed of oven baked bricks cut in standard sizes, unlike the simpler, irregular, sun-dried mud bricks of the Mesopotamians. Early engineers also created sophisticated plumbing and sewage systems. These systems could rival any urban drainage systems built before the 19th century. The uniformity in the cities' planning and construction suggests that the Indus peoples had developed a strong central government (Wright, 2010). So far as the origin of the word Indus is concerned, Scholars are of the opinion that the name 'Indus' is the origin of the word 'Hindu.' The original Indian name of the river is Sindhu. The ancient Iranians had difficulty in pronouncing an initial sibilant 's' and changed it to an aspirate sound 'h', hence, 'H-indu' instead of 'S-indhu.' The Greeks referred to the river as the 'Indos,' and the later Arabs referred to it as 'al-Hind.' Eventually the name came to be applied to the people of the subcontinent, namely, the 'Hindi,' the 'Hindus' and the 'Indians' (Kosambi, 1964). It is believed that the Indus Valley Civilization belonged to the copper Stone Age as the presence of iron tools and implements has not yet been established at any part of this civilization.

The Indus civilization is best known through the excavation of Moen-Jo-Daro which literally means 'The mound of the dead,' which is located in Sind, about 40 kilometres from Larkana town, and another almost due north of Hyderabad city about 209 kilometres (Gray, 1940). A similar and larger ruin, Harappa, is located some 644 kilometres northeast of Moen-Jo-Daro in Punjab province, and in this general area between Moen-Jo-Daro and Harappa there are numerous mounds indicating other cities of same Indus civilization, and in addition there are mounds of an even earlier and distinct Amri Culture. The third site lothal (dockyard of the Indus Valley Civilization) was discovered some 60 years ago, located about 7 kms away from the Ahmedabad-Bhavnagar highway, India (Agarwal, 2009).

The upper levels of Moen-Jo-Daro are dated about 2550 BC. The lower levels probably go back much further, but high groundwater has prevented effective excavation into these earlier cities (Kenoyer, 1998). The mounds above the ruins average 2 to 3 meters in height (with maximum of 7 m) and cover an area of about one kilometre (Gray, 1940), but floods and erosion by Indus river have probably greatly reduced the original extend of the ruins (Figure 2.2).

2.2 PHYSICAL SETTINGS

As the name denotes, the greater Indus region was home to the largest of the four Ancient urban civilizations of Egypt, Mesopotamia, South Asia and China and was spread along river Indus and its tributaries that is Jhelum, Chenab, Ravi, Beas and Sutlej. Geographically, the civilization was covered an area of one million square kilometres (Sharma, 1992), bounded by great Himalayas in the North, Arabian Sea in the South, Rajasthan desert in the East and rugged hills and plateaus of Balochistan in the West. The most out standing sites are Moen-Jo-Daro, Harappa, Kot Diji, Dhulavira, Rakhigarhi, and Lothal (Figure 2.1). The remains of the Harappan settlements are located in a vast desert region of Cholistan, Thar, Nara, and Kharan deserts in the lower Indus basin. The Indus civilization sites have arid continental climate (precipitation less than 254 mm) with hot long summers and short winters (Hasan & Khan, 2010). Some of the sites like lothal and Sutkanjen Dor have marine climate and under the influenceof the Indian ocean (Figure 2.1).

There are two theories about the climate of Indus civilization. According to the first theory the climate has been changed. It used to be wet, rainy and severe. There were a lot of floods as well especially during snowmelt in March. These floods resulted in the change of the river course. The agricultural growth mainly depended on the rains. Moen-Jo-Daro was destroyed nine times by the heavy floods. The second theory says that the weather was hot and dry and in support of this argument they say that the baked bricks were only used for construction purposes on the riverbanks and not used in Baluch areas away from the river (Admin, 2005).

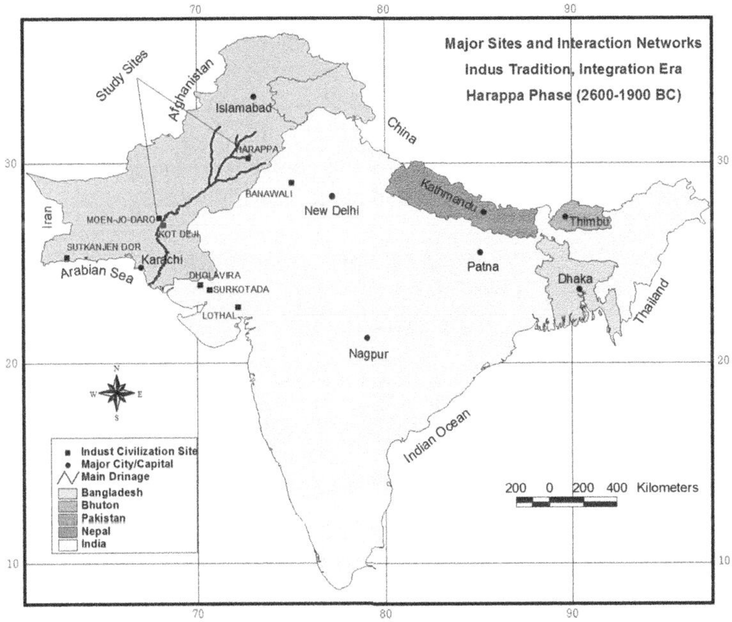

Figure 2.1 Indus valley Civilization, Major sites and Interaction Networks (*ca.* 2600–1900 BC).

2.3 HISTORY, CULTURE, AND TOWN PLANNING OF INDUS VALLEY CIVILIZATION

2.3.1 Introduction

The current work deals with the evaluation of sanitation and wastewater management and town planning of the Indus Valley Civilization with special reference to Harappa, Moen-Jo-Daro, Dholavira and Lothal. The work is based on the personal visits of the sites and literature. Some of the information regarding the Indus Valley Civilization have been collected from the field, whereas most of the historical information from the work of the previous workers. The work explains historical background, town planning, culture, drainage, sanitation, toilets, baths, water reservoirs, and dockyards of the Indus Valley Civilization. Harappa located at 72°–75′ east longitude and 30°–53′ north latitude at the eastern bank of Indus river in Upper Indus Basin. The second major site of the Indus civilization Moen-Jo-Daro, located at 68° east longitude to 27°–42′ north latitude at the western bank of Indus river in Sind province, Lower Indus Basin, Pakistan. The third site of Dholavira is at 70°–17′ east longitude to 24° north latitude at the southeastern border of Pakistan near coast of Indian ocean in Gujrat district of India. The forth site, Lothal is located at 72°–20′ east longitude to 22°–78′ east latitude on the coast of Indian ocean, district Gujrat, India (Figure 2.1).

2.3.2 History

During the 1980's, while the construction of railway line was going on in Pakistan, French archaeologists found the remains of Indus valley Civilisation at Harappa and Moen-Jo-Daro. They excavated it further to discover systematic housing colonies built by mud of the Mehargarh people. Further excavation leaded to discovery of their amazing irrigation and drainage system. Unearthed ornaments, plates and dishes, drinking glasses, tools made up stones, painted bowls and jars further established the facts about their prosperity and development (Kenoyer, 1998). Anthropologists believe that these centres of Indus Valley Civilization were epitome of development and one of the finest examples of flourishing trade and agriculture based economy. The people of Indus Valley Civilization made the clever and resourceful use of rivers present in their area surrounding them. Mohen-Jo- has a sophisticated system of water supply & drainage and its brickwork, is highly functional and completely waterproof. The granaries are also intelligently constructed, with strategic air ducts and platforms (Figure 2.2).

Figure 2.2 Town planning of Indus Valley Civilization (Kenoyer, 1998).

2.3.3 Culture

The civilisation of Indus valley was mostly an urban culture maintained by additional agricultural production, trade and commerce. Craft of the Indus Valley Civilization was very advanced. Iron as a metal was unknown to the Harappan people but copper and bronze were used in making statues. Music and dance appear to be the main sources of entertainment (Figure 2.3).

Figure 2.3 Music and Dance of Indus Valley Civilization (Kenoyer, 1998).

Agriculture was their main occupation. There has been enough evidence of the cultivation of wheat, barley, peas, mustard, cotton and rice. Enough evidence of religious practices was not found as no temples have yet been discovered. From the Pashupati seal, it appears that they worshipped Lord Shiva. Several earthenware figurines of Mother Goddesses have also been found. It is also said that nature worship had been a significant part of their ritual as discovered in the seals (Kenoyer, 1998). The people of Indus Civilization had highly developed chalcolithic (stone implements contemporary with copper or bronze) culture which was similar to and in some respect superior to the contemporary Mesopotamian and Egyptian civilization (Gray, 1940). They were organized into towns, with a well developed commerce in all directions. They had wheeled vehicles for transport. They were skilled metal workers, using gold, silver, copper and lead, and tin in bronze. They made cotton into textiles. They had domesticated the dog, pig, sheep, buffalo, humped zebu, camel, and elephant, but probably not the horse or cat. They had well built and commodious houses, and baths, with drainage system built of well burned brick, and the common people probably enjoyed a degree of comfort unknown in other parts of the world, for in other contemporaneous civilizations the skill of the artist and builder was lavished upon the temples and places, but the homes of the ordinary people were apparently inadequate and ephemeral hovels.

2.3.4 Town planning

Indus Valley Civilization had a proper town planning with well laid out streets, separate living quarters, flat-roofed brick houses, and efficient drainage system and ventilation. Town Planning is one of the most outstanding and remarkable features of the Indus Valley Civilization. The Town Planning of Indus Valley Civilization reveals that the civic organisations of the city were highly developed and even at the present age, the sites are special attractions for archaeologists. Archaeological surveys have revealed several interesting facts about Harappa, Moen-Jo-Daro and lothal town planning. In the 3rd decade of the present century, archaeological investigations had been conducted at Moen-Jo-Daro, Harappa and lothal, which have brought out many interesting observations. The people at that time were technologically advanced and very knowledgeable in the laying out of the construction of the city as a whole (Rothermund & Kulke, 1998). The roads, dwelling houses, large buildings and forts (dockyards) were very well executed. People followed a system of centralised administration (Figure 2.2). The houses were even protected from sounds, odours, and thieves. The streets were formed in grid system. Similar sized bricks were used for construction of buildings; wood and stone were also used in buildings. Municipal authorities maintained effective drainage system. Harappan, Moen-Jo-Daro and Lothal town planning had the insertion of many travelling houses, which ranged from two roomed to large buildings. Houses were properly placed on both sides of the roads, and also in the lanes. Moen-Jo-Daro has a sophisticated system of water supply and drainage and its brickwork, is highly functional and the amazing part of it is the great bath (Figures 2.4 and 2.5), streets, bathrooms, and drainage and sanitation system. The granaries are also intelligently constructed, with strategic air ducts and platforms are divided into units.

Figure 2.4 Great bath at Moen-Jo-Daro (Kenoyer, 1998).

Figure 2.5 House Bathroom and outlet, House Bathing Platform at Moen-Jo-Daro, and row of private Bath at Lothal (Kenoyer, 1998).

The doors of the houses opened in the lanes and not on the roads. The houses were built on plinths that rose above the street level with stairs recessed at the wall at the front door. The house planning did not let any hindrance on the roads so everything was well organized (Figures 2.7 and 2.8). There were the government houses which were differently executed, dwelling houses which ranged from single to several stories with many rooms, public baths (Figures 2.4, 2.5, 2.6, and 2.7) common wells, and sanitation (Kuthiala, 2009). At Harappa many rooms consisting of two chambers around the courtyard of a dwelling house have been discovered, which are meant for the staying of the laborers (Fayaz, 2012).

Figure 2.6 Great Bath outlet Moen-Jo-Daro and Lothal Sanitation (Kenoyer, 1998).

Figure 2.7 Drainage and Sanitation system of Indus Valley Civilization (Kenoyer, 1998).

The Mohen-Jo-Daro ruins present a picture of a community in which both personal and community cleanliness was quite effectively practiced, and the water supply reasonably safeguarded from contamination as a rule.

Harappa town planning has stunned the archaeologists worldwide. It has become a landmark for the contemporary civilisation when technological advancements have been made which is helping to achieve great heights. It has inspired the contemporary generation. The concept of bathing pools and granaries gives a glimpse of the modern day swimming pools and storehouses where grains can be stored. It was a proper furnished city. This facilitated the Harappa dwellers to live a luxurious life with proper sanitation and regulation (Saleem, 2012).

Lothal (Gujarat) is one of the most prominent cities of the ancient Indus valley civilization. Located in Bhāl region of the modern state of Gujarāt and dating from *ca.* 2400 BC. Lothal town planning is represented a structure of dockyard, industrial and trade center of the Indus Valley Civilization (Figure 2.2). The whole town was situated on a patch of high ground. Rising from the flat alluvial plains of Bhal, a wall was erected to encircle the town, and a platform was built where goods were checked and stored. The warehouse was divided into 64 rooms of around 3.5 square meters each, connected by 1.2 meter wide passages (Mulchandani & Shukla, 2010).

2.4 WATER AND SANITATION

The Harappan town had very good drainage and sanitary system. The main drain was associated with each and every house ensuring the proper dumping of waste. In order to check the maintenance, inspection holes were provided. The drains were covered and connected to the bigger sewerage outlets, which ensured the channel of dirt out of the city (Figure 2.7). For water, the big houses had their own wells, other wells would serve groups of smaller houses. Almost every house had a bathroom, usually a fine sawn burnt brick pavement, often with a surrounding curb. The house drains start from the bathrooms of the houses and join up to the main sewer in the street, which was covered by brick slabs or corbelled brick arches (Figures 2.5, 2.7, and 2.8). On the streets there are manholes for cleaning; some drains flow to closed seeps, others flow out of the city (Jansen, 1985). These water wells and well planned sanitation and sewerage system is one of the great signs that lead to the well developed Indus Valley Civilization (Jones, 1967).

The bath and kitchen waters, as well as drainage from the latrines, and the roof drainage, usually did not run into the street drains direct, but entered them via tightly brick-lined puts, with outlets to the streets drains about three-quarters of the distance above the bottom. Apparently these pits were cleaned out from time to time, as were the setting basins or soakage pits located along the street drains. These pits may have been the ancient precursors of our present day septic tanks and grit chambers. In some houses the drainage water discharged into large pottery jars places in the street at the foot of the vertical drains in the street walls (Gray, 1940). Houses also had rubbish chutes built into the walls and descending from the upper floors, at the foot of which chutes there were sometimes provided bins at the street level which could be cleaned out by the scavengers. Public rubbish bins were also provided at convenient places (Gray, 1940).

The most unique aspect of planning during the Indus Valley civilization at Lothal was the system of underground drainage. The main sewer, 1.5 meters deep and 91 cm across, connected to many north-south and east-west sewers. It was made from bricks smoothened and joined together seamlessly. The expert masonry kept the sewer watertight. Drops at regular intervals acted like an automatic cleaning device (Figure 2.6).

A wooden screen at the end of the drains held back solid wastes. Liquids entered a cess poll made of radial bricks. Tunnels carried the waste liquids to the main channel connecting the dockyard with the river estuary. Commoner houses had baths and drains that emptied into underground soakage jars. The rooms of the upper town were obviously built for upper classes. They had private pathed brick baths (Figure 2.5) and a remarkable network of drains and cesspools (Mulchandani & Shukla, 2010).

2.4.1 The great bath

The 'great bath' is without doubt the earliest public water tank in the ancient world located at the archeological site of Moan-Jo-Daro. The tank itself measures approximately 12 m north-south and 7 m wide, with a maximum depth of 2.4 m (Deonarine *et al.* 2001). Two wide staircases lead down into the tank from the north and south and small sockets at the edges of the stairs are thought to have held wooden planks or treads. At the foot of the stairs is a small ledge with a brick edging that extends the entire width of the pool. People coming down the stairs could move along this ledge without actually stepping into the pool itself (Figures 2.4 and 2.6).

The floor of the tank is water tight due to finely fitted bricks laid on edge with gypsum plaster and the side walls were constructed in a similar manner. To make the tank even more water tight, a thick layer of bitumen (natural tar) was laid along the sides of the tank and presumably also beneath the floor.

Brick colonnades were discovered on the eastern, northern and southern edges. The preserved columns have stepped edges that may have held wooden screens or window frames. Two large doors lead into the complex from the south and other access was from the north and east. A series of rooms are located along the eastern edge of the building and in one room is a well that may have supplied some of the water needed to fill the tank. Rainwater also may have been collected for this purposes, but no inlet drains have been found (Saffy, 2011).

The principal community bath was a structure of considerable size, conforming somewhat with our ideas of a swimming pool, though perhaps being used rather as a place for religious ceremonials than for either mere pleasure or for only the cleansing of the body. The structural features of the pool indicate an excellent ability in construction, considering the building materials available at that time and place. For example, waterproofing was accomplished by a membrane or coating of asphaltum between the inner and outer walls of the pool or tank (Gray, 1940).

2.4.2 Water treatment

The Indus Valley Civilization was known for their water management. Most of the excavations have been found around the areas of the cities of Harappa, Moen Jo Daro and Dholavira. They were known for their obsession with water (Jansen, 1989). They prayed to the rivers everyday and gave them a divine status. They had well-constructed wells, tanks, public baths, a wide drinking system and a city sewage system. Each city had two regions – a higher ground, which contained the 'Citadel', was the main administrative area and the lower city where the houses were situated. All the important areas were situated on the higher ground. The baths and wells were situated there, which suggests the importance they were given (Nambiar, 2006).

The inhabitants of Moen-Jo-Daro were masters in constructing wells. It is estimated that about 700 wells have been built within their city, an average of one well for every third house. They were constructed with tapering bricks that were strong enough to last for centuries (Ann, 2009). The cities too had strong walls to resist damages due to floods. One reason for this large number is that Moen-Jo-Daro received less winter rain and was situated further from the Indus River than the other prominent cities. Hence it was necessary to collect and store water for various purposes (Figures 2.6 and 2.7).

2.4.3 Baths and wells

One of the best-known excavations is the Great Bath of Moen-Jo-Daro, which has before discussed. In addition to wells, archaeologists have also found remains of giant reservoirs for water storage. Reservoirs were situated around the metropolis which was fortified with stonewalls. The Archaeological Survey of India has revealed that one third of the area of the city of Dholavira in the Rann of Kutch, was devoted to collection and distribution of fresh water (Figure 2.2). The city was situated on a slope between two streams. At the point where one of the streams meets the city's walls, people carved a large reservoir out of rock. This was connected to a network of small and big reservoirs that distributed water to the entire city all year round. All the reservoirs together could hold about 248480 cubic meters of water. Such was the importance they gave for water storage. According to Gray (1940), many of the houses in Indus civilization had their individual wells within buildings. These wells were usually circular in plan, thought at time oval, and had copings of stones or bricks at the floor level, and brick lining for a moderate depth below the surface (Figure 2.8). In a few instances the streets drains ran rather too close to the wells, and it is possible that some contamination of the well occurred. But in most cases the wells were located at adequate distances from the drains.

Figure 2.8 Different types of wells and water tanks in Indus civilization (Kenoyer, 1998).

Generally, the Moen-Jo-Daro ruins present a picture of a community in which both personal and community cleanliness was quite effectively practiced, and the water supply reasonably safeguarded from contamination as a rule. Practically every house in Moen-Jo-Daro had its bathroom, always placed on the street side of the building for the convenient disposal of waste water into the street drains. Where latrines have been found in the houses, they were placed on the street wall for the same reason. Ablution places were set immediately adjacent to the latrines, thus conforming to one of the most modern of sanitary maxims. Where baths and latrines were located on the upper floor, they were drained usually by vertical terra-cotta pipes with closely fitting spigot joints, set in the building wall (Figure 2.5).

In the bathroom, people stood on a brick 'shower tray' and tipped water over themselves from a jar. The clean water came from a well. Dirty water drained through a pipe out through the wall into the drain in the street (Lofrano & Brown, 2010). These ancient terra-cotta pipes, still sound after nearly five thousand years, are the precursor of our modern verified clay spigot-and-socket sewer pipe, and are an excellent guarantee of the durability of this material (Figure 2.7).

Figure 2.9 Dockyards at Lothal and Courtesy of Professor Kenoyer, 1998.

2.4.4 Drainage system

The Indus civilization had an elaborate sanitary and drainage system, the hallmark of ancient Indus cities. The authorities maintained a highly efficient drainage system. Each and every house had a connection with the main drain. These even had inspection holes for maintenance. The conduits to the main drains ran through the middle of the streets below the pavement level and were covered with flat stones and sturdy tile bricks. The covered drain was connected to the larger sewerage outlets which finally led the dirty water outside the populated areas. The urban plan found in these cities included the word's first urban sanitation system. The elaborate brick-linked drainage system for the removal of rainwater is of unparalleled engineering skill (Rothermund & Kulke, 1998).

With such an extensive domestic water storage system, the associated problem that arises is that of drainage. Town planners of Moen-Jo-Daro had built the worlds first known main drainage system. It was a central system that connected every household in the city (Figure 2.6). Every house had a drinking water well with a private bathroom. Earthenware waste pipes carried sewage from each home into covered channels that ran along the centres of the city's main streets into the nearby agricultural fields, rivers, or streams. The drains took waste from kitchens, bathrooms, and indoor toilets. The main drains even had movable stone slabs as inspection points. The houses had excellent plumbing facilities for provision of water (Rothermund & Kulke, 1998). Toilets had brick seats. The toilet was flushed with water from jars. The waste flowed out through clay pipes into a drain in the street. Waste was carried away along the drains to 'soak pits' (cesspits), Cleaners dug out the pit and took the waste away. They also took away rubbish from bins on the side of houses. Each street and lane had one or two drainage channels, with brick or stone covers which could be lifted to remove obstructions in the drains. The drains were usually ranged from 46 to 61 cm below the street level, and varied in dimensions from 30.50 cm deep and 23.00 cm wide (Gray, 1940). When the drain could not be covered by flat bricks, or stone slabs, the roof of the drain was corbelled.

2.4.5 Irrigation system

The Indus Valley Civilization in Pakistan (from *ca.* 2600 BC) also had an early canal irrigation system. Large scale agriculture was practiced and an extensive network of canals was used for the purpose of irrigation. Sophisticated irrigation and storage systems were developed, including the reservoirs built at Girnar in *ca.* 3000 BC (Shirsath, 2009). Beside, some of the toy pictures of the Indus Valley

Civilization indicate that they were a proper system of water supply into different houses and places. Mostly, women had the responsibility to supply water into different places. Farmers made good use of water from the rivers. They sowed seeds after the rivers had flooded the fields, as flood water made the soil rich. They planted different crops for winter (which was mild and wet) and summer (which was hot and dry). They were probably the first farmers to take water from underground wells. They may have used river water to irrigate their fields. Their main cultivation products, amongst others, were peas, sesame seed and cotton. They also domesticated wild animals in order to use them for harvesting their farms (Lawler, 2008).

2.4.6 Rainwater harvesting and storage system

The Indus Valley Civilization, that flourished along the banks of the river Indus and other parts of western and northern India about 5000 years ago, had one of the most sophisticated urban water supply and sewage systems in the world. The fact that the people were well acquainted with hygiene can be seen from the covered drains running beneath the streets of the ruins at both Moen-Jo-Daro and Harappa. Another very good example is the well-planned city of Dholavira, on Khadir Bet, a low plateau in the Rann in Gujarat. One of the oldest water harvesting systems is found about 130 km from Pune along Naneghat in the Western Ghats (Kenoyer, 1991). A large number of tanks were cut in the rocks to provide drinking water to tradesmen who used to travel along this ancient trade route. Each fort in the area had its own water harvesting and storage system in the form of rock-cut cisterns, ponds, tanks and wells that are still in use today. A large number of forts like Raigad had tanks that supplied water (Figure 2.10).

Figure 2.10 Water tank at Moen-Jo- Daro (Kenoyer, 1998).

'The kind of efficient system of Harappans of Dholavira, developed for conservation, harvesting and storage of water speaks eloquently about their advanced hydraulic engineering, given the state of technology,' (Subramanian, 2010). One of the unique features of Dholavira is the sophisticated water conservation system of channels and reservoirs, the earliest found anywhere in the world and completely built out of stone, of which three are exposed. Dholavira had massive reservoirs. They were used for storing the freshwater brought by rains or to store the water diverted from two nearby rivulets. This clearly came in wake of the desert climate and conditions of Kutch, where several years may pass without rainfall. A seasonal stream which runs in north-south direction of the site was dammed at several points to collect water (Figure 2.11). The great bath at Moen-Jo-Daro is also an evidence of the water conservation and storage system (Violett, 2007).

Figure 2.11 Hydraulic Engineering Indus Civilization at Dholavira, Rann at Gujrat (Wales, 2010).

The inhabitants of Dholavira created sixteen or more reservoirs of varying size. Some of these took advantage of the slope of the ground within the large settlement, a drop of 13 m from northeast to northwest. Other reservoirs were excavated, some into living rock. Recent work has revealed two large reservoirs, one to the east of the castle and one to its south, near the Annexe , shown in Figure 2.12 (Wales, 2010).

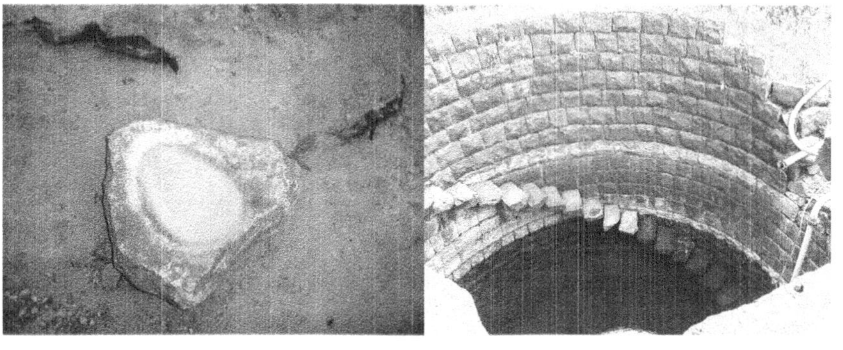

Figure 2.12 Storm water storage in rock tank and dry well at Dholavira and Lothal (Wales, 2010).

Reservoirs are cut through stones vertically. They are about 7 meters deep and 79 meters long. Reservoirs skirted the city while citadel and bath are centrally located on raised ground. A large well with a stone-cut through to connect the drain meant for conducting water to a storage tank have also been found. Bathing tank had steps descending inwards as in Figure 2.12.

A large number of tanks (Figure 2.12) were cut in the rocks to provide drinking water to tradesmen who used to travel along the trade route. Each fort in the area had its own water harvesting and storage system in the form of rock-cut cisterns, ponds, tanks and wells that are still in use today. A large number of forts like Raigad had tanks that supplied water for domestic use (Wales, 2010).

2.4.7 Public toilets

In each society from time to time the administration felt the need to provide public toilet facilities to those who could not afford to have individual toilets. The public toilets have a long history in number of countries and most of these were constructed and managed by municipalities. But there was all around disgust with their poor maintenance, vandalism and lack of basic facilities (Pathak, 1995). In the absence of proper toilet facilities, people perforce had to defecate and urinate wherever they could. Defecating on the road, open spaces, or just easing themselves in the river was very common (Figures 2.13 and 2.14).

Figure 2.13 Public toilets, Open toilet and house hold toilet in Indus Valley Civilization, (Kenoyer, 1998).

Figure 2.14 Indus Valley Civilization Open toilets Commodes, (Kenoyer, 1998).

The third millennium BC was the 'Age of Cleanliness.' Toilets and sewers were invented in several parts of the world, and Moen-Jo-Daro *ca.* 2800 BC had some of the most advanced, with lavatories built into the outer walls of houses. These were primitive 'Western-style' toilets made from bricks with wooden seats on top. They had vertical chutes, through which waste fell into street drains or cesspits (Pathak, 1995).

The toilets at Moen-Jo-Daro, built about 2600 BC, were only used by the affluent classes. Most people would have squatted over old pots set into the ground or used open pits (Figure 2.14). The people of the Indus Valley Civilization in Pakistan and north-western India had primitive water-cleaning toilets that used flowing water in each house that were linked with drains covered with burnt clay bricks. The flowing water removed the human wastes (Hooper, 2011).

Toilets would have been an essential feature in Moen-Jo-Daro, but the early excavators identified most toilets as post-cremation burial urns or sump pots. This brick structure had a hole in the top that was connected to a small drain leading out of the base into a rectangular basin (not reconstructed). Early excavators suggested that structure with a hole and drain located are thought to have been toilets (Figure 2.14). For the human urinate, they may have used a hole in the ground at open places that connected to near by drain. The toilets of Indus Valley Civilization were different than the Roman and Greek Civilizations. This difference is the main evidence of the cultural difference between them (Hooper, 2011). Figure 3.13 reveals that the Indus Valley Civilization in Pakistan has the concept of toilet and latrine and a well established waste system at that time.

2.4.8 Dockyard at Lothal

The dominant sight at Lothal is the massive dockyard which has helped make this place so important to international archaeology. Spanning an area 37 meters from east to west and nearly 22 meters from north to south, the dock is said by some to be the greatest work of maritime architecture before the birth of Christ (Mulchandani & Shukla, 2010). To be sure, not all archaeologists are convinced that the structure was used as a dockyard and some prefer to refer to it as a large tank that may have been a reservoir (Figure 2.9).

It was excavated besides the river Sabarmati, which has since changed course. The structure's design shows a thorough study of tides, hydraulics and the effect of sea water on bricks. Ships could have entered into the northern end of the dock through an inlet channel connected to an estuary of the Sabramati during high tide. The lock gates could then have been closed so the water level would rise sufficiently for them to float (Figure 2.9).

An inlet channel 1.7 meters above the bottom level of the 4.26 meter deep tank allowed excess water to escape. Other inlets prevented siltation of the tanks and erosion of the banks. After a ship would have unloaded its cargo, the gates would have opened and allowed it to return to the Arabian sea waters in the Gulf of Combay (Mulchandani & Shukla, 2010).

Archaeological finds from the excavations testify to trade with ancient Egypt and Mespotamia. The hydraulic knowledge of the ancient Harappans can be judged by the fact that boats could dock at Lothal in the 1850's. In 1942 timber was brought from Baruch to nearby Sagarwala. It is said that then the dockyard could hold 30 ships of 60 tons each or 60 ships of 30 tons each. This would be comparable to the modern docks at Vishakapatnam (Mulchandani & Shukla, 2010).

A long wharf connected the dockyard to the main warehouse, which was located on a plinth some 3.5 meters above the ground. The first concern of the Harappan engineers might have been to ensure against floods and tides (which may have been their undoing at Mohenjo-daro & Harappa).

2.5 CONCLUSION

By about 1700 BC, the Indus Valley Civilization was on the verge of decline. The causes of its decline are not certain. The physical existence of the civilization ended due to various factors as given below.

(a) Ecological changes led to the decline of land and agriculture, thereby enforcing the need to evacuate to other areas might have been the reason for the disintegration of the Indus valley civilization. Shifts in the monsoon pattern and changes in temperature led to the area become even more arid (Carr, 2012).

(b) Increase in population, excessive deforestation, decline in agriculture might have created economic problems leading to the gradual decay of the culture. The marked decline in the quality of building and town planning indicates that the authorities were losing control with passage of time.
(c) The changes in the Indus flow and correspondent widespread flooding would have disrupted the agricultural base and destroyed the civilization.
(d) The invasion of the Aryans is the other view that is said to be another reason which might have also led to the decline of the Indus valley. Thus ended the most brilliant civilization of the ancient world.
(e) Hypothesis is made here that the destruction of the Indus Valley Civilization is the result of an earthquake caused by the plate tectonic moment of the Indian plate. This earthquake was caused at night time when most of the people were asleep. The dead bodies that buried in the Harappa and Moen-Jo-Daro sites are the evidences of the earthquake disaster as most of them lay properly along with juvenile. The people who were safe and a live shifted to other areas and constructed new sites for survival. However, this needs further research to study the evidence of earthquake disaster that destroyed the Indus valley civilization.

The first basic concept embodied in the Indus civilization was the belief that no one individual had the right to usurp the wealth and resources of the land and use them for his or his family's benefit. The wealth of the cities was distributed among all segments of the society. There were rich people and poor people, but the executive hold of one individual or family on the wealth of the cities is nowhere found.

The second basic concept ingrained in the Indus civilization is the separation of the clergy from the state administration. Religion was a very important part of the spiritual, social and cultural life of the Indus people. The role of the priests was to provide spiritual comfort and to promote cleanliness and purity of thought and actions. There is no evidence to suggest that they craved for power or to indicate their involvement in the affairs pertaining to state administration.

The third basic concept, which characterized the Indus valley civilization, was that of unified culture and decentralized form of government. The Indus civilization was spread far and wide over widely different regions and terrain, yet these regions were knit together by common bonds of religion and culture. The design of the artifacts and the layout of the towns were all very similar. All this uniformity was achieved without the direct interference of the central regime in the administration of the city states.

The fourth concept which characterized the Indus civilization was that of well planned drainage, sanitation system, dockyards, and hydraulic engineering. The houses had their own wells, bathroom and toilet. Almost every house had a bathroom, usually a fine sawn burnt brick pavement, often with a surrounding curb. The house drains start from the bathrooms of the houses and join up to the main sewer in the street.

The proud people of Indus were docile, peace loving and accommodative. The Indus person demonstrated tolerance and broadmindedness. Our quest to search for our identity has taken us to the land of Mighty Indus. There is absolutely no doubt that the Pakistani are the people of the forgotten Indus civilization, who were docile, peace loving, accommodative, moderate and open minded, traits that we have lost. It is time to rediscover and restore Pakistan as a liberal, progressive, modern Muslim state with its rightful place into the community of nations.

2.6 REFERENCES

Story of Pakistan.com (2005). Climate of Indus Civilization. http://storyofpakistan.com/climate-of-indus-civilization/ (accessed 26th May 2014).

Agarwal A. (2009). Lothal – A Port of Harappan Civilization. http://photos.abhinavagarwal.net/2012/02/lothal-port-of-harappan-civilization.html

Ann C. (2009). History of Water Pumps, eHow Contributor. http://www.ehow.com/facts_5031932_history-water-pumps.html

Carr T. J. (2012). The Harappan Civilization, Archeology. Online. http://www.archaeologyonline.net/artifacts/harappa-mohenjodaro.html

Deonarine A., Jim M. and Khan O. (1995 and 2001). 'Great Bath,' Mohenjo-daro. Online. http://www.harappa.com/indus/8.html

Fayaz A. (2012). Indus valley civilization, Pakistani Perspectives. Online. http://www.articlesbase.com/history-articles/indus-valley-civilization-pakistani-perespective-5884227.html

GOI (2004). History of the Toilet, From Ancient Indian Drainage Channels to a Three-Pipe Water Closet, Explore the History of the Toilet. http://www.motherearthliving.com/green-living/history-of-the-toilet.aspx

Gray H. F. (1940). Sewerage in ancient and mediaeval times. *Sewage Works Journal* **12**(5), 939–949.

Hooper A. (2011). What is the Complete History of the Toilet? http://answers.yahoo.com/question/index?qid=20120220121010AARe9nX

Jansen M. (1985). Mohenjo-Daro, city of the Indus Valley, Endeavour, New Series. Volume 9, No. 4, Pergamon Press, Printed in Great Britain, London, UK.

Jansen M. (1989). Water supply and sewage disposal at Mohenjo-Daro. *World Archaeolgy* **21**(2), 177–192.

Jones Jr. D. E. (1967). Urban hydrology – a redirection. *Civil Engineering* **37**(8), 58–62.

Kenoyer J. M. (1991). The Indus Valley tradition of Pakistan and Western India. *Journal of World Prehistory* **5**(4), 331–385.

Kenoyer J. M. (1998). Mohenjo-Daro., an ancient Indus Valley Metropolis. University of Wisconsin, Madison. http://www.harappa.com/indus3/kenoyer.html

Khan S. and Hasan M. (2010). Climate of Pakistan, BALWOIS 2010. Ohrid Macedonia, p. 28.

Kosamb D. D. (1964). The Culture and Civilization of Ancient India in Historical Outline. House 803, Poona 4, India, p. 188.

Kuthiala P. (2009). Indus Valley Civilization-A Summary. http://www.thoughtdots.com/2009/09/11/indus-valley-civilization-a-summary/

Lawler A. (2008). Boring no more, a trade-savvy Indus emerges. *Science* **320**, 1276–1281. http://www.academicroom.com/article/boring-no-more-trade-savvy-indus-emerges

Lofrano G. and Brown J. 2010. Wastewater management through the ages: a history of mankind. *Science of Total Environment* **408**, 5254–5264. http://www.ncbi.nlm.nih.gov/pubmed/20817263

Mulchandani A. and Shukla D. (2010). A Walk Through Lothal. http://www.harappa.com/lothal/index.html

Nambiar A. (2006). Indus Valley – How They Managed Their Water Resources. Water Management in the Ancient Indus Valley. http://voices.yahoo.com/indus-valley-they-managed-their-water-resources-82580.html

Pathak B. (1995). The History of Toilets. In: International Symposium on Public Toilets, Hong Kong. http://www.plumbingsupply.com/toilethistoryindia.html

Rothermund D. and Kulke H. (1998). A History of India. 4th edn, New Delhi, India. http://www.fsmitha.com/h1/ch05.htm

Saffy H. (2011). The Great Bath at Moenjodaro, Wikipedia. Online, en.wikipedia.org/wiki/Great_Bath,_Mohenjo-daro

Saleem S. (2012). Harappa and Mohenjo Daro. http://ugghani.blogspot.in/2012/06/harappa.html

Sharma L. P. (1992). Ancient India History, Indus Valley Civilization. http://www.culturalindia.net/indian-history/ancient-india/indus-valley.html

Shirsath P. B. (2009). Irrigation Development in India: History & Impact. Water Technology Centre, WTC, IARI, New Delhi, http://indiairrigation.blogspot.com/2009/01/history-of-irrigation-development-in_01.html

Smith M. L. (2006). The archaeology of South Asian cities. *Journal of Archaeology Research* **14**, 97–142.

Subramanian T. S. (2010). The rise and fall of a Harappan city, Frontline 27(12). http://www.frontline.in/navigation/?type=static&page=archive

Tekwani N. (2009). How rainwater harvesting was practised in the ancient times? http://in.answers.yahoo.com/question/index?qid=20110521034619AAUCCPJ

Violett P. L. (2007). Water Engineering in Ancient Civilizations: 5000 Years of History, (Translation into English by Forrest M. Holly). International Association of Hydraulic Engineering and Research (IAHR), Madrid. CRC Press, Boca Raton FL.

Wales J. (2010). Dholavira, Wikipedia. http://en.wikipedia.org/wiki/Dholavira#cite_note-frontline-1#cite_note-frontline-1

Wright R. P. (2010). The Ancient Indus: Urbanism, Economy, and Society. Cambridge University Press, New York.

Chapter 3

Sanitation and water management in ancient South Asia

Hilal Franz Fardin, Annick Hollé, Emmanuèle Gautier-Costard, and Jacques Haury

3.1 INTRODUCTION

This chapter focuses on the geohistory of the early water and sanitation management techniques in South Asia. The Indian subcontinent is constituted by the lands located south of the Himalaya-Karakoram-Hindu Kutch ranges.[1] This region has hosted one of the first urban cultures which had developed various techniques of water management, including wastewater management, namely the Indus valley civilisation. But water and wastewater have been managed in other parts of the subcontinent during early history. In this paper, we will describe the main wastewater management techniques practiced in ancient South Asia. We also will inventory the types of wastewater techniques according to some of the major archaeological sites of this area. Thus, the chapter follows a chronological structure through the regions of South Asia: the greater Indus Valley during the urban phase of the Harappan period (2600 BC–1900 BC); the Northern and Central South Asia in the Northern Black Polished Ware period (600 BC–200 BC); and the Southern South Asia during the Sangam period (300 BC–300 AD). It will also discuss in a critical point of view the potential ways of diffusion of wastewater techniques in South Asia during Antiquity (2600 BC to 300 AD).

3.2 WASTEWATER MANAGEMENT DURING THE HARAPPAN PERIOD

The first human settlements in South Asia have been dated to about two million years ago. Then that region has witnessed various Palaeolithic cultures, including Acheulian in the entire peninsula (Misra, 2001). Approximately during the same period as in the Middle East, populations of the greater Indus valley developed Mesolithic techniques around 7000 BC (Jarrige, 2006). Protoagriculture and protopastoralism appeared in Mehrgarh, Baluchistan, in the north-western part of the sub-continent (Fuller, 2006). This South Asian protoagricultural phase has many similarities with the culture of the very same period of

[1] In this chapter, 'South Asia' means Afghanistan, Bangladesh, Bhutan, India, Nepal, Pakistan, and Sri Lanka. Maldives Islands, although they are included in South Asia, are not mentioned on the maps of the present paper.

the eastern part of Mesopotamia (Jarrige, 2006), and with Kashmir as well (Mani, 2006). Thus, the foundations of the first South Asian urban culture, that is the Indus civilisation, are considered to be in the western part of Baluchistan range. We will here explore further the Harappan civilisation, since it is the first to have developed sanitation techniques in the Indian subcontinent.

3.2.1 Indus Valley palaeogeography and culture

Harappan culture spread all over the Indus valley region and on the banks of the palaeo-river Ghaggar-Hakra, considered as the Vedic Saraswati (Misra, 2001). The economy was based on agriculture, handicraft and trade. During Early and Mature Harappan period, agriculture was based on the wheat/barley 'package' (Fuller, 2006). With the late Harappan period, cultivation shifted to the rice/millet 'package', in the middle of the 3rd millenary BC. According to the biophysical settings, the shift in agricultural pattern was due to an adaptation to some climatic changes (Madella & Fuller, 2006), but also probably to socio-cultural factors (diffusion of Harappan culture to the south of the sub-continent; aggregation with the southern local culture who used to cultivate rice and millet). Indeed, an aridification of the climate seems to have occurred about 2000 BC (Ponton *et al.*, 2012). When summer monsoonal crops (rice and millets) started to take the place of the winter rainfall based crops (wheat and barley), the whole social systems had to change. Because the Indus writing has not been deciphered yet, we cannot mention anything about religious or cultural aspects, apart from the remains of material culture.

Thus, the decline of the Harappan culture is due to multiple factors: climatic changes, demographical growth, emergence of a cultural diversity (Aryans/Dravidians; Tribal/Urban), and maybe water-related issues. Indeed, because most of the Harappan sites were based on water management systems, a change in the hydro system would have made a change in the whole society. It has been demonstrated that the palaeo-river Ghaggar-Hakra disappeared during the 2nd millennium BC, while it was the main residential plain of the Harappan culture. Moreover, it has been pointed out that some sanitarian issues could have contributed to the decline of this civilisation. Effectively, the possible contamination of the wells by the domestic wastewater could have produced various health and environmental damages, as the development of water born diseases (Ewald, 1991). Wastewater management is a characteristic of the urban and semi-urban settlements of the Harappan civilisation. Sewage and drainage were composed by complex networks, especially in Mohenjo-daro and Harappa. Latrines, soak-pits, cesspools, pipes and channels were the main elements of wastewater disposal. The Harappan civilisation is considered to have 'declines' after about 2000 BC, yet wastewater management has spread all over the South Asian region.

In this part, we will present the two main types of wastewater management systems during the urban phase of the Harappan period, namely the centralised and the decentralised systems.

3.2.2 Sanitation centralised systems

According to some authors, the development of sanitation is linked to the development of urbanisation because 'guaranteeing pure water for people became a prerequisite for successful urbanisation (Vuorinen, *et al.*, 2007, p. 50). Drainage systems are a key-factor in the development of Indian urbanism (Smith, 2006, p. 130). It has been pointed out that drainage and sanitation are major characteristics of the first urban sites of the Harappan civilisation (Kenoyer, 1991, p. 354). Jansen (1989) did a descriptive overview of wastewater management at Mohenjo-daro (Figure 3.1). According to Wright (2010, p. 122), there were few houses with latrines in this city. These toilets were of two types: made of clay bricks with a seat, or consisting of a simple hole in the ground. Domestic outflows from toilets and bathing platforms were transported to the drains of the streets through terracotta pipes. In some cases, the effluent was directed to pits made of clay bricks (Jansen, 1989; Wright, 2010). When these pits were three-quarter filled, the outflow ran into the drains of the streets (Mackay, 1936, p. 51). These pits were probably emptied, and the sludge was sent to sites dedicated to that purpose, as it has been suggested about solid wastes (Jansen, 1989). To prevent the drainage systems from getting clogged, pits were located at the junction of several drains, or in certain places where the drains extended over a long distance. These pits had the function of manholes. Downstream of these manholes, a wider drain was implemented (Wright, 2010), receiving water partially treated through these pits (sedimentation). To reduce the friction of the water on the walls, bends were curved using key-bricks (Mackay, 1936, p. 52), the same characteristic was used for the sewer system in Lothal (Rao, 1979, p. 78).

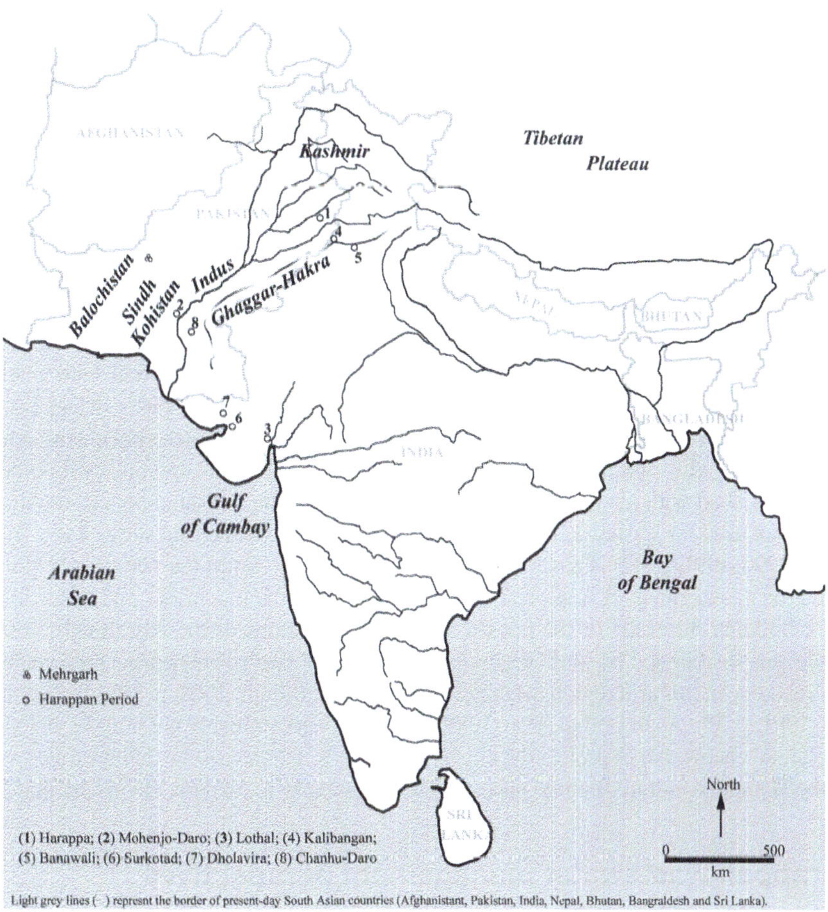

Figure 3.1 Map of the sites of the Harappan period cited in the text (Map by H. F. Fardin).

Almost every house of Mohenjo-daro was connected to the sewer system, and the network was facilitated by the urban morphology in checkerboard. The same type of centralised wastewater collection was used at Harappa (Singh, 2008, p. 149). Recent excavations have revealed that the houses of the periphery had bathrooms and sewage pipes (Singh, 2008, p. 152). But because the remains of the site were used as building material for the towns and villages around, the plan of Harappa contains many gaps (Mackay, 1936, p. 7). Lothal is located several hundred kilometres from the Indus Valley, and a complex network of wastewater and storm water collection systems had also been implemented (Rao, 1979, pp. 77–81; Kirk, 1975, p. 21; Leshnik, 1968, p. 912). Similarly, each house had a bathroom (Rao, 1979, p. 77), but only one house is described as having latrines (Rao, 1979, p. 81).

3.2.3 Sanitation decentralised systems

Another type of wastewater collection system was also practiced. For example at Kalibangan, toilets and bathrooms outflows were collected in U-shaped channels made of wood or terracotta bricks. These effluents poured into a jar placed in the main street (Chakrabarti, 2004, p. 92). Sometimes the jar was perforated so that contaminated water was allowed to flow out (Mackay, 1936, p. 51). The same model of wastewater collection was used in Banawali, where effluents were channelled into drains made of clay bricks, before reaching the jars (Bisht, 1984, p. 96). As highlighted by Bisht, although a centralised system for collecting rainwater was implemented, it was not used for wastewater collection. Scavengers might have collected the sludge from the jars, which was then poured in a space provided for this purpose (Bisht, 1984, p. 96). It was argued that such a system was in operation at Surkotadu (Chakrabarti, 2004, p. 102), and in some houses of Mohenjo-daro. In the later case, the wastewater did not join the sewer system, which is certainly due to the remoteness of these houses in relation to the network of drains (Jansen, 1989, p. 189). According to data available on the subject, it seems that sites of smaller size did not have a centralised sewer system. Another difference with the larger sites of the Harappan culture, it is rare to find well within houses in Kalibangan (Chakrabarti, 2004, p. 92).

It has been demonstrated that even the poorest Chanhu-daro households had bath platforms, which wastewater flowed through channels made of baked bricks, or through clay pipes (Mackay, 1937, p. 531). Although this site is much smaller than Mohenjo-daro and Harappa wastewater system was centralised. This feature leads to wonder why some Harappan sites had a centralised sewer system, while other sites of the same period were satisfied with jars to collect wastewater. Also during the same period, the drainage system of Dholavira was not used for wastewater management, but only for channelling storm water (Archaeological Survey of India, 1990–1991, p. 32). Thus, Dholavira is an exception within the Harappan culture, both in its architecture and the management of water resources. In addition to the raw bricks and terracotta, stones were widely used for building the walls of the houses and the fortifications of the city (Singh, 2008, p. 155). In addition, the site was surrounded by tanks receiving the waters of two rivers nearby, presumably to store the precious water resource in this arid region, where rainfall is scarce (Singh, 2008, p. 155; Archaeological Survey of India, 1990–1991, p. 35). As in Lothal (Rao, 1979), the walls were probably not built for defensive purposes in view of a potential enemy, but to protect the city against floods. Dholavira site demonstrates once again that the people of the Harappan culture seem to have certain adaptability to extreme environmental conditions.

3.2.4 Harappan wastewater management perspectives

However, at Mohenjo-daro, the sewer system enveloped the seven hundred individual or collective wells that had the city (Jansen, 1989). Thus, domestic water could be contaminated by sewage, which seeped into the basement through pits and drains (Mackay, 1936, p. 53). Due to the potential contamination of wells by sewage, Ewald (1991) argues that these poor sanitary conditions were likely to cause epidemics, such as cholera, or other enteric diseases. This author goes further by stating that this gap in sanitation systems was the origin of the desertion of the Harappan urban settlements (Ewald, 1991, pp. 11–14). Furthermore, nothing is said in the specialised literature about a possible link between wastewater drainage and irrigation in the Harappan sites.

Since at least 3200 BC, it seems that during the dry season, most of the Harappan settlements in the flood plain used to irrigate agricultural parcels (Wright, 2010, p. 71). Only few sites had developed storing water facilities. *Gabarbands* are among those exceptions, which are located in the current Baluchistan and Sindh Kohistan. They have been dated from the early Harappan period, or Kot Diji phase (2800–2600 BC). The *gabarbands* are 'stone built dams construction designed to control and store water' (Wright, 2010, p. 31), probably used for irrigation. It has been argued that rainwater tanks were installed at Lothal to irrigate agricultural land during the dry season (Wright, 2010, p. 167). At Lothal, one can witness the spaces where wastewater was channelized. The wastewaters of the eastern part of the city flowed into what is considered as a dock; those of the south-eastern zone into the river; and those of the northern into the seasonal stream (Rao, 1979, p. 81). Because of the trapezoidal shape of the 'dock', this space considered as a mooring area for boats has long been discussed. Thus, taking into account present-day lagooning system, it is reasonable to ask whether this 'dock' did not have a function, at least informally, of urban effluents treatment.

3.3 WASTEWATER MANAGEMENT DURING EARLY HISTORIC TIMES

Some archaeological records on wastewater management have been discovered in some non-urban settlements in the Northern and the Central parts of the subcontinent for the Chalcolithic period, as well as in Southern South Asia for the late few centuries BC. Because we are now focusing on a period from about 500 BC to 300 AD, the social-cultural characteristics are almost similar to the ones of Hindu-Buddhist cultures of present-day South Asia. Many authors have studied the relation between the various categories of the society, that is the castes, and pollution and/or water (Dumont, 1966; Alley, 1994; Mosse, 1997; Douglas, 2001). In various sites of Ancient South Asia, wastewater was canalised into cesspools, which could have been ring-wells, as in Ujjain (~500 BC), as it will now be studied in further details.

3.3.1 Northern and Central South Asia

Around 500 BC, the drainage system of Ujjain (Figure 3.2) included 'soak-pits built of pottery-rings or pierced pots' (Kirk, 1975, p. 32). It is considered that the ring-wells received domestic wastewater (Mate, 1969, p. 244). This point of view can help understand why Dhirajlal and Bhalchandra (cited in Narr, 1961, p. 311) questioned whether some buildings were circular wells or soak-pits, in 150 BC Nasik. We can notice the superimposition between the drainage and the sewer systems, as it was in Europe until the 19th *ca.*, before the development of hygienism theories.

In the 3rd *ca.* BC, Taxila domestic sewage flowed from terracotta pipes to pits that can be considered as septic tanks or soak-pits (Singh, 2008, p. 335). Pushkalavati, the ancient capital of Gandhara (Stein, 1927, p. 437), is located near the present town of Charsadda, at the confluence of Swat and Kabul rivers, the two main routes linking South Asia to Central Asia. The road through Charsadda, called Peukelaotis by the ancient Greeks, was the one followed by Alexander the Great to enter the Indus Valley (Codrington, 1944), who settled a garrison in the city (Stein, 1927, p. 437). As in Taxila, the same type of wastewater management system was commonly used in that very site: drains, soak-pits and/or cesspools (Singh, 2008, p. 389). In ancient Delhi, still during the 3rd *ca.* BC drains channelled wastewater into 'wells, which may have functioned as soak-pits' (Singh, 2006, p. 119). Once again, we can note with interest the correlation between well and sewage disposal.

3.3.2 Southern South Asia

In the southern part of South Asia, the development of water resource management began to spread during the Sangam period (300 BC to 300 AD): tanks (*ery* in Tamil) were used to store water in order to irrigate paddy fields (Sita, 2000, p. 35). Aquaculture was practiced in the lotus ponds (*tamarai kulam* Tamil) (Sita,

2000, p. 36). Ery were used to collect the runoff for irrigation purpose. The Grand Anicut, built around the 1st *ca.* AD, was mentioned as the oldest dam for irrigation in the region (Bijker, 2007, p. 111).

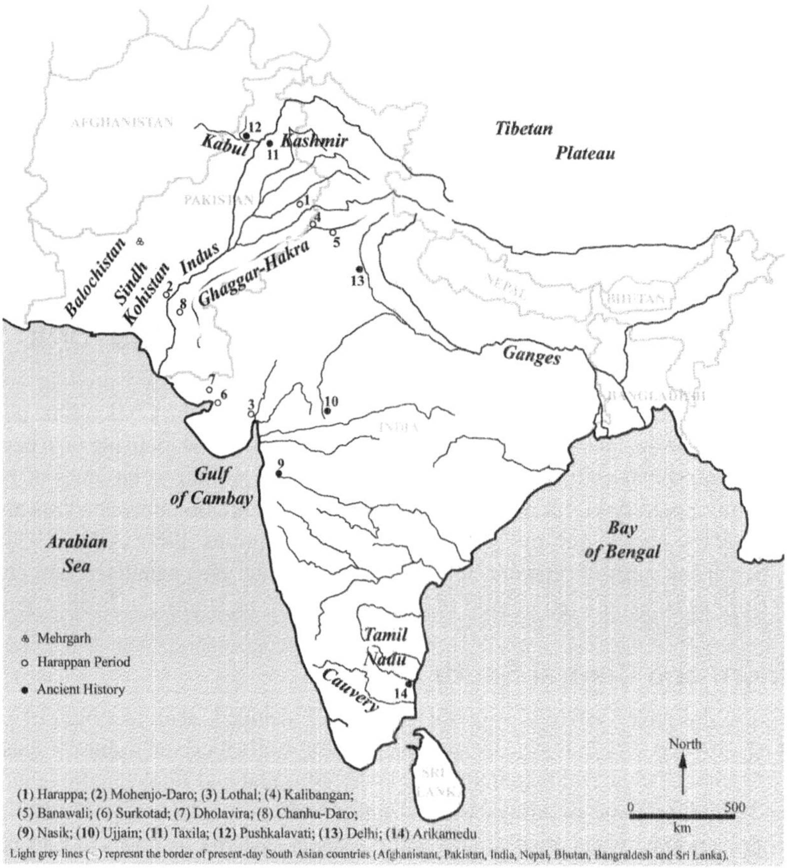

Figure 3.2 Map of the sites of the 'Ancient History period' (500 BC–300 AD) cited in the text (Map by H. F. Fardin).

Arikamedu site is considered to have been occupied from the Northern Black Polished Ware (NBPW or NBP, from 600 to 200 BC; Singh, 2006, p. 99), and most likely from the late 1st *ca.* BC to the middle of the 1st *ca.* AD (Begley, 2004). This site is an ancient port town located in the south-east of India, a few kilometres south of the present-day Pondicherry. Several ceramics have been dated to about 300 BC by archaeomagnetic methods (Ramaswamy & Duraiswamy, 1990). But we have to keep in mind that 'it is frequently not possible to postulate when an object arrived at a site, when it was discarded and when and how a fragment from it was deposited in a given location' (Begley, 2004). Most authors consider this site as a Roman colony (Kirk, 1975), since many elements of Roman origin were discovered at Arikamedu, as coins and pottery, and because the presence of the Romans and Greeks has been demonstrated in this region for the same period (Ray, 1988). The presence of these Mediterranean populations gave the opportunity to local blacksmiths to improve and diversify their techniques (Sita, 2000, p. 32). What interests us particularly are the drains, probably used for the disposal of wastewater since the beginning of Arikamedu urban development, that is from the 1st *ca.* BC (Casal, 1949). The wastewater management was improved around 150 AD with corbelled drains that seem unique to this region during this period (Begley, 1983). This system drained water of the basins. These spaces included adjacent rooms (Bharadwaj, 1997), believed to be elements of a 'protoindustrial' textile production and dyeing (Begley, 1983). Whereas until that time, in the rest of India, drainage systems and other wastewater management equipment were related to domestic effluents, at Arikamedu wastewater management was about industrial effluents.

A correlation can be stated with some ancient sites in the Mediterranean area. Indeed, the management of wastewater of Pompeii fullonicae (De Feo & De Gisi, 2012) was based on the same system as Arikamedu textile production site. Again, no author mentions a possible link between wastewater treatment and irrigation in Arikamedu. However, canals and ditches for irrigation were commonly used in this region at the same time (Bijker, 2007).

3.4 DISCUSSION: CULTURAL AND TECHNICAL INNOVATIONS AND DIFFUSIONS IN ANCIENT SOUTH ASIA

In south-east India, the development of drainage techniques coincides with the period when the Mediterranean populations settled in this part of South Asia. Since there were technical inputs to pottery and metallurgy, why would it not have the same technical inputs about runoff and wastewater management? In addition, can the corbelled drains of Arikamedu be considered as techniques borrowed to the Romans? Indeed, at the same period, the Romans created the biggest sewage system of antiquity: the Cloaca Maxima. This work, started in the late 7th *ca.* BC (Hopkins, 2007), is considered to be almost completed at the end of the 1st *ca.* AD, when it became necessary to improve sanitation because of population growth (Cooper, 2001). It was during the same century that the vaulted drains of this equipment were probably built (Adam, 1999, p. 319 and p. 321). But it has been shown that corbelled vaults were also used at Mohenjo-daro for the drainage of the Great Bath (Jansen, 1989, p. 185). Thus, we cannot confirm that Arikamedu corbelled drains have a Roman origin. In addition, it has been proved that an ancient trade route linked Taxila to Ujjain (Figure 3.3), at least since the time of the Emperor Ashoka (Eggermont, 1966), that is to the 3rd *ca.* BC (268–231 BC; Robb, 2003, p. 40). We suggest that this trade route could have been a possible route of diffusion of wastewater management techniques in South Asia, at least from the NBP. Indeed, it has been demonstrated that trade routes are the major ways by which technology is spread (Keller, 2000).

About the NBP period, Magee *et al.* (2005) report that the diffusion of techniques was most likely from west to east, and that this culture could be earlier than 500 BC. This trade route was also well-known by Mediterranean populations of the 1st *ca.* AD, as mentioned by Pliny and Ptolemy (Eggermont, 1966, p. 262) or in the *Periplus of the Eritrean Sea* (Ray, 1987, p. 98). The contact between the peoples of northern South Asia and those of the Mediterranean probably took place during the Achaemenid and Gandhara periods, and we know that the management of wastewater was carried out during the Harappan and NBP periods. Kenoyer (1997, p. 277) notes that 'many of the technologies first developed in the Indus cities provided the foundation for later technologies used in South Asia and other regions of the Old World.' In southern India, sanitation appeared during the NBP, it means when commercial and cultural contacts between South Asia and the Mediterranean region were already important (Kirk, 1975). But we have to take into consideration the link between the South Asia populations and Eastern Asia populations. Ancient Chinese cultures also have developed sanitation techniques, but are considered to appear around 2300 BC (De Feo *et al.* 2014), thus after the Harappan ones. And no sign of trade between these two parts of Asia during these ancient periods has been discovered until now. But later, when China and South

India started having intense commercial relations (around 300 BC, Francis and Francis, 2002) irrigation techniques through pond system (named 'multi-pond system' in China, and 'tank system' in South India) have been developed in both regions of Asia. Further investigation should be done in order to determine whether there has been a technological transfer of irrigation system.

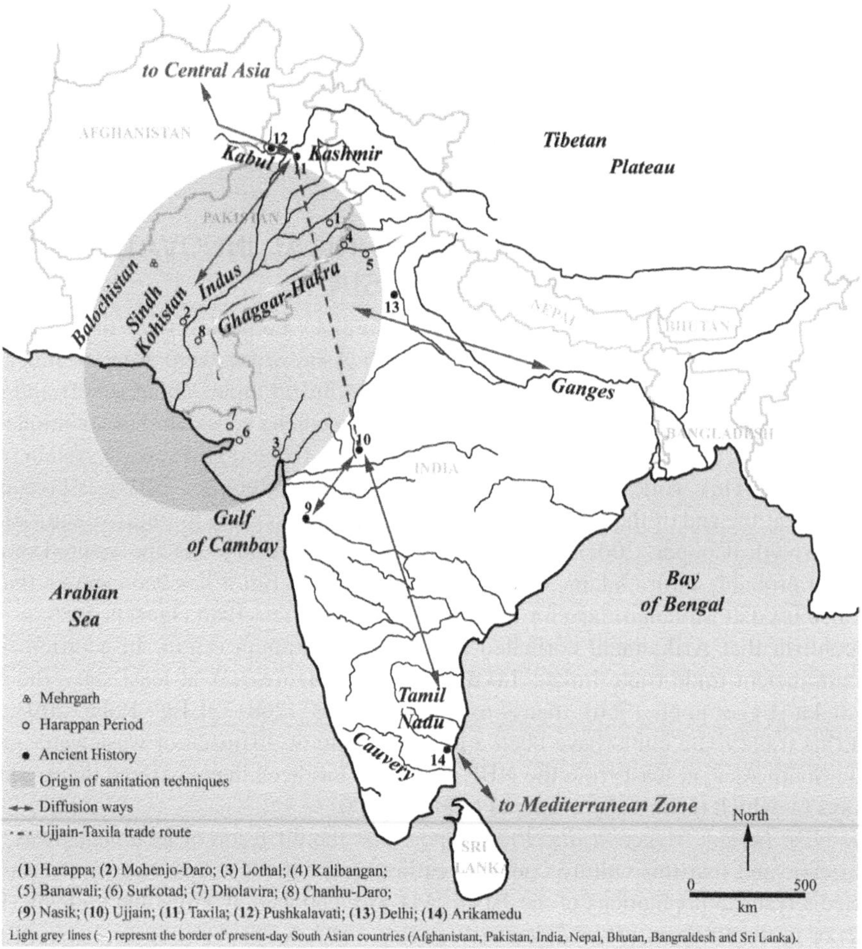

Figure 3.3 Map of the probable diffusion of wastewater techniques in South Asia during early History (Map by H. F. Fardin).

After Kroeber (1940), technical innovation is mainly due to a stimulus created when a single person or a group of persons is in contact with an exogenous technique. But, diffusionism theories have been criticized for having a europeanocentric point of view (Blaut, 1993), and indeed technology innovation can appear in any society. The current research in social sciences points out the heterogeneity of social networks (Law, 1992), as well as how humans and non-humans ('nature', techniques, sciences, knowledge etc.) are coproduced by each other (Latour, 1988, 2011). These various studies about the relation between society and technology cannot be ignored when we analyse any kind of technical evolution. Indeed, though these views of society are about contemporaneous society, it can be applied to ancient cultures. Thus, we have to be critical about the idea of technical diffusion. The transfer of sanitation techniques from Harappan culture to NBP cultures seems to be wrong, because according to archaeological discoveries not any culture has developed sanitation techniques from 1600 BC to 500 BC. Moreover, Kenoyer has pointed out that some Harappan technologies 'disappeared with the end of the Indus cities and, with the breakdown in trade networks' (Kenoyer, 1997). So, this remark questions the absence of sanitation techniques from 1600 BC to 500 BC. Is this absence of sanitation correlated to the absence of urbanity during this period? But, for example, centralised sanitation techniques have been discovered in Chanhu-Daro, which covered only 6 ha (Kenoyer, 2002), and such a built-up area cannot be considered as urban. Thus, Harappan sanitation techniques are not a characteristic of urban agglomeration only, and centralised sanitation is probably linked to socio-political factors rather than land use (De Feo *et al.*, 2014).

3.5 CONCLUSION AND PERSPECTIVES

Wastewater is managed in South Asia for about 5000 years. Harappan civilisation (3200 BC–1900 BC) has been the first South Asian culture to develop various techniques to collect domestic effluents. Two main types of systems were used: a centralised one, with sewage and drainage networks, and a decentralised one, where outflows were managed at the household level through jars and soak-pits. During the second urban phase, which occurred in the Northern Black Polished Ware period (600–200 BC), only decentralised wastewater management was practiced in South Asia. But it is in this very period that wastewater management techniques diffused in the whole subcontinent, and seems to follow the main trade route of the region, which joined Ujjain to Taxila. (Table 3.1). It was also at the same time that South Asia had intensive trade with the rest of the old world, especially with the Middle East and the Mediterranean zone, and with East Asia. But, still it is considered that wastewater management techniques used in the subcontinent found their origin in South Asia. We have to keep in mind that early history techniques cannot be transposed in any present-day society, since socio-political, demographical and environmental context are very different.

Table 3.1 Wastewater management techniques in South Asia during Ancient History.

Historical Periods	**3200 BC**	**1900 BC**	**500 BC**	**200 BC**	**150 BC**	**150 BC – 150 AD**
	Harappan periods		**Northern Black Polished ware (NBP)**			**Sangam**
Sites	Greater Indus Valley: Mohenjo-daro, Harappa, Dholavira etc.	Gulf of Cambay/ Gujarat: Lothal	Central India: Ujjain	Northern South Asia: Taxila, Pushkalavati and Delhi	Central India: Nasik	South India: Arikamedu
Wastewater Management Techniques	– Latrines – Drainage and Sewage disposal: • Soak-pits • Pipe network • Drains with sedimentation cesspits	– Drainage and Sewage disposal: • Soak-pits • Pipe network • Drains with sedimentation cesspits	– Drainage and Sewage disposal: • Soak-pits • Pierced pots (latrines?) • Ring-wells = soak-pits?	– Drainage and Sewage disposal: • Drain-pipes • soak-pits • cesspools • Ring-wells = soak-pits?	– Drainage and Sewage disposal: • Ring-wells = soak-pits?	– Covered drains for industrial effluent disposal

3.6 REFERENCES

Adam J. P. (1999). Roman Building: Materials and Techniques. Routledge, London, p. 360.

Alley K. D. (1994). Ganga and Gandagi: interpretations of pollution and waste in Benaras. *Ethnology*, **33**(2), 127–145.

Archaeological Survey of India (1990–1991). Excavation at Dholavira, District Kutch. In Indian Archaeology. A review, 1990–1991. Archaeological Survey of India, Delhi, pp. 26–35.

Begley V. (1983). Arikamedu reconsidered. *American Journal of Archaeology*, **87**(4), 461–481.

Begley V. (2004). The Dating of Arikamedu and its bearing on the archaeology of early historical South India. In: South Indian Horizons, J. L. Chevillard and E. Wilder (eds), Institut Français de Pondichéry/Ecole Française d'Extrême-Orient, Pondicherry, pp. 513–537.

Bharadwaj H. C. (1997). Town planning, building and building materials. In: History of Technology in India, A. K. Bag (ed.), volume 1, Indian National Science Academy, New Delhi, pp. 498–523.

Bijker W. E. (2007). Dikes and dams, thick with politics. *Isis*, **98**(1), 109–123.

Bisht R. S. (1984). Structural remains and town-planning of Banawali. In: Frontiers of the Indus civilisation, B. B. Lal and S. P. Gupta (eds), Books and Books, New Delhi, pp. 89–97.

Blaut J. M. (1993). The Colonizer's Model of the World. Geogrqphicql Diffusionism and Eurocentric History. The Guilford Press, New York/London, p. 246.

Casal J. M. (1949). Les fouilles de Virapatnam-Arikamedu. *C.R. séances Acad. Inscr. Belles-Lettres*, **2**, 142–147.

Chakrabarti D. K. (2004). The Archaeology of Ancient Indian Cities. Oxford University Press, Delhi, p. 296.

Codrington K. de B. (1944). A geographical introduction to the history of Central Asia (continuation). *Geography Journal* **104**(3–4), 73–91.

Cooper P. F. (2001). Historical aspects of wastewater treatment. In: Decentralised Sanitation and Reuse, P. Lens, G. Zeemann and G. Lettinga, (eds), IWA Publishing, London, pp. 11–36.

De Feo G. and De Gisi S. (2012). Water and wastewater management in the treatment process of a Roman *Fullonica*. In: Proceeding of the 3rd International Conference on Water and Wastewater Technologies in Ancient Civilisation, mars 22–24, 2012, Istambul, pp. 396–402.

De Feo G., Antoniou G., Fardin H. F., El-Gohary F., Zheng X. Y., Reklaityte I., Butler D., Yannopoulos S. and Angelakis A. N. (2014). Historical development of sanitary sewers worldwide. *Sustainability* (in Press).

Douglas M. (2001). De la Souillure – Essai sur les Notions de Pollution et de Tabou (Purity and Danger: An analysis of Concepts of Pollution and Taboo). La Découverte & Syros, Paris, p. 207.

Dumont L. (1966). Homo Hierarchicus – Le Système des Castes et ses Implications (Home Hierarchicus: An Essay on the Caste System). Gallimard, Paris, p. 449.

Eggermont P. H. L. (1966). The Murundas and the ancient trade-route from Taxila to Ujjain. *Journal of Economics, Soc. Hist. Orient*, **9**(3), 257–296.

Ewald P. W. (1991). Transmission modes and the evolution of virulence, with special reference to Cholera, Influenza, and AIDS. *Hum. Nat.* **2**(1), 1–30.

Francis P. Jr, and Francis P. (2002). Asia's Maritime Bead Trade: 300 BC to the Present. University of Hawaii Press, Honolulu, p. 313.

Fuller D. Q. (2006). Agriculture origins and frontiers in South Asia: a working synthesis. *J. Wld. Prehist.*, **20**, 1–86.

Hopkins J. N. N. (2007). The Cloaca Maxima and the monumental manipulation of water in archaic Rome. *Wat. Rome*, **4**, 1–15.

Jansen M. (1989). Water supply and sewage disposal at Mohenjo-Daro. *Wld. Archaeol.*, **21**(2), 177–192.

Jarrige J. F. (2006). Mehrgarh neolithic. *Prāgdhārā*, **18**, 135–154.

Keller W. (2000). Geographical localization of international technology diffusion. Working Paper 7509, National Bureau of Economic Research, Cambridge, USA, p. 58.

Kenoyer J. M. (1991). The Indus Valley tradition of Pakistan and Western India. *J. Wld. Prehist*, **5**(4), 331–385.

Kenoyer J. M. (1997). Trade and technology of the Indus Valley: new insights from Harappa, Pakistan. *Wld. Archaeol.*, **29**(2), 262–280.

Kenoyer J. M. (2002). Mature Harappan. In: Encyclopedia of Prehistory, N. Peregrine and M. Ember (eds), South and Southwest Asia, Springer, New York, USA, **8**, pp. 271–283.

Kirk W. (1975). The role of India in the diffusion of early cultures. *Geogr. J.*, **141**(1), 19–34.

Latour B. (1988). Science in Action: How to Follow Scientists and Engineers through Society. Harvard University Press, Cambridge, p. 288.

Latour B. (2011). Networks, societies, spheres: reflection of an actor-network theorist. *Internat. J. Comm.*, **5**, 796–810.

Law J. (1992). Notes on the theory of the actor-network: ordering, strategy, and heterogeneity. *Sys. Pract.*, **5**(4), 379–393.

Leshnik L. S. (1968). The Harappan 'port' at Lothal: another view. *Am. Anthropol.*, **70**(5), 911–922.

Mackay E. J. A. (1936). La Civilisation de l'Indus: Fouilles de Mohenjo-Daro et d'Harappa. Payot, Paris, p. 206.

Mackay E. J. A. (1937). Excavations at Chanhu-Daro. *J. Roy. Soc. Arts*, **85**(4405), 527–545.

Madella M. and Fuller D. K. (2006). Palaeoecology and the Harappan civilisation of South Asia: a reconsideration. *Quat. Sc. Rev.*, **25**, 1283–1301.

Magee P., Petrie C., Knox R., Khan F. and Thomas K. (2005). The Achemenid empire in South Asia and recent excavations in Akra in northwest Pakistan. *Am. J. Archaeol.*, **109**(4), 711–741.

Mani B. R. (2006). Kashmir Neolithic and early Harappan: a linkage. *Prāgdhārā*, **18**, 229–247.

Mate M. S. (1969). Building in ancient India. *Wld. Archaeol.*, **1**(2), 236–246.

Misra V. N. (2001). Prehistoric human colonization of India. *J. Biosci.*, **26**(4), 491–531.

Mosse D. (1997). The symbolic making of a common property resource: history, ecology and locality in a tank-irrigated landscape in South India. *Dev. Change*, **28**, 467–504.

Nair K. J. (1961). Review on Report on the excavations at Nasik and Jorwe 1950–51. *Anthropos*, **56**(1–2), 310–311.

Ponton C., Giosan L., Eglinton T. I., Fuller D. Q., Johnson J. E., Kumar P. and Collett T. S. (2012). Holocene aridification of India. *Geophy. Res. Let.*, **39**(3), L03704.

Ramaswamy K. and Duraiswamy D. (1990). Archaeomagnetic studies on some archaeological sites in Tamil Nadu, India. *Phys. Earth Planet. Inter.*, **60**, 278–284.

Rao S. R. (1979). Lothal, a Harappan Port Town (1955–62). Vol I, Archaeological Survey of India, New Delhi, p. 268 + CXXVII (Plates).

Ray H. P. (1987). Early historical urbanization: the case of the Western Deccan. *Wld. Archaeol.*, **19**(1), 94–104.

Ray H. P. (1988). The Yavana presence in ancient India. *J. Econ. Soc. Hist. Orient*, **31**(3), 311–325.

Robb P. (2003). A History of India. Palgrave, New York, p. 344.

Singh U. (ed.) (2006). Delhi Ancient History. Social Science Press, New Delhi, p. 231.

Singh U. (2008). History of Ancient and Early Medieval India: from the Stone Age to 12th century. Pearson Education/ Dorling Kindersley, New Delhi, p. 704.

Sita K. (2000) Irrigation system of the Sangam Tamils. In: Aspects of South Indian History, J. Alasiar, K. K. Kusuman, R. Rajalingom, P. Sarveswaram and B. Sobhanan (eds), Rajesh Publications, Nagercoil, pp. 29–36.

Smith M. L. (2006). The archaeology of South Asian cities. *J. Archaeol. Res.*, **14**, 97–142.

Stein A. (1927). Alexander's campaign on the Indian North-West frontier: notes from explorations between upper Swat and the Indus. *Geogr. J.*, **70**(5), 417–440.

Vuorinen H. S., Juuti P. S. and Katko T. S. (2007). History of water and health from ancient civilizations to modern times. *Wat. Sci. Tech.: Water Supply*, **7**(1), 49–57.

Wright R. P. (2010). The Ancient Indus: Urbanism, Economy, and Society. Cambridge University Press, New York, p. 396.

Chapter 4

Evolution of sanitation and wastewater technologies in Egypt through centuries

Fatma A. El-Gohary

4.1 INTRODUCTION

As societies moved from nomadic cultures to building more permanent sites, the disposal of human excreta became an important concern. As will be presented, sanitation and wastewater management is an issue that has been dealt with in many different ways and knowledge has been lost and regained. This chapter intends to highlight the evolution of sanitation and wastewater treatment technologies through centuries and how it has influenced the development of this sector in Egypt.

4.1.1 Historical perspective

Modern humans (Homo sapiens) have dwelled on this earth for some 200,000 years, most of that time as hunter-gatherers and gradually growing in number (Vuorinen *et al.*, 2007). Approximately 50,000 years ago modern man began to inhabit every corner of the world and people were constantly on the move. The first human communities were scattered over wide areas and waste produced by them was returned to land and decomposed using natural cycles. Disposal problems were limited primarily because they were small communities of nomadic hunter–gatherers (Giusy & Jeanette, 2010). Occasionally, people were troubled by pathogens transmitted by contaminated water, but it has been postulated that the waterborne health risks of hunter-gatherers were small.

Some 10,000 years ago, when people adopted an agrarian way of life, mankind established permanent settlements. This new type of livelihood spread everywhere and the population began to expand faster than ever before. Sedentary agricultural life made it possible to construct villages, cities and eventually states, all of which were highly dependent on water. This created a brand new relation between humans and water. Pathogens transmitted by contaminated water became a very serious health risk for the sedentary agriculturists.

The earliest known permanent settlement, which can be classified as urban, is Jericho from *ca.* 8000–7000 BC, located near springs and other bodies of water. In Egypt there are traces of wells, and in Mesopotamia of stone rainwater channels, from *ca.* 3000 BC. From the early Bronze Age city of Mohenjo-Daro, located in modern Pakistan, archaeologists have found hundreds of ancient wells, water pipes and toilets. The first evidence of the purposeful construction of the water supply, bathrooms, toilets and drainage in Europe comes from Bronze Age Minoan (and Mycenaean) Crete in the second millennium BC.

The first urbanization in Europe occurred during antiquity (*ca.* 500 BC–500 AD) around the Mediterranean region. The share of urban population reached some 10–20% in the centuries around the birth of Christ. The most urbanized areas were the Eastern Mediterranean, Egypt, North Africa (modern Tunisia), the Apennine Peninsula (modern Italy), and the southern part of the Iberian Peninsula, most of which were areas of quite modest rainfall. In this period the archaeological and written sources become richer, and consequently improved our possibilities to study the relationship between water and health of people.

In general, guaranteeing pure water for people became a prerequisite for successful urbanization and state formation. Evidence of activities concerned with human health and water supply has been found in civilizations throughout human history. The human search for pure water supplies must have begun in prehistoric times. The experience of humankind from the very beginning testifies to the importance and safety of groundwater as a water source, particularly springs and wells. The way in which water supply and sanitation was organized was essential for early agricultural societies. If wells and toilets were in good shape, health problems and environmental risks could be avoided.

The realization of the importance of pure water for people is evident already from the myths of ancient cultures. Religious cleanliness and water were important in various ancient cultures. Ideas of the salubrity of water were connected to the general 'scientific' level of the society. The first known Hellenic philosophical thinkers and medical writers also recognized the importance of water for the public health.

Furthermore, it should be kept in mind that the salubrity of the water supply must have differed markedly in accordance with the social status of people in the ancient towns. The rich had running water in their homes; the poor had to fetch their water from public fountains. The rich had their own baths and toilets, while the poor had to use public toilets and baths. All this must have lead to different health conditions and levels among rich and poor people.

4.1.2 Evolution of sanitation in ancient time

4.1.2.1 Wastewater management

Sanitation is a term primarily used to characterize the safe/sound handling and disposal of human excreta as well as other waste products (Avvannavar & Mani, 2008). It is well known that the relationship between humans and sanitation has seen substantial changes, due to the influence throughout the ages by cultural, social and religious factors (Sorcinelli, 1998; Wolfe, 1999; De Feo & Napoli, 2007; Avvannavar & Mani, 2008). It is incredible how much history is found at the end of a sewage system; from food to hygienic habits, from the use of pharmaceuticals and birth control pills to more intimate sexual habits. There is no more reliable source of customs and behavior of a society than its waste products and this fact is beyond the perception of the civilization. A sociological analysis that is more truthful than the analysis of a wastewater does not exist (Giusy *et al.*, 2010). 'The history of men is reflected in the history of sewers' wrote Victor Hugo (1892) in Les Miserable's, 'it has been a sepulcher, it has served as asylum, crime, cleverness, social protest, the liberty of conscience, thought, theft, all that the human law persecute or have persecuted is hidden in that hole'.

Although the importance of good quality drinking water for urban populations was realized since the antiquity. Yet the importance of proper sanitation for the protection of public health was not understood by modern cities until the 19th century (Brown, 2005; Vuorinen *et al.*, 2007; Cooper, 2001). For centuries wastewater management was not given much, if any, consideration. Much has been written about the history of water supply systems, yet, there is a lack of corresponding information on wastewater management

(Giusy *et al.*, 2010). Sadly, when it came to waste management and sanitation, countries, even those that suffered epidemics, tended to have short memories. This is surprising since the lack of sanitation affects human development to the same or even greater extent as the lack of clean water. In most cultures, wastewater was disposed of in the streets and near population centers creating serious impacts on public health and the environment. This is evident by the numerous epidemics which occurred throughout Europe until the nineteenth century (Lucking, 1984; Brown, 2005; HDR, 2006; Aiello *et al.*, 2008). At the same time, the idea of wastewater management is as old as man himself. Simply put, man has struggled through the ages with the problem of what to do with his waste. Man knew instinctively, even in his earliest existence, the importance of allowing animal and human waste to go downstream, yielding to the natural flow of things (Eddy, 2000). He may not have known all of the consequences, but he surely found the prospect of harvesting drinking water from the same area of the stream used for waste disposal (Eddy, 2000).

Throughout history wastewater management has presented people and governments with far reaching technical and political challenges. The story of waste and wastewater management is at once a story of human ingenuity and human frailty (Sorcinelli, 1998; HDR, 2006). A number of keystone events defined the speed at which environmental management evolved through the ages. Some of these events were scientific, such as stream purification models, while others were socio-economic such as two World Wars (Seeger, 1999; Shifrin, 2005; Cooper, 2007). However according to the recent Human Development Report (2006), the lesson from the past is that progress in wastewater management and sanitation was driven above all by political coalitions uniting industrialists, municipalities and social reformers. This means that if on one side developing new technologies as well as appropriate strategies for wastewater management is required, on the other side there must be an urgent need to overcome the stigma of a polluted environment.

4.1.2.2 *Wastewater management in ancient time*

Our earliest archeological records of central water supply and wastewater disposal date back about 5000 years, to Nippur of Sumeria. In the ruins of Nippur there is an arched drain with each stone being a wedge tapering downward into place. Water was drawn from wells and cisterns. An extensive system of drainage conveyed the wastes from the palaces and residential districts of the city.

In the 3rd millennium BC time period the Indus civilization had bathrooms in houses and sewers in streets (Adams, 1981). The sewage system was a network of effluent drains (built of brick masonry) constructed along the streets. These drains, located along one side of the street, were U-shaped approximately 50–60 cm deep. The drains were built of bricks set in clay mortar with covers made of loose bricks, flagstones or wooden boards. Covers could be removed for cleaning purposes. Wall drain chutes were used through which effluent flowed into the public drain or into a catchment basin.

Two millennia BC, the Hellenes and Egyptians had adequate supplies of drinking water for their cities, drained streets, had bathrooms in their houses, and, in Crete, water flushing arrangements for toilets. The Incas also had impressive sewerage systems and baths. Sites excavated in the Indus Valley and in Punjab show that bathrooms and drains were common in Indian cities 4 millennia ago. Streets were drained by covered sewers 0.61 m deep and made of molded bricks cemented with a mortar of mud. Within the houses, drain pipes were made of pottery embedded in gypsum .

Historical records show that the Mesopotamian Empire (*ca.* 3500–2500 BC) was the first civilization to formally address sanitation problems arising from community living (Giusy *et al.*, 2010). The Mesopotamian Empire states of Assyria and Babylonia marked great advances in civilization during the second millennium BC. The ruins from Mesopotamian cities contain well-constructed storm drainage and sanitary sewer systems. In the ruins of Ur and Babylonia, there are remains of homes which were connected to a drainage system to carry away wastes (Jones, 1967) as well as latrines leading to cesspits. For example, the ancient cities of Ur and Babylon, located in present day Iraq, had effective drainage systems for storm water control (Jones, 1967). The systems contained vaulted sewers and drains for household waste and gutters and drains specifically for surface runoff. The material of choice was baked brick with an asphalt sealant. Rainwater was also collected for household and irrigation uses. The Babylonians were partially motivated to construct urban drainage systems by their desire to remain clean. The Babylonians, like other ancient civilizations, viewed uncleanness as a taboo; not because of the physical uncleanness but the moral evil it suggested (Reynolds, 1946). In retrospect, the Mesopotamians viewed urban runoff as a nuisance flooding concern, waste conveyor, and a vital natural resource (Giusy *et al.*, 2010).

Close to *ca.* 4000 years ago, approximately *ca.* 1700 BC, the Minoan Palace of Knossos on the isle of Crete featured four separate drainage systems that emptied into great sewers constructed of stone (Angelakis *et al.*, 2012). The palace latrine (Figure 4.1) was the world's first flushing toilet with a wooden seat and a small reservoir of water (Eddy, 2000). The efficient separate urban drainage system integrated with rainwater collection devices suggests that the Minoans viewed urban runoff as a nuisance flooding concern, a waste conveyance mechanism, as well as a vital natural resource. Also, the first known bathrooms, in the sense of a room that is devoted to bathing and personal care were constructed in Minoan Crete (*ca.* 3200–1100 BC). Lavatories (or toilets) were often associated with baths.

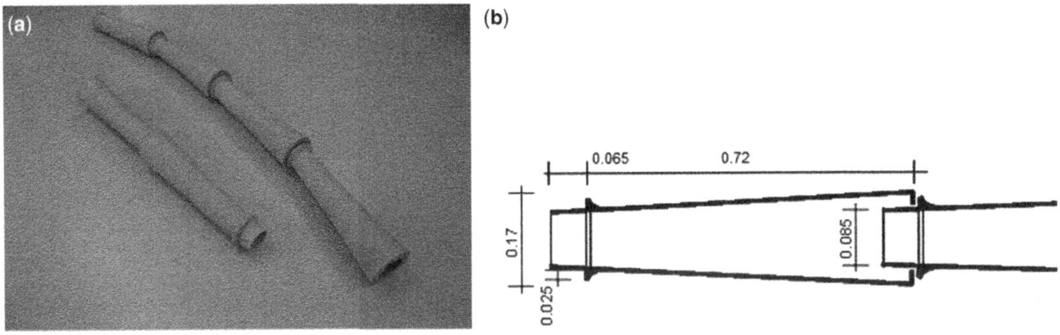

Figure 4.1 (a) Fitted terracotta pipes, of conical geometry with their dimensions, and (b) recently constructed pipes of same material and geometry looking very much like modern ones (with permission of A. N. Angelakis).

4.1.2.3 Ancient urban drainage systems

Urban drainage is defined to include two types of fluids: wastewater and storm water (Butler & Davies, 2000). Wastewater is water that after use for life support, industrial processes, or life enhancement must be collected and disposed of appropriately to prevent nuisances and polluted conditions from developing in urban areas. Storm water is runoff produced by precipitation. Both wastewater and storm water must be considered during urban drainage system planning. Historically the two waters have either been combined into a single conduit (i.e., combined sewers), or have been kept separate during collection and disposal (e.g., separate sewers).

Historically, urban drainage systems have been viewed with various perspectives. During different time periods and in different locations, urban drainage has been considered a vital natural resource, a convenient cleansing mechanism, an efficient waste transport medium, a flooding concern, a nuisance wastewater, and a transmitter of disease. In general, climate, topography, geology, scientific knowledge, engineering and construction capabilities, societal values, religious beliefs, and other factors have influenced the local perspective of urban drainage. For as long as humans have been constructing cities these factors have guided and constrained the development of urban drainage solutions.

From *ca.* 3000 to 1500 BC, early plumbers laid sewage and drainage systems. Archaeologists have discovered underground channels that remained virtually unchanged for centuries. Ancient gravity sewers were developed in response to the density of populations living in close proximity or in cities. These large central systems were actually analogous to sewers developed in the 1800s in London and other large cities. During these times, there were many outbreaks of various diseases, such as dysentery, cholera, infectious hepatitis, typhoid and paratyphoid, and various other types of diarrhea Eddy, 2000).

Historical accounts of ancient civilizations (e.g., Indus and Minoan) suggest urban drainage systems were constructed with great care and that the objectives of the systems were to collect rainwater, prevent nuisance flooding, and convey wastes. The systems that eventually met their objectives likely did so after trial-and-error modifications. In general, planning and design were limited. Few numerical standards existed for urban drainage and engineering calculations were not used during design. Despite the lack of optimization and the use of trial-and-error construction methods, numerous ancient urban drainage systems can be rated very successful. Lewis Mumford summarized the state of ancient urban infrastructure when he stated that ancient sewer systems were an 'uneconomic combination of refined technical devices and primitive social planning' (Mumford, 1961).

Webster (1962) and Kirby *et al.* (1956) described the technologically advanced urban drainage systems that the Indus civilization constructed for several of their more important cities. Ruins from two cities in particular provided a detailed glimpse of the Indus urban drainage systems. The ruins from Harappa and Mohenjo-Daro, two Indus cities separated by about 350 miles, suggest that they were arranged according to a plan and that the urban drainage system was coordinated with the layout of the town sites. Connections were built from most residences to open channels constructed in the center of the streets. The channels were either excavated into the ground or constructed above ground of burnt brick (Figures 4.2 and 4.3). Although houses were connected to the drainage channels, wastewater was not permitted to flow directly to the street sewers. First, the wastewater was passed through tapered terra-cotta pipes into a small sump. Solids settled and accumulated in the sump, while the liquids overflowed into the drainage channels in the street when the sump was about three-fourths full. The drainage channels could be covered by bricks and cut stones, which likely were removed during maintenance and cleaning activities. Another interesting feature of the channel was the inclusion of a cunette (Webster, 1962). The cunette was probably constructed to convey the smaller flows associated with daily wastewater discharges, while the entire channel would only be used during wet weather events. Overall, the Indus civilization viewed urban drainage systems as providing the dual purposes of waste and storm water conveyance.

The Minoan civilization flourished on the Island of Crete from about *ca.* 2800 BC to 1100 BC. The ruins from this civilization located on the Aegean Sea revealed elaborate systems of well-built stone drains (Figure 4.4), which carried sanitary sewage, roof runoff, and general surface drainage (Gray, 1940). The drains emptied into a main sewer that disposed of the sewage a considerable distance from the origin of the wastes. The frequent and torrential rains in Crete resulted in excellent flushing of the system. Ruins from the palace-city of Knossos indicate that a two-conduit system was installed, where one conduit collected sewage and the other rainwater (Angelakis *et al.*, 2012). The efficient separate urban drainage system integrated with rainwater collection devices suggests that the Minoans viewed urban runoff as a nuisance flooding concern, a waste conveyance mechanism, as well as a vital natural resource (Burian & Edwards, 2002).

Figure 4.2 Mohenjo-Daro excavated sewer channel (Hodge,1992).

Figure 4.3 Harappan above ground sewer channel constructed of burnt brick (Kirby *et al.*, 1956).

Figure 4.4 Minoan storm drain: (a) at Knossos palace and (b) at Tylissos houses in Crete, Hellas (with permission of A. N. Angelakis).

The Mesopotamian Empire states of Assyria and Babylonia marked great advances in civilization during the second millennium BC. The ruins from Mesopotamian cities contain well-constructed storm drainage and sanitary sewer systems. For example, the ancient cities of Ur and Babylon, located in present day Iraq, had effective drainage systems for storm water control (Jones, 1967). The systems contained vaulted sewers and drains for household waste and gutters and drains specifically for surface runoff (Maner, 1966). The material of choice was baked brick with an asphalt sealant. Rainwater was also collected for household and irrigation uses. The Babylonians were partially motivated to construct urban drainage systems by their desire to remain clean. The Babylonians, like other ancient civilizations, viewed uncleanliness as a taboo, not because of the physical uncleanness but the moral evil it suggested (Reynolds, 1946). In retrospect, the Mesopotamians viewed urban runoff as a nuisance flooding concern, waste conveyor, and a vital natural resource.

4.2 SANITATION IN ANCIENT EGYPT

Proper sanitation is an important factor in any community in order to protect the health of the people and the environment. These issues were also important in ancient Egypt. By *ca.* 2500 BC the Egyptians were pretty adept with drainage construction, accentuated by the significance that water played in their priestly rituals of purification and those affecting the burial of the kings. According to their religion, to die was simply to pass from one state of life to another. If the living required food, clothing and other requirements of daily life, so did the dead. Thus, it's not surprising that archaeologists have discovered bathrooms in some Egyptian tombs.

Excavators of the mortuary temple of King Suhura at Abusir discovered niches in the walls and remnants of stone basins. These were furnished with metal fittings for use as lavatories. The outlet of the basin is closed with a lead stopper attached to a chain and a bronze ring. The basin emptied through a copper pipe to a trough below. The pipe was made of 1/16″ beaten copper to a diameter of a little less than 2″. A lap joint seam hammered it tight. Also found within a pyramid temple built by King Tutankhamen's father-in-law at Abusir, was a brass drain pipe running from the upper temple along the connecting masonry causeway to the outer temple on the river.

Excavators have discovered a tomb which supposedly contains the body of Osiris before he became a god. It contains the dividing line between Life and Death, for example, a deep moat containing water that surrounds all sides of the figure of the god on his throne. After *ca.* 5000 years, water still fills the canal through underground pipes from the River Nile.

The sanitation methods of the ancient Egyptians may seem crude when compared to the modern conveniences available in the 21st century. However, they did have what appears to have been a workable, viable sanitation system. The degree of sanitation available to certain individuals varied according to their social status. A sociological analysis that is more truthful than the analysis of a wastewater does not exist.

According to the description of Herodotus (Histories II), in the Egyptian city of Herakopolis (BCE, 2100), the average person treated their wastes much like those in Ur, they threw the wastes into the streets. However, 'in the elite and religious quarters, there was a deliberate effort made to remove all wastes, organic and inorganic to locations outside the living and/or communal areas, which usually meant the rivers.' The homes of the wealthy were airy and roomy, literally. There were bedrooms, servants' quarters, halls, dining rooms – and bathrooms. Actually, a 'bathroom' was usually a small recessed room with a square slab of limestone in the corner. The bathroom would be fitted with a slightly inclined stone-slab floor and the walls were typically lined to a certain height (about half a meter) with battered stone slabs to protect against dampness and splashing (Breasted, 1906). Drainage of wastewater was provided by setting a basin beneath the spout of the floor slab in the bathroom, or sometimes by drainage channels running through the outer wall into a vessel or straight into the desert sand. Remains of early earth closets with limestone seats have been also discovered. The disposal of the human waste was most probably in the sandy soil. Certain more well-to-do homes had toilets. Their seats were made of limestone. The less wealthy who could not afford to have a limestone toilet, used toilet stools, under which a ceramic bowl was placed. Furthermore, toilet stools with a hole in the middle and a clay pot beneath were also used as portable toilets and they were often buried with senior officials (Breasted, 1906). The excrement, which was collected in jars containing sand, was emptied into pits outside the walls of the house, in the river and even in the streets. Instead, poor people used wooden stool with a hole on it. The toilets used beds of sand to catch/contain the wastes. Servants cleaned the sand regularly (Giusy & Jeanette, 2010).

At the same time, ancient Egyptians were early developers of pipes and the techniques of making copper alloys. In the beginning, of course, their pipes and fittings were very crude. Like the Mesopotamians, they used clay pipes made from a combination of straw and clay. First it was dried in the sun, and then baked in ovens. As they improved their clay sewer pipes, the Egyptians were able to drain the low-lying portions of the Nile Valley, and gradually the entire region evolved into a fertile garden (the plumber.com). The Helenes made their earthenware pipes in curved sections as well as straight, and of tapering shape so that each fitted into the next. This method was adopted by the Romans. In the sixth century BC the city of Athens was served by two aqueducts terminating in a reservoir from which water was distributed to the city, initially by a stone masonry channel (Figure 4.5), and later by pipes of earthenware and lead materials. The Egyptians were also skilled in working metals. They melted metal in a crucible over a super-hot fire, provided by men fanning the fire with blowpipes made of reeds tipped with clay. The molten metal was poured out and allowed to cool, then beaten out with smooth stones into sheets of the required thickness. It was then cut to shape. One explanatory picture in a tomb chapel describes the process as 'causing metal to swim' (the plumber.com). Very ancient metal pipes are known to have been used in Egypt. They were of hammered copper 1.4 mm thick. These pipes were cemented into grooves cut in solid flagstones.

Figure 4.5 Stone channels (http://www.ethanholman.com/history/egypt/dailylife/sanitation.html).

4.3 DOMESTIC WASTEWATER TREATMENT IN THE MODERN WORLD

4.3.1 International level

What we must understand is that what flows in sewers at that time differs from that in Hugo's description. Wastewater is typically classified according to its origin; domestic, industrial, commercial or urban (WEF, 2009). Domestic wastewater comes from residential sources including toilets, sinks, bathing, and laundry.

It can contain virtually anything from cleaning chemicals, soaps, and detergents to bacteria and other pathogenic organisms. Industrial wastewater is discharged by manufacturing facilities and commercial wastewater from offices, hotels, stores and other enterprises. Municipal or Urban wastewater is typically a mixture of domestic, industrial and commercial wastewater. If a community has a combined sewer system rather than a separate system, then storm water is also included in the mixture.

The 20th century witnessed a revolution in wastewater management, environmental science and societal views towards pollution. Scientific discovery, debates on societal priorities and government interest evolved through the century beginning with unhindered pollution and ending with attempts an increasing control (Shifrin, 2005). A milestone was the Eighth Report (1912) of the Royal Commission on Sewage Disposal which introduced the concept of biochemical oxygen demand (BOD) and established standards and tests to be applied to sewage and sewage effluents which were copied by many other countries (Giusy, 2010). Streeter and Phelps (1925) and Imhoff and Mahr (1932) pioneered aeration/deaeration models that allowed scientists to predict allowable BOD loads to surface waters. Governments began to mandate waste treatment. Before the First World War led to the interruption of installation of wastewater treatment facilities, they were constructed in the main cities of Europe (Seeger, 1999; Cooper, 2007).

The Second World War also delayed development of wastewater treatment until 1948 causing increasing pollution to the waters (Giusy & Jeanette, 2010). In addition many wastewater plants were damaged during the war and not rebuilt for many years (Seeger, 1999). After the end of the war there was rapid progress in wastewater treatment in the United Kingdom and the United States, but not Europe (Cooper, 2007).

The period from 1981 to 1990 was officially deemed International Drinking Water Supply & Sanitation Decade (IDWSSD), during which an estimated $133.9 billion dollars was invested worldwide. Approximately 1.6 billion people were served with safe water and close to 750 million with adequate excreta disposal facilities. At the end of the decade – according to WHO estimates – 1 billion people still lacked access to safe water and 2 billion lacked sanitation services. Moreover, only about 25% of the $133.9 billion went to rural areas, where 75% of the poor in developing countries reside.

Unfortunately, the *International Water Decade* paid insufficient attention to the issue of sanitation and wastewater reuse in the developing world (Alaerts *et al.*, 1993; WHO, 1987; Munasinghe, 1992; Black, 1994; Giles & Brown, 1997). Although fresh water systems have been increasingly developed for the urban poor, urban drainage and sanitation systems have not been scaled-up proportionally, because it was largely supply-driven and did not respond to demand, and it did not focus on sustainability. Evidence gathered from India toward the end of the IDWSSD showed that while many rural systems had been installed, maintenance of those systems was either assigned to overburdened state agencies, to local governments with little technical and financial capacity, or not assigned at all.

Additionally, the global initiative was not able to overcome the weaker voice in policymaking that has plagued rural populations due to their poverty, their distance from policymakers, and their wide dispersion across a large area, which inhibits them coming together as a unified lobby (WHO Fact Sheet NO 112).

4.3.2 Trends in the development of wastewater services in Egypt

Several social, cultural, economic and environmental factors have shaped the development of sanitation services in Egypt and will influence the development of wastewater management services in the future. Historically, private household sanitation facilities in Egypt were built to satisfy households demand for privacy and convenience.

On-site disposal systems were commonly used in rural areas, un-sewered small communities and un-sewered developments in urban areas. The Sanitary pit privy or earth pit privy were the most commonly used systems in rural Egypt. A pit capacity of 1.5 m^3 is considered adequate for four or five persons. In locations where the soil is heavy and impervious or where there is no room to establish the sanitary pit privy at a safe distance from drinking water supply wells, the concrete vault privy is used. Where indoor plumbing has been installed, and the house is provided with drinking water supply system, the indoor toilet or privy is connected to an out-door septic tank. On-site pit latrines and soak away pits are not a viable solution for high density urban areas as they depend on the permeability of soil and multiple systems can overload the infiltration capacity of the local strata (Alaerts, 1996; Giles & Brown, 1997). Septic tank systems and vault toilets are effective in containing wastes, provided they are properly lined. However, these systems require frequent servicing, depending on the size, and are often maximized in their capacity to the degree of overflowing across streets and yards, thus contributing to non-point pollution sources. The cost to regularly service on-site septic systems is expensive. Consequently, regular servicing does not occur, and the function of the system becomes inefficient (Black, 1994). Another problem associated with septic tanks, is the number of vehicles needed to adequately maintain and service household-level tanks; the costs associated with the consumption of fossil fuels can be very high (Strauss *et al.*, 1998).

High overall population growth and even higher urban population growth have not only increased pressure on water resources, but also created a tremendous additional demand for water and sanitation infrastructure and on-site systems have begun to threaten the groundwater resources. Efforts financed by National and International Organizations allowed the development of improved on-site sanitation systems in Egypt.

4.3.2.1 Development of Cairo's wastewater system

In order to gain understanding of the development of wastewater treatment technologies in Egypt it was found appropriate to trace the development of the system in Cairo city from its inception at the beginning of the century until the present day.

In 1885 a commission was nominated by the Ministry of Public Works to study all questions related to the health of the people of Cairo and the best means of improving the sanitary conditions. During the period between 1885 and 1906 various sanitary improvements were made and a number of comprehensive sewerage schemes were proposed by consultants appointed by the commission. The projected plan was prepared to cover a period of twenty five years (1907–1932) and to provide the city (which at that time lay entirely on the east bank of the Nile) with a sewerage system and sewage treatment facilities. The treatment plant consisted of six sedimentation tanks. One fifth of the settled sewage was directed to trickling filter before rejoining the main flow from the works. It has been estimated that the population of Cairo will increase during this period from 700,000 to nearly one million and that the ultimate flow of the wastewater would be 50,000 CMD.

By mid nineteen twenties it was recognized that there was a need to establish a sewerage system for Giza and the other rapidly expanding communities on the west bank of the Nile and a scheme to serve this area was prepared. The scheme was commissioned in 1939 and comprised gravity sewers draining to a pumping station to treatment facility at Abu Rawash consisting of primary sedimentation.

In the mid nineteen fifties a new Commission was appointed to prepare a Master Plan for sewerage facilities for Cairo for the remainder of the century. These included the construction in 1940 of an emergency canal from Ein Shams to Gabal el Asfar to replace the emergency reservoir located near Ein Shams. The construction of sub-collectors all over the city and additional sedimentation tanks in Gabal el-Asfar continued up to 1952. Over the years, additional sewerage systems and sewage treatment facilities are added and now over 90% of the population of Cairo are served by six activated sludge wastewater treatment plants (Figure 4.6).

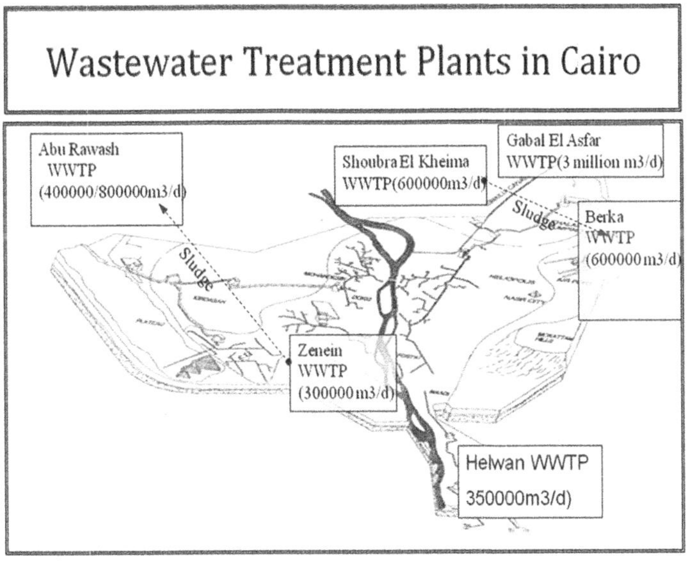

Figure 4.6 Six activated sludge wastewater treatment plants (Greater Cairo Wastewater Organization).

4.3.2.2 Development of sanitation on the national level

To fulfill the requirements of the first International Drinking Water Supply & Sanitation Decade (IDWSSD) launched during the period from 1981 to 1990, the Egyptian Governments have made significant strides towards improving water and wastewater services in the country. However, most sector investment was directed to the provision of drinking water. Also, construction of sewage networks and treatment plants was directed to Cairo and Alexandria. Sanitation in small towns and rural areas remained inadequate. In Egypt access to improved sanitation is less than half that for water. This gap matters, not just because access to sanitation is intrinsically important, but also because the benefits of improved access to water and to sanitation are mutually reinforcing. In Egypt, the rate of sewerage connection varies, with impressively high rates (more than 90%) in some cities, such as Cairo to less than 10% in rural areas. High levels of pollution in areas without appropriate sanitation systems undermine the potential health and environmental benefits of near universal access to water.

On the sanitation side, it became clear that investment has been primarily directed to large sanitation infrastructure. The high capital, operation, and maintenance costs of this infrastructure have made it more difficult to recover the full cost from users and to make wastewater utilities financially sustainable. High costs are often the result of implementation of inappropriate technologies. Large scale conventional treatment systems have been viewed as the only option for dealing with domestic wastewater. A large proportion of the cost of these systems is tied up in the pipes, pumping equipment and earth works required to install this infrastructure, making it an expensive option.

To achieve the Millennium Development Goals (MDGs) for water and sanitation, the Government of Egypt has embarked on an ambitious program more focused towards maintaining sustainable sanitation services for peri-urban and rural areas. In light of the growing realization that weak institutional framework is a major contributing factor to the poor level of water and sanitation services , substantial efforts have been undertaken to develop a nationwide sector reform program.

In general, community wastewater management services in Egypt evolved in three phases. During the first phase, wastewater services were provided to large cities and urban centers. In the second phase, wastewater collection services and some treatment works were provided to secondary towns. Centralized conventional wastewater collection systems were typically provided to large cities and secondary towns. The third phase, currently underway will provide wastewater services to smaller towns and rural areas. At the beginning, the trend was to provide conventional centralized sewerage systems to individual communities and to transport the waste from several individual communities to centralized treatment facilities. Resistance to application of non-conventional wastewater management systems often stems from the wastewater management agencies preference to conventional systems for the following reasons:

(a) Establishment of well design standards and operating procedures for conventional systems;
(b) The security and safety of selecting familiar solutions;
(c) The general negative perception of planners, engineers, and the public towards non-conventional systems;
(d) sub-quality service; and
(e) The lack of knowledge of non-conventional options and technologies.

Recently, there is a growing realization between experts and decision makers in Egypt that decentralization of sanitation programs into smaller-scale projects can bring benefits at an affordable cost to those in greatest need. Centralized water based sanitation is in general not economically feasible. Moreover, it produces wastewater which does not always meet the criteria required for sustainability and environmental security. New configurations employing the best practices of sanitation technology and management for rural and urban contexts are now under consideration. The challenge facing decision makers is therefore, to go beyond traditional classification between 'small, appropriate' and 'modern/advanced technologies and to develop rural and peri-urban sanitation with a mix of scales, strategies, technologies, payment systems and decision-making structures, that better fits the physical and human systems for which they are designed. This does not mean that the macro picture should not be considered. On the contrary, the decentralization should take place after an adaptable strategic macro framework has been defined to sketch out the overall direction for sanitation service provision. Several options have recently been proposed and appear feasible, but necessitate further development. There is a wide range of innovative sanitation technologies to choose from. In some cases it may be preferable not to install sewers, but to continue to use existing on-site sanitation technologies such as cesspools and septic tanks. In other cases, sewers may be installed only for a block of houses connected to a communal septic tank. Under certain circumstances, however, the high-cost solution of connecting to a citywide sewerage network is the only feasible technical solution.

On the national level, the number of municipal wastewater treatment plants Egypt increased by 10 times between 1985 and 2005. By the end of 2012, Egypt had 372 municipal wastewater treatment treating an average of 10.1 million m^3/d, serving more than 18 million people.

The largest wastewater treatment plant in Egypt is located in Gabal el Asfar to the Northeast of Cairo, serving about 9 million people and treating 2 million m^3/d in 2009. The plant discharges into the Belbeis Drain and then into Bahr El Baqar Drain (BBD), which in turn drains to Lake Manzala 170 km away from Cairo. The drain and Lake Manzala had been identified as 'black spots' by the Egyptian Environmental Action Plan back in 1992. The quality of water improved substantially after the completion of the first stage of the plant in 1999, but the drain and the lake are still environmentally fragile. Tendering for further expansion of the plant with financing from the African Development Bank began in 2011. It would increase its capacity to 2.5 million m^3/d. A planned third stage would bring capacity to 3 million m^3/d, serving 12 million people and making Gabal al Asfar one of the largest wastewater treatment plants in the world.

4.4 CONCLUSIONS

Appropriate sanitation and good hygiene are fundamental to health, survival, growth and development. However, these basic necessities are still a luxury for many of the world poor people. Over 2.6 billion of our fellow citizens, mostly in developing countries lack basic sanitation. Efforts to prevent death from diarrhea or to reduce the burden of such diseases as *ascaris*, hookworm, *schistosoiasis* and trachoma are doomed to failure unless people have access to basic sanitation. Lack of basic sanitation indirectly inhibits the learning abilities of millions of school-aged children who are infested with intestinal worms transmitted through inadequate sanitation facilities and poor hygiene.

There are still a great deal of unresolved problems related to sanitation principles. In the developing world, such problems were intensified to an unprecedented degree. Moreover, new problems have arisen such as the contamination of surface and groundwater. Naturally, intensification of unresolved problems led societies to revisit the past and to reinvestigate the successful past achievements. To their surprise, those who attempted this retrospect, based on archaeological, historical, and technical evidence were impressed by two things: the similarity of principles with present ones and the advanced level of lavatories engineering and urine separation practices.

However, it is important to note that water and sanitation is a complex sector that involves multiple stakeholders ranging from the community level to the international level The role and responsibilities of each stakeholder will vary depending on the nature of the service being delivered and on the level of water availability and scarcity in the area. The entry point, depends on the local socio-economic and environmental conditions

4.5 REFERENCES

Adams R. (1981). Heartland of cities: surveys of ancient settlements and land use on the central floodplain of the Euphrates. University of Chicago Press, Chicago, USA.

Aiello A. E., Larson M. S. and Sedlak R. (2008). Hidden heroes of the health revolution. Sanitation and personal hygiene. *American Journal of Infection Control*, **36**, 128–51.

Alaerts G. J., Veenstra S., Bentvelsen M. and van Duijl L. A. (1993). Feasibility of anaerobic sewage treatment in sanitation strategies in developing countries. *Water Science & Technology*, **27**(1), 179–186.

Angelakis A. N., Koutsoyiannis D. and Papanikolaou P. (2012). On the geometry of the Minoan water conduits. In: Proc. of the 3rd IWA International Symposium on Water and Wastewater Technologies in Ancient Civilizations, March 22–25, 2012, Istanbul, Turkey, pp. 172–177.

Avvannavar M. S. and Mani M. 2008. A conceptual model of people's approach to sanitation. *Science of the Total Environment*, **390**, 1–12.

Black M. (1994). Mega – Slums: The Coming Sanitary Crisis. Water Aid, London, UK.

Breasted J. H. (1906). Ancient Records of Egypt, vol. V, Historical Documents from the Earliest Times to the Persian Conquest. University of Chicago Press, Chicago, Illinois, USA.

Brown J. (2005). The early history of wastewater treatment and disinfection. Impacts of Global Climate Change; pp. 1–7. doi:10/061/40792(173)288. American Society of Civil Engineers. ASCE/Library.

Burian S. J. and Edwards F. G. (2002). Historical Perspectives of Urban Drainage. Global Solutions for Urban Drainage. Proceedings of the 9th International Conference on Urban Drainage, 8–13, September 2002, Portland,OR. pp. 1–16. doi: 10.1061/40644(2002)284

Butler D. and Davies J. W. (2000). Urban Drainage. E & FN Spon, London.

Cooper P. F. (2001). Historical aspects of wastewater treatment. In: Decentralised Sanitation and Reuse: Concepts, Systems and Implementation, P. Lens, G. Zeeman and G. Lettinga (eds), IWA Publishing, London, UK.

De Feo G. and Napoli R. M. A. (2007) Historical development of the AugustanAqueduct in Southern Italy: twenty centuries of works from Serino to Naples. *Water Science Technology*, **7**, 131–8.

Eddy N. (2000). Wastewater management, surfaces as an important issue in the new Millennium. *Small Flows Quarterly Winter 2000*, **1**(1), 14–19.

Giles H. and Brown B. (1997). 'And not a drop to drink'. Water and sanitation services in the developing world. *Geography*, **82**(2), 97–109.

Giusy L. and Jeanette B. (2010). Wastewater management through the ages: a history of mankind. *Science of the Total Environment*, **408**, 5254–5264.

Gray H. F. (1940). Sewerage in ancient and mediaeval times. *Sewage Works Journal*, **12**, 939–946.

HDR (Human Development Report) (2006). Beyond Scarcity: Power, Poverty and the Global Water Crisis. United Nations Development Programme, New York.

Hodge A. T. (1992). Roman Aqueducts & Water Supply. Gerald Duckworth & Co. Ltd., London.

Hugo V. (1982). Les Misérables, Trans. Norman Denny. Penguin Classics, New York, 2001.

Imhoff K. and Mahr G. (1932). Required degree of treatment of sewage prior its disposal by dilution. *Sewage Works Journal*, **4**, 892–8

Jones Jr, D. E. (1967). Urban hydrology – a redirection. *Civil Engineering*, **37**(8), 58–62.

Kirby R. S., Withington S., Darling A. B. and Kilgour, F. G. (1956). Engineering in History. McGraw-Hill Book Company, Inc., New York, NY.

Lucking B. (1984). Evaluating the Sanitary Revolution: Tyohus and Typhoid in London 1851–1900. Urban Disease and Mortality in Nineteenth Century England, London-New York.

Maner A. W. (1966). Public works in ancient Mesopotamia. *Civil Engineering*, **36**(7), 50–51.

Mumford L. (1961). The City in History: its Origins, its Transformations, and its Prospects. Harcourt, Brace & World, New York, NY, USA.

Munasinghe M. (1992). Water Supply and Environmental Management. Developing World Applications. West View Press, Boulder, CO.

Reynolds R. (1946). Cleanliness and Godliness. Doubleday and Company, Inc., New York, NY, USA, pp. 198–204.

Seeger H. (1999a). The history of German wastewater treatment. *European Water Management*, **2**(5), 51–56.

Shifrin N. S. (2005). Pollution management in twentieth century. *Journal of Environmental Engineering-ASCE*, **131**, 676–91.

Sorcinelli P. (1998). Storia Sociale Dell'acqua. Riti e Culture. Mondadori, Milano.

Strauss, M., Heinss, U. and Montangero, A. (2000). On-Site Sanitation: when the pits are full – planning for resource protection in faecal sludge management. In: Proceedings, Int. Conference, Bad Elster, 20–24 November 1998. Schriftenreihe des Vereins fuer Wasser-, Boden- und Lufthygiene, 105: Water, Sanitation & Health – Resolving Conflicts between Drinking – Water Demands and Pressures from Society's Wastes, I. Chorus, U. Ringelband, G. Schlag and O. Schmoll (eds). IWA Publishing, London, UK.

Streeter H. W. and Phelps E. B. (1925). A study of the pollution and natural purification of the Ohio River III. Factors concerned in the phenomena of oxidation and reparation. *Public Health Bulletin*, **146**, 1–75.

Vuorinen H. S., Juuti P. S. and Katko T. S. (2007). History of water and Health from ancient civilizations to modern times. *Water Science & Technology*, **7**, 49–57.

Webster C. (1962). The sewers of Mohenjo-Daro. *Journal Water Pollution Control Federation*, **34**(2), 116–123.

WEF Water Environment Federation (2009). Design of Municipal Wastewater Treatment Plants, Manual of Practice No. 8. Water Environment Federation, Alexandria, VA.

Wolfe P. (1999). History of wastewater. World of water 2000–the past, present and future. Water World/Water and Wastewater International Supplement to Penn Well Magazines, Tulsa, OH, USA.

World Health Organization (WHO) (1987). Wastewater stabilization ponds: Principles of planning and practice. WHO EMRO Technical Publication No. 10. Regional Office for the Eastern Mediterranean, Alexandria developing world. *Geography*, **82**(2), 97–109.

World Health Organization (WHO) (Reviewed November 1996). Fact Sheet No. 112, www.who.ch (accessed December 2013).

Chapter 5

Sanitation and wastewater in the Central Andean Region, Peru: An overview from Pre-Columbian and Colonial times to nowadays

A. Reyes-Knoche

5.1 INTRODUCTION

In South America and especially in Peru, the catchment and secure provision of water has been and still remains challenging. Reasons are particular geographic, hydrologic and climatic conditions, such as diverse and sometimes adverse environments, irregular availability of water, and immanent natural hazards, among others. Sanitation as well as returning the used waters into the natural water cycle also have been and still are main components of the implemented water resources management strategies.

The present chapter intends to give an overview of the sanitation and wastewater situation in the Central Andean Region with focus on Peru's Coastal and Andean regions, from Pre-Columbian times (*ca.* 3000 BC to 1532 AD) and Colonial times (1532–1821 AD) to nowadays, complementing a previous publication on sustainable water supply in Pre-Columbian civilizations in Ancient Peru and South America (Reyes-Knoche, 2012). Since the vast subject spans about 5000 years of human development and history in an eclectic natural and social environment (Figures 5.1, 5.2 and 5.3), the challenge has been to select and bring out summarized representative examples, cases and available facts, and still provide the reader with an overall impression. This chapter, supported by diagrams, tables and carefully chosen historic images and own photographs from different epochs, encompasses examples from Pre-Columbian civilizations, includes European and foreign influences brought in since the Spanish Colonial times, and concludes with an insight into the actual situation, development and trends.

5.1.1 The environment

5.1.1.1 Geography and hydrology

Peru is located in the central and western part of South America, neighboring with Ecuador and Colombia to the north, Brazil to the east, Bolivia to the south-east, Chile to the south and the Pacific Ocean to the west (Figure 5.2). It covers a surface of 1,285,216 km^2 including 8,8% of water surface. The most critical limitation affecting development is the irregular distribution of water (Reyes-Knoche, 2012). The water availability in Peru versus today's water demand is shown in Figure 5.5.

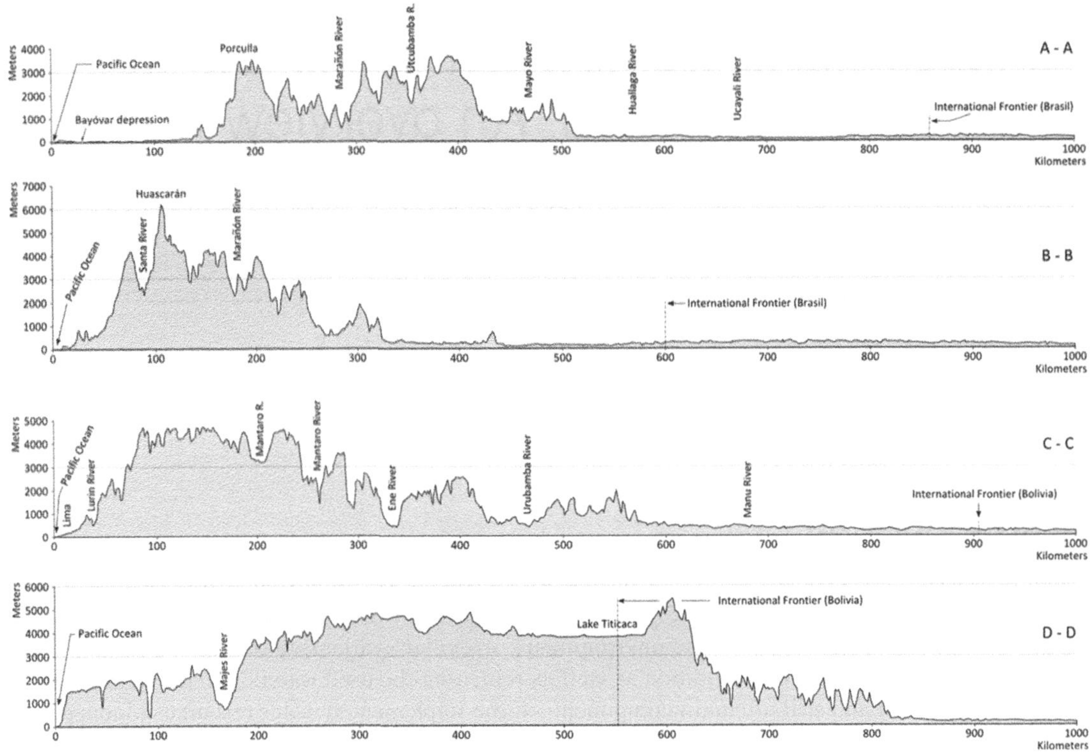

Figure 5.1 Topographic profiles/transects of the Peruvian territory: A-A at 06°10′ (Bayóvar), B-B at 09°11′ (Samanco), C-C at 12°08′ (Callao), and D-D at 16° parallel (see Figure 5.2). Own elaboration based on USGS GTOPO30 and SRTM-3 digital elevation models, and data of the Instituto Geográfico Nacional of Peru (Reyes-Knoche, 2011, 2012).

The Andes Mountains lie as a continuous chain of highland along the Pacific Ocean, dividing the territory into three geographic regions, the '*costa*' (coast), the '*sierra*' (highlands), and the '*selva*' (amazon forest/jungle, lowlands). The *costa* is a narrow littoral of 80 to 150 km width, largely arid except for valleys created by seasonal rivers originating in the Andes. The *sierra* is the region of the Andes, the world's longest exposed mountain range, 200 to 700 km wide and an average height of about 4000 m above sea level. The highest peak of the country is the Huascara´n at 6768 m. The *sierra* includes also the '*altiplano*' (high plain), a plateau with an average height of 3300 m, and Lake Titicaca, the world's highest navigable lake, and largest lake in South America. The *selva* is the flat and wide Amazon rainforest that covers almost 60% of the country's surface (Reyes-Knoche, 2012). The above-described varied environment is visualized in Figure 5.1 and 5.2.

The Peruvian rivers originate mainly in the Andes (Figure 5.2). They drain into three basins, the Pacific Ocean basin (283,600 km^2, 22%), the Atlantic Ocean basin (952,800 km^2, 74%) and the Lake Titicaca basin (48,800 km^2, 4%).

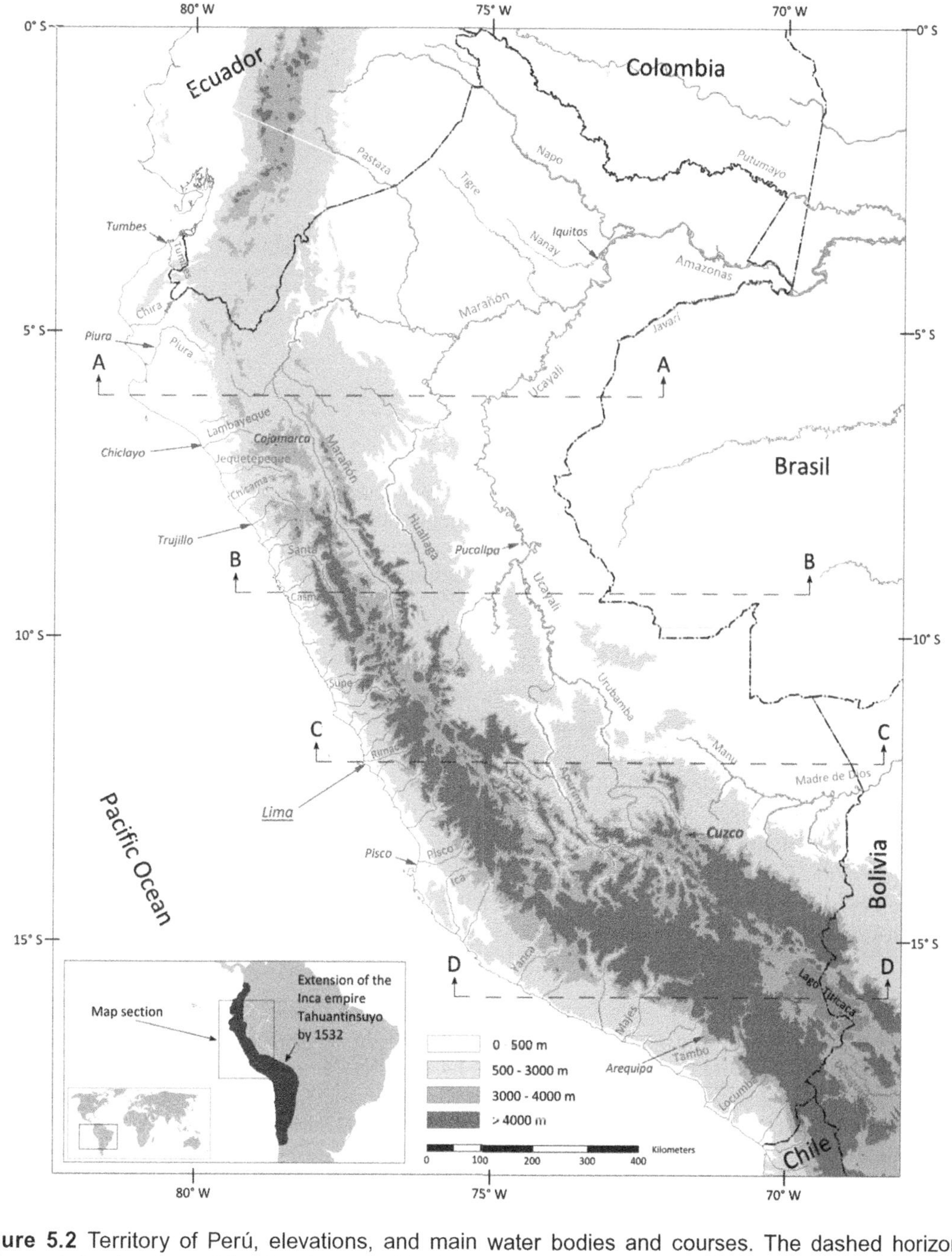

Figure 5.2 Territory of Perú, elevations, and main water bodies and courses. The dashed horizontal lines mark the location of the four topographical profiles/transects A-A, B-B, C-C and D-D contained in Figure 5.1. The small map on the left shows the territorial extension of the Inca Empire *Tahuantinsuyo* by 1532. Elaboration based on USGS GTOPO30 digital elevation model and data of the Instituto Geográfico Nacional of Peru (Reyes-Knoche, 2011, 2012).

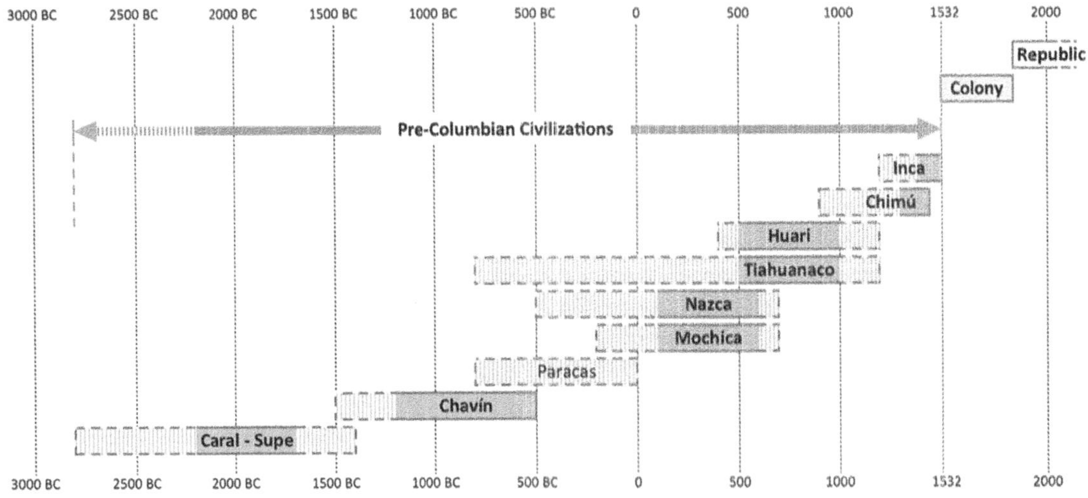

Figure 5.3 Indicative timeline representing the chronological appearance of selected pre-Columbian civilizations within *ca.* 4500 years of known pre-Columbian history, followed by about 300 years of the Spanish Viceroyalty and close to 200 years after Independence till nowadays. The shaded segments till the year 1532 indicate the most probable periods for each representative pre-Columbian civilization. The dashed lines indicate further possible periods of existence.

The rivers of the Pacific Ocean basin, 53 in total, are steep and short. Most of them carry water only during the rainy season and remain dry the rest of the year. Only a few of them keep significant flow rates along the year, such as the rivers Tumbes, Santa, Cañete and Majes.

The rivers of the Atlantic Ocean basin, 44 in total, are much longer, have a much larger flow, and are less steep once they exit the '*sierra*' (highlands). The main rivers of this basin are the Urubamba, Mantaro, Apurímac, Ucayali, Huallaga, Marañón, Napo, the Amazonas River, the world's largest river by volume and length, and the Madre de Dios, Putumayo and Yavarí rivers.

The rivers of the Titicaca Lake basin, 9 in total, are generally short, non-navigable, with small gradients and sparse flow variations. The main rivers of this basin are the Ramis, Illave and Coata rivers (Reyes-Knoche, 2012).

5.1.1.2 Climate

The Peruvian territory has a great climatic diversity due to the influence of the Andes and the cold Humboldt Current. The *costa* (coast) is characterised by moderate temperatures, low precipitations, and high humidity, except for its warmer, moist northern region. In the *sierra* (highlands), rain is frequent during summer months, and temperature and humidity diminishes with altitude up to the frozen peaks of the Andes. The *selva* (amazon forest) features heavy rainfall and high temperatures, except for its southernmost part, which has cold winters and seasonal rainfall (Reyes-Knoche, 2012). Table 5.1 shows meteorological data for representative cities, as well as their populations in 2012.

5.1.1.3 Natural ocurrences

The global coupled ocean-atmosphere phenomenon '*El Niño*', also known as 'ENSO' (El Niño Southern Oscillation) excerts its influence in Peru, causing in exceptional years heavy rains in the northern coast and intense droughts in the southern *sierra*, particularly in the *altiplano*. The strengths and consequences of the *El Niño* phenomenon are extremely variable and very difficult to predict. However, based on historical data a certain periodicity could be ascertained. While 'normal' *El Niño* occurences seem to appear every 3 to 4 years, 'very strong' *El Niño* phenomena with strong temperature anomalies and heavy rains in the north coast seem to appear every 9 to 12 years. The last two strong ones occurred in 1982–83 and 1997–98. To illustrate the magnitude of 'very strong' *El Niño* effects, the River Piura in the north coast reached a flow rate of 4424 m^3/sec compared to the average annual flow rate of 29.1 m^3/sec (SENAMHI). It is believed, that 'Mega' *Niños* appear every 500 to 1000 years, and that they could have induced the sudden fall or rise of some Pre-Columbian civilizations (Reyes-Knoche, 2102).

Table 5.1 Altitude, average mean temperature, and precipitation of representative cities in the Peruvian coast, Sierra and Selva, and their populations in the year 2012. Own compilation based on data of SENAHMI and INEI, Lima, Peru.

Region	City	Altitute [m above sea level]	Average mean temperature [°C]	Average mean precipitation [mm]	Population (2012)
Coast	Piura	55	24.4	72	417,892
	Chiclayo	34	21.0	62	583,159
	Trujillo	26	19.2	12	765,495
	Lima	30	19.2	15	9,437,493
Sierra	Arequipa	2508	15.4	99	844,407
	Cuzco	3249	12.5	736	405,842
	Cajamarca	2750	14.5	736	204,543
	Chachapoyas	2435	15.3	796	27,986
Selva	Tingo María	665	24.8	3302	70,742
	Yurimaguas	184	26.9	2047	70,575
	Pucallpa	154	26.3	1567	211,591
	Iquitos	126	26.2	2853	422,055

Peru is also particularly exposed to earthquakes, as the country is located in a seismic zone. The interference between the Nazca and the South American tectonic plates is located near the Peruvian coast. Minor earthquakes occur constrantly. The most recent severe earthquakes causing major damages took place in 1946, 1970, 2001, 2005, and 2007 with different epicentres (IGP).

5.1.2 Indicative timeline

The indicative timeline shown in Figure 5.3 encompasses the time period covered in the present chapter. It summarises the chronological appearance of selected pre-Columbian civilizations within *ca.* 4500 years of known pre-Columbian history, followed by about 300 years of the Spanish Viceroyalty and close to 200 years after Independence till nowadays.

The timeline section for the Pre-Columbian civilizations has been prepared and updated taking into account several timeline versions and/or chronology attempts from various archaeologists indicating differing dates. Due to the lack of written evidence the exact dates are mostly unknown. The shaded segments till the year 1532 indicate the most probable periods for each representative pre-Columbian civilization. The dashed lines indicate further possible periods of existence.

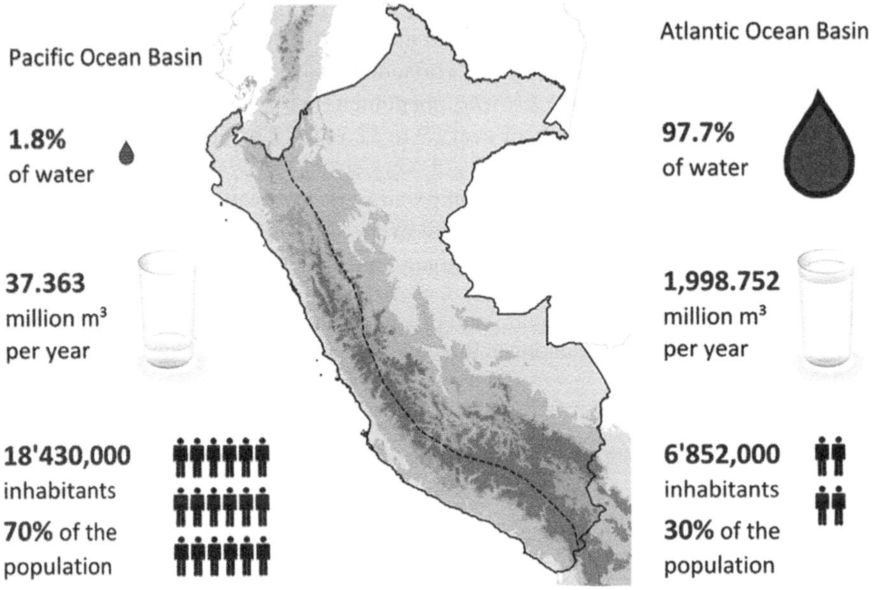

Figure 5.4 Water availability in Peru versus today's water demand. In today's Peru most of the population is concentrated in the Basin with considerably less water resources. While roughly 2% of Peru's water is available for 70% of the population, about 98% of Peru's water is available for the remaining 30% inhabitants. Figure prepared based on data of Oxfam International.

5.1.3 Population development

The population development in Peru within the last 600 years can be appreciated in Figure 5.5. The qualitative estimation from about the year 1450 to present days takes into account available census data from colonial times, from the early republican times as well as modern times census data (INEI).

In the early years of the Spanish colony the native population decreased considerably, mainly due to diseases brought in by the colonists (e.g., influenza, pox, measles, typhus, pest) that turned into epidemic diseases within the native population due to the lack of the correspondent antibodies. But also the effects of heavy forced labours (e.g., in gold and silver mines) contributed to the considerable diminution of the native population till about the middle of the sixteenth century. However, the census figures assorted during the colonial times can be considered as conditionally accurate. They sometimes resulted from estimations based on counting representative portions of the population. Up to what extent the slaves of African origin have been counted as well as the Amazon jungle inhabitants has been taken into account, for example, remains in some cases unknown. Nevertheless, the census figures become more accurate towards present days. The population development of Peru's capital for illustration purposes is shown in Figure 5.5. Note that presently about 1/3 of Peru's population live in Lima.

5.1.4 Water qualities over time

The diagram shown in Figure 5.6, originally conceived in the context of water reuse (Asano, 2002, 2007), has been chosen herein to visualize the development of sanitation and wastewater in the time span and region covered by the present chapter. The text onwards will refer to this diagram at some stages.

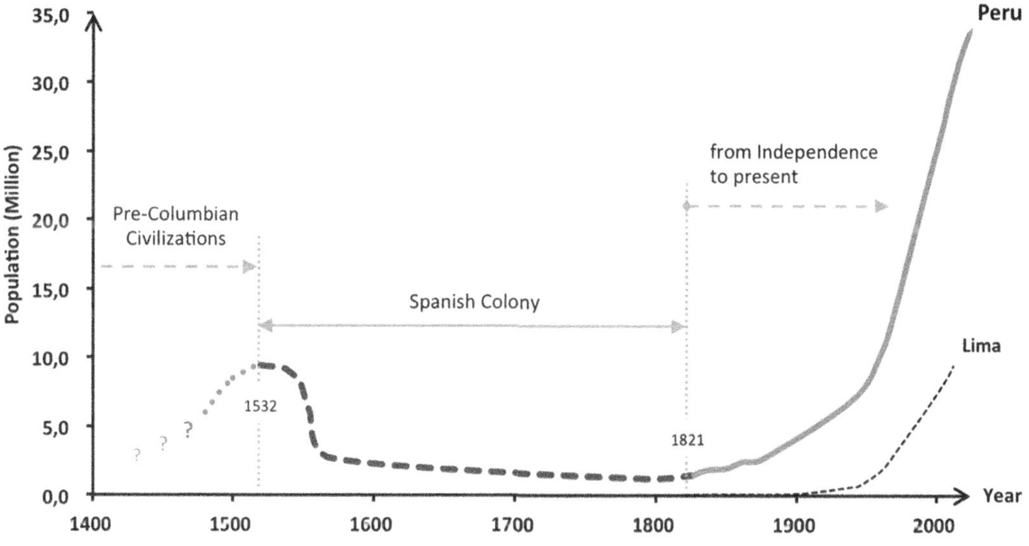

Figure 5.5 Population development in Peru within the last 600 years, from about 1450 to present days. Own estimation taking into account available census data (INEI) as well as historical information from primary and secondary sources.

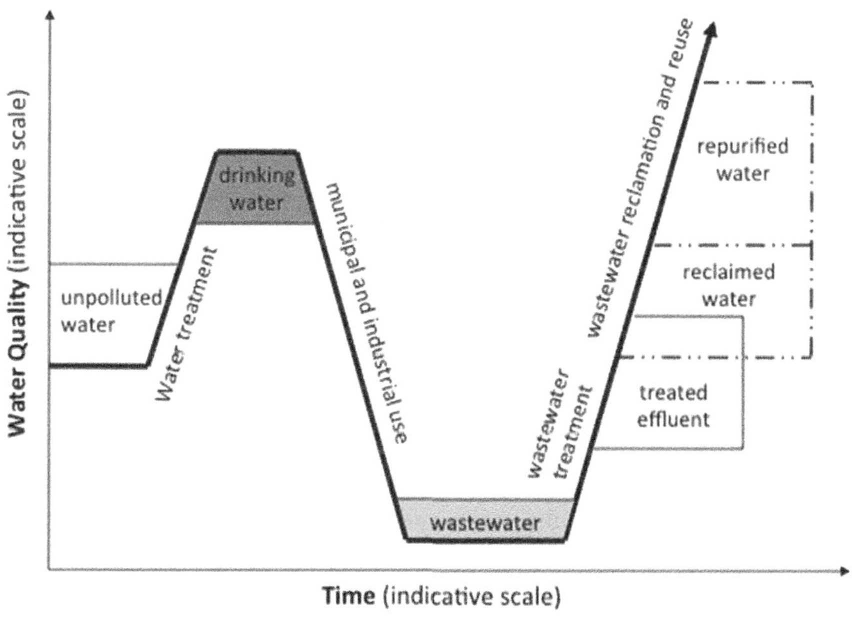

Figure 5.6 Water quality changes with water uses and treatments, drawn after Asano (2002, 2007), indicative scales.

The graph shows the water quality changes along different uses and respective treatments in an indicative timeframe. The quality of unpolluted water as found in nature is enhanced with water treatment to become drinking water. After for example, municipal and industrial use the quality decreases and the water becomes wastewater. Depending on the further use the wastewater shall ideally undergo an appropriate treatment before being returned to the water cycle or being reused for agricultural or other purposes requiring sometimes up to drinking water qualites or even higher water qualities.

5.1.5 Water availability versus demand

As mentioned further up the most critical limitation affecting development in the Central Andean Region is the irregular distribution of water. This circumstance applied for Pre-Columbian, Colonial and early Republican times and remains unchanged till nowadays.

In today's Peru most of the population is concentrated in the basin with considerably less water resources. About 70% of the population has to subsist in the Pacific Ocean Basin with roughly 2% of Peru's water availability, while the remaining 30% inhabitants are generously spread out in the Atlantic Ocean Basin, the area offering the remaining 98% of available water (Figure 5.5, see also Table 5.1). The comparatively small figures of the Titicaca Sea Basin have been disregarded in this appreciation.

The demographic situation during Pre-Columbian times was different. The habitat for most of the population was the *sierra* (Andean region), the region with comparatively ample water ressources. The population in the *costa* (coastal region) lived mainly along the coastal valleys of today's northern Departments of *Lambayeque* and *La Libertad*, and the Department of *Ica* in the southern coast. Even though the coastal population also reached a considerable number, the settlements remained within the limits of sustainable development. The settling of the *selva* (Amazon region), the region with abundant water ressources but uncomfortable living conditions, began mainly in modern times especially due to the exploitation of different natural ressources for industrial purposes (e.g., caoutchouc, petroleoum, timber, agriculture, cattle-breeding).

5.2 PRE-COLUMBIAN ERA (*ca.* 3000 BC TO 1532 AD)

During Pre-Columbian times today's sources of industrial water contamination didn't exist. The constituents of domestic wastewater were less polluting than those of nowadays and the volumes returned to the water cycle were also comparatively smaller. The water was used in its natural appearance and the used water cleaning happened by natural means (mother earth, father sun), and was sufficient (Figure 5.4). Water was generally taken in upper parts of the catchment areas and conducted by gravity to the cities, settlements, agriculturally used areas, and to other water users, before being returned to the natural water bodies or courses at lower parts of the water catchment areas. The water intakes were located at places with prudent distances to wastewater discharge points and away from places of deposition of excreta.

One spectacular example of this principle can be observed at the archaeological site of *Chavín de Huantar* (Chavín culture, Figure 5.3). The water source was the *Mosna* River where the water was catched at a high-lying point. The water was then conducted to the site by gravity for the various water uses and sanitary purposes, and returned afterwards by the network of drainage conducts to the lower laying *Mosna* River, a tributary of the *Marañon* River, the principal source of the *Amazonas* River. The archaeological site contains also underground channel systems constructed at different superposing levels properly designed to drain also regular and even intense storm water (Reyes Carrasco, 1973). As the stone build structures on the surface were built in different stages the network galleries and underground channels had to be planned in advance, and constructed following a sewer system master plan. Hydraulic details and cross sections of the underground channel system as well as the system elements visible on surface prove the quite good hydraulic engineering capabilities of the Chavín master builders.

Chroniclers of the early Spanish colonial times also give testimonials of pretty skilled provision and utilization of water sources in the different environments of the *costa* and *sierra*, in such a way that 'no drop of the precious resource was wasted' (e.g., Cobo, 1653, among others). The *Racarumi* intervalley channel build by the *Mochica* in the northern coast of Peru is certainly one of those examples (Figure 5.7).

Figure 5.7 View of a section of the pre-Columbian *Racarumi* intervalley channel (*Mochica*, northern coast) normally submerged under the water body of the Tinajones reservoir since 1968. The pre-Columbian channel continues its course behind the earth dam in the picture, placed perpendicular to its axis. The reason why the reservoir was considerably empty could not be ascertained with exactitude. Most likely high water demands for agricultural needs (e.g., Ethanol production) coupled with sporadic water distribution usages of the competent water resources management authorities and/or delay of the replenishing rains at the water source in the nearby Andean Region may have been the reasons. (Picture taken during investigation field works by Reyes-Knoche, November 2011; Reyes C. & Reyes-Knoche, 2012).

Figure 5.8 Two artisanal water wells built and operated in the traditional *Mochica* pre-Columbian manner. The wells are located in the course of one section of the pre-Columbian *Taymi* channel between the *Chancay-Lambayeque* and *La Leche* valleys in the north coast of Peru. In the picture, also A. Reyes-Knoche taking note of the position (S06°42.348 W079°40.732) during investigation field works in November 2011 (photograph courtesy of Reyes Carrasco).

In rural areas pre-Columbian inhabitants obtained their water also from artisanal carved wells, as it was the case in the northern coast of Peru. The same tradition is still being applied nowadays (Figure 5.8).

The 'Stone of *Sayhuite*' (13°32′50.21″S 72°48′10.29″W) may be another proof for the holistic and careful water resources management planning approaches of pre-Columbian master builders (Figure 5.9). The stone is most likely a scale hydraulic and territorial allocation model. The natural environment as well as different water use and drainage elements is carved on its surface. Some archaeologists believe that the stone may rather have been manufactured for water cult purposes as it contains also animal representations. But the usage as ceremonial stone for water rituals may have been the latter purpose, after serving as integral water resources planning instrument. Although the dating is not known, it is believed, that the stone has been manufactured either by *Huari* or *Inca* stonemasons (Figure 5.3).

Figure 5.9 View over the 'Stone of *Sayhuite*' (picture on the left). The stone is believed to be a scale model representation of a natural environment, for example, catchment area(s), with different water use elements carved also on it, probably a precursor of today's hydraulic models. The picture on the right shows details of the same stone. Both photographs taken at the *Sayhuite* archaeological site by A. Reyes-Knoche, October 2011.

During the 4500 years of currently known Pre-Columbian history several highly developed civilizations emerged mainly in the western coast and Andean regions (Figure 5.3). They developed and implemented remarkable sustainable water supply and sanitation solutions within the same challenging geographic, hydrologic and climatic conditions Peruvian's have to face nowadays. Examples of this Pre-Columbian water resources management heritage are the *Chavín de Huantar* underground channel system (Reyes Carrasco, 1973; Reyes-Knoche, 2012), the *Cumbe Mayo* aqueduct, the *Mochica* intervalley irrigation schemes, channels and aqueducts, the *Nazca* underground aqueducts and lines (Reyes-Knoche, 2009, 2012), a Pre-Columbian surveying instrument (Reyes Carrasco, 1980; and Reyes-Knoche, 2012), the *Tiahuanaco* and *Huari* raised-field farming technique, the *Huari* and *Inca* terraces, the *Chimú* city of Chan Chan, the Chimú territory and water resources in comparison to those from Babylonia and the ancient Egypts in the Nile Valley, and selected examples from the *Valle Sagrado de los Incas* (Sacred Valley of the Incas) and *Machu-Picchu* (Reyes-Knoche, 2009, 2012). Figure 5.10 shows another representative example built by the *Incas*.

Figure 5.10 Panoramic view on top of the vestiges of the Inca fortress *Sacsayhuamán* with partial view over the city of Cuzco, the ancient capital of the Inca Empire *Tahuantinsuyo* (picture on top). The bottom picture gives a detail view of the basement of the *'Torreón Muyuc Marca'* (*Muyuc Marca* tower), visible also in the picture on top, which is believed to be the basement of a circular water reservoir with radial outlets and concentric channels. Both panoramic photographs taken and composed by A. Reyes-Knoche, October 2011.

Around the year 1530 almost the total of the territories, people and knowledge of former Pre-Columbian civilizations formed part of the Inca Empire *Tahuantinsuyo*, the pre-Hispanic state with the largest extension in America, with a population of about 15 million inhabitants of different cultures, languages and aspects, a surface area of about 900,000 km^2, and a length of more than 5000 km, encompassing territories of the coast and the Andes of actual Peru, Bolivia, Ecuador, Colombia, Chile and Argentina.

Even though the Inca Empire *Tahuantinsuyo* existed only the last 100 years before the arrival of the Spanish colonists, the Inca managed to apply the knowledge gained by almost all other previous pre-Columbian civilizations along *ca.* 4500 years of development (Figure 5.3). Their economy was based on agriculture; hence water catchment, distribution and technical use of water were of vital importance.

5.2.1 Eradication of hunger

Besides their firm administration, notable knowledge and remarkable achievements in many disciplines, probably the most outstanding success of the Incas of Peru was the eradication of hunger (Reyes-Knoche, 2012). In contrast, according to the actual report on the status of achievements of the Millenium Development Goals, the corresponding target of the Goal 1 'Eradicate extreme poverty and hunger' is considered to lie 'within reach if recent slowdowns in progress can be reversed'. According to those statistics up to 8% of the human beings living in Latin America & the Caribbean are suffering today from hunger (UNSTATS and MDGs). The target is considered as achieved if the proportion of people who are undernourished lie within 7% of the total population in the Region, with population figures of 2012 (World Bank, INEI) about 40.7 million people in Latin America & the Caribbean, and about 2, 1 million in Peru.

The level of civilization, eradication of hunger and sustainability the Incas achieved was only possible with the consequent implementation of their institutional framework and organisation, strong leadership and skilled experts and workers. However, with the forced colonization by the Spaniards the Incas could not reach their own development climax.

5.2.2 No water-related diseases

To date there are no evidences of major diseases associated with inadequate water supplies and poor sanitation and hygiene, such as cholera or hepatitis during pre-Columbian times. Also, for example, there are no evidences of existence and outbreaks of Pest epidemics or pandemics in pre-Columbian Peru. Cleaning, personal hygiene and even water cult are believed to be fundamental components of pre-Columbian cultures and, as described above, their water usage and disposal followed holistic water resources management strategies conceived in harmony with the natural environment.

5.3 FROM THE SPANISH COLONY (1532–1821) TO EARLY REPUBLICAN TIMES AROUND THE BEGINNING OF THE XX CENTURY

With the arrival of the Spanish colonists and the creation of the Viceroyalty of Peru the *Inca*, and all pre-Columbian civilizations they once ruled, were forced to undertake abrupt cultural, economic and political changes. Their territories and populations sudden became part of the Spanish Kingdom. The *Inca* State was exchanged by the viceroyal administration. The colonists imposed Spanish institutions and customs supressing the indigenuous traditions and ideologies, establishing their regime with the aid of military force and legitimation by the Catholic Church. Mining of gold and silver, exploitation of land, natural ressources and labor force of the conquered became the new base for the economy.

The majority of the Spanish colonists that arrived especially at the early times of the colony, were adventurer, impecunious and illiterate individuals motivated by the chance to obtain a portion of the prospective richness the new world offered. Their idiosyncracy was determined by the late medieval European and Spanish culture at the early beginning of the cultural changes of the Rennaissance. The native individuals were considered primitive pagans that had to be civilized under Christianity. The further developments during the colony came along with the developments and progresses that happened in Spain and Europe during those times.

5.3.1 Colonial settlements

The Spanish colonists settled they cities following Spanish and European patterns of that times. Early settlements along the coast were among others Piura (1532), Lima (1535) and Trujillo (1535). At the beginning of the colony the Spaniards settled also cities in the Andes, the main habitat of the Incas, such as Cajamarca (pre-Hispanic *ca.* 1320, no Spanish foundation), Cuzco (pre-Hispanic *ca.* XIII century, Spanish 1534) and Ayacucho/Huamanga (prehispanic, Spanish 1540), all three already important pre-Columbian settlements, as well as Jauja (1534), Chachapoyas (1538), Moyobamba (prehispanic *ca.* 1420, Spanish 1540), Huánuco (1539), and Arequipa (1540), among others. Nevertheless, very soon Lima, located close to the Pacific Ocean harbor of Callao, became the seat of the administrative and economic power of the Spanish Viceroyalty of Peru, instead of Cuzco, the former capital of the Inca Empire *Tahuantinsuyo* located in the Andes (Figure 5.2). Gradually the new founded cities became the center of life, and accordingly, the Andean population began to migrate from the rural areas to the cities, from the Andean Region to the Coast, and ultimately to the major coastal cities and the capital Lima, a trend that was intensified with the course of time till nowadays (Figures 5.4 and 5.5).

5.3.2 Water resources management

Basically, the Spaniards continued to use the existent pre-Columbian water infrastructure they encountered with the knowledge and practices they knew from Europe. There is no evidence of major water infrastructure undertakings during colonial times with comparable dimensions to those realized by the pre-Columbian civilizations. On the contrary, the existing water infrastructure began to be disregarded due to the shift of priorities from agriculture towards gold and silver mining and less people to feed due to the demographic collapse in the middle of the XVI century (Section 5.1.3, Figure 5.5).

5.3.2.1 Mining

Gold and silver mining became the most important economic activity, especially during the first half of the Spanish colony. Initially the colonists applied the existing pre-Columbian mining procedures (e.g., Cobo, 1653), and introduced further on extraction processes made available through the technical progresses along the course of time, marking the begin of industrial use of water with its consequent deterioration of quality (Figure 5.5).

5.3.2.2 Agriculture

During colonial times the principal crops along the coast were sugar cane, cotton, grapes, olives and later on also rice. The labor force in the fields consisted mainly of slaves, mostly of African origin. The production centres were the *haciendas*, large estates that evolve from the early colonial times *encomiendas*.

At the time sugar cane and cotton production became a profitable business, the optimal use of water resources in agriculture got importance again. Step-by-step the *hacendados* (landlords) rehabilitated parts of the ancient irrigation systems, such as the intervalley channels along the north coast, and started to think about solutions to ensure and expand their profitable agriculture production. With the course of time and increase of water demands this initiatives led to the realization of the actual irrigation schemes in the valleys of the north coast, the former territories of the *Mochica* and *Chimú* (Figures 5.3 and 5.7), and along the south coast, the former habitat of the *Nazca* (Figure 5.3). Most of the actual irrigation systems, especially the ones in the north coast, follow the irrigation schemes conceived, implemented and operated by pre-Columbian civilizations.

In the *sierra* the crops produced during colonial times were potatoes, camote, oca, quinua, maize, wheat, among others, whose main purpose was to secure the subsistence of the mineworkers and related population. Even though the Spaniards introduced European agriculture procedures, the native population kept using in parallel their ancient agicultural technologies, producing their traditional crops, pasturing South American camelids and working in their ancient forms of collective work.

5.3.3 Urban water supply and sanitation

The urban water supply and sanitation arrangements of colonial settlements followed Spanish and European patterns of those times, and adopted with the course of time the technologies and progresses made in Spain, Europe and the rest of the occidental world. While the emphasis was set towards water supply, the evacuation of used waters as well as the removal of waste was not really taken into account.

5.3.3.1 Urban water supply system

The case of Lima shall illustrate the urban water supply and sanitation situation and development during colonial times till the commencement of the XX century. At the beginning water was taken mainly from the *Rimac* River, the nearby water source, and transported in big clay jars to the city end users. The construction of the system to supply Lima with clean water was commenced between 1561 and 1564 and finalized and put into service by the end of 1578. The system consisted of a water intake at *La Atarjea*, and an aqueduct of about 6 Km to a city reservoir from which water was distributed via fired clay pipes mainly to public fountains and throughs and to the fountain of the *Plaza de Armas* (main square).

The water supply network grew with the times. The exchange of clay pipes into cast iron pipes began in 1834 with one line, and continued from 1857 onwards. Around 1861, 40 years after independence, Lima had 165 fountains distributed within the city perimeter for its *ca.* 100,000 inhabitants: 6 big public fountains, 25 public throughs, 10 fountains in public buildings, 13 fountains in monasteries, 6 fountains in convents, 5 fountains in hospices, 7 fountains in schools, 3 fountains in barracks, and 90 fountains in private houses (Bromley, 1961).

5.3.3.2 Aguadores

The water distribution in the city of Lima was complemented by the *aguadores* (water porters). They filled up *pipas* (water recipients, jars) with water at the public fountains and delivered the water to the home premises of people who could afford their services. The aguadores were generally released Afro-american slaves and were organised in a body of water porters. They delivered the water to their customers by foot or with the aid of a mule (Figures 5.11, 5.12 and 5.13).

Figure 5.11 Four watercolour paintings showing *aguadores* (water porters) from Peru's colonial and early republican times. The paintings on the left and right show *aguadores* fulfilling their weekly duty of watering the main public plazas of Lima, the capital of the former Viceroyalty of Peru (Fierro, 1840 and 1850). The second picture from left to right shows an *aguador* (water porter) wearing his typical working clothes, mounted on a mule and carrying two *pipas* (water storage-jars) (Fierro, 1854). The third image from left to right shows another colonial times *aguador* fulfilling his bi-monthly duty of killing street dogs as a measure of preventing rabies (Fierro, 1850).

Besides their principal activity of water home delivery the *aguadores* had to fulfil their weekly duty of watering the main public plazas of Lima (Figure 5.11). Additionally, as a measure of preventing rabies, twice a month they had to kill dogs they found in the streets without proper dog tags, a task they carried out initially with the sole aid of a wooden stick with plumb weighting applications (Palma, 1833–1919).

In Lima the *aguadores* continued to offer their services after colonial times till around 1900. In provincial cities the services of these *aguadores* were sometimes required even till around the mid of the XX century, according to relations of eyewitnesses among older fellow men (Figure 5.13).

Figure 5.12 Further impressions of past means of water home delivery in Peru. From left to right: The portrait of an *aguador* of Lima (Fuentes Delgado, 1866); an *aguador* of Cuzco wearing comparatively elementary clothes, and carrying one small water recipient instead of the two bigger ones of his coastal colleague that counted on the support a beast of burden (Cuzco, around 1920, postcard published by Bazar Pathé around 1920); an *aguador* filling his *pipas* in a public fountain (Lima, around 1900, unknown photographer); drawing of an *aguador* transporting water in Lima (Cazeaux & Charton, 1854).

Figure 5.13 Three views to means of water home delivery in Peru around 1900. On the left side, a water porter ('*aguador*') with appurtenance and two wooden water storage-jars ('*pipas*') carried by a mule. The public fountain, his water source, can be seen behin the mules neck (Lima around 1900, unknown photographer, postcard published by E. Pollack-Schneider around 1900 in Lima). The picture on the bottom right shows another *aguador* fixing one *pipa* on the mules back while his other *pipa* is being filled with water in a public fountain (Lima around 1900, unknown photographer). The picture on the top right shows an artisanal carriage loaded with several *pipas* and ready to be hauled by a mule (Chiclayo around 1910, unknown photographer).

5.3.3.3 Urban sanitation and hygiene

The water supply systems constructed in the new founded colonial cities generally didn't include corresponding proper drainage systems to evacuate wastewater, as mentioned above. In Lima as well as in other coastal cities due to the lack of rain also no storm water sewer were foreseen (Table 5.1). And, following the common practices of Spanish and European cities of those times waste as well as used waters were disposed in the nearby streets or accumulated in dumping places in the outskirts.

Open ditches along main streets transported every sorts of waste along with the portions of wastewater reaching them. By 1861 Lima had 196 open ditches along the streets. The canalization of these ditches began in 1868 and continued till 1901 when all ditches within the city limits disappeared (Bromley, 1961). Waste kept accumulating in the cities and city outskirts. Most of the streets were dusty and without any sorts of pavement such as cobblestone. The quadruped means of transportation left their excrements and emictions in the streets and plazas.

As a consequence the sanitary situation in the cities was not quite ideal for the human health, gently expressed, offering rather optimal environments for the proliferation of all kinds of diseases, such as those brought into the country as a result of maritime commerce (e.g., cholera). The oil painting in Figure 5.14 depicts a typical scene in Lima few years after the Independence from the Spanish Kingdom, and gives an idea about the prevailing urban hygienic conditions of those times.

Figure 5.14 The oil painting represents a typical scene in Lima few years after the Independence from the Spanish Kingdom, and gives an idea about the prevailing hygienic conditions. Waste thrown away on the public street is being selectively collected by turkey vultures while the inhabitants perform their normal course of life (Rugendas, oil con canvas, between 1842 and 1845).

5.3.4 Rural water supply and sanitation

Through colonial times and into the mid-twentieth century most of the people lived in the countryside. Even though Lima as capital of the Viceroyalty of Peru and later capital of the Republic of Peru was already the biggest city in the country, its population around 1850, close to 30 years after Independence, was about 100,000 inhabitants (Figure 5.5).

The villages and housings in the countryside were frugal compared to those of the cities, and were also the result from the mixture between the colonist influences and native traditions. Countryside villages also developed according to Spanish arrays and included native traditional components. The housing in remote rural areas were even more elementary in comparison, a trend that can be observed even till today, as if time had stand still for centuries. Already during colonial times the rural areas were handled with low priority by the different administrations. Thus, water supply and sanitation in rural areas dragged behind the development in the cities, a tendency that lasted till nowadays (Figure 5.22).

The living conditions for the workers in the *haciendas* (land estates) were more precarious. During the colony and till the abolition of slavery in 1854 especially the costal *haciendas* employed considerable amounts of slaves, most of them of African origin. The housing and sanitary conditions put at their disposition were harsh and sometimes comparable to those of workhorses and other plodders at the *haciendas*, as their value was seen in comparison to them. There were times that four African slaves were traded for one horse and vice versa. With the abolition of slavery workers from Asia, mainly from Canton in China, replaced the gap as cheap labourers, initially working for refunding their passages from China to Peru. Figure 5.15 gives a view into the *ranchería* (residential area of *hacienda* workers) of one of the better-conducted sugarcane *haciendas* of the Peruvian North coast during the mid of the twentieth century. Although the water supply and sanitation facilities shown in the picture were very elementary, the situation does not correspond to the general situation of the haciendas at those times. During the mid of the twentieth century the *haciendas* were commonly equipped with less sanitation facilities. During colonial times the reality was worse.

Figure 5.15 The first image shows two rural *aguadores* (water porters) at a ditch nearby Chiclayo, Peru (Brüning, 1903). The second picture gives a view into the *ranchería* (residential area of *hacienda* workers) of a sugarcane *hacienda* in the Peruvian North coast. The housing of each family consisted of one room without windows and an access door. The contiguous habitations were located on both sides of long unpaved ways with discontinuous foot-walks. Latrines for all inhabitants were located inside wooden cabins installed in the middle of the ways at certain distances. A water tap for drinking and cooking purposes was located behind each of those cabins (photograph by Kosok, between 1948 and 1949).

The living circumstances of the mineworkers in the Andes were even worse. In view of the extreme hard working conditions in the mines, the main cause for illnesses and even early deaths, the precarious sanitary conditions they had to deal with were probably the easy part.

5.4 FROM AROUND THE BEGINNING OF THE XX CENTURY TILL TODAY

The trend set during colonial times continued after the Independence from the Spanish Kingdom till nowadays. The administrative, economic and cultural center was cemented in Lima. The cities continued to attract the rural population. The uncontrolled migration from the rural areas to the cities continued to grow without planning. People moved from the Andean Region to the Coast, to the major coastal cities and to the capital city of Lima (Figures 5.4 and 5.5). Furthermore, the rural areas were still handled with low priority by the different governments. And as a consequence, the development of water supply and sanitation in rural areas still drag behind the development in the cities (Figure 5.22).

Along its Hispanic history Peru experienced different periods of bonanza mainly due to the exploitation of valuable natural resources, such as gold and silver, sugarcane, guano, salpeter, caoutchouc, fish meal, natural gas, and mining as well as gold and silver again today. During those periods of bonanza the very profitable exports also increased government revenues, enabling major infrastructure investments, eventually also in water supply and sanitation. However the investments didn't follow a holistic and sustainable long-term development master plan for the country, and in the case of water and sanitation, as in many other cases, the priority was always given to Lima, the capital of the country.

5.4.1 Snap-shots of water supply and sanitation development

5.4.1.1 First chlorination plant in the country

The first chlorination plant in the country was installed in the water supply system of Lima in 1917. It is documented that from that moment on Lima was supplied with potable water, a statement that cannot be proved today. However, today's tap water in Lima and in most of the cities in Peru is not suitable for human consumption.

5.4.1.2 Water supply coverage

With the course of time also other cities in the country installed water treatment facilities and networks to supply its populations with drinking water. However, as the water supply network didn't reach all the inhabitants, alternative means to obtain the vital element had to be contrived.

According to relations of eyewitnesses among older fellow men the *aguadores* (5.3.3.2) continued to offer their services in provincial cities in the old-fashioned way, sometimes even till the mid of the twentieth century, while in Lima the the traditional services of the *aguadores* ceased around 1900 (Palma, 1833–1919). Nevertheless the service of home water delivery is still required in cases of water connection malfunctions, or especially in the marginal zones where customers are not connected to the water network. This is the case in the Provinces (Figure 5.21) as well as in the Megacity Lima.

Another option to get water was, and sometimes still is particularly in rural areas, to informally drill a well either within the own premises or at the most creative places (Figure 5.8). The rural areas continued to be handled with low priority by the different administrations. In consequence the water supply coverage in rural areas continued at comparatively low levels (Section 5.4.1.7).

5.4.1.3 Wastewater collection

Around the beginning of the twentieth century various provincial cities did neither count on dense water supply networks nor central sewerage systems, despite their colonial flairs and tidy appearances (Figure 5.16). Urine and feces were sometimes held in cesspits. While the priority was given to the development of the water supply systems, the evacuation of wastewater was not given the same importance (Figure 5.22).

Figure 5.16 The image on the left gives a view into a street of the former colonial town Lambayeque (Brüning, September 1916). The photograph on the right shows stagnant stormwater in the center of Chiclayo due to El Niño torrential rains and the lack of sewers (*Elías Aguirre* Street at the Cathedral), (Brüning, March 1925).

The water supply systems and the sewerage systems were not conceived and/or implemented as part of integral master plans. The results were partially implemented systems followed by different sporadic and inconsistent patchworks over the time. This phenomenom happens till today in many Peruvian cities. Such is the case of the actual sewerage network of the city of Chiclayo. The right picture in Figure 5.16 shows stagnant stormwater in the *Elías Aguirre* Street at the Cathedral, very close to the main plaza, after a rare *El Niño* torrential occurrence (5.1.1.3). The rare intense precipitation (Table 5.1) could not be drained due to the lack of sewers at those times. Today even though the city counts on a sewerage network, the sewerage system collapses frequently at several points. For example, about 300 m further down the street shown in the right picture the sewerage system collapses frequently, gushed out wastewater stagnates in that street crossing, and pedestrian as well as motorised vehicles have to overcome that malodorous wet obstacle.

5.4.1.4 Water drainage in Cuzco's city center

The colonial city center of Cuzco was build over the basements and walls of the ancient *Inca* capital city, following its street courses. While several colonial constructions were damaged during various earthquakes the Inca structures remained intact. Different street views of Cuzco's historical center capturing impressions from three different epochs are shown in Figured 5.17 and 5.18. The original shape of the streets including their water drainage provisions remained mostly unchanged.

Figure 5.17 Two street views during the XIX century, city of Cuzco, the former capital of the Inca Empire *Tahuantinsuyo* (Squier, 1877). Note the foot-walk and the adjacent indentation along the street (left picture) and the conduit in the middle of the street (right picture). Stormwater from heavy rains could also be discharged making use of the whole street cross-section, as building entrances have a stone echelon. The Spanish colonial buildings in the center of Cuzco were build on top of the stone Inca basement and walls, following the original street courses.

Figure 5.18 Four modern times street views from Cuzco's *centro histórico* (historic center), showing similar characteristics than those of the precedent elderly street views contained in Figure 5.17. From left to right: View to the *Calle Loreto* (section of a photograph by Martín Chambi, around 1930), and three views to different streets in the proximity (photographs by Reyes-Knoche, 1996). Note the stone echelons at the building entrances (second picture from the left, right wall). The two pictures from the right give also an impression of the slopes the streets of Cuzco's *centro histórico* can reach. The street shown in the right picture is the *Hatun Rumiyoc* Street where the famous *piedra de los doce ángulos* (twelve angle stone) is located (second stone in the second row, from the bottom right of the picture).

5.4.1.5 The drinking water supply source of the city of Tumbes around the year 2000

Figure 5.19 documents the situation of the water supply system of the city of Tumbes at the source around the year 2000. Damages occurred during the heavy El Niño phenomenon of 1997/98 was still visible along with obvious operation and maintenance deficiencies. Since 2005 a private operator under the modality of a concession operates the water supply company. From 2005 to 2010 the water supply coverage decreased from 70 to 67% and the sewerage coverage decreased from 45 to 44%, according to SUNASS figures.

In the same period the payment behaviour fall off from 1.5 to 3.4 months, and the average tariff increased from 1.40 to 1.53 PEN/m^3. On the other hand the number of service hours increased from an average of 7.5 to 15.5 hours/day, along with an increase of the minimum pressure in the water supply network (SUNASS). The actual concessionaire still operates the same water treatment plant depicted in Figure 5.19, cleaned up and apparently arranged with some cosmetic changes. The plant was built in 1954.

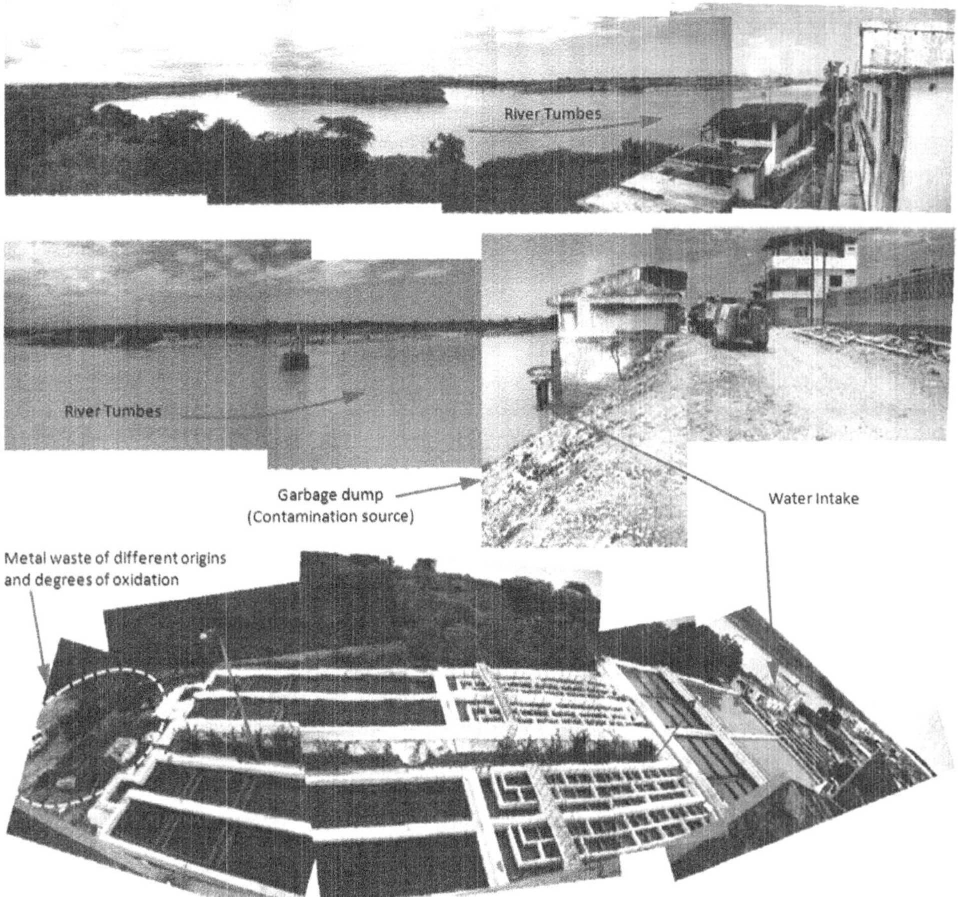

Figure 5.19 Water source, water intake and treatment plant of the city of Tumbes in the year 2000. On top: Panoramic view of the river Tumbes, the main drinking water source of the city of Tumbes, as seen nearby the water intake and the treatment plant. In the middle: Tumbes river and drinking water intake. Note the garbage dump (contamination source) just before the water intake. Bottom: Photocomposition showing the conventional water treatment plant of Tumbes. Note the metal waste of different origins and degrees of oxidation (on the left) and the overall carelessness of the water treatment plant (photographs and compositions by Reyes-Knoche, February 2000).

5.4.1.6 Waste disposal and mobile water supply devices in Pisco around 2010

Figure 5.20 shows a typical informal waste disposal site as it can be found in many public places all along the country, especially in provinces and outskirts. Figure 5.21 shows typical contemporary mobile water tanks.

Figure 5.20 Informal waste diposal site located in the outskirts of Pisco between the Pacific Ocean shores and the road to the *Reserva Nacional de Paracas* (Paracas National Park). The text written on the white concrete panel surrounded by litter (right picture) says: 'Provincial Municipality of Pisco, the Pisco Ecological and Turistic circuit of Pisco will be build here. Intangible Zone. Prohibited to dispose rubbish' (photographs taken by Reyes-Knoche, October 2010).

Figure 5.21 Mobile water tank trucks in the provincial city of Pisco, modern times means of transportation of the contemporary *aguadores*. Left: Two water tank trucks from the city's water company waiting to supply water to customers not connected to the water network or without working water connections. Right: A municipal water tank truck watering a public grassed area. Similar vehicles of different manufacturers and conservation conditions can be seen circulating in the whole country, in rural areas, in provincial towns as well as in the capital city of Lima, especially in the marginal zones (photographs by Reyes-Knoche, October 2010).

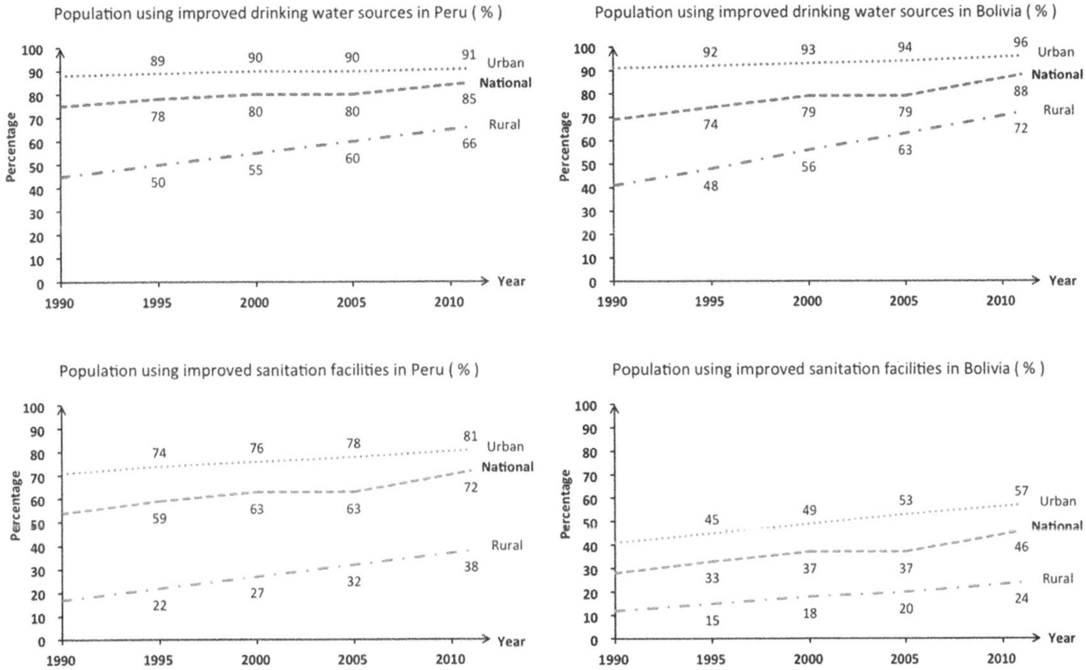

Figure 5.22 Water, sanitation and Hygiene in Peru and Bolivia between 1990 and 2011. Percentages of population using improved drinking water sources (upper diagrams) and sanitation facilities (bottom diagrams). Graphs elaborated based on the Global Health Observatory Data Repository, WHO, 2013.

5.4.1.7 Water, sanitation and hygiene between 1990 and 2011

While the urban population using improved drinking water resources in Peru according to WHO's definition increased from 89% to 91% between 1990 and 2011, 34% of the rural population and about 15% of the national population still remain without access to improved drinking water resources (Figure 5.22).

On the other hand, the urban population using improved sanitation facilities in Peru increased between 1990 and 2011 from about 70% to 81%. But with about 62% of the rural population still needing improved sanitation facilities, the gap between rural and urban development is even bigger.

In comparison to the Bolivian neighbors the Peruvian sanitation situation looks healthier in the rural and urban areas. But on the contrary, the percentage of people in Bolivia having access to improved drinking water sources is slightly higher in the urban areas and higher in the rural areas (Figure 5.22, WHO).

5.4.1.8 Water supply and sewerage in today's Lima

Since mid of the last century the population of Lima grew rapidly to a Megacity of *ca.* 9.44 million inhabitants, about 1/3 of the total population of Peru. The growth has been uncontrolled and followed a seed that was planted early in colonial times.

Today's water supply situation in Lima is quite challenging. About one million inhabitants have no access to water. The limited water resources coming from the Andes seem to diminish with the effects of climate change, such as glacial shrinkage as well as falling rain intensities. The water losses in the existing overaged water network lie in the order of 40%. Consequently, the scarce water resources are being used inefficiently. In addition, the water demand still increases due to the persisting uncontrolled migration, informal settlements and augmenting industrial water demand.

The difficulties in sanitation and wastewater are challenging as well. Around 1.5 million inhabitants are not connected to a sanitary sewer system. The existing sewer system has not been implemented following a long-term masterplan, and consists of a conglomeration of sewer lines of different materials, diameters, ages and degrees of conservation. It may happen that subsystems of the sewerage network work properly alone but fail in combination, causing frequent collapses of the sewer networks. And, leaking sewers become sources for groundwater contamination. Taking into account these facts the statistical figures related to the access to improved sanitation facilities can be questioned (e.g., Figure 5.22).

Figure 5.23 gives an impression of the infrastructure level in Miraflores, one of the modern urban districts of Lima that counts with satisfying water supply and sewerage services, in contrast to many peripheral areas of the capital city of Lima and provincial cities that do not entirely count on these basic sanitary services (see in comparison previous Figures contained in the present chapter).

Figure 5.23 Panoramic view over the green ceiling of the multi-story commercial center '*Larcomar*' in Miraflores, Lima, located between the *Circuito de Playas* coastal highway (not in the picture, sight covered by *Larcomar*) and the *Malecón de la Reserva* street (in the picture), as seen from the swimming-pool terrace of the JW Marriott Lima Hotel (panoramic photograph by Reyes-Knoche, January 2013).

5.4.1.9 Wastewater treatment and reuse

Till the beginning of 2013 almost 84% of the collected domestic and industrial wastewater of Lima was discharched directly into the River Rímac and into the Pacific Ocean without any previous wastewater treatment. Only 16% of Limas wastewater was treated with small plants and oxidation ponds. This situation began to change in February 2013 with the start of operation of the first phase of the new wastewater treatment plant (WWTP) *Taboada*. The inauguration of the second phase was scheduled to happen in July 2013. The *Taboada* WWTP is designed to pre-treat an average of 14 m^3/s and a maximum of 20,3 m^3/s with the completion of phase 2, in total the equivalent population of more than 4.3 million, which represents around 56% of the population of Lima and Callao and about 75% of the wastewater generated by these two adjacent cities (SEDAPAL). The pre-treated water is being sent through a 3.9 Km long underwater outfall to the Pacific Ocean for its final clarification. The remaining wastewater collected in the southern zones of Lima shall be treated with the same principle in the WWTP La Chira, currently under construction.

The reuse of treated, or as the case may be of pre-treated wastewater is being contemplated as an option in the future by SEDAPAL, for example, for watering parks and green areas within the city. However, untreated wastewater of different origins, compositions and qualities is already being used informally for the irrigation of agricultural land for growing crops destined for human consumption, in Lima as well as in the province.

With regard to wastewater treatment provincial cities are at large behind the developments of the capital. Wastewater is often discharched untreated into the ocean, nearby waterbodies or seepage surfaces.

5.4.2 Protection of water sources, pollution hazards, and water qualities

Figure 5.5 summarises the water quality changes along different uses and the respective ideal treatments the water shall obtain in an indicative timeframe. The graph helps to visualize what needs to be done in order to stay within permissible environmental and health limits and offers a qualitative guideline for the responsible use of the water resources. Accordingly, and taking into account the above described snapshots of water supply and sanitation development (5.4.1) lots needs to be done in Peru. But probably, the most prominent cases for water pollution hazards can be found in the mining sector due to the toxicity levels involved, the consequences the contaminations can cause, and due to the fact that Peru is among the world's leading producers of gold, silver, copper, zinc, tin, lead, molybdenum, quicksilver, selenium, cadmium, among others. The following actual example illustrates the risk exposure and dangers the mining industry bare.

The Yanacocha open pit gold mine is located in the northern sierra of Peru at an altitude of *ca.* 3800 m and 48 km north of Cajamarca (6°59′29.09″S 78°31′47.45″W). Yanacocha is considered to be the second largest gold mine in the world, and the largest and more profitable gold mine in Latin America. The gold is obtained by separating microscopic gold fragments from large amounts of soil with the aid of large quantities of a dilute alkaline cyanide solution (heap leaching). The huge amounts of earth movements in the gigantic open pit mine irreversibly transform the environment (Figure 5.24). The alkaline cyanide solution requires large amounts of water. And finally, the biggest hazard, if adequate protective measures are not taken, the highly poisonous cyanide can persistently pollute water, cause adverse effects on the environment surface and subsurface, as well as cause adverse health effects to human beings and animals.

Up to what extent the consortium operating the *Yanacocha* gold mine is actually taking the necessary environmental precautions to avoid dangerous contaminations remains controversial. In any case, the corresponding treatment of the highly toxic acid wastewater is costly and reduces profits. Local communities detected severe contaminations, fish and frog deaths, as well as unusual sediments in the riverbeds. In the year 2000 a consortium transport truck spilled 151 kg of mercury along 40 km of a road contaminating the town of Choropampa and two neighboring villages, and poisoning about 900 persons. In the year 2004 due to massive protest of the local population the consortium had to cancel their initiative to expand their operations to the nearby *Cerro Quillish*, a mountain that supplies water to the city of Cajamarca.

However, the consortium actually foresees the exploitation of copper-gold-and-silver bearing ores at the projected *Conga* open pit mining site, located nearby *Yanacocha*, at altitudes between 3700 and 4270 m, and comprising an area of about 2000 ha. The plant nominal capacity is dimensioned to process 92.000 tons per day, allowing processing the mineral content of 3.1 billion pounds of copper and 11.6 million ounces of gold. The mining would be completed in aprox. 19 years, and the ore would be processed during the last 17 years (Minera Yanacocha, feasibility study). But, the *Conga* project site is located within a fragile ecosystem, and on top of five populated and agriculturally active sub-watersheds.

Figure 5.24 Satellite view over a section of the Yanacocha open pit gold mining site, showing a leaching pond on the left, and an open pit on the right. The image encompasses an area of about 2.400 m × 1.400 m. The total area of the Yanacocha mining site surpasses the urban area of the nearby city of Cajamarca. Source of satellite view: Google earth.

5.4.3 Actual developments and trends

During the last years Peru is experiencing another epoch of wealth due to the high demand and high prizes of its natural resources such as gold, silver, copper, and so on. This situation is increasing also government revenues. Macroeconomic indicators developed quite good, placing Peru among the exemplary emerging economies. And the investment conditions are gaining attractiveness. As a result, Peru has again the chance to develop and implement a holistic and sustainable long-term development master plan for the country, giving water supply and sanitation infrastructure the corresponding priority.

Within the last decade lots of efforts have been undertaken to reform the Peruvian water sector. In 2009 the *Autoridad Nacional del Agua* (National Water Authority) was created and the *Ley de Recursos Hídricos* (Law of Water Resources) was passed. The *Superintendencia Nacional de Servicio de Saneamiento* (The Water Regulator Agency) has been consequently introducing means of monitoring the performance of the water supply and sanitation entities across the country and making efforts to adjust the tariffs accordingly. Recently, in June 2013 the *Ley de Modernización de los Servicios de Saneamiento* (Law of the modernization of Sanitation Services) was passed. The objective of the law is the establishment of measures to increment coverage, and ensure quality and sustainability of the sanitation services across the country, promoting the development, environmental protection and social inclusion. The law includes also provisions destined for ameliorating the participation of the private sector.

Multilateral donors such as The World Bank, Interamerican Development Bank and the Corporación Andina de Fomento as well as Bilateral Cooperation Agencies, such as KfW and GIZ from Germany, AECID from Spain, CIDA from Canada, JBIC from Japan, among others, as well as the European Commission are supporting for many years the development of the Peruvian water sector financing and co-financing infrastructure investment measures and providing technical assistance to improve the efficiency and sustainability of the water and sanitation services.

5.5 CONCLUSION

In Peru the catchment and secure provision of water has been and still remains challenging. Reasons are particular geographic, hydrologic and climatic conditions, such as diverse and sometimes adverse environments, irregular availability of water, and immanent natural hazards, among others. Sanitation as well as returning the used waters into the natural water cycle played and still play a major role in the quest for sustainable development.

Peru counts on a very long water resources management tradition. In the past there were already civilizations that managed to find and consequently implement forms of sustainable development in almost the same living environment of today. With the colonization the further development of their civilizations stopped and their form of living ceased to exist. Instead, the late medieval occidental, Spanish and catholic cosmovision defined henceforward the new order. About 300 years of Spanish Viceroyalty followed along with the cultural and technological developments of the occidental world. And, to some extent the development of the country stagnate, and the level of water resources management and sanitation once achieved was buried in oblivion. The population concentrated more and more in the coast, instead of the Andes, the former preferred habitat. The time after Independence from the Spanish Kingdom didn't change that much in terms of development, as the structures within the country remained. Today's situation is somehow the yield of the seeds sowed during colonial times.

The water resources management, water supply and sanitation to-do list of today's Peru is long. But ancient civilizations such as the *Inca* already proved the feasibility. The level of civilization, eradication of hunger and sustainability they achieved was only possible with the consequent implementation of a long-term holistic strategy along with a firm institutional framework and organisation, strong leadership and skilled experts and workers.

5.6 REFERENCES

ANA, Autoridad Nacional del Agua (National Water Authority). Water Resources Management Data. Lima, Peru. http://www.ana.gob.pe (accessed 22 August 2013) (Website contents in Spanish).

Asano T. (2002). Water from (waste)water – the dependable water resource. *Water Science and Technology*, **48**(8), 23–33.

Asano T., Burton F. L., Leverenz H. L., Tsuchihashi R. and Tchobanoglous G. (2007). Water Reuse – Issues, Technologies, and Applications. Metcalf & Eddi/AECOM, McGraw Hill, USA.

Bromley Seminario S. (1961). *Las Viejas Calles de Lima* (The old streets of Lima), manuscript published 2005 by the *Municipalidad Metropolitana de Lima* (Metropolitan Municipality of Lima). Lima. Peru (in Spanish).

Brüning H. H. (1848–1928). In: H. H. Brüning and C. Raddatz (eds), (1990). *Fotodokumente aus Nordperu von Hans Heinrich Brüning, 1848–1928* (Photographic documents of Northern Peru, Hans Heinrich Brüning, 1848–1928), Hamburgisches Völkermuseum, Hamburg, Germany (in German and Spanish).

Bustíos Romaní C. (2004). *Cuatrocientos años de la Salúd Pública en el Perú, 1533–1933* (Four hundred years of public health in Peru, 1533–1933), Fondo Editorial Universidad Nacional Mayor de San Marcos, Lima, Peru (in Spanish).

Cazeaux E. and Charton E. (1854). *L'aguador, porteur d'eau de Lima, Pérou* (The 'aguador', water porter of Lima, Peru), Le Magasin Pittoresque, Livraison n°14, Bureaux d'abonnement et vente, Paris, France (in French).

Cobo B. (1653). *Historia del Nuevo Mundo* (History of the New World), three volumes published 1890, 1891 and 1892. Sociedad de Bibliófilos Andaluces, Imp. de E. Rasco, Sevilla, Spain (in Spanish).

Fierro P. (1807–1879). Public Domain Replicas of Watercolour Paintings.

Fierro P. (1854). *Tipos limeños vistos por Pancho Fierro* (Types of Lima as seen by Pancho Fierro). Editorial de Artes Aurora, 1979, Leningrad, USSR/Saint Petersburg, Russia (in Spanish).

Fuentes Delgado M. A. (1866). Lima: or, Sketches of the Capital of Peru, Historical Statistical, Administrative, Commercial and Moral. Trübner & Co., London, UK.

IGN, *Instituto Geográfico Nacional* (National Geographical Institute) (1989). *Atlas del Perú* (Atlas of Peru). Instituto Geográfico Nacional, Lima, Peru (in Spanish).

IGP, *Instituto Geofísico del Peru* (National Geophysical Institute). Seismological and Geophysical Data, Lima, Peru. http://www.igp.gob.pe (accessed 22 August 2013) (Website contents in Spanish).

INEI, *Instituto Nacional de Estadística e Informática* (National Institute of Statistics and Informatics), Lima, Peru. http://www.inei.gob.pe (accessed 22 August 2013) (Website contents in Spanish).

KfW Entwicklungsbank, German Development Bank, Frankfurt am Main, Germany. http://www.kfw-entwicklungsbank.de (accessed 20 August 2013) (in German).

Kosok P. (1965). Life, Land and Water in Ancient Peru. Long Island University Press, Brooklin, New York, USA.

Kroß E. (1992). *Die Barriadas von Lima: Stadtentwicklungsprozesse in einer lateinamerikanischen Metropole* (The marginal urban zones of Lima: City development processes in a Latin American metropolis). F. Schöningh, Paderborn, Germany (in German).

Lossio J. (2003). *Acequias y Gallinazos, Salud ambiental en Lima del Siglo XIX* (Ditches and turkey vultures, Environmental Health in Lima in the XIX century), Instituto de Estudios Peruanos (IEP), Lima, Peru (in Spanish).

MDGs, Millennium Development Goals Indicators. The Millennium Development Goals Report 2013, United Nations Statistics Division, mdgs.un.org, New York, USA (accessed 28 August 2013).

Minera Yanacocha S. R. L., (Yanacocha Mining Company), Lima, Peru. http://www.yanacocha.com.pe (accessed 02 September 2013) (in Spanish).

Municipalidad Metropolitana de Lima (2007). *Acuarelas de Pancho Fierro y seguidores: Colección Ricardo Palma* (Watercolor paintings of Pancho Fierro and followers). Pinacoteca Municipal Ignacio Merino, Lima, Peru (in Spanish).

Oxfam America (2009). Mining conflicts in Peru: Condition critical. Briefing paper, Boston, USA.

Oxfam International, Oxford, UK. http://www.oxfam.org (accessed 15 March 2012).

Palma R. (1833–1919). *Tradiciones Peruanas completas* (Complete Peruvian Traditions). 5th edn, 1964, Aguilar, Madrid, Spain, pp. 391–393 (in Spanish).

Regal Matienzo A. (1970). *Los trabajos hidráulicos del Inca en el antiguo Perú* (The hydraulic works of the Inca in ancient Peru). Lima, Perú (in Spanish).

Reyes Carrasco L. V. (1973). *Análisis hidráulico del sitio arqueológico de Chavín de Huantar* (Hydraulic analysis of the archaeological site of Chavín de Huantar). Universidad Nacional de Ingeniería, Lima, Peru (in Spanish).

Reyes Carrasco L. V. (1980). *Reflexiones sobre un dispositivo precolombino, precursor de los niveles y teodolitos actuales* (Pre-Columbian device, precursor of actual levelling instruments and theodolites). III Congreso Nacional de Ingeniería Civil, Cuzco, Perú (in Spanish).

Reyes Carrasco L. V. (2005). *Mecánica de Fluidos, Hidráulica y Diseños Hidráulicos, Tomo I & II* (Fluid Mechanics, Hydraulics and Hydraulic Designs, Volume I and II). Chiclayo, Perú, p. 410 and 413 (in Spanish).

Reyes Carrasco L. V. (2009). *Sobre la alimentación de agua de las poblaciones que habitaron las pirámides de Túcume* (About the water supply of the former populations around the Pyramids of Túcume). In: XVII Congreso Nacional de Ingeniería Civil, Chiclayo, Perú (in Spanish).

Reyes Carrasco L. V. and Reyes-Knoche A. (2012a). The water supply of the population that once inhabited the Pre-Columbian Pyramids of Túcume and today's Water Supply system in the same Region. In: 3rd IWA Specialized Conference on Water and Wastewater Technologies in Ancient Civilizations, March 22–24, 2012, Istanbul, Turkey.

Reyes Carrasco L. V. and Reyes-Knoche A. (2012b). About a Pre-Columbian surveying instrument, precursor of today's levelling instruments and theodolites. In: Third IWA Specialized Conference on Water and Wastewater Technologies in Ancient Civilizations, March 22–24, 2012, Istanbul, Turkey.

Reyes-Knoche A. (1991). *Traditionelle Bewässerung am Beispiel des alten Perus* (Traditional Irrigation in Ancient Peru). Seminar 'Ökologische und soziale Folgen von Stauanlagen', Institut für Hydromechanik, University of Karlsruhe (today's KIT Karlsruhe Institute of Technology), Germany (in German).

Reyes-Knoche A. (2009a). Water supply and sanitation in ancient Peru: Pre-Columbian cultures and the Incas. In: 2nd IWA Symposium on Water Technologies in Ancient Civilizations, May 28–30, 2009, Bari, Italy.

Reyes-Knoche A. (2009b). Water Solutions in ancient Peru – Overview and options to solve today's and future water challenges. In: 1st IWA Development Congress, November 15–19, 2009, Mexico City, Mexico.

Reyes-Knoche A. (2011). *Nachhaltige Wasserwirtschaft und Anpassung an den Klimawandel im Präkolumbischen und heutigen Peru – Überblick und Rekonstruktion als Beitrag zur Lösung heutiger und zukünftiger Herausforderungen am Beispiel des Siedlungsraumes der Mochica in Nordperu, Arbeitstitel* (Sustainable water management and adaptation to climate change in Precolumbian and today's Peru, working title). Synopsis of Ph.D. thesis, University of Heidelberg, Germany (in German).

Reyes-Knoche A. (2012a). Sustainable Water supply in Pre-Columbian Civilizations in Ancient Peru and South America, Chapter 11 In: Evolution of Water Supply Through the Millennia, A. N. Angelakis, L. W. Mays, Dimitris Koutsoyiannis, Nikos Mamassis (eds), IWA Publishing, London, UK.

Reyes-Knoche A. (2012b). Sustainable water supply – The case of Pre-Columbian Peru. In: 3rd IWA Specialized Conference on Water and Wastewater Technologies in Ancient Civilizations, March 22–24, 2012, Istanbul, Turkey.

Rugendas J. M. (1802–1858). *Escena de una calle de Lima, entre 1842–1845* (Scene in a street in Lima, between 1842–1845). In: Oil on canvas, in *El Perú Romántico del Siglo XIX* (The Romantic Peru of the XIX century), Carlos Milla Batres, 1975, Lima, Peru (in Spanish).

SEDAPAL, *Servicio de Agua Potable y Alcantarillado de Lima* (Drinking water and sewerage services of Lima), Lima, Peru. http://www.sedapal.com.pe(accessed 30 August 2013) (in Spanish)

SENAMHI, *Servicio Nacional de Meteorología e Hidrología* (Meteorological and hydrological historical data), Lima, Peru. www.senamhi.gob.pe (accessed 22 August 2013) (Website contents in Spanish).

Squier E. G. (1877). Incidents and Explorations in the Land of the Incas, Hurst & Company, New York, USA.

UNSTATS, United Nations Statistics Division, Millenium Development Goals Indicators. unstats.un.org (accessed 22 August 2013).

Valencia Carpio M. (1997). *Historia del abastecimiento de agua potable de Lima 1535–1996* (History of the water supply of Lima), Servicio de Agua Potable y Alcantarillado de Lima (SEDAPAL), Ministerio de la Presidencia, Lima, Peru (in Spanish).

WHO, World Health Organization (2006). Guidelines for the Safe Use of Wastewater, Excreta and Greywater. Geneva, Switzerland.

WHO, World Health Organization, Global Health Observatory Data Repository. apps.who.int/gho/data/node.main (accessed 15 Agust 2013).

WHO, World Health Organization and UNICEF, United Nations International Children's Emergency Fund (2013). Progress on Sanitation and Drinking Water. 2013 update, Geneva, Switzerland, New York, USA.

WSP, Water and Sanitation Program (2008). *Agua y saneamiento para las zonas marginales urbanas de América Latina* (Water and Sanitation for the marginal urban zones of Latin America). LEDEL S. A. C., Lima, Peru (in Spanish).

WSP, Water and Sanitation Program, The World Bank (2013). Desafiando al paso del tiempo. La experiencia de gestión de servicios de agua y saneamiento en cuatro pequeñas ciudades del Perú (Challenging the pass of time. Experiences in management of WS and sanitation services in four small Peruvian towns). Lima, Peru (in Spanish).

The World Bank, Open Data, data.worldbank.org, Washington, D.C., USA (last accessed 31 August 2013).

Chapter 6

History of urban wastewater and stormwater sanitation technologies in Hellas

G. P. Antoniou, G. Lyberatos, E. I. Kanetaki, A. Kaiafa, K. Voudouris, and A. N. Angelakis

6.1 PROLEGOMENA

In contrast to other ancient civilizations (Egyptians, Mesopotamians, and Indus), developed in regions with high water availability, all major Hellenic cities during the several phases of the Hellenic civilization, which lasted for millennia, were established in areas that had low water availability. This is the case both on the continental and the insular country, since the Bronze Age (Zarkadoulas *et al.*, 2008). Moreover, safety reasons and the efforts to avoid occupying fertile lands, resulted in the construction of the settlements on the top of hills or on rocky areas. It is probable that these factors decreased the availability of water.

Ancient Hellens lived in harmony with nature and the environment and they appear to have had some knowledge of sanitation since the early Bronze Age (*ca.* 33,000). The Minoan civilization is universally considered one of the grandest and most brilliant ones (Koutsoyiannis & Angelakis, 2003). The architectural and hydraulic functions of storm water and wastewater sewerage systems were of special significance in the construction of the principal Minoan centres. Archaeological and other evidences indicate that, during the Middle Bronze Age, advanced water management and sanitary techniques were practiced in Minoan settlements. These include the construction and use of bathrooms and other sanitary facilities, relevant to wastewater and storm sewer systems. The hydraulic and architectural functions of sewer systems in palaces and cities are regarded as one of the salient characteristics of the Minoan civilization (Angelakis & Spyridakis 1996a; De Feo *et al.*, 2011). Thus, it is no accident that the main technical and hydraulic operations associated with catchment basins, surge chambers, manholes, urinals and toilets, laundry slabs and basins and sewerage and drainage systems, including effluent disposal sites, have been practiced in varying forms since *ca.* 3000 BC (Angelakis & Spyridakis, 1996b).

The status of urban wastewater and stormwater technologies in Ancient Hellas is reviewed, based on the results of archaeological studies of the 20th century. Emphasis is placed on the construction, operation and management of wastewater and storm water sewerage systems during the Minoan times (2nd Millennium BC). The achievements of this period in supporting the hygienic and functional requirements of palaces and cities were so advanced, that they could be paralleled only to modern urban water systems. Many remains of sanitary structures have been found in Minoan sites (e.g., Knossos, Phaistos, Malia, Gournia, and Tylissos). The hygienic installations are considered as a characteristic factor of the standard of living and economic prosperity, in both domestic and public uses (Lyrintzis & Angelakis, 2006; Angelakis *et al.*, 2012a). Due to these reasons, lavatories had become, and often still are considered luxury items (Antoniou & Angelakis, 2008; Antoniou & Angelakis, 2009). These advanced technologies were further developed during the Hellenistic, Roman, and Venetian periods. In parallel, the advanced Minoan technologies expanded all over Hellas in later periods of the Hellenic civilizations, for example, in the Mycenaean, Archaic, Classical, Hellenistic, and Roman periods (Angelakis & Koutsoyiannis, 2003; De Feo *et al.*, 2012).

In Ancient Hellas, the realization of the importance of pure water availability for people is evident already from the early Classical period. The importance of water and 'religious cleanliness' were known in Classical Hellas. The ancient City States were created in two fashions: the older ones through natural growth, incorporating quite often a rectangular grid, and the newer ones, systematically, following the 'Hippodamian system'. In the Hippodamian cities, for example the original form of the Agora in Athens, like that of the city itself, the layout was very simple. A flat open space with suitable sewerage and drainage systems were the first requirements (Doxiadis, 1964). Despite their differences, the concept hidden behind both building processes was the same: To take advantage of the natural landscape and to create both public and private spaces, according to rational and functional considerations with man at the center.

Sewers in Athens delivered storm water and human wastes to a collection basin outside town. From the basin, storm water and wastes were conveyed through brick-lined conduits to the fields, in order to irrigate and fertilize fruit orchards and field crops. However, the first known epidemic of 430–426 BC in Athens, responsible for the death of the great statesman Pericles, decimated the population and contributed significantly to the decline and fall of Classical Hellas. The enclave held 155,000 inhabitants out of the 400,000 total population of the Attica region (Schladweiler, 2002). It is characteristic that, during that time, sanitary installations became a necessary space, even for ordinary middle class houses (e.g., Olynthos, Delos, and Dystos).

In addition, Romans were famous for their interests, craftsmanship, and ability of implementing engineering projects, for example, aqueducts, amphitheatres, civic centres', and of course, sewers and drains. They developed very advanced technology for sanitation, including baths with flowing water and underground sewers and drains. In rich and well developed urban regions of the Roman civilization, a typical street included: (a) a burier lead water pipe and (b) an underground sewer. Some historians believe that lead pipes caused lead poisoning and contributed to the eventual fall of major Roman cities. Sewage from the nearby buildings and homes was discharged into the sewerage system through pipes. Most of such sewerage systems were eventually drained into the rivers or to the sea, in the cases of coastal urban areas (Schladweiler, 2002).

In general, Ancient Hellens contributed to the historical development of sewers and drains in urban areas. These achievements can be compared with the modern systems, which were established only in the second half of the 19th century in European and American cities. However, very little progress was made from *ca.* 300 AD through the middle of eighteenth century (during the so called Dark Ages). Starting from 1850 (Industrial Revolution), the 'state-of-the-art' for sewage disposal advanced considerably beyond sewage conveyance. Cities and towns had the benefit of knowing the actual connection between sewage and the sources of drinking water (e.g., the adverse potential impact on their health/welfare). That knowledge resulted in great strides being made in collecting and conveying sewage away from people's homes, and in treating the sewage, prior to its discharge either into their source of potable water or at a point near their source (Paranychianakis *et al.*, 2009).

In this Chapter, a review of the basic status of the wastewater and storm water management in Ancient Crete, Hellas is attempted. It is evident that extensive systems and elaborate structures for urban wastewater and storm water were designed, constructed, operated and managed properly, in order to support the hygienic and functional requirements of palaces, cities, and other settlements, since the early Bronze Age. Significant developments relevant to the hygienic lifestyle in the Minoan era, and the Archaic, Classical, Hellenistic, and Roman civilizations, the Middle Ages, and the Ottoman period are considered. The evolution of the major sanitary achievements during the long history of Hellas is considered a significant part of the world history in this field. Thus, the main objectives of this Chapter are: (a) to review the health and environmental risks arising from the water supply and wastewater and stormwater management, (b) to briefly present the trends and the developments in sanitation technologies in Ancient Hellas, (c) to provide information on the status and to review and compare sanitation technologies among several Hellenic civilizations, and (d) to develop a timeline for urban wastewater and stormwater sanitation technologies in Hellas. The information provided is expected to contribute to '*how to learn from the past*'.

6.2 PREHISTORIC TIMES

6.2.1 Minoan crete

Crete is an island, considered as a solid and uniform region with variable climate and anomalous geography, since the Neolithic era. Crete was inhabited by scattered populations that lived partly in caves at some distance from the coast, but also in organized settlements. Although very little is known about the origin of the first settlers, a preoccupation with pottery technology and various artifacts indicate their origin from Anatolia and possibly from Egypt, rather than from mainland Hellas (Angelakis *et al.*, 2005; Angelakis *et al.*, 2012b). The population of Crete was reinforced at the beginning of the Minoan period, for example, immediately after the *ca.* 3000 BC, with the arrival of new settlers, probably from Asia Minor (Angelakis & Spyridakis, 1996a). Furthermore, linguistic affinities and appellations indicate affinity with the Cretan population Cow, which in the mid-late Minoan period, settled in Asia Minor (Huxley, 1961). Contacts of Minoans with other prehistoric civilizations (e.g., Mediterranean's, Europeans, Asians, and even in North American) have been reported (Mariolakos, 2010).

Important cultural progress was observed in Crete throughout the third and second Millennium BC, while an unprecedented cultural technological development was observed in the mid-Minoan period (*ca.* 2100–1600 BC), when the population of the island in the central and southeastern regions increased significantly. At that time the first well organized settlements were developed, the first palaces were built, and generally a thriving cultural and uniform civilization was developed (Angelakis *et al.*, 2012b). By the end of this period, the arts, technology, manufacturing, and commerce with the Aegean islands, Egypt and the Near East flourished. In the first phases of the Late Minoan period (*ca.* 1600–1400 BC), Crete appears to be flourishing, as evidenced by the size urban areas and residences, and sumptuous palaces of this period (History of the Greek Nation, 1970). The thriving arts, the development of metallurgy, the construction of advanced and equipped palaces and an excellent road system, prove the existence of a rich, highly cultivated, well-organized society and governance in Crete. The fall of the Minoan civilization, with the destruction of the palace, is placed after *ca.* 1400 BC (Angelakis & Spyridakis, 1996a).

The cultural development that occurred at different periods in Crete, covers many aspects, characteristic of the modern world, such as architecture, art, technology, navigation, agriculture, forestry, and environmental protection. In Minoan Crete, knowledge of 'hydraulics' was quite evolved. Of the most remarkable features of the Minoan civilization was the construction and use of sanitary and storm sewer systems and wastewater in several palaces and cities, with an advanced architecture and hydraulic operation. Perhaps the most striking feature of the palace of Knossos is the sewerage and drainage system that crosses the entire interior of the palace (Angelakis *et al.*, 2005). In the following sections an attempt of reviewing the sanitation and health practices and technologies in the Minoan civilization is undertaken.

Baths, toilets, and others sanitary infrastructures (or systems)

The first sophisticated sewerage and drainage systems and bathrooms, including lavatories that were constructed in Minoan Crete, date from *ca.* 2500 BC to 1600 BC. Thus, lavatories were flushed and connected to the drainage and sewerage systems. It should be noted that in the Minoan times, bathrooms were not considered necessary, merely convenient, and most palaces did not have them. Although the exact function of Minoan rooms is difficult to define, in a number of houses, the lavatory was quite clearly located in the private living rooms (e.g., Knossos, Phaistos, Tylissos, Malia, and Gournia). In most cases, the evidence for the identification of a lavatory is the existence of a sewer at the floor level, passing through the exterior wall and connecting with the outside central sewerage and drainage system. However, in some homes, there are also traces of some sort of provision for a stone or wooden seat (Antoniou *et al.*, 2013).

Generally, the function of the palace sites is difficult to determine. Evans (1921–1935) identified three rooms at Knossos that were used as baths. One of the most interesting rooms in the ground floor in the residential quarter of the Knossos palace was identified as a lavatory (Figure 6.1). Remains of a clay tube were found just outside the door of the room. It is thought that water was poured through a hole on the floor immediately outside the lavatory door, while an under-floor channel linked the hole with the vertical clay pipe under the lavatory seat (Castleden, 1993). The lavatory could thus be flushed even during a rainless summer, either by an attendant outside the lavatory or by the user. The lavatory in the residential quarter of the Palace of Minos in Knossos (Angelakis *et al.*, 2005) is probably the earliest flush lavatory in history.

Figure 6.1 North 'lustral basin' at Knossos palace (with permission of A. N. Angelakis).

At certain times of the year, the drains at the Knossos palace may have been adequately flushed through by the rainfall that fell into the light-wells, but in general it was supposed that water was poured into the lavatories in order to flush them. It was also observed that there was sufficient space at the end of the seat at Queen's Hall lavatory in Knossos for a large pitcher (Graham, 1987). As with today, toilet flushing systems were available (Figure 6.2). The rinse and drain tube, off from the exterior of the entrance, crossed along the toilet-room, passed under the seat and ended in the main sewer (Castleden, 1993). The toilet could thus be rinsed with the emptying of a container with water at the mouth of the conduit outside the door. Evans (1921–1935) states emphatically the provision of a satisfactory space for placing the container on one side of the seat and points out that many human societies have not advanced so even nowadays.

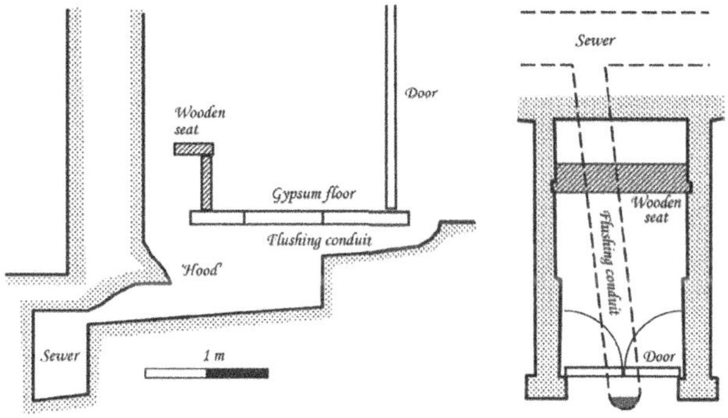

Figure 6.2 Section and plan of ground-floor toilet in the Queens' department in the palace of Minos, Knossos, Crete (adapted from Angelakis *et al.*, 2005; Graham, 1987).

The lavatory is similar in function to that of the so-called Queen's Hall, those found in the Phaistos and Malia palaces as well as in some of Minoans cities and houses. However, there is a difference of the one in Knossos from those in Phaistos and Malia palaces. The point of distinction is that it is at the level of the floor with consequent absence of steps. Evans (1921–1935) reported the latter as 'lustral chambers'. Also, Graham (1987) made the assumption that a room, which started out as a 'lustral chamber', later became an ordinary bathroom. In fact, as a result of investigations by Platon (1990), it must be assumed that this also happened twice in the houses at Tylissos, located to the Southwest of Iraklion.

In several bathrooms, clay tubs were used. A variety of such tubes have been discovered in Minoan sites. The clay tubes in the Minoan bathrooms must have been filled and emptied manually, rather than being directly connected to the sewers for discharging. However, on the 'Caravansarai', a rest house just to south of the palace, a footbath for the weary travelers was supplied by a direct pipe, and the overflow discharged through another conduit (Figure 6.3); a branch of the water channel also served as a drinking trough (Angelakis & Spyridakis, 1996a).

Figure 6.3 The 'Room of the Spring' in Caravanserai which was used for footbaths and the overflow discharged conduit (with permission of A. N. Angelakis).

Similar facilities have been reported elsewhere in Minoan Crete. The study of Platon (1990) provided some statistics on water tanks, baths and other sanitary structures, and concluded that most of them were constructed during the mid-Minoan period. It indicated 16 adjacent structures to the inner rooms, seven near the holy shrines and two at the palace entrances. In two cases, there were bathing facilities, seven were occupied by soil material, and two had been reconstructed and converted into toilets. Also in 14 cases, various sacred objects were found. Finally, frescoes associated with sacred subjects, were found in two 'sanitary' reservoirs in the southern home of the Knossos palace, as well as in a cistern-bath in the northeastern part of the palace of Zakros. Alexiou (1964) states that the absence of bathhouses in some 'sanitary' cisterns and their existence in others, should not be considered coincidental. Platon (1974) and Graham (1987) report that cisterns were used for cleansing the body, and were symbolic for the soul. Most Minoan baths and toilets were connected associated with independent external septic systems, a practical indicator of advanced management of wastewater and the environment in that Era (Panagiotakis, 1987).

The clay bathtubs of Minoan Crete were filled and emptied by hand. However, one feet-tube the house 'Caravansarai' in the southern part of the palace of Knossos, filled with water from the water supply pipe and the overflow was drain through another conduit in the central sewer (Angelakis & Spyridakis, 1996b).

Sewerage and drainage systems

Evans (1921–1935) and MacDonald and Driessen (1988, 1990), describing the state of sewerage in the Minoan palace of Knossos, provided a plan of the original structure. The total spread of the system, including secondary pipelines and drains exceeds 150 m.

The best-explored part of the drainage and sewerage system is what lies beneath populated parts of the palace. It results to one big loop with the highest point at the well (*φρεάτιο or πηγάδι in Hellenic*) near the grand staircase, and it discharged through a connected conduit that follows the slope to the east of the palace (Evans, 1921–1935). In the departments of 'Hall of the Double Axes' and 'Hall Queen' with the side apartments, the sewage was discharged through at least five wells. The wastewater of the so called Queens toilets discharged through the same systems. The system drains also rainwater from the roof and is probably associated with the toilets in the upper floors. The main dimensions are 79 cm × 38 cm and it is made of stone, plastered with mortar. The fairly large size allowed the passage for cleaning or maintenance. Finally, there were holes at intervals for ventilation (Graham, 1987). Parts of the drainage and sewerage system are shown on Figure 6.4.

(a) (b)

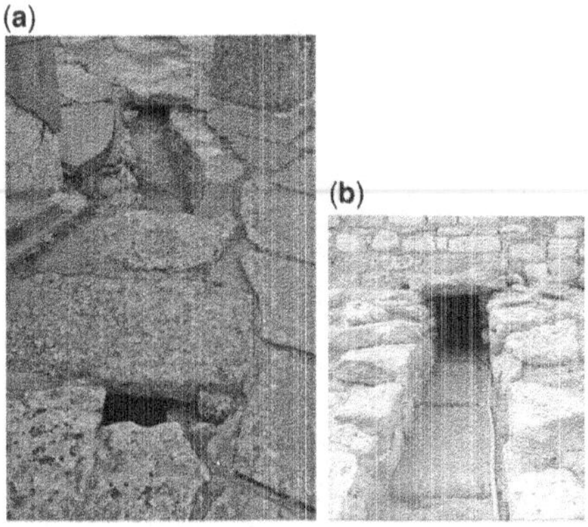

Figure 6.4 Parts of Knossos sewerage and drainage system: (a) inside the palace and (b) central system (with permission A. N. Angelakis).

Sophisticated drainage and sewerage systems existed in other cities and Minoan palaces, as in Phaistos (Figure 6.5) and Zakros. These were very advanced, quite dense, and very well protected (Platon, 1974). As at Knossos and Zakros, the main channels were built with stones, fairly large diameter to permit passage for cleaning and maintenance (Figure 6.6a). The lower section was made of terracotta pipes. There is evidence that the Knossos' system did not effectively function in times of intense rainfalls, although the location of the area in a natural slope, is favorable for the drainage of rainwater hypothesis. Another such sophisticated sewerage system is at the Vathipetro Minoan village, located South of Iraklion. It is one of the first grapes and/or olives crasher-press in the whole history of mankind (Figure 6.6b).

Figure 6.5 Parts of the sewerage system in the palace of Phaistos: (a) within the palace and (b) the main drain (with permission A. N. Angelakis).

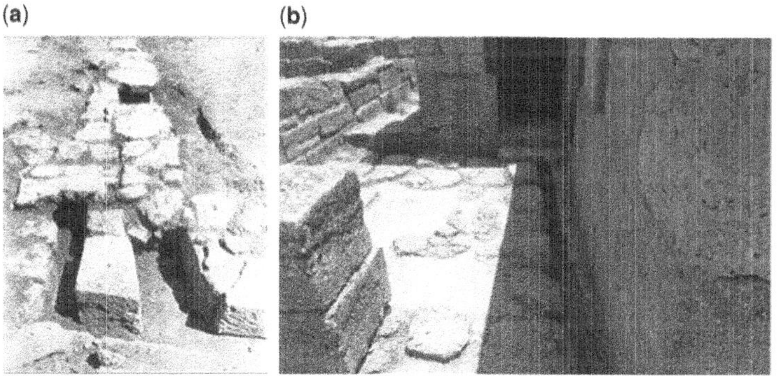

Figure 6.6 Parts of Minoan sewerage and drainage systems (a) in the palace of Zakros and (b) in the village of Vathipetro (with permission A. N. Angelakis).

The most advanced Minoan sewage system seems to be that of the villa of Agia Triada. This system caused the admiration and surprise of several visitors, among them the Italian writer Angelo Mosso (1907), who visited the villa in the early 20th century. During a heavy rain, he noticed that the whole system functioned perfectly and recorded the incident saying: '*... all the sewers were still working! It was very interesting for me to see the water in the drainages and sewers so big that a man could enter. I doubt if there are other examples of ancient sewerages working after 4 thousand years*'. The American H. F. Gray (1940) that carries the story adds: '*Perhaps it may be allowed to doubt whether modern drainage and sewerage systems will operate at least one thousand years*'.

In addition, wastewater spreading to the soil, also known as 'land treatment' has a long history, as demonstrated by the elaborate sewerage systems in Minoan palaces and cities since *ca.* 2600 BC (Angelakis

& Spyridakis, 1996b). Indications of wastewater application to agricultural land goes back to the Ancient Hellenic civilizations on Crete, Sparti, Athens, and Cyprus, approximately 4000 years ago (Angelakis *et al.*, 2005; Crouch, 1996). More on this is given on Chapter 24.

6.2.2 Other locations

Apart from Crete, both in the Hellenic island and in the south mainland a high level technology and expertise for water management have been also evidenced as far back as the 3rd Millenium BC. This includes dams, irrigation projects and water supply and drainage installations. A large number of hydraulic projects and especially sewerage systems have been found in many sites since the Bronze Age. Urban sewerage technology has been developed in parallel with the water supply, definitely connected with the existence of well organized settlements with high cultural level and a central government. Under these conditions, advanced drainage systems have been found in palaces and town houses, such as in Polichni of Lemnos, where a main underground sewer was leading storm and wastes out through the city walls (Brea, 1964). Drainage and sanitary installations are also attested at Acrotiri, Thera, dating in the middle of the 2nd Millennium BC, when the city was reconstructed after a major disaster, probably an earthquake (Palyvou, 1997, 2005).

Wastewaters from the houses have been channeled in built or earthen pipes, vertical inside the building walls or under the floors, which were evacuating to a central stone built sewer under the main road. This was a straight sided inlet section, covered by stoneslabs, and pouring out directly into the sea. The drainage system was established at a level higher than ground floors of most houses, in order to protect them from moisture and rain waters and to ensure a better functioning of the sewerage and sanitary installations, located on the upper floors. A well-preserved lavatory has been found in the city of Akrotiri, situated at the upper floor of the West house. It consists of a twin bench, built within a niche in the wall (Thera). A hole in the floor between the two benches was essentially the rim of a clay pipe embedded in the wall. The pipe ended to a pit outside the house, through which the main sewer has been running (Fig. 6.7). An advanced system of oblique and vertical slabs within the pit was both ensuring an even flow of the wastewaters and trapping gases and nasty odors.

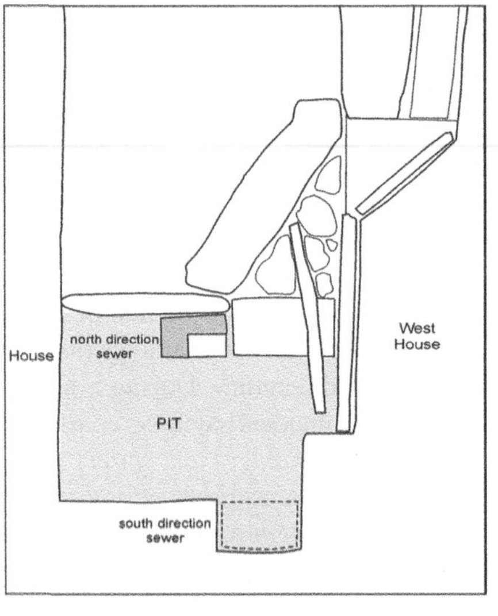

Figure 6.7 Thera, Akrotiri. Sanitation installation (adapted from Palyvou, 2005 with modifications).

During the Mycenaean era, in southern mainland Hellas the existence of permanent and well-organized establishments with palaces favored hydraulic technology. Apart from urban drainage infrastructures, hydraulic technology was also used for major water projects such as dams, built of earth and roughly shaped blocks in Tiryns (Balcer, 1974). Dams were built either to divert streams towards arable land for irrigation, or to channel waters away from fields that might be damaged. In Boeotia, on the other hand, at Orchomenos, large quantities of surface waters were controlled and channeled in dikes and canals, during the Late Bronze Age (Knauss, 2003; Argoud, 1988). The development of hydraulic engineering during the Late Bronze Age, in the Hellenic mainland is also evidenced by the existence of bathtubs associated with drainage infrastructures in Pylos. Engineers were able to drain stormwaters and control and dispose the wastes. A little further north, in Dimini, probably Mycenaean Iolcos, a permanent settlement classified as urban, three of the houses had bathtubs with drains in separate rooms, implying advanced knowledge of water management and high living standards (Adrymi-Sismani, 1990).

In contrast to the Bronze Age settlements of Southeastern Hellas, where water supply and sewerage networks had been a major concern, Macedonia gives a completely different picture. During prehistory, in Macedonia, water was definitely the most crucial reason for inhabitation and remaining in one place. However, water management projects are completely missing. Both neolithic and especially Bronze Age sites were located next to springs, streams, or along rivers and lakes and had no supply or sanitary facilities at all. Water for everyday needs had been stored in large jars and cisterns opened in impermeable soils. In prehistoric Macedonia, urban hydraulic networks, both for supply and drainage, did not exist, as in the Cretan-Mycenaean world, because settlements there had not been classified as urban. Well organized settlements, with central government, appeared in Northern Hellas since 7th century BC and developed during 6th century BC.

Dimini township in central Hellas dating of both the Neolithic period (end of fourth millennium BC) and the Mycenaean period (*ca.* 14–13th centuries BC) has been discovered, northeastern of ancient Iolkos, the city of Jason. A well-constructed wide road with 'Megaroid' houses were built with the same orientation on either side of a wide street have been excavated. In addition a very well preserved drainage system has been recently discovered.

6.3 HISTORICAL TIMES

6.3.1 Introduction

Alcmaeon of Croton (flourish *ca.* 470 BC) was the first Hellene doctor to state that the quality of water may influence the health of people. Also, the Hippocratic treatise 'Airs, Waters, Places' (*ca.* 400 BC) deals with the different sources, qualities and health effects of water. Thus, the importance of water for public health was recognized for the first time and the first well-organized baths, toilets, and sewerage and drainage systems appeared. Pipes of lead (of lengths of 3 m or more) and bronze were used by the Hellenes to distribute water. In the Hellenic cities, there was a system of aqueducts, but for the most part, few above-ground structural arches were incorporated. However, a lot of tunnels through hills, siphons under valleys and rivers and other structures were implemented. Also, the first public fountains were constructed in the Athens Agora.

Besides the care for drinking water supply, the organized settlements of the Classical period improved the wastewater management as well. In Athens, there were sewers which delivered storm water and human wastes to a collection basin outside the town. From that basin, the storm water and wastes were conveyed through brick-lined conduits to fields, in order to irrigate and fertilize fruit orchards and field crops. The enclave held 155,000 inhabitants out of the 400,000 total population of Attica. These results could be also related with the coexistence of older structures, as cesspits, during that time (Antoniou, 2007). The evolution of the town planning along with the implementation of the grown, by the time, skills in hydraulic structures, were combined in the new settlements, for example, Olynthus and the reconstructed towns, such as Miletos. It is characteristic that during that time, sanitary installations became necessary, even for ordinary middle class houses (e.g., Olynthos, Delos, and Dystos).

In the Hellenistic era, the achievements and the knowledge of the Classical period on hydraulic matters were improved. The raised living standards combined with the wealth in several cities of Hellas led to more articulated constructions of lavatories and sewerage systems. The establishment of the Hippodamian town planning as a well studied, designed and applied system, contributed to the high quality urban hygienic drainage.

Also the Romans were proud for their 'rooms of easement' (e.g., lavatories or latrines, which were an essential part of urban communal places. Usually, lavatories were associated with public baths. Public baths included such rooms, adjacent to gardens. There, Roman officials would sometimes continue discussions with visiting dignitaries while sitting on the lavatories. Elongated rectangular platforms with several adjacent seats were utilized (some with privacy partitions, but most without). These lavatory rooms were often, as were the baths. Water was directly transported from aqueducts or through the baths and was continuously flowing under the lavatory seats. The sewage from both the baths and the lavatories was discharged to the underground sewers and eventually to the rivers or to the sea.

6.3.2 Classical period

During the Classical period, the improvement of the urban planning and the implementation of the Hippodamian system in the new cities of the *ca.* 5th and 4th centuries BC was adjoined with the construction of sound wastewater and storm water networks, fulfilling not only the contemporary public hygienic requirements, but also the increasing domestic hygienic and comfort standards. These standards were referring to both baths and lavatories, and thus to enlarged needs for wastewater management.

At the beginning of the period, the preexisting structures, such as private cesspits, and less articulated drainage networks were predominant, but later, due to the factors mentioned above, several improved structures were applied in the beginning in the new towns, and afterwards in the existing ones.

The urbanization of Macedonian settlements in the Classical Age was definitely combined with the appearance of water transfer projects and the construction of storm and wastewater systems. Thus, the implementation of the Hippodamian system and the stabilisation of architectural types within the urban tissue, have played a major role in water management. There is no doubt that Hippodamian urban planning, with the regular blocks and vertical routes encouraged further development of hydraulic infrastructures, which have been implemented and enriched systematically over the Hellenistic and Roman times (Kaiafa, 2008).

Classical Olynthus has been built on a flat-topped hill, according to the Hippodamian city plan. In addition to the long distance subterranean pipeline, serving water supply, the city was also equipped with storm drainage and sewer systems, most of them beneath floors and roads. Many remains of sanitary facilities have been found in many of the houses of the classical town, testifying the high living standard of its inhabitants. Approximately 30% of the houses had a separate bathroom, with clay bathtubs placed on a cement floor with drains. In many cases, wastewaters were discharging from the court by sloping the surface to a point. From there, they were simply conducted away from the houses into the middle of the adjacent streets or alleys, either directly passing through the walls or by means of stone drains, rectangular or round in cross-section. These were open channels of cobblestones, or of channeled stone blocks or of cement faced rubble or of roof tiles. Frequently, terracotta pipes were evacuating the domestic sewage to a drainage path between the two rows of houses on each block (Robinson, 1938). In many cases there was no effort to control waste water that ended up directly through the pipes to the streets (Figure 6.8).

Figure 6.8 Olynthus sanitation facilities (adapted from Robinson, 1938 with modifications).

The technological abilities of that period were applied also to the construction of the domestic and public lavatories of these periods. The limited evidences from the Classical times refer to cesspits in Athens or elsewhere (Filimonos, 2000) and the individual defecation appliances, consisted of clay vessels, as those found in Olynthos (Robinson, 1938; Antoniou, 2007). The archaeological evidences which can date formations of a typical lavatory in the Classical period are limited. The most sound one refers to the small lavatory (Figure 6.9) of the building named Gymnasium at Minoa on Amorgos island (Antoniou, 2007, 2010). This is dated in the late Classical period.

Figure 6.9 Restored view of the Gymnasium's lavatory in Amorgos (Antoniou, 2007).

6.3.3 Hellenistic period

The main innovations on the hydraulic structures in the Hellenic antiquity appeared and evolved during the Hellenistic period, along with the establishment of the Hippodamian system of town planning. This time, the well planned drainage became an essential feature in most settlements. Additionally, the typical lavatory incorporated flushing with running or not water and was established in most Hellenistic residencies, from the 'cosmopolitan' Delos to the mountainous Dystos (Antoniou, 2007).

After Philip's death, during the period of his successors the harmonious and functional Hippodamian city planning was continued and favored various technologies for collecting transporting, using water and disposing storm and wastewater, always based on the principles of aesthetics, safety and hygiene. Scientific and engineering progress enabled the construction of more sophisticated hydraulic infrastructures, which were much more necessary, as sanitation standards were advanced by the appearance and extended use of baths and toilets.

Bathing facilities are also found scattered in Macedonia more in the form of individual bathtubs and less as public structures. In Amphipolis, bathrooms were arranged in the north wing of the Palaestra of the Gymnasium of the city, a complex structure in the second half of the 4th century BC, equipped with integrated hydraulic systems (Figure 6.10). The drainage network was organized with various channels passing their used water quantities on to the central collector, which crossed the complex evacuating in a nearby stream (Lazaridi, 1990).

Figure 6.10 Amphipolis. Gymnasium, central stone built drain (with permission of A. Kaiafa).

In Pella, baths have been found in many structures. An organized form of baths for public use was found to the South of the Agora, constructed in the last quarter of the 4th century BC, with a sophisticated drainage system, consisting of drainage elements, which were achieving immediate and continuous disposal of wastewaters. Moreover, an impressive wastewater management, fulfilling all the conditions of hygiene, has been found in a small bath integrated into the NE corner of the Palaestra building of the palace in Pella of the late 4th century BC. This included a spacious stone-built pool (Figure 6.11). Its drainage infrastructure included a hole piercing the wall, which was evacuating into a stone drain nearby, in addition to holes on the floor around, enabling percolation of overflows (Chrisostomou, 1996).

Figure 6.11 Pella. Pool with drainage infrastructure in Palaestra (Chrisostomou, 1996).

Apart from bathing structures, urban drainage systems have been also advanced during the Hellenistic times, all over the Hellenic world. The Hellenistic city of Rhodes had an advanced drainage network under the roads, in addition to a well-organized supply system. A complex web of stone-built drains equipped with wells at intervals, running underground, was receiving the wastewaters from smaller earthen, lead, rockcut or built pipes from adjacent buildings. Drains were all stone-built channels, coated with clay, usually covered with slabs forming either a flat or a pitched cover. Elsewhere in the town arched drains were also found. The town had also small sanitary facilities, organized baths, with advanced drainage infrastructure, an undeniable evidence of a high hydraulic engineering (Doumas 1973–1974) (Figure 6.12). Relevant drains were recently discovered in the area of the great Hellenistic Stoa, in the town of Kos.

Figure 6.12 Rhodes. Urban drainage networks (Doumas, 1973–1974).

Similar applications have been observed in Delos (Figure 6.13), despite the lack of a harmonious urban plan. A well organized sewerage network was feeding underground slab-covered drains. This well-formatted ditch network was recorded during the excavations of the town (Chamonard, 1924), and recent researches provided significant information about its engineering (Moretti & Fincker, 2011). Possibly, this network was managing mainly wastewater rather than storm water, since most of the rainwater was carefully collected in the numerous private and public cisterns of the town (Antoniou *et al.*, 2013). Besides, the strong relation of the inhabitants with water and water management structures is reflected in numerous inscriptions found on the sacred island (Hellmann, 1988).

Figure 6.13 Delos. Drains under the roads (Chamonard, 1924).

In other smaller settlements on the Aegean islands, there were relevant but simpler systems, as in Minoa on the island of Amorgos. There, a complex drainage network was managing in some branches storm water, but mostly wastewater, providing also a significant amount of reuse for the operation of the lavatories in the residences of the settlement (Antoniou, 2010).

At the same time, the general development of hydraulic science and technology is evident in cities of Macedonia, which was the center of the Hellenistic world. In urbanized Macedonian settlements, built mostly according to the Hippodamian system, water supply and sanitation for everyday needs was a primary concern of the authorities. Drainage and sewerage systems were planned and built simultaneously with the development of roads and the construction of the buildings.

Pella is a typical example in the Macedonian Kingdom; built in a form of building blocks in a chessboard network of roads. An impressive feature of the town-planning at Pella is the drainage system. Beneath the streets of the city, there was an elaborate sewerage and drainage system, which enhanced the quality of life and health of the inhabitants. Water was brought to the houses from systems of water pipes, cisterns and cleaning shafts. Drains took excess and refuse water from the houses, and emptied it into sewers beside the horizontal and vertical earthy streets, which, using the slope of the hill (gravity), carried the wastewater and rainwater away.

Similar articulation was found in all Hellenistic Macedonian cities, central drains were the most crucial part of drainage networks, discharging wastes through city walls. Drain openings in fortification walls were hydraulic elements confirming a high level of hydraulic knowledge even before mid 4th century BC in Macedonia (Kaiafa, 2008). In the city of Thassos, wastewaters were outpouring from arched exposures, piercing the entire width of the sea walls. These exposures were closed by iron bars since the end of the 5th century. The same infrastructures were also used during all Hellenistic and Roman times ensuring cleanliness and public hygiene (Figure 6.14).

Figure 6.14 Thassos. Drainage opening on city walls (Grandjean, 1994).

In Amphipolis, instead of the usual small-sized drain holes in the fortification, an impressive sewerage system was chosen on the north walls (Lazaridi, 1989, 1990). Located along the bank of the river, very close to the city, north walls have been always in danger of collapsing under the pressure of the inner or outer flood water. The project consisted of drainage vertical slits, rectangular in shape, ensuring the safety of the walls in case of river overflowing or large amounts of rainwater gathering there. The whole system was based on a similar one, of the Classical period. The much more cared Classical drainage project has been also found on the northern city wall, arranged in repeats of two or three. It consisted of vertical trapezoidal slits in the rampart, of 1.65–1.90 m high, with the narrow width side 0.22 m outwardly. The width of classical slits was further reduced with metal vertical rods, so as not to allow the passage of an intruder, even if it was a small kid.

As far as the lavatories of the Hellenistic period are concerned, they present clearly the typical formation. The typical ancient Hellenic lavatory consists of single or multiple benches, wooden or stone-made, having defecation keyhole shaped openings (Antoniou, 2007). The benches are supported by cantilever stone blocks, and were situated over the main ditch, where flushing water was running, continuously or periodically. According to the type and the layout of the benches, the shape of the supporting beams and the keyhole openings, a kind of typology can be categorized.

The more usual layouts are the L shaped, as in Ithidikis residence in Minoa Amorgos (Figures 6.15 and 6.16) and at the Lecheon street in Corinth (Figure 6.17), and the U shaped, as in Epidauros and Philippoi. Depending on their use, they are distinguished in domestic lavatories and public ones, usually related with public buildings or complexes.

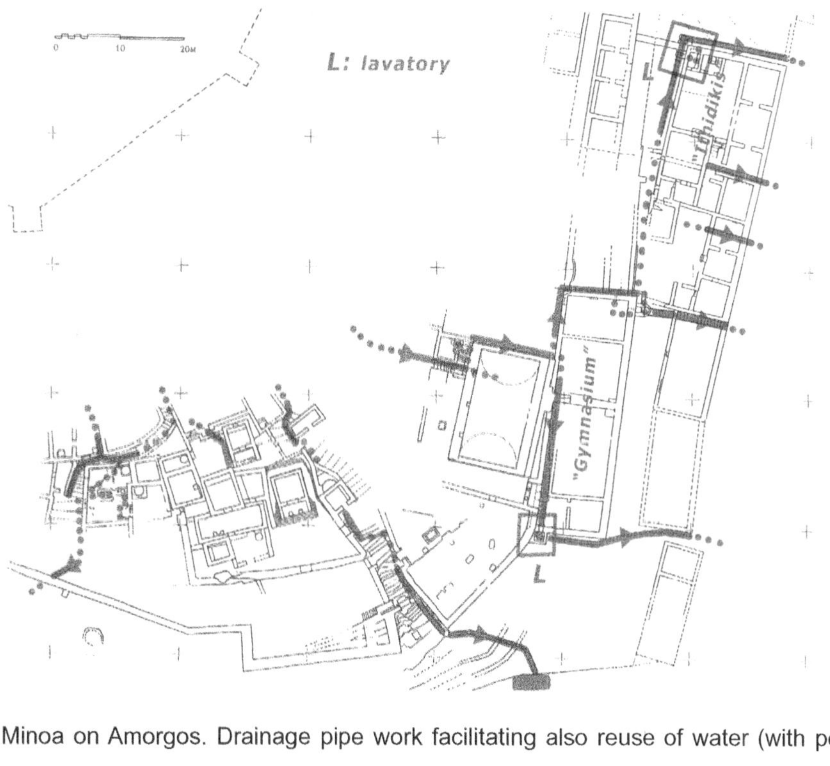

Figure 6.15 Minoa on Amorgos. Drainage pipe work facilitating also reuse of water (with permission of G. Antoniou).

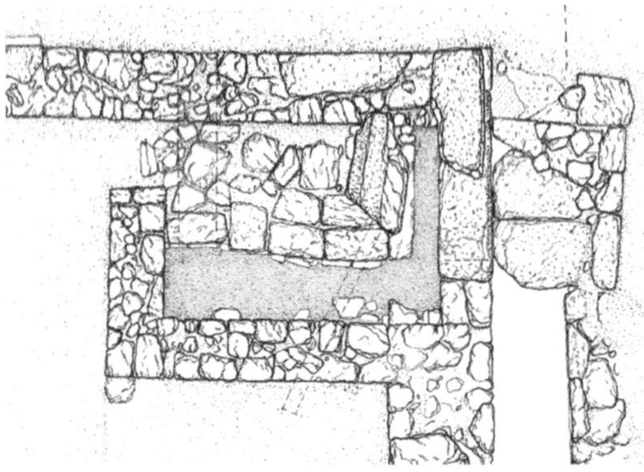

Figure 6.16 Lavatory of Ithidiki's residence in Minoa Amorgos (with permission by G. Antoniou).

Figure 6.17 Corinth. Lavatory along the Lecheon street (with permission by G. Antoniou).

In Delos public lavatories have been found in the Palaestra and the Gymnasium. In the Palaestra of the Lake, there were three spaces, believed to have been public toilets, formed after the rearrangement of the original classical building (Chamonard, 1924). The neighbouring, smaller and newer Palaestra has a lavatory as well. In both buildings, lavatories have been placed in the perimeter, and particularly near the path of some drainage. The north-eastern lavatory of the Lake's Palaestra was probably supplied by the water from the bath or even the colonnaded atrium. In the south-western lavatory, there was probably a small rectangular reservoir in the middle.

The public lavatory of Asclipieion of Pergamon was characterized by its ground plan, which is more complicated than the usual rectangular form as is evident from its ground plan. In the Asclipieion of Kos, the public lavatory was part of a later extension of the lower perimetric portico towards the west. Here, it is remarkable that the monolithic reservoir that drained the water from the small perimetric half pipe (similar in Corinth; Figure 6.17 and the later one in Ephesus; Figure 6.18) for the washing of sponghia (σπογγιά in Hellenic) to the main conduit, also provided water that was reused inside the lavatory itself (Schatzmann 1932; Antoniou 2007). Also the placement resembled that of the lavatory of Kotyo's Portico in Epidaurus.

Figure 6.18 Ephesus. Lavatory at the Thermae. Small perimetric half pipe (with permission of Antoniou 2007).

Lavatory capacity can be classified according to the number of the toilet seats, which correspond to the maximal number of users at any one time:

(a) The very small domestic lavatories, used by two or three people of the house (e.g., Figure 6.19, Hoepfner 1999).
(b) Moderate sized domestic lavatories with more than four defecation seats.
(c) Small public lavatories with evidence for at least four users at a time (e.g., 'Gymnasium' of Minoa in Figure 6.15 and in Palaestra of Delos in Figure 6.20).
(d) Large public lavatories used by more than ten or twenty people. These were generally constructed during the Roman period (Figures 6.11, 6.17, and 6.18) (Orlandos, 1940).

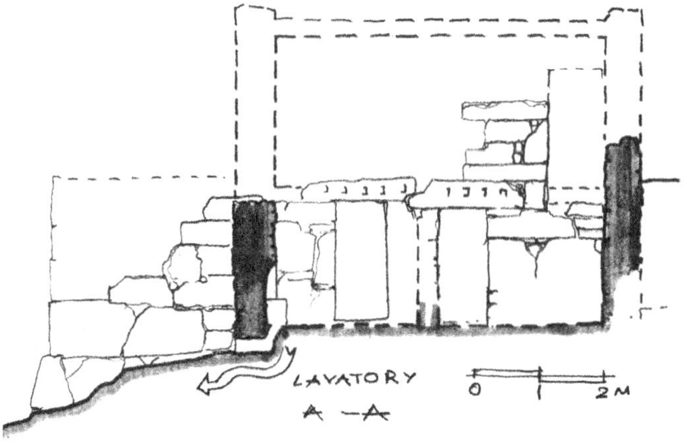

Figure 6.19 Dystos. Section of a ruined house (Hoepfner, 1999).

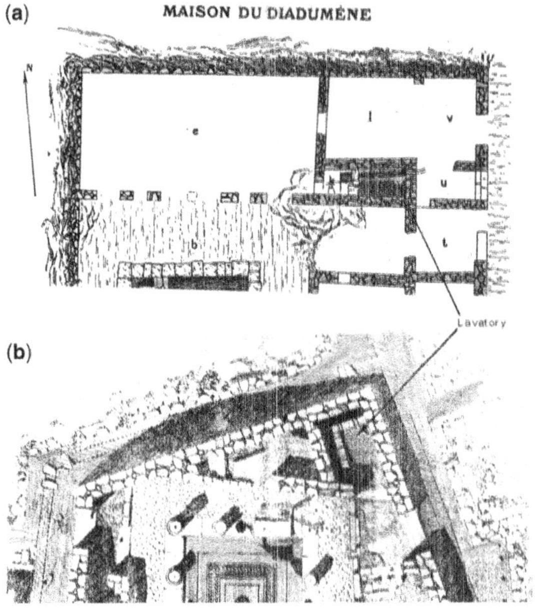

Figure 6.20 Delos. Lavatories of houses: (a) with flushing hole and (b) L-shaped ditch (Chamonard, 1924).

Most of the ancient names for toilets mentioned at the ancient written sources as apopatos, apochorisis and so on refer to a private space (the part –από -apo). At the earlier examples the single defecation vessels (e.g., in Olynthos) or cesspits (e.g., in Athens), justify that privacy, referred to that terminology of the classical Era. Despite this, the excavation at many private lavatories clearly demonstrates evidence for their simultaneous use by more than one person. Even in residences where the inhabitants numbered 5–10 people, there were lavatories with two up to four defecation openings, concluding the simultaneous usage by more than one person! Research on relative aspects (Trümper, 2012) approaches the use of private and public spaces by the genders, but it could be considered as unclear whether there was a simultaneous usage of the domestic lavatory by residents of different sexes. For the public lavatories, there is also no doubt that they were used by dozens of people, often more than fifty. This was a practice which expanded during the Roman era and survived in many Byzantine and medieval lavatories of the Eastern Mediterranean area, including monasteries (Myriantheos, 1987) and castles such as Mytilene (at the 16th century semi subterranean refuge).

6.3.4 Roman period (*ca.* 67 BC–330 AD)

In Roman times, water management urban systems, based on the Hellenic knowledge, has been characterized by a great technical progress. Romans developed high engineering skills and were able to expand these technologies in large scale projects and to improve supply and drainage infrastructures. Throughout the Roman times, in Macedonian cities, no big changes have been identified in material or types of structure of drainage channels. However, sewage networks over the years have been enriched and become more complicated according to the growing needs. Earthen or lead pipes, sometimes piercing private and public building walls, visible rockcut tunnels, underfloor stone or brick built channels, constituted complex sewerage networks, discharging to large, invisible, vaulted or flat covered sewers. This is definitely related to the organization and extended use of more complex public baths and latrines, with increasing demands for continuous water flow and the absolute need for channeling away of wastewaters and sewage continually. This was exactly the chief innovation of Roman times, the constant washing of drains by overflow from aqueduct-fed fountains and cisterns, thus improving sanitation and public health. This actually allowed the construction of multiple-seat public latrines.

Baths and latrines in Macedonia were playing a major role in the life style, all over the empire. Macedonian engineers under these conditions developed water supply sewerage and sanitation systems, constructing impressive sanitary achievements, such as public toilets and drains (Kaiafa, 2008). At least five latrines have been found in Dion, Thessaloniki and Philippoi. They all had rectangular ground plan and were usually embedded into more complex buildings, baths, thermae, palestrae. The necessary drainage infrastructure was an integral part of their drainage network. Yet in Dion, public latrines are also found outside the walls joining the Sanctuary of Demeter, during the 2nd century AD. The ditch under the defecation bench was ensuring uninterrupted cleanliness, supplied continually by an underground stone-built conduit, probably filled with overflow water from a source in the sanctuary of Asclepius nearby (Pingiatoglou, 2003). All the latrines in Macedonia were positioned in the corner of the buildings or across the streets. The location was facilitating waste management, as waste could flow into the urban sewerage system directly, through a single discharge channel.

Engineers in latrines, positioned on the southeast corner of Great Thermae of Dion, exploited the very practical technical advantages of sharing drainage and using discharge water for flushing (Figures 6.21). The channel beneath the circumferential marble bench with the defecation holes was essentially the continuation of a stone-built vaulted sewer, crossing the complex. It was accepting wastewaters from tanks and basins, and stormwater from the roofs, via an elaborate system of clay, lead pipes, stone or brick built conduits, lying vertical and horizontal inside the walls, under the floors or piercing masonries. Rainwaters gathered in a sewer pit in the middle of the courtyard enriching the arched drain which was flushing the latrines, reducing offensive odors and contributing to the public hygiene, by leading all the wastes and impurities directly outside the city through the walls nearby (Pantermalis, 1989; Karadedos, 1986, 1990, 2000; Kaiafa, 2008).

(a)

(b)

Figure 6.21 Dion: (a) Great baths and (b) latrines (with permission of Karadedos, 2000).

The standalone public toilets in Dion, with a rectangular groundplan, had an impluvium at the centre, which was improving the illumination, providing continuous ventilation, hence making the interior somewhat more pleasant for the users. Its hydraulic infrastructure was connected directly with the urban supply network, which was feeding a small constantly flow fountain in the middle of the western side wall of the building. Water from the fountain at first was rushing in a shallow ditch in front of the defection seats so as people could clean themselves, rinsing a sponge on a stick. Simultaneously, the flowing waters were pouring into a deeper ditch, passing under the bench type seats along the walls with the usual keyhole shaped defecation openings, evacuating into the urban sewerage system (Pantermalis, 1989).

In Athens two public lavatories dating from the Roman period have also been preserved. In the south-eastern corner of the Attalos' Stoa (Travlos, 1982) and a stand-alone one, to the east of the Roman Agora (Figure 6.22). Both ground plans have an almost square shape. The Roman market lavatory was built after the Agora. Apart from the rectangular entrance lobby it was also characterized by the deep conduit underneath the benches and the impluvium at the centre of the room. From the surviving remains, it appears to have had 62 defecation openings, which correspond to matching urinal holes on the floor (Orlandos, 1940).

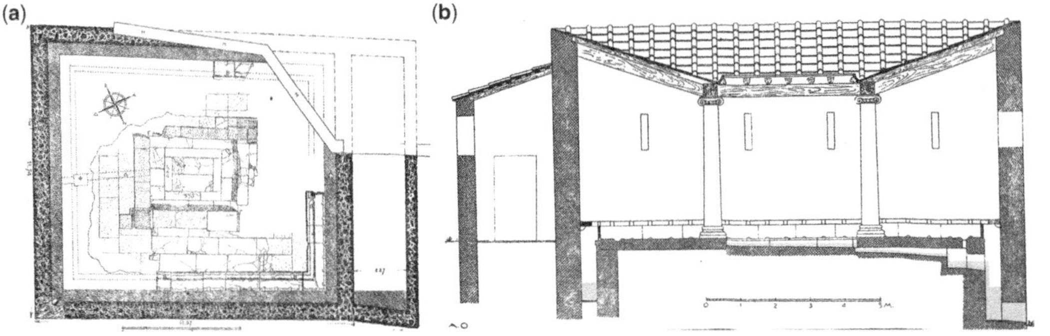

Figure 6.22 The lavatory outside Roman Agora, Athens: (a) plan and (b) restored longitudinal section (Orlandos, 1940).

Latrines in Palaestra in Philippoi in East Macedonia were equipped with flushing systems connected to urban storm drainage conduits, almost like those of the independed latrines in Dion (Figure 6.23). A conduit running under the side street was supplying uninterruptedly the sewage duct, pouring in via a small square cross hole positioned in the northeast corner of the room (Lemerle, 1937). The typical Hellenic layout was predominant here, as in other lavatories in the Hellenic region built during the Roman period, as those in Epidauros (Figure 6.24).

Figure 6.23 Philippoi. Palaestra, Latrines (with permission by G. Antoniou).

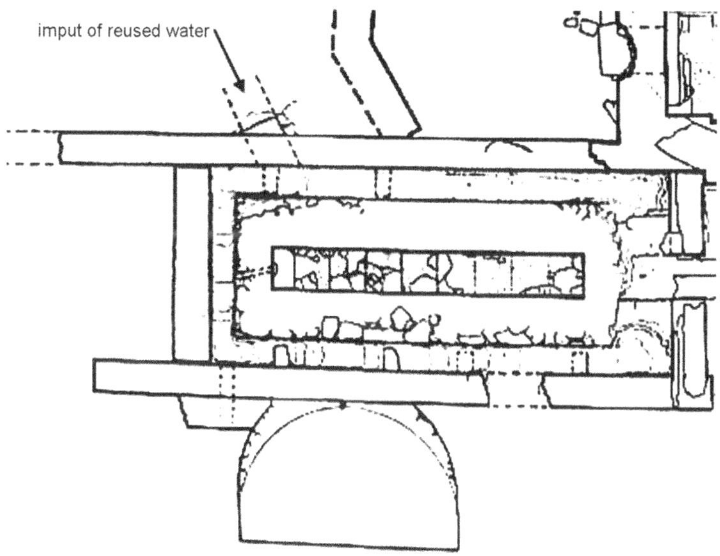

Figure 6.24 Epidaurus. Plan of the lavatory in Kotyo's Stoa (with permission by G. Antoniou).

Elsewhere in the same city, latrines were an integral part of a small bath structure (βαλανείο), of the 1st century, with a simple technological solution for continuous cleanliness and elimination or at least reduction of unpleasant odors (Gounaris, 1990, 2004). Dirty waters from a shallow basin were running off into the ditch in front of the seats and then flowing into the deep conduit along the walls underneath the benches, which was enriched with the wastes from natatio frigita right next to the toilets room. Impurities of urination and defecation under the seats were in this way flushed by the wastewaters of the bath tanks and were channeled into the main collector of the urban sewerage network under the side street.

At the Ventio's Thermae in Efessos the traditional Hellenic typology was maintained inside a typical Roman building (Figure 6.18). They were characterized by the elongated rectangular impluvium that was quite monumental for the size of the chamber. It is situated close to the entrance of the Thermae, and it can be used not only by the users of the baths but also by the people walking outside the complex (Scherrer, 2000).

The public lavatory at Epidaurus, at the east end of Kotyos' portico, appears to have been one of the later buildings of this type in Hellas. It had a rectangular ground plan and was supplied with naturally flowing water, thought to have been channelled from the north-eastern baths (Figure 6.24). It is believed to have been built when the portico was partly standing, and the poor construction includes stones of other collapsed buildings of the sanctuary. The elongated shallow tank in the middle, made of tiles, had a small sewerage pipe ending at the main perimetric conduit. The similarities of the structures in that lavatory with the Roman lavatory at Vercovicium Fort, a Legion's camp by the Hadrian's Wall in Scotland, proves the well spread techniques all over the Empire.

It is evident that in many cases of Roman baths, the lavatories were being used not only by the users of the baths, but also by people of the area near the baths. On the other hand, it is not clarified if this combined usage was original or if it was implemented later on. In Thessaloniki, latrines embedded in 4th century baths of the Galerios complex, were equipped with drainage infrastructure, connecting also with the sewerage system of the palace (Figure 6.25). Wastewaters from the pools were used to flush the conduit of the rectangular section under the bench seats, evacuating directly into main underground sewers, as parts of the urban hydraulic network (Athanasiou *et al.*, 1999).

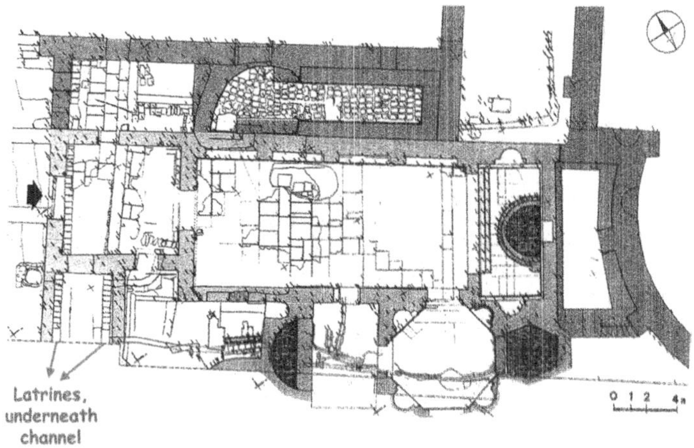

Figure 6.25 Thessaloniki. Galerius palace, latrines in palace baths (with permission of Athanasiou 1999).

Stone-built central drains in Macedonia have been developed under vertical and horizontal streets, along the central axis, ensuring storm and wastewater sanitation. They were all of rectangular section, made of materials of second use (Kaiafa, 2008). Their walls were made of stones and mortar, while slabs were used for the drainbeds. In the Roman cities of Veria, Dion and Philippoi, central drains had no extra coverage, as they were covered directly with road plaques, which had slits in the joints, enabling the percolation of rainwater. Sewers of Philippoi had impressive dimensions, ranging from 0.55 m–1.00 m in width and 0.90 m–1.70 m in height. Somewhat smaller were the sewers in Veria. In Roman Thassos, as well, massive slabs of the road surfaces were set onto the rims of underground sewers functioning simultaneously as covers (Grandjean, 1994) (Figure 6.26).

Figure 6.26 Thassos. Drains (Grandjean, 1994).

Sometimes vaulted collectors, instead of horizontal covered, were developed under paved or foot-stepped-soil roads. Vaulted sewers of large dimensions were used long before the Hellenistic period, since the end of the 8th century BC in Assyria. In Roman Hellas, they were preferred for urban sanitation in several cities, such as Samos, Elis, Patras, Kos and Rhodes. In Macedonia, vaulted channels, with arches made of bricks and beds coated with tiles or bricks, composed the wastewater and sanitation network of Thessaloniki, in the period of Tetrarchy (293–313 AD), reaching 1.20–1.40 m in width and 1.05 m–1.80 m in height, following the steep southern slope. They were located along the main axes of horizontal and vertical streets of the city, built mostly of stone and mortar, just like the arched drains under the road decks in Samos, of the same time. Elsewhere, the side walls were constructed of bricks in successive layers, such as notified contemporary examples of the drainage network of Patras (Kaiafa, 2008). At intervals, manholes on top for maintenance and cleaning, had an average size of 0.75 m × 0.45 m. Similar openings have been identified in sewers of Elis or Rhodes.

During the Roman era, central underground sewers are the most important part of an organized sewerage system of a city, as they were collecting and evacuating the stormwaters and impurities through drainage openings at the fortification. On the late Roman SE walls of Dion, the rectangular run-off mouth had a width of 0.80 m and a height of 1.15 m and a decorative arched upper side, made of bricks was placed vertically next to each other, ensuring the discharging of the wastewaters into the river protecting the building from flooding and ensuring public health (Figure 6.27).

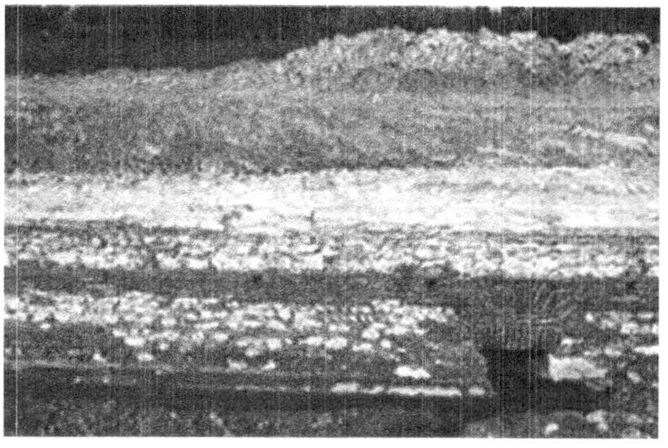

Figure 6.27 Dion. Drain hole on the city wall (Stefanidou-Tiveriou, 1998).

In the Roman period, the underground sewers and drains were designed on such a large scale, that in certain section's wagons loaded with hay could drive through. The Roman sewerage and drainage systems have been over praised. Despite their longevity, they ignored basic sanitary principles. Sewers and drains in Rome and other major cities discharged directly into the rivers, the polluted state of which must have been a constant pollution problem. They carried sewage, urban runoff, and drainage water together. This made it necessary to have large openings along the streets (Hansen, 1983).

It is known that Roman elaborate latrines were not a haven for the lazy, the slaves, the poor, or the invalids. These groups had to resort to chamberpots (Hansen, 1983). These were emptied into containers of garbed placed under the stairwells or, if containers were not available, jars could be emptied into a nearby cesspool or into the central sewer through openings provided (Figure 6.28). There were many who found the distances to cesspools or sewers too far, and who found it more efficient to empty the contents of their chamberpots from windows onto the streets below. During the Roman period, several auxiliary elements of the typical lavatory as the keyhole shape openings in the perimetric ditch, etc become quite well formed and articulated (Figure 6.29).

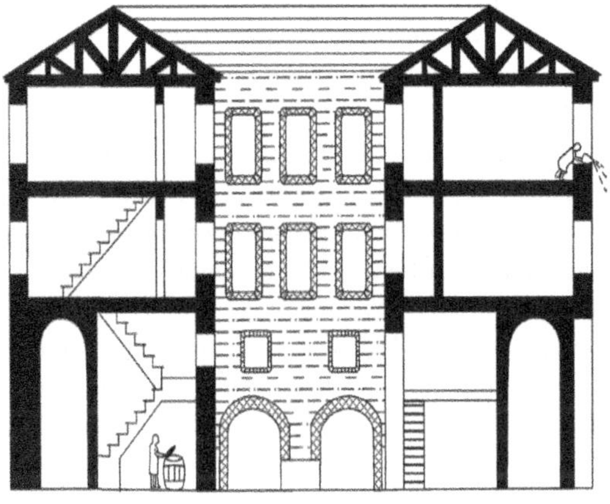

Figure 6.28 A hypothetical Roman tenement building with a chamberpot being dumped on the street below in the left-corner and another chamberpot being emptied into a barrel located under a staircase in the lower-right corner (adapted from Hansen 1983 with modifications).

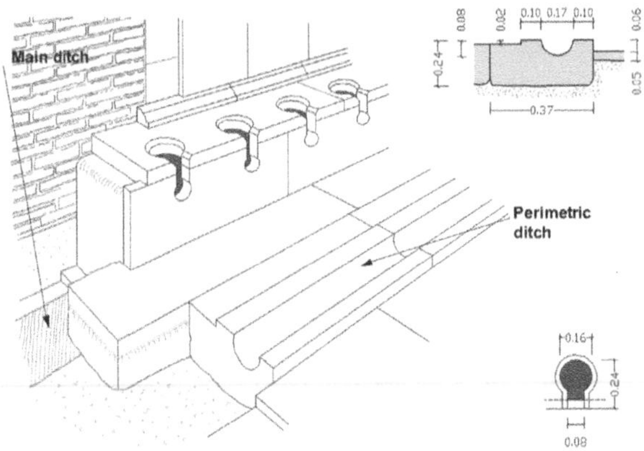

Figure 6.29 Latrine in Miletos with dense arranged holes (adapted from Schatzman, 1932 with modifications).

With the spread of the Roman culture throughout the empire, awareness of the design of effective private and public latrines spread too. Public lavatories flourished during Roman Era, and these can be classified into three typical forms with the following examples: (a) the lavatory of Pompei's Palaestra, (b) the complex of Triklinon's in Ostia (Hoepfner, 1999), and (c) the Largo Argentina lavatory in Rome. Therefore, in the 2nd and centuries BC, lavatories were built in monumental forms and sizes, equivalent to other constructions of the Romans. It is clear that the Roman influence on such technologies was significant. Roman engineers, with their devotion to useful public works, were critical agents for the construction of numerous public lavatories all across their Empire. Moreover, the size of these latrines was adjusted to accommodate the growing cities of that era and their increasing population. The typical lavatory form was incorporated, not only in most Roman Thermae, but also in other public buildings, such as Gymnasia and Palaestra. This was similar to that of the later Hellenistic period, but in larger scale. In many cases, larger lavatories were added in complexes with a high number of visitors, as at the Askleipieion of Pergamon (Figure 6.30). Finally, all Roman mansion houses and villas had a proper lavatory, usually operated with running water.

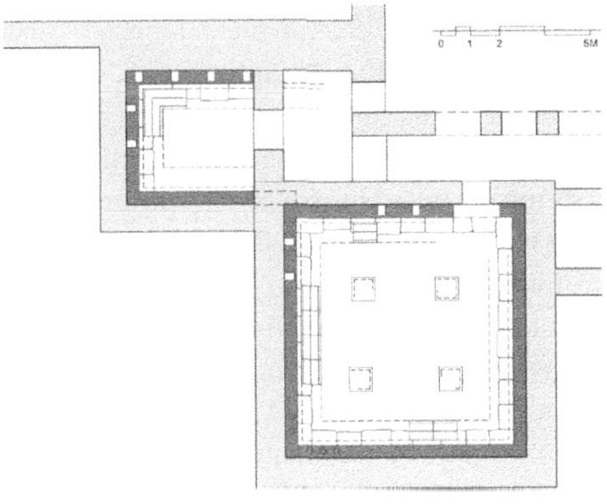

Figure 6.30 Double public latrine at Askleipieios' Sanctuary at Pergamum (adapted from Hansen, 1983 with modifications).

The element of running water, provided to the Roman cities by their famous aqueducts, is critical for the expansion of the size of the lavatories. In many cases, it replaced the reused water supply that was a quite familiar water source for the earlier centuries. The more widespread existence of toilets, the larger size to cater a greater proportion of the population, and the running water from aqueducts to flush the facilities all year round, were technological improvements to the typical forms of the ancient Hellenic lavatory. This was typical of the Roman approach towards improving the conditions of everyday life. During the years of Emperor Vespasian's rule, lavatories became an important source of income for the imperial funds, since they began to charge an entrance fee (Suetonius XXIII, Kline *et al.*, 2010).

6.4 MEDIEVAL TIMES

6.4.1 Introduction

The decline of the ancient world influenced not only the technological achievements accumulated by that time, but also the relevant skills and constructional abilities. In addition to that, the morals and the social habits introduced by Christianity influenced the practices in the formation of the lavatories and particularly the importance of privateness, not only referring to defecation, but also to bathing. On the other hand, the vivid tradition of the ancient world, especially in the East, permitted the continuation of earlier relevant habits. In the early centuries of the Byzantine Empire, there were several common lavatories, mainly in monasteries, facilitating multiple users at the same time, as it will be presented below. Despite that, wastewater management seems somehow neglected by the communities and was mostly resolved privately. The private character of lavatories resulted to the reduction of their size and to the possibility for their placement in spaces closer or next to main rooms (Orlandos, 1937).

6.4.2 The Byzantine period (*ca.* 330–1453 AD)

As it was mentioned above, the wastewater and lavatory technical practises and the relevant applications during the Byzantine period present interesting characteristics related mostly to a combination of various historical aspects, as the partial continuation of the ancient tradition and practises, the barbaric raids and their results, the social reformation due also to the Christianity, the power of the Church within the state, and so on.

The continuation of the Roman Empire was the Byzantine state, which preserved the Roman name and inherited the Greco-Roman civilisation of the antiquity. Thus, the new state adapted at its beginning not only administrative institutions, but also the constructional and technical achievements of the pre-existing empire. The Greco-Roman civilization survived mostly in the eastern part, since after the 5th century AD, the western part was lost due to the barbaric invasions. Therefore the medieval characteristics of the west remained away from the east for quite long time.

Several typical ancient Hellenic and Roman formations can be traced in the applied wastewater and storm water technologies, as well as in the lavatories built till the 8th century. From the majority of the archaeological findings, it seems that the wastewater and storm water networks were inferior to the relevant Hellenistic and Roman ones. Later information about mature Byzantine settlements as Mystras, refer to both piping networks (Orlandos, 1937, Figure 6.31) and individual residential water-tight cesspits (Velenis, 1978).

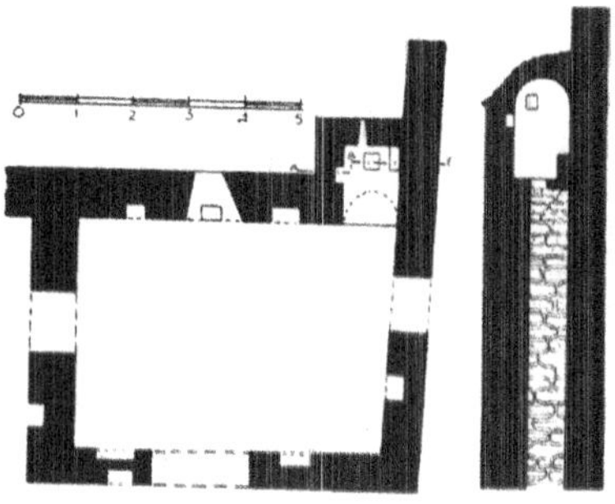

Figure 6.31 Mystras. Chamber of mansion house with toilet (Orlandos, 1937).

Typical toilets with perimetrical ditch (Figure 6.32) can be found in baths of the first Byzantine centuries as in Marea (*ca.* 6th–8th century AD) (Szymanska & Babraj, 2009) and the double bath at the pilgrimage complex of Abou Mina (*ca.* end of 6th century AD) (Müller-Wiener, 1966) in Egypt. They are mentioned here as it can be considered that they had been created under the Greco Roman tradition. In both cases, the toilet is adjacent to the bath, but closer to the entrance, or the disrobing hall, or the entrance patio. According to the written sources about the life in the Byzantine era, which states that the users in the baths were visiting the toilet after undressing and before entering the bath (Koukoules, 1952), that placement of the toilet is well testified. Moreover, it can most certainly be concluded, out of the surviving remains, that lavatories were used by many people simultaneously, despite their use 3–4 centuries after the prevalence of Christianity which promoted privacy. Construction wise, these toilets were supplied by water through pipes or the ditch system of the bath, possibly reusing water effluent from the main bath as well. In Marea, an entry chamber (Figure 6.32) resembles similar plans found during the end of the Hellenistic period, for example, in Athens (Orlandos, 1940; see Paragraph 6.3.1) Their placement at the outer zone of the building served as well for the sewage, like the way it was usually applied at the ancient lavatories, for example, in Delos, Dystos, Kos, and so on (see Paragraph 6.3.1; Antoniou, 2007).

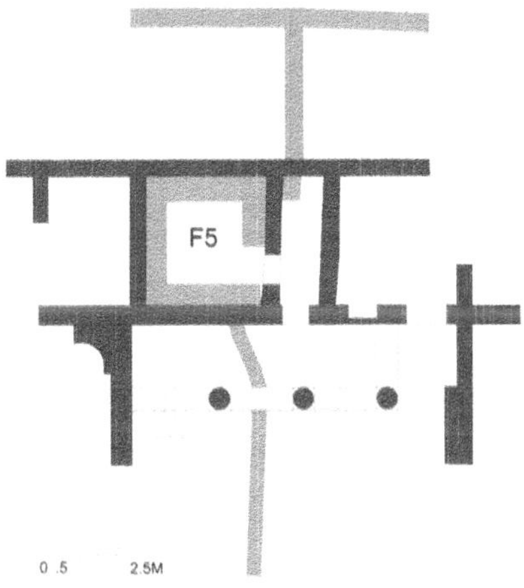

Figure 6.32 Lavatory at Marea baths Egypt. Water ditches in gray (adapted from Szymanska & Babraj, 2009 with modifications).

It seems that the collective use, the connection or attachment to bathing facilities and the semi-perimetrical ditch, supplied by the bath's water or separate stream water (as in Aksaray Sultan Han & Incir Han) are characteristics, which survived up to the 12th–13th centuries and are found at Selcuk, or Rum Selcuk according to Prof. Kiel, Hans-Caravansarais in Anatolia (Yavuz, 2011).

Monastic lavatories with a collective character are not rare in the greater region of the Eastern Mediterranean, during the first Byzantine centuries. The row of the collective latrines at St Symeon monastery in Assuan, Egypt (Figure 6.33), and their placement, lead to the existence of a kind of sewage (Monneret & Villard, 1926). On the other hand, a similar row at the great eastern tower (*ca.* 6th–7th century AD) of St. Catherine's monastery at Sinai (Myriantheos, 1987) had undoubtedly a collective use, but the waste was drained exactly outside the walls, via pipes running through the thickness of the towers' wall.

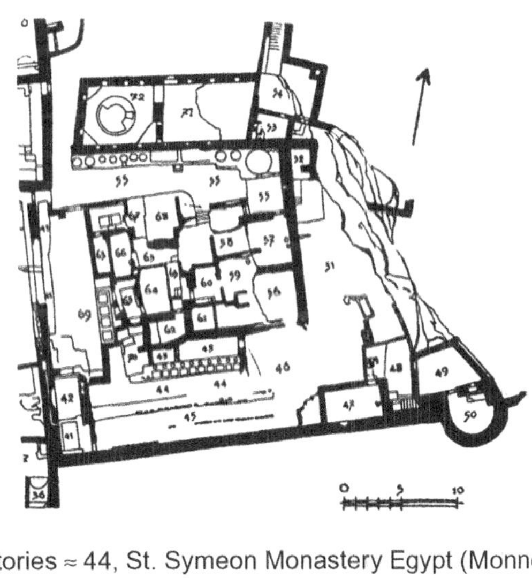

Figure 6.33 Collective lavatories ≈ 44, St. Symeon Monastery Egypt (Monneret & Villard, 1926).

Similarly, the lavatories of Zygos monastery in Mountain Athos, in Hellas – dated *ca.* 10th century AD – (Papangelos, 2005) have a relevant layout, but are placed outside the walls, attached to them. On the other hand, it is not certain if it was operating as a collective toilet or not, forming just a row of toilets. The wastewater also drained directly outside via inclined pipes.

Despite the numerous examples of collective use lavatories, there are surviving structures of private or semiprivate toilets in monastic buildings. A characteristic case is the combination of private and semiprivate toilets in the QR195 hermitage dated early *ca.* 7th century (Henein & Wuttman, 2000; Gaubert, 2009). There, a chair type toilet was applied in single or twin layout.

The private toilet became part of the typical layout of the Byzantine Monastery and was constructed not only at the wings of the cells, but also in other buildings of the complexes, as the Estia (Figure 6.34), the Hospitals, and the Guest Houses. In some cases – as in the Hospital of Mega Meteoron monastery – where a separating wall is absent, there was probably a wooden partition (Orlandos, 1958). Doors resembled a kind of curtain, the so called velothyron (Orlandos, 1958). In many cases, there are pipes draining out the waste, inclined as in Vrontiani monastery in Samos (Figure 6.35), or vertical as at the lavatory of the Mega Meteoron Hospital (Figure 6.36). On the other hand, there are monastic literal references about cleaning or removing the waste '... *σαρώσατω τα λύματα* ...' (Orlandos, 1958), which leads to the conclusion that there was no sewage system in every case.

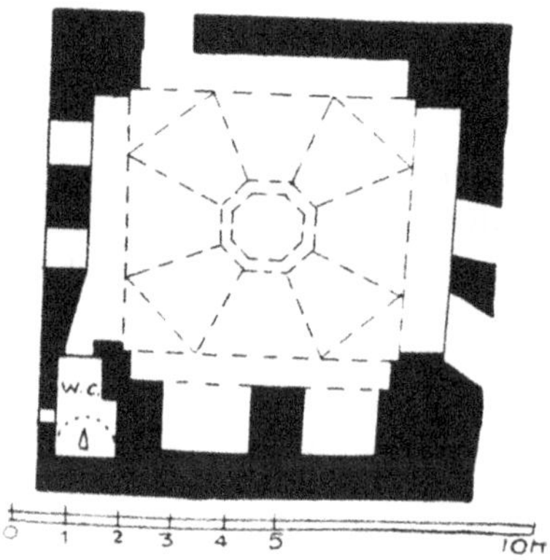

Figure 6.34 Estia (fireplace) of Helandariou monastery (Orlandos, 1927).

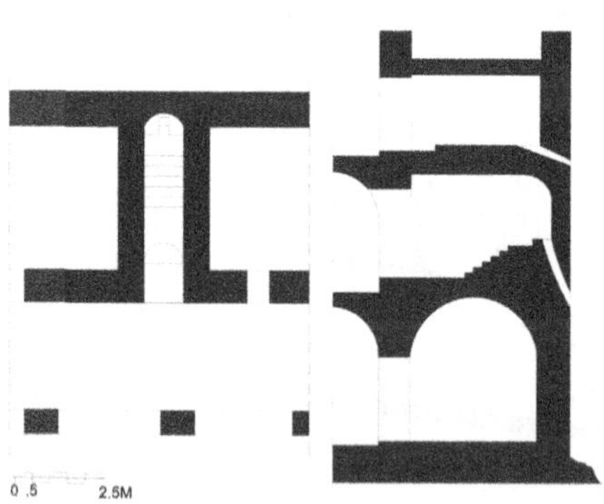

Figure 6.35 Samos, Vrontiani. Toilets among monks' cells (Antoniou after Orlandos, 1958).

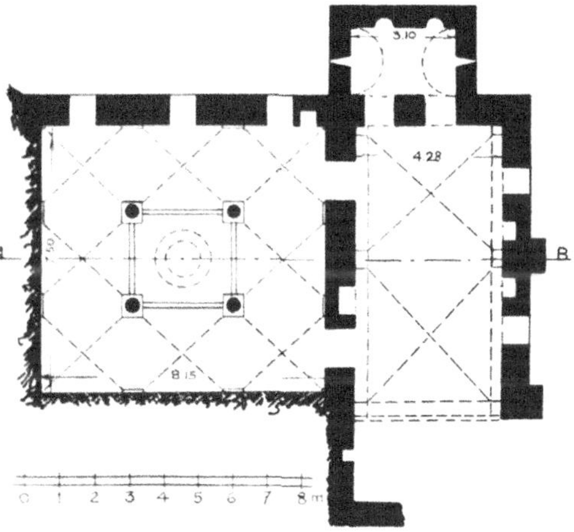

Figure 6.36 Hospital of the Mega Meteoron (Orlandos, 1927 with modifications).

Apart from the fact that there are some references to using water out of the monastery's cistern like in Nea Moni in Chios (Orlandos 1958), not much information exists about the way the toilets were supplied with flushing water. In most cases a bucket would be used.

Most probably the residential toilets were flushed in a similar way. On the other hand many surviving structures, like the residences of Mystras in Hellas, testify that a waste pipe network existed. It was consisted not only of vertical pipes, but also of clay pipes under the streets (Orlandos, 1937). These pipes coming from each house were centralized in a junction point, and were led to the town's sewage network. Moreover the owners were responsible for the maintenance of the pipes up to the junction point (Orlandos, 1937). The residential byzantine toilet was protruding (Figures 6.37 and 6.38) and situated at the corner or side of the main chamber of triclinion, found often in each floor. Their usual semicircular edge, rather than their vaulted ceiling, was the reason of the term exedra, one of the names that had been given to the toilets by that time (Orlandos, 1937). Traces testifying domestic twin or triple toilets are very limited) and in Thessaloniki they coexist with typical single user toilets (Karydas, 1999).

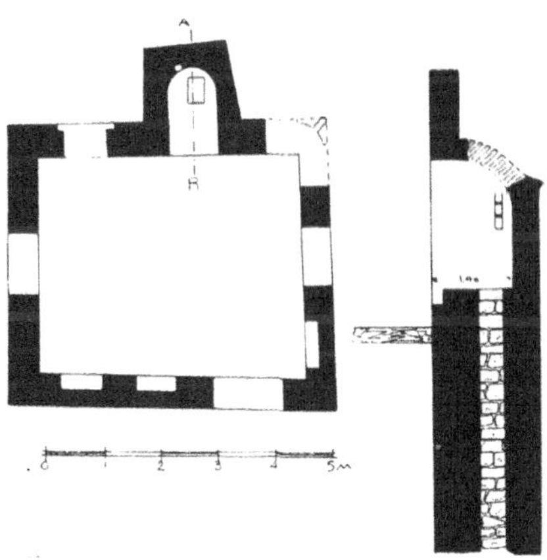

Figure 6.37 Mystras. Chamber with toilet (Orlandos, 1937).

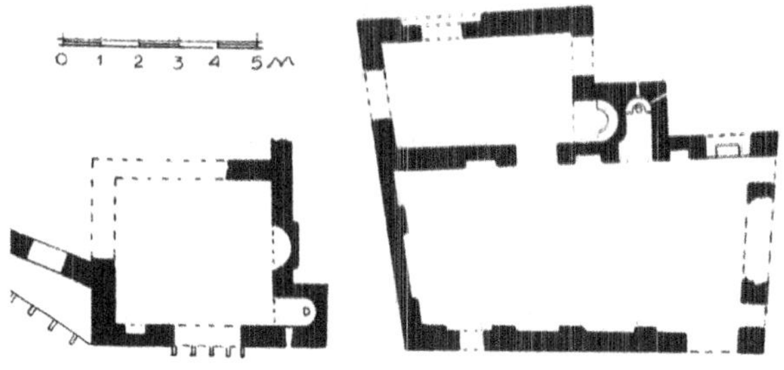

Figure 6.38 Mystras. Chambers with toilets. Regular & mansion house (Orlandos, 1937).

Similar protruding residential type toilets with or without a sewage pipe- dated in mid and late byzantine period, survive in some monasteries of Mt. Athos (Katsaros, 2002). Illumination was provided by oil lamps inside small niches on the walls. Written sources about 24 h burning oil lamps in monasteries' Typikon, refer also to the toilet oil lamps (Orlandos, 1958).

6.4.3 The Venetian period (*ca.* 1204–1669 AD)

Byzantium weakened from Ottoman and crusader attacks and lost its ability to retain Hellenic provinces. Consequently, the Byzantines sold Crete and other islands of Southern Hellas to the Venetians receiving a low, rather symbolic payment (1212 AD). The Venetian Republic became dominant and ruled the trade in the Eastern Mediterranean. Crete and other islands flourished during the Venetian period. They let people live at ease, speaking Hellenic and maintaining their religion. As a result this period of Hellenic history was rather prosperous. Under the rule of Venetians many fortresses, which were later improved by the Turks, were built. Also, aqueducts, irrigation windmills (not improved later by the Turks) and other water infrastructures were built. However, during this period water availability, quality, and sanitation were at a low level. Several cities suffered from the lack of water. The water supply depended on water cisterns, in which storage rainwater was collected. The Black Death (plague) of 1348 hit Crete particularly hardly. Plagues followed in 1398, 1419, 1456, 1523, 1580, 1592, 1678, 1689, 1703, and 1816, and some of these were responsible for the killing of one third of the population. Many Cretans migrated overseas during difficult periods on the island. Some acquired great fortunes abroad.

Venetians, however, built advanced wastewater and drainage systems in several castles (fortresses), usually located in coastal areas. Such castles are in: Preveza, Parga, Kalamata, Koroni, Argos, Methoni, Zakynthos, Chania, Gramvousa island in Crete, Frangokastello in Crete, Rethymnon (Fortetsa), Iraklion, Palaiokastro in Iraklion, and Keryneia in Cyprus. Inside the castles. the Venetians constructed beautiful buildings and water supply and drainage systems for protection from stormwater. These constructions were very modern at the time. In some cases, the rainwater and the sewage from the baths and the lavatories were directly discharged to the sea.

In addition, during the Venetian period, just like during the late Byzantine times, the construction of lavatories reflected the need for privacy, which was imposed by social and religious factors. In addition, they reflected the lack of spacious housing within fortresses. The result was small spaces for the defecation (the absolutely necessary), similar in most cases to the examples of the Byzantine era, presented above, that refer to. Even in countryside houses (like in Chromonastiri in Crete, Figure 6.39) and other spacious buildings, the size of the lavatories was limited. In addition, their location was next to the most private sectors of the residences, such as bedrooms, in contrast to the ancient houses, in which they were placed closer to the entrance or commonly accessed areas.

Figure 6.39 Protruding lavatory at the 2nd floor of a rural residential complex in Chromonastiri on Crete (with permission by G. Antoniou).

During the Early Christian period, in orthodox monasteries, several common lavatories, accommodating multiple users, have been recorded (e.g., in Sina and in Egypt see 6.4.2). On the other hand, from that period only small private toilets can be testified. In contrast in several early dated monasteries in Northern Europe latrine or reredorter (or necessarium from the East) chambers, were accommodating many monks. Examples may be found in Scotland, such as in the Saint Andrew Cathedral Priory, the Dumferline and the Inchcolm Benedictian and Augustinian Abbeys (Wright, 2004). These evidences suggest the continuation during these times of the Roman tradition, but adjusted to the ethics of the era, by using separating screens (Green, 1992).

Small protruding niches on the higher floors of the buildings, with free open air drainage, were quite common practice for the toilets in the late Byzantine and Venetian periods in Hellas, and several lavatories of that type are preserved at the monasteries of Mountain Athos (Theocharidis, 1991). This is in contrast to typical Byzantine buildings. which had a built-in drainage conduit (Orlandos, 1937; Katsaros, 2002). Despite that, there are cases, especially in fortresses, such as in Lemessos, where a built closed vertical pipe conveyed the wastes and wastewaters to a cesspit located on the ground level, (Figure 6.40).

Figure 6.40 Lavatory at the Castle of Lemessos (with permission by G. Antoniou).

6.5 MODERN TIMES

6.5.1 The Ottoman period (*ca.* mid 14th–1923 AD)

The emergence of the Ottoman Turkish tribe from western Turkey and their conquest of the Balkan lands occurred ca the middle of the 14th century AD. The urban infrastructure of many former Byzantine cities was then enriched with various constructions necessary for their society's basic needs, financed mostly by the institution of the Vakıf (Kanetaki, 2012b), such as buildings of: a) religious character (e.g., mosques), b) secular ones (of social-public character and for domestic use, including commercial buildings, e.g., bedestens), and c) works of military architecture, such as fortifications and their towers (Cahill, 1991; Kanetaki, 2011).

As Islam prescribes ablution before prayer, (Wudu, The Encyclopedia of Islam, 2001) (the duty of washing is mentioned in the Kur'an: in Surah V, '... *when you get ready for the worship, wash your face and hands up to the elbows* ...') and the use of flowing water is indicated for ablution before prayers in the mosques, as well as for total cleansing of the body, this statement encouraged Ottoman rulers to construct fountains, public steam baths ('*ḥammām*' or '*hamam*'), as well as baths over thermal springs (*kaplıca*). Apart though from the Koran's guidance, the lack of hygienic conditions, as far as washing installations were regarded during these times (Kanetaki, 2012a), rendered the above mentioned establishments important, since people's everyday needs had to be satisfied.

Buildings of public use, as well as private dwellings, were equipped also with lavatories (toilets), which although of minor importance -in regards to their size, compared to the general layout of the main construction, (and practically, showing no artistic, architectural or aesthetic value ...), were necessary installations, situated usually somewhere outside of the main groundfloor plan. As they were often shed-like and weak constructions, most of them have been turned into ruins, a fact which nowadays doesn't allow us to descriptively present the latrine technology. It is interesting to notice, that a few private residences dating from that chronological period do preserve distinctive examples of survived toilets.

Sanitary installations (e.g., toilets), were incorporated in most Ottoman architectural types of religious and secular buildings, such as mosques (camı), medrese, türbe, hospitals (darüşşifas), hammams (Kanetaki, 2004; Õnge, 1990), thermal baths (kaplıca), as well as their grouped complexes (külliyes) (see Figure 6.41 the Beyazit complex in Edirne). Lavatories were situated at a rather remote part of the complex, and squat toilets were placed one next to the other, separated through partitions for privacy. Exceptions can be traced in the case of smaller scale buildings, such as kaplicas (thermal baths), (see Figure 6.42 the Eski Kaplica of Bursa) and hammams, as the toilet was either in an intimate section of the tepidarium incorporated in it, or close to the disrobing hall of the building close to the depilation room, instead of being located outside of the building.

The examples above, used as evidence of the existence of latrines in the Ottoman period, are chosen from the first two capitals of the Ottoman Empire, Bursa and Edirne, as only a few buildings once belonging to a külliye dating from the Ottoman period have nowadays survived in the Hellenic lands, while the Haci Evrenos Kaplica in Traianoupolis (late 14th century), is in a bad state of decay and doesn't offer such constructional details (Kanetaki, 2012a). Examples of latrines have been evidenced in a few Ottoman hammams still existing in Hellas, where sanitary installations actually consist of a small cell just to fit a man who may squat down on his heels, located close to the disrobing hall, just before entering the warm part, for example, Bey hammam, Thessaloniki (Figure 6.43), Zambeliou and Douka str. hammam of Chania, Crete (Figure 6.44), Hammam of the Winds, Athens (Figure 6.45), and Karavangeli hammam, Lesvos (Figures 6.46).

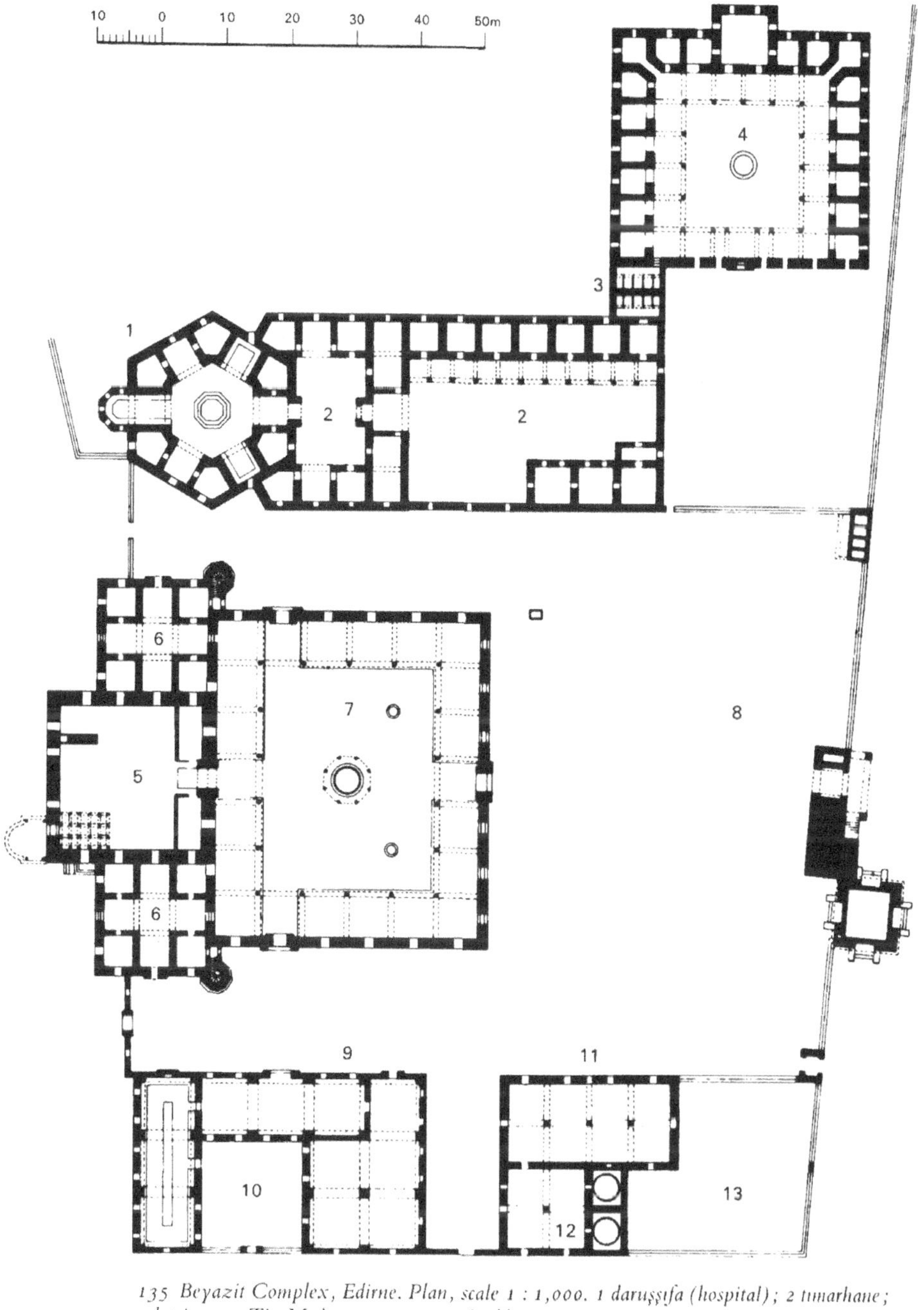

135 Beyazit Complex, Edirne. Plan, scale 1 : 1,000. 1 daruşşifa (hospital); 2 tımarhane; 3 latrines; 4 Tip Medrese; 5 mosque; 6 tabhanes; 7 mosque courtyard; 8 precinct; 9 imaret; 10 courtyard; 11 store; 12 bakery; 13 courtyard.

Figure 6.41 Latrines in Beyazit's complex, Edirne (Goodwin, 1971).

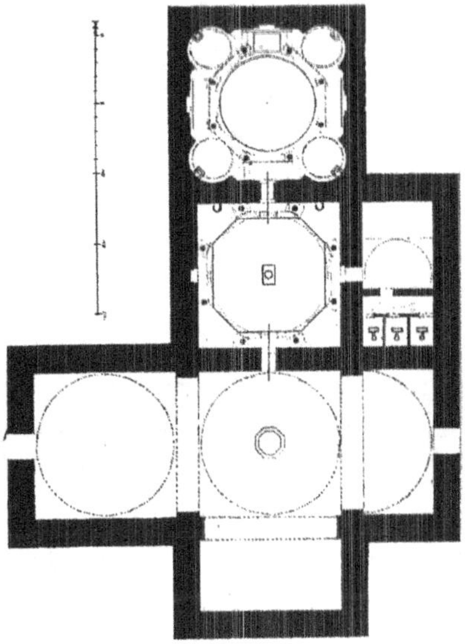

Figure 6.42 Latrines in Bursa's Eski Kaplica (Glűck, 1921).

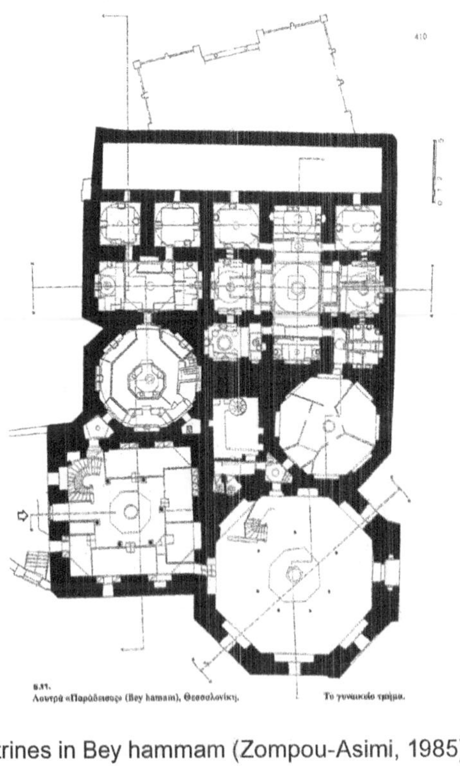

Figure 6.43 Thessaloniki. Latrines in Bey hammam (Zompou-Asimi, 1985).

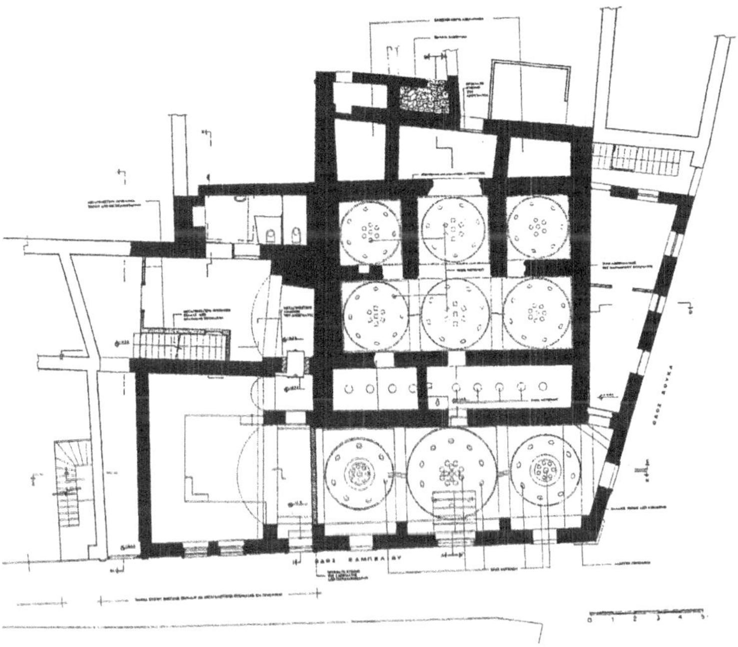

Figure 6.44 Chania, Crete. Latrine of Zampeliou and Douka hammam (Kanetaki, 2004).

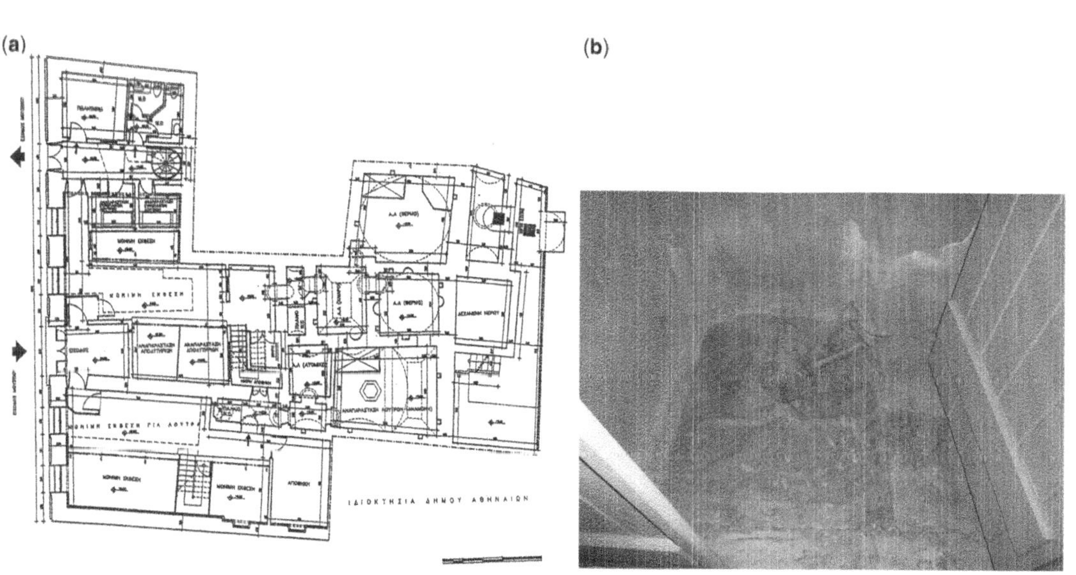

Figure 6.45 Baths and latrines: (a) Abid Efendi or Hammam of the Winds, Athens (plan: Directorate for Cultural Buildings and Restoration of Recent Monuments, Ministry of Culture, 2000), and (b) Latrine in the men's section, Abid Efendi or Hammam of the Winds, Athens, Hellas (with permission of E. Kanetaki, 2013).

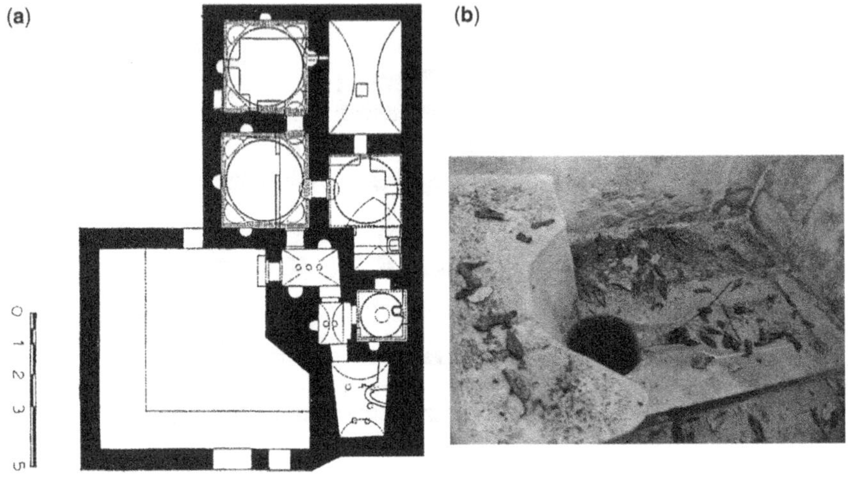

Figure 6.46 Lesvos, Mytilene. Latrine in Karavangeli hammam (with permission of E. Kanetaki, 2004).

The paving of the latrines was usually covered with marble slabs, which were also easier to clean, while a special key-hole shaped monolithic marble structure was located in each compartment. That specially curved slab was adequately detached from the wall, in order that the user's body would not come in contact with the wall, while at the same time its vertical surface showed a convex profile, in order to avoid splashing by human waste. There is a tap or a full sewer inside the cell for cleaning purposes (Figure 6.47). Usually openings set on the roof provided the necessary ventilation of the latrine. The direction of the cells in Ottoman toilets is very important, because they are oriented as much as possible towards the opposite direction of the Mecca. No flushing system has been traced, and the waste was removed from the toilet through ceramic pipes, placed under the key-hole shaped slab, connected most probably with a sewage pit (Dimitriou, 2002).

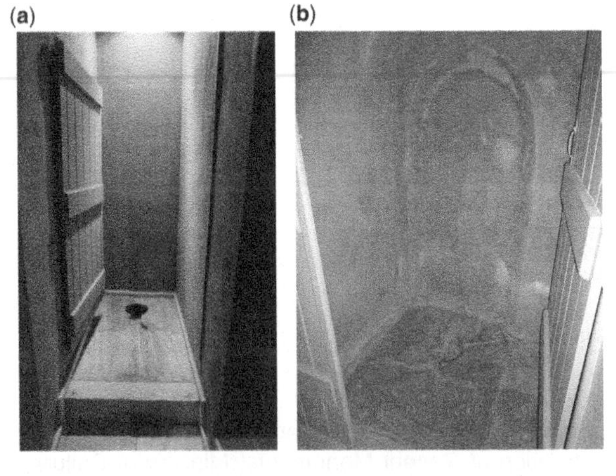

Figure 6.47 Ottomann latrines: (a) In the women's section, Abid Efendi or Hammam of the Winds, Athens and (b) in the men's section, Abid Efendi or Hammam of the Winds, Athens (with permission of E. Kanetaki).

6.5.2 The Present times (1900 AD to the present)

With the industrial revolution, new possibilities opened up in the area of community hydration and wastewater management. Ancient water supply systems relied heavily on gravity as the main force for supplying water to a city. With the development of modern pumping systems, it has been possible to deliver water from the underground aquifer and from long distances either up- or down-hill. It has also been possible to have a continuous water supply in high rise buildings, by simply securing a high enough pressure in the water supply network. Increasing water demands are now largely met through wastewater reuse and through sea water desalination.

Waste and wastewater treatment is the other very important area, in which scientific and technological advances in the last 150 years have completely changed the picture as far as waste (water) management is concerned. Here it must be mentioned that in the ancient times, solid waste was to a large extent handled along with wastewater. Thus, water was used not only as a source for drinking, washing and irrigation, but also as the main carrier for solid waste. In our days, we are collecting and managing waste separately from wastewater. In addition, wastewater is on many occasions handled separately from rainwater runoff. Separate sewer systems secure collection of more uniform quantities and characteristics.

In addition, wastewater treatment facilities have been developed, that secure an effluent of such a quality that it may be disposed or even reused, without any adverse consequences on public health and the environment. These advances have been of course applied in Hellas just as in all other areas of the world. In the last two centuries, technological achievement is commonly available to every country and thus we may no longer talk about technological achievements of 'different civilizations'. We all share a common scientific and technological civilization.

Today in Hellas, more than 95% of the population is connected to wastewater collection services. Of this latter figure, more than 80% is also connected to a wastewater treatment plant. It is also worth noting the importance of investments in infrastructure assets, with more than 26,000 km of wastewater networks and almost 350 wastewater treatment plants (Paranychianakis *et al.*, 2009). Also, the rainwater in urban areas is discharged: (a) through the combined wastewater sewerage system 10%, (b) through a separate surface drainage system 75%, and (c) on to roads and then directly to surface water or infiltration to the soil (SUDS) 15%.

Remote areas and establishments require, of course, a distributed and small-scale wastewater management system. Indeed, there has been substantial progress in developing such systems. Constructed wetlands, among others, present often a viable economic solution to effectively cleaning and possibly reusing the generated wastewaters for irrigation purposes.

6.6 FUTURE TRENDS

It is known that in the future, urbanization will worldwide continue to increase with 80–90% of the future world growth (~145,000 inh/d) anticipated in urban areas. In a recent article (Howe *et al.*, 2012), the City of the Future and the key water-related issues associated with it are discussed. Among other issues, the unavoidable interplay between energy and water, the need for an integrated approach when considering centralised versus decentralised systems, the need for novel approaches to provide water resilience to sudden changes due to climatic and other stressors and the need for a greener approach to water and wastewater management are deliberated.

In Hellas also, the urban population is estimated to be 65% (it was 42% in 1960) of the total population with an annual increase of 0.70% (Paranychianakis *et al.*, 2009). In these new urban developments, new specific sanitation technologies should be developed to regulate and manage wastewater and stormwater. Interesting late developments include the effluent reuse locally, the possibility to separate urine from the rest of the wastewater using appropriate toilet equipment, and other advanced achievements. This is important, as urine contains most of the nitrogen typically found in wastewater. Such a separation of wastewater to 'grey' and 'black' fractions facilitates its treatment and opens new reuse possibilities.

Another interesting possibility has to do centrally with the main wastewater treatment technologies that are being employed. Starting in the early 20th century and until today, the activated sludge process and its variations, represent the main wastewater treatment technologies used in the developed world. When this process was conceived, it represented great progress, as it allowed biodegradation of dissolved organic matter, along with production of sludge, purifying thus the received wastewater. The sludge is then used to produce biogas, through the anaerobic digestion process and hence energy is recovered. The digested sludge may be used as soil amendment. A new possibility that is being developed these days, is the direct production of electricity while treating the dissolved organic matter in a microbial fuel cell. This may well improve the overall economics of wastewater treatment (Krishna Lamichhane & Babcock, 2013; Tremouli *et al.*, 2013; Zhuwei Du *et al.*, 2007).

Finally, the development of proper decentralized systems technology is anticipated to become very important in the future. Septic tanks are to be replaced with efficient small-scale systems that will allow environmental protection, while at the same time producing energy, water and/or sludge for reuse. Treated water can be easily reused locally for various purposes such as toilet flushing, watering gardens or car washing or even for direct potable use (Leverenz *et al.*, 2011). Sludge from decentralized plants can be used as fertilizer even in urban landscape areas (Lyberatos *et al.*, 2011).

Also, measures and technologies for harvesting of rainwater in order to reduce the flood risk and increase water availability should be developed. Thus, much we learn from the past technologies and practices implemented (e.g., design philosophy, adaptation to the environment, and decentralization management of water and wastewater projects, architectural and operation aspects, diet habits, and sustainability as a design principle); *probing the past, forging the future.*

The following recommendations are proposed for stormwater management:

(a) Construction of small interception dams in the main torrents of the hilly region, aiming at retardation of wintertime torrential flows and increased groundwater recharge is proposed, and
(b) Construction of rainwater tanks to reduce the storm water flow and consequently the floods risk. Furthermore, rainwater tanks should be used as source of alternative water supply and a means of stormwater management. It is pointed out that the rainwater quality does not meet the drinking water guidelines on many instances and should be disinfected before drinking (Van der Sterren *et al.*, 2012).

6.7 DISCUSSION AND CONCLUSIONS

In contrast to earlier ancient civilizations (Egypt, Mesopotamia, and Indus) that flourished in water-abundant environments (large river valleys), ancient Hellenes preferred to establish their settlements in dry, water-scarce sites (Mays *et al.*, 2007). In Ancient Hellas, there exist some medium-scale rivers and lakes and there have been no major cities close to them in the Hellenic antiquity. It can be argued that in such choices, climate and health have been the main criteria: dry climates are generally more convenient to live and healthier as they protect the population from water-related diseases, since the Bronze Age (Zarkadoulas *et al.*, 2008).

It is well documented that most technological developments relevant to water supply and wastewater are not achievements of present-day engineers but date back to more than two to three thousand years ago. These developments were driven by the necessities to make efficient use of natural resources, to make civilizations more resistant to destructive natural elements, and to improve the standards of life, both at public and private level. Hellenic civilizations have been closely connected to hygienic living standards and a comfortable lifestyle. To achieve these, both technological infrastructures and management solutions were developed. In Crete, hygienic technologies were practiced as early as in the Minoan period of the island (*ca.* 3200–1100 BC) and were followed in several other cases in mainland Hellas, the Aegean islands and Cyprus. The historical development of sanitation and wastewater and stormwater management through the centuries in Hellas is shown in Table 6.1. The technological frame created comprised of:

(a) bathrooms, toilets (resembling modern day ones with flushing devices) and other sanitary facilities,
(b) urban wastewater and storm water management systems, and
(c) water supply facilities that ensure superior water quality and safety against pollution and sabotage.

The Minoan civilization is universally considered to be one of the grandest and most brilliant ones. The architectural and hydraulic function of storm water and wastewater sewerage systems were of special significance in the construction of the principal Minoan centres. Archaeological and other evidences indicate that, during the Middle Bronze Age, advanced water management and sanitary techniques were practiced in Minoan settlements. These include the construction and use of bathrooms and other sanitary facilities relevant to wastewater and storm sewer systems. The hydraulic and architectural function of sewer systems in palaces and cities are regarded as one of the salient characteristics of the Minoan civilization (Angelakis & Spyridakis, 1996a). Thus, it is not by chance that the main technical and hydraulic operations associated with catchment basins, surge chambers, manholes, urinals and toilets, laundry slabs and basins and sewerage and drainage systems, including disposal sites of the effluent, have been practiced in varying forms since *ca.* 3000 BC (Angelakis & Spyridakis, 1996b). These systems were so advanced that can be compared with the modern systems, which were established only in the second half of the 19th century in European and American cities.

There is no evidence suggesting the connection between the Minoan era and the Classical and Hellenistic periods regarding the water and wastewater systems. However, someway, hydraulic engineers' knowledge was not lost with the Minoans and was re-gained later in the Hellenistic and then the Roman period. In Classical and Hellenistic Hellas an advanced, comfortable, and hygienic lifestyle was developed, as manifested from long-term very efficient sewerage systems, bathrooms and flushing toilets, which can only be compared to the modern ones, re-established in Europe and North America a century and half ago.

The progress in the Hellenic civilizations has been closely connected to hygienic living standards and a comfortable lifestyle. To achieve these, both technological infrastructures and management solutions were developed. The importance attached to the hygienic use of water in ancient Hellas is highlighted in the case of Athens, a city established in one of the driest places of Hellas. The entire Peisistratean aqueduct (*ca.* 6th century BC), which transferred water from the Hymettos Mountain to the city center, was constructed as an underground channel. There were bathrooms, latrines and other sanitary facilities, both public and private. Finally, an extended wastewater management network connected every single building of the Athenian Agora to the so-called Great Drain. The amazing evolution and development of structures for bathing, sanitary installations can be traced from the Classical cities up to the cities of the Hellenistic period and the public and private similar facilities during the Roman period. This period, the scale of baths and toilets, as well as the sewerage and drainage systems in Roman cities were further increased. Also, the status of urban sewerage and stormwater drainage systems in Hellas during the medieval times is reviewed, based on the results of archaeological and historical studies. It is also evident that such structures and installations have survived until the end of the ancient world and have been implemented at least during the beginning of the Byzantine period.

Table 6.1 A Timeline for historical development of sanitation and wastewater management in Hellas.

Period	Achievements	Comments
ca. 6500–3300 BC	First confirmed evidence of habitation and first farmers. A first successful effort in wastewater management was the wastewater drainage.	In Crete and several Aegean islands
ca. 2100–1100 BC	Sewerage and drainage systems available in Crete Hellas. Also terracotta pipes drained to stone sewers and possibly the first 'flushed' toilet was implemented at Knossos palace (Crete, Hellas).	Minoan setlements (e.g., Knossos, Zakros, Agia Triada, and Tylissos).
ca. 1500–800 BC	The realization of the importance of pure water for people is evident already from the myths of ancient cultures. Religious cleanliness and water important in various ancient cults.	In various Mycenaean cities (in south Hellas).
ca. 800–480BC	Sanitary dark ages	In several areas
ca. 480–67 BC	Alcmaeon of Croton (floruit *ca.* 470 BC) was the first Hellene doctor to state that the quality of water may influence the health of people. Also, Hippocratic treatise *Airs, Waters, Places* (*ca.* 400 BC) deals with the different sources, qualities and health effects of water. Thus, the importance of water for the public health for first time was recognized and the first well organized baths, toilets, and sewerage and drainage systems appeared. Also, the first public fountains were implemented. However, the first known epidemic of 430–426 BC, happened in Athens, caused the death of the great statesman, Pericles, decimated the population and contributed significantly to the decline and fall of classical Hellas. The enclave held 155,000 inhabitants out of the 400,000 total population of Attica. It is characteristic that during that time sanitary installations became a necessary space even for ordinary middle class houses (e.g., Olynthos, Delos, and Dystos).	Hellas, Asia Minor, south Italy, and northern African states
ca. 67 BC–330 AD	The importance of water for the public health was widely recognized in several parts of the world. Pliny the Elder in the first century AD had in his works a long section concerning the different opinions on what kind of water is the best. Also, urban sewerage and drainage of large scale were recognized. In addition, the first baths and toilets both private and public of large scale appeared in several urban areas. The first sewerage network system in Rome connected to houses at *ca.* 100 AD.	In eastern Mediterranean, Egypt, north Africa (modern Tunisia), the Apennine peninsula (modern Italy).

	However, in public toilets facilities were common to all; they were cramped, without any privacy, and had no decent way to wash one's hands. The private toilets most likely usually lacked running water and they were commonly located near the kitchens. The rich had running water in their homes; the poor had to fetch their water from public fountains. Water-borne infections must have been among the main causes of death. Dysentery and different kinds of diarrheas must have played sever problems to the populations.	
ca. 330–1700 AD	There was no improvement in methods of waste removal in Europe for many centuries. With the advent of the Dark Ages little progress was made for 1400 years, from the late forth century. Baths, toilets and sewerage and drainage systems were further improved by Byzantines and Venetians. Medieval cities, castles and monasteries had their own wells, fountains or cisterns. Usually towns built a few modest latrines for the inhabitants, but these were mostly inadequate for the size of the population. The lack of proper sanitation increased the effects of epidemics in medieval towns in Europe. Also, in several Asian countries (e.g. China, India, and Vietnam) were implemented various types of drainage systems in the religious temples developed under several dynasties.	The transition from the ancient world is not certain.
ca. 1700–1900 AD	The great epidemics of cholera and typhoid fever occurred in England during 1830–1850 and their association with the pollution of water sources with raw wastewater made clear the need for sanitation and the protection of water resources and motivated health agencies to set sanitation rules and environmental policies to protect public health. Thus, the first large-scale projects of unintended however effluent reuse were set at the beginning of 1800s when 'sewage farms' were developed as an attempt to protect public health and to control water pollution. This practice was mainly developed from 1840 to 1890 in England, while in the 1870s the first systems were appeared in the United States, France, Germany, Hellas, and other European countries.	After the beginning of the Industrial revolution.

(*Continued*)

Table 6.1 A Timeline for historical development of sanitation and wastewater management in Hellas (*Continued*).

Period	Achievements	Comments
1900–1990 AD	Developing the basic treatment processes in developed world. The age of process development (industrial revolution). Decentralized waste management (DWM) concepts (e.g., privy vaults, cesspools, and dry sewage collection) were predominantly used in urban and rural areas up to the middle of the 19th century, mainly in the USA and in central Europe. Decentralized dry sewage systems were more common in Europe and Asia than in the USA because Europeans and Asians had more experience using human excrement as fertilizer and doing so cost effectively.	
1990–present	Towards even stringent environmental standards (Process refinement). An early attempt for centralized wastewater management in the USA was the construction of public and private combined sewers to transport the cumulative wastes from a city block or from several city blocks to a nearby water body. Afterword the DWM systems became inadequate and was gradually replaced with centralized water-carriage sewer (CWM) systems. CWM systems were rapidly expanded in European countries and the directive 91/271/EC begin to be implemented. CWM systems remain the preferred wastewater management option in newly urbanizing areas since today. These new DWM technologies were entirely different than those used before including several small mechanical (e.g., attached and suspended biomass) and natural systems (e.g., wetlands, land application, and on site) included recycling and reuse of wastewater.	From the globalization of the technological civilization of our times.

Venetians accomplishments in hydraulics are worth noting, such as the construction and operation of cisterns and drains. Many of these technologies were developed and used in the famous castles constructed during that period in several places. Water was connected to Islam, so that during the Ottoman period, there was a water tap in all mosques. Hammams, which are presently also referred to as Turkish baths, and were established in all the regions of the Ottoman Empire, played an important role in the Ottoman culture and served as places of social gathering, ritual cleansing, and as architectural structures, institutes and so on. The cleansing of the body symbolises the cleansing of the soul, according to the Koran. Modern water and wastewater technologies were derived from ancient Hellenic technological achievements, which were not totally forgotten during the Dark Ages. However, the evolution of sanitary engineering not entirely linear but rather characterized by short discontinuities and regressions.

In conclusion, through the ages, innovation has played a key role, in ensuring the progress required to meet the emerging challenges. There is a lot to be learnt from studying the technological progress that has been historically the result of the need to address water distribution and sanitation. The stormwater and wastewater hydraulic works in the Minoan, Hellenic, Roman, and Venetian civilizations are sometimes not too different from the modern ones, since present technologies descend directly from the engineering of those times. In Ancient Crete, wastewater and storm water public works, including disposal practices, are characterized by simplicity, robustness of operation, and absence of complex controls. In the future, water and wastewater management systems based on reapplication of old practices and philosophical approaches, using new equipment, in order to effectively meet the modern emerging challenges could be of great significance.

Finally, it is obvious that ideas, technologies, and practices developed during most periods of Hellenic civilization greatly influenced our today knowledge; as Will Durant (1939, p. vii–viii) put it: 'Excepting machinery, there is hardly anything secular in our culture that does not come from Hellas (Koutsoyiannis & Patrikiou, 2013). In addition, more than 2.6 billion people do not use improved sanitation thus there is a huge need for sustainable and cost-effective water supply and sanitation facilities, particularly in cities of the developing world (Bond *et al.*, 2013). Applicability of selected ancient Hellenic sanitation systems for the contemporary developing world should be seriously considered.

6.8 REFERENCES

Adrimi-Sismani B. (1990). Excavations in Magnesia, Dimini. AD 45, B1 *Chronicle*, pp. 194–197 (in Hellenic).

Alexiou S. (1964). *Minoan Civilization*. Alexiou Sons, Iraklion, Crete, Hellas (in Hellenic).

Angelakis A. N. and Spyridakis S. V. (1996a). The status of water resources in Minoan times: A preliminary study. In: Diachronic Climatic Impacts on Water Resources with Emphasis on Mediterranean Region, A. N. Angelakis and A. S. Issar (eds), Springer-Verlag, Heidelberg, Germany, pp. 161–191, Chapter 8.

Angelakis A. N. and Spyridakis S. V. (1996b). Wastewater management in Minoan times. In: Proceedings of the Meeting on Protection and Restoration of Environment, August 28–30, Chania, Hellas, pp. 549–558.

Angelakis, A. N. and Koutsoyiannis, D. (2006). Wastewater management in Minoan civilization. In: Second Conference on Ancient Technologies, October 17–21, 2005, Athens, Greece. Technical Chamber of Greece, Athens, Hellas, pp. 751–756 (in Hellenic).

Angelakis A. N. and Koutsoyiannis D. (2003). Urban water resources management in ancient Greek times. In: The Encycl. of Water Sciences, Markel Dekker Inc., B. A. Stewart and T. Howell (eds), Madison Ave. New York, USA, pp. 999–1008.

Angelakis A. N., Koutsoyiannis D. and Tchobanoglous G. (2005). Water resources technologies in the Ancient Greece. *Water Research*, **39**(1), 210–220.

Angelakis A. N., Lyrintzis A. G. and Spyridakis S. V. (2012a). A brief History of water and wastewater technologies in Minoan Crete, Greece. In: Proceedings of the 3rd IWA International Symposium on Water and Wastewater Technologies in Ancient Civilizations, March 22–25, 2012, Istanbul, Turkey, pp. 208–216.

Angelakis, A. N., Dialynas, M. G. and Despotakis, V. (2012b). Evolution of water supply technologies in Crete, Greece through the centuries. In: Evolution of Water Supply Throughout Millennia, A. N. Angelakis, L. W. Mays, D. Koutsoyiannis and N. Mamassis (eds), IWA Publishing, London, UK, pp. 227–258.

Antoniou G. P. (2007). Lavatories in Ancient Greece. *Water Science and Technology, Water Supply*, **7**(1), 155–164.

Antoniou G. P. (2010). Ancient Greek lavatories. Operation with reused water. In: Ancient Water Technologies, Mays (ed.), Springer, Dordrecht, Heidelberg, London, UK, and New York, USA, pp 67–86.

Antoniou, G. and Angelakis, A. N. (2008). Sanitary and purgatory structures in Ancient Greece. In: Fifth Symposium on Archaeometry of the Hellenic Society of Archaeology, October 8–10, 2008 in Athens, Hellas, no 19.

Antoniou G. and Angelakis A. N. (2009). Historical development bathrooms (toilets) and other sanitary and purgatory structures in Greece. In: Proceedings of the Second IWA International Symposium on Water and Wastewater Technologies in Ancient Civilizations. May 28–30, 2009, Bari, Italy.

Antoniou G., Kathijotes N., Angelakis A. N. and Spyridakis D. S. (2014). Historical Development of Technologies on Water Resources Management and Rainwater Harvesting in the Hellenic Civilizations. Inter. *Journal of Water Resources Development*, **14**(08), 1–14.

Argoud G. (1988). Amenagements hydrauliques en Grece, des Myceniens a l'epoque Hellenistique. In *l' Eau et les Hommes en Mediterranée et en mer Noire dans l' Antiquité*, De l'époque Mycénienne au règne de Justinien (Man and water in Mediterranean and Black sea in Antiquity, From the Mycenaean times till Justinian era), Actes de Congrès International, Athènes, Greece, 20–24 Mai 1988, pp. 35–55.

Athanasiou F., Malama V., Miza M. and Sarandidou M. (1999). The baths of the palace of Galerius. *The archaeological work in Macedonia and Thrace* 13, pp. 191–206 (in Hellenic).

Balcer J. (1974). The mycenaean dam at tiryns. *AJA*, **78**, 141–149.

Bond T., Roma E., Foxon K. M., Templeton M. R. and Buckley C. A. (2013). Ancient water and sanitation systems – applicability for the contemporary urban developing world. *Water Science and Technology*, **67**(5), 935–941.

Brea L. B. (1964). *Poliochni: città preistorica nell' isola di Lemnos* (Poliochni: Prehistoric city in the island of Lemnos), I, Roma, Italy (in Italian).

Cahill J., Reinhard K., Tarler D. and Warnock P. (1991). It had to happen: scientists examine remains of ancient bathroom. *Biblical Archaeological Review*, **17**(3), 64–69.

Castleden R. (1993). *Minoans: Life in Bronze Age Crete*. Routledge, 11 New Fetter Lane, London, UK.

Chamonard J. (1924). Le Quartier du Theatre. Contient , Vol. 1: Etude sur l'habitation délienne à l'époque hellénistique, Vol. 2: Construction et technique, Vol. 3: Planches. Paris, France: 1922–1924 (The Quarter of the Theatre: Vol. 1: Study on the Delian home in the Hellenistic period, Vol. 2: Construction and technique, Vol. 3: Boards. Paris, France: 1922–1924). Collection Exploration archéologique de Délos, 8, 1–2–3.

Chrisostomou P. (1996). Palace of Pella. *The archaeological work in Macedonia and Thrace*, **10a**, 105–142 (in Hellenic).

Crouch D. C. (1996). Avoiding water shortages: some Ancient Greek solutions. In: Diachronic Climatic Impacts on Water Resources with Emphasis on Mediterranean Region, A. N. Angelakis and A. S. Issar (eds), Springer-Verlag, Heidelberg, Germany, pp. 129–159, Chapter 7.

De Feo, G., Mays, L. W. and Angelakis, A. N. (2011). Water and wastewater management technologies in Ancient Greek and Roman civilizations. In: Treatise on Water Science, P. Wilderer (ed.), Academic Press, Oxford, UK, Vol. 1, pp. 3–22.

De Feo, G. Antoniou, G. P., Mays, L., Dragoni, W., Fardin, H. F., El-Gohary, F., Laureano, P., Kanetaki, E. I., Zheng, X. Y. and Angelakis, A. N. (2012). Historical development of waste water management. In: Handbook of Engineering Hydrology, S. Eslamian (ed.), Taylor & Francis Group, 6000 Broken Sound Parkway NW, Boca Raton, Florida 33487, USA, Vol. 2, Chapter 67: (in press).

Dimitriou O. (2002). Latrines and sanitation facilities in Greek houses of ancient, byzantine and modern times, Doctoral Dissertation, Dept. of History-Archaeology, Aristotle University, Thessaloniki, Hellas.

Doumas Ch. (1973–1974). Antiquities and monuments of Dodecanese, Rhodes. AD 29, B3 *Chroniques*, **29**, 946–965.

Doxiadis C. (1964). The ancient Greek city and the city of the present. *Ekistics*, **18**(108), 346–364.

Durant, W. (1939). The Life of Greece (The Story of Civilization, Part II). Simon & Schuster, New York, USA.

Evans S. A. (1964). The Palace of Minos at Knossos: A Comparative Account of the Successive Stages of the Early Cretan Civilization as Illustrated by the Discoveries. Vol. I, The Neolithic and Early and Middle Minoan Ages. Biblo and Tannen, New York, USA.

Evans, S.A. (1921–1935). The Palace of Minos at Knossos: A Comparative Account of the Successive Stages of the Early Cretan Civilization as Illustrated by the Discoveries. Vols. I–IV, Macmillan and Co., London (reprinted by Biblo and Tannen, New York, USA, 1964).

Filimonos M. (2000). Municipal waste in agricultural production. The testimony of ancient Rhodes. In: Archaeology and Environment in the Dodecanese. Research and Cultural Tourism. International Conference, Department of Mediterranean Studies, Laboratory of Archaeometry, I. Liritzis and A. Sampson (eds), Rhodes, Hellas (in Hellenic).

Galanaki M. (2014). Waterborne diseases in the Hippocratic treatise: Airs, Waters, Places. In: Evolution of Sanitation and Wastewater Management through the Centuries, A. N. Angelakis and J. B. Rose (eds), IWA Publishing, London, UK (this book).

Gaubert M. (2009). Les Bains et les Moines in *Le bain collectif en Egypte* (Baths and monks in the public baths in Egypt). In: IFAO, M. Boussac *et al.* (eds), Caire, Egypty, pp. 297–303, Figs. 1–2.

Glück H. (1921). Probleme des Wőlbungsbaues I: Die Bäder Konstantinopels, Arbeiten des Kunsthistorischen Instituts der Universität Wien, Band XII, Wien, Austria.

Goodwin G. (1971). A History of Ottoman Architecture, Thames and Hudson Ltd., London, UK (reprinted 2003).

Gounaris G. (1990). The balnaeum and structures north of Octagon of Philippoi, Athens, Hellas (in Hellenic).

Gounaris G. (2004). Philippoi. University Studio Press, Thessaloniki, Hellas, p. 62.

Graham J. W. (1987). The Palaces of Crete. Princeton University Press, Revised Edition. Princeton, New Jersey, USA, pp. 108–110.

Grandjean Y. (1994). *L' eau dans la ville de Thassos.* L' eau, la santé et la maladie dans le monde grec. *Bulletin de Correspondance Hellénique* Supplément, **28**, 283–296 (in French).

Gray H. F. (1940). Sewerage in ancient and medieval times. *Sewage Works J.,* **12**(5), 939–946.

Green J. P. (1992). Medieval Monasteries, Archaeology of medieval Britain, Leicester University, Leicester, New York, USA.

Hellmann M.-Chr. (1988). Le vocabilaire de l' eau dans les inscriptions de Delos. In l' Eau et les Hommes en Mediterranée et en mer Noire dans l' Antiquité, De l' époque Mycénienne au règne de Justinien, Actes de Congrès International, Athènes, Hellas, Mai 20–24 1988, pp. 181–196.

Hansen R. D. (1983). Water and waste water in imperial Rome. *Water Resources Bulletin,* **19**, 263–269 and www.waterhistory.org.

Henein N. and Wuttman M. (2000). Kellia II. L' ermitage copte OR 195. I Archaeologie et architecture, FIFAO 41 (in French).

Howe C., Skinner R. and Ewert J. (2012). Implementing the city of the future: tackling the key issues. *Water,* **21**, 12.

Huxley G. L. (1961). Crete and the Luwians. Oxford University Press, Oxford, UK.

Hoepfner, W. (ed.) (1999). Geschichte des Wohnens. VoL. I. Stuttgart: Deutsche Verlags-Anstalt, Germany. Thessaloniki, Hellas, 2005 (in Hellenic).

History of the Greek Nation (1970). Publisher Ekdotiki Athinon, Athens, Hellas (in Hellenic).

Kaiafa A. (2008). Urban water supply and drainage systems in Hellenisic and Roman Macedonia. Doctoral dissertation, Dept. of Architecture, Aristotle University, Thessaloniki, Hellas (in Hellenic).

Kanetaki E. (2004). Othomanika loutra ston elladiko xwro (Ottoman baths in the Greek territory), Technical Chamber of Greece, Athens, Hellas.

Kanetaki E. (2011). Bathhouses in the former ottoman province of the Greek Lands: a contribution to the study of their history and architecture. In: Acts of Conference Bathing Culture, N. Ergin (ed.), 2007 Istanbul, Ancient Near Eastern Supplement Series 37, Peeters, Leuven, Belgium, pp. 221–255.

Kanetaki E. (2012a) The use of water in historic balnear buildings of Greece. In: Acts of the 3rd IWA Specialized Conference on Water and Wastewater Technologies in Ancient Civilizations, (IWA WWTAC 2012). Istanbul, Turkey, pp. 279–287.

Kanetaki E. (2012b) Architectural and Technical Aspects Regarding the Construction of Hammams in Ottoman Greece. In: Acts of the 2nd International Conference Balneorient, Thermae, Hammam, Th. Fournet and M.-Fr. Boussac (eds), Damascus, Syria.

Karadedos G. (1986). Water supply and Sanitation in ancient Dion. Archaeologists talk about Pieria, pp. 33–41 (in Hellenic).

Karadedos G. (1990). Hydraulic expertise and technology of the Ancestors. *Scientific Annals of the Faculty of Architecture*, **XII**, 143–172 (in Hellenic).

Karadedos G. (2000). Hydraulic expertise and technology – The case study of Dion. Water relations, Water as a source of life in antiquity, pp. 51–68 (in Hellenic).

Karydas N. (1999). Excavations in Street Glafkos. AEMEΘ 13, Thessaloniki, Hellas, pp. 370–374 (in Hellenic).

Katsaros A. (2002). The tower of the transfiguration of the vatopedi monastery. In: The Towers of Mount Athos, KEDAK, Centre for the Preservation of the Athonite Heritage, Thessaloniki, Hellas, pp. 45–49.

Kline A. S. (2010). Suetonius: The Lives of the Twelve Caesars, VIII, Vespasian XXIII, http://www.poetryintranslation.com/PITBR/Latin/Suethome.htm.

Knauss J. (2003). Arcadian and boiotian orchomenos, centres of mycenaean hydraulic engineering. *Irrigation and Drainage Systems,* **5**, 363–381.

Koukoules F. (1952). Byzantine life and culture (*Βυζαντινών Βίος* και *Πολιτισμός),* Vol. 4, CIFA, 73, 76 Athens, Hellas, pp. 309–311, 435 (in Hellenic).

Koutsoyiannis D. and Angelakis A. N. (2003). Hydrologic and hydraulic sciences and technologies in Ancient Greek times. In: The Encycl. of Water Sciences, Markel Dekker Inc., B. A. Stewart and T. Howell (eds), Madison Ave, New York, USA, pp. 415–418.

Koutsoyiannis D. and Patrikiou, A. (2013). Water control in Ancient Greek cities. In: Water and Urbanization, T. Ostigard, (ed). pp. 130–148.

Krishna M., Lamichhane K. M. and Babcock R. W. (2013). Survey of attitudes and perceptions of urine-diverting toilets and human waste recycling in Hawaii. *Science of the Total Environment*, **443**, 749–756.

Lazaridi K. (1989). The excavation of the Gymnasium of ancient Amphipolis. PAE, pp. 205–215 (in Hellenic).

Lazaridi K. (1990). The gymnasium of amphipolis, city and country in ancient macedonia and thrace. In: Memory of Lazaridis D., Proceedings of Archaeological Conference. Kavala May 9–11 1986, pp. 241–273.

Lemerle P. (1937). Palestre a Philippes. *Bulletin de Correspondance Hellénique* (BCH) 61, pl IX, XI, De Bocard Paris, France (in French).

Leverenz H. L., Tchobanoglous, G. and Asano, A. (2011). Direct potable reuse: a future imperative. *Journal of Water Reuse and Desalination,* **1.01**, 2–10.Water Reuse and Desalination.

Lyberatos G., Sklyvaniotis M., and Angelakis A. N. (2011). Sewage bio-solids management in EU countries: Challenges and prospective. *Fresenius Environmental Bulletin,* **20**(9A), 2489–2495.

Lyrintzis A. and Angelakis, A. N. (2006). Is the 'Labyrinth' a Water Catchment Technology? A Preliminary Approach. In: IWA Specialty Conference: *1st International Symposium on Water and Wastewater Technologies in Ancient Civilizations*, A. N. Angelakis and D. Koutsoyannis (eds), Iraklion, Hellas, pp. 163–174.

MacDonald C. F. and Driessen J. M. (1988). *The Drainage System of the Domestic Quarter in the Palace at Knossos.* British School of Athens, Hellas, 83, 235–358.

MacDonald C. F. and Driessen J. M. (1990). The Storm Drains of the East Wing at Knossos. In: L' habitat égéen préhistorique, P. Darcque and R. Treuil (eds), [Bulletin de Correspondance Hellénique, Supplément 19], pp. 141–146.

Mariolakos I. (2010). www.ekke.gr/estia/Cooper/Mariolakos/New_Mariolakos_greek.pdf (last accessed, October 2012).

Mays L. W., Koutsoyiannis D., Angelakis, A. N. (2007). A brief history of urban water supply in antiquity. *Water Science & Technology, Water Supply*, **7**(1), 1–12.

Monneret D. and Villard H. (1926). *Annales du Service de Antiques de l' Egypte* XXVI pl. 2 (in French).

Mosso A. (1907). Escursioni nel Mediterraneo e gli Scavi di Creta, Fratelli Treves, Editori, Milano, Italy, p. 290 (in Italian).

Moretti J. C. and Fincker M. (2011). Les Reseaux d' Eau Courante a Delos in Les Reseaux d' Eau Courante Dans L' Antiquite, Pr Univ. de Rennes.

Muller-Wiener W. (1966). *Abu Mena 4. Vorlaufiger Bericht,* MDAIK 21: pp. 171–187, fig. 3.

Myriantheos M. (1987). The Bastion at the SE side of the Sinai monastery wall, In: Proceeding of 7th Symposium of Byzantine Archaeology and Art, Athens, Hellas, XAE p. 55 (in Hellenic).

Őnge Y. (1990). Toilets in Turkish Ottoman architecture, in: Acts of the Seventh International Congress of Turkish Art, (T. Majda Ed.), Polish Scientific Publishers, Warsaw, pp. 185–190.

Orlandos A. (1927). *Monasterial Architecture* (Μοναστηριακή Αρχιτεκτονική), 1st Ed. Estia, Athens, Hellas.

Orlandos A. (1937). Palaces and Houses of Mystras. Archive of Byzantine Monuments in Greece, Vol. C, No. 1, Athens, Publ. Estia.

Orlandos, A. (1940). The Role of the Roman Building Located northern of Horologe of Andronikos Kiristos. *Proceeding of Athens Academy*, Athens, Hellas, pp. 251–260 (in Hellenic).

Orlandos A. (1958). *Monasterial Architecture* (Μοναστηριακή Αρχιτεκτονική), 2nd Ed. BAE, Athens, Hellas.

Panagiotakis N. M. (1987). Crete: History and Civilization. Vikelea Library. Association of Municipalities and Communities of Crete, Iraklion, Hellas. Vol. 1, p. 416 (in Hellenic).

Palyvou K. (1997). Sewage and sanitary installations in the Aegean during the 2nd Millennium BC. Ancient Greek Technology, 1st National Conference of Ancient Technology, pp. 381–389 (in Hellenic).

Palyvou K. (2005). Akrotiri Thera: An architecture of affluence 3500 years old, pp. 41–42, 51–53. INSTAP Philadelphia, USA.

Pantermalis D. (1989). The great thermae. *Archaeology* **33**, 29–34.

Pantermalis D. (1998). Dion 'Εκατόμβες και σωτήρια', *The archaeological work in Macedonia and Thrace,* **12**, 1998, 291–298 (in Hellenic).

Papangelos I. (2005). Zygos monastery in Mountain Athos (Αθωνική Μονή Ζυγού). Publisher J.A. Papangelos. Thessaloniki, Hellas, ISBN 9606306119, p. 53, Figs 9 (in Hellenic).

Paranychianakis N. V., Kotselidou O., Vardakou E. and Angelakis A. N. (2009). Greek Regulations on Wastewater Reclamation and Reuse. Hellenic Union of Municipal Enterprises for Water Supply and Sewage. Larissa, Hellas, pp. 160 (in Hellenic).

Pingiatoglou S. (2003). The sanctuary of Demeter in Dion. *The archaeological work in Macedonia and Thrace*, 425–434 (in Hellenic).

Platon N. (1974). *Zakros, The new Minoan Palace*. The Athens Archaeological Society, Athens, Hellas (in Hellenic).

Platon M. (1990). New indications for the problems of purgatory cisterns and bathrooms in Minoan World. In: Proceedings of the 6th International Cretologic Congress. Literary Association Chrysostomos, Chania, Hellas, A2, 141–155 (in Hellenic).

Robinson M. D. (1938). Excavations at Olynthus, Part VIII, The Hellenic house, a study of the houses found at Olynthus with a detailed account of those excavated in 1931–1934. Baltimore, Johns Hopkins press, Maryland, USA.

Schladweiler J. C. (2002). Tracking down the roots of our sanitary sewers. Historian: Arizona Water & Pollution Control Association (Member Association – WEF, Section – AWWA).

Schatzman P. (1932). Kos. Berlin, Germany. A.I.D.R. pp. 68–69, Tfl 34.

Scherrer P. (ed.). (2000). Ephesus, The New Guide. Ö.A.I, Vienna, p. 168.

Stefanidou-Tiveriou Th. (1998). Fortification, excavation of ancient Dion, 1, Thessaloniki, Hellas.

Szymanska H. and Babraj K. (2009). Le bains de Marea in: *Le bain collectif en Egypte*. M. Boussac *et al.* (eds), IFAO, Caire, Egypt (in French).

Theocharidis P. (1991). Examples of residence in Mount Athos Forts, castles and monasteries cells of the early Ottoman period. The Monastery of Simonos Petra. In: G. Lavas and D. Philippidis (eds), Greek Traditional Architecture Macedonia B, Thrace, Vol. 8, Publ. Melissa, Athens, Hellas, pp. 271–276.

Thompson H. and Wycherlay R. (1972). The Agora of Athens XIV, ASCSA Princeton.

Travlos I. (1982). Pictorial Dictionary of Ancient Athens Hacker New York, USA.

Tremouli A., Antonopoulou, G., Bebelis, S. and Lyberatos, G. (2013). Operation and characterization of a microbial fuel cell fed with pretreated cheese whey at different organic loads. *Bioresource Technology*, **131**, 380–389.

Trümper M. (2012). Gender and Space, 'Public' and 'Private'. In: A Companion to Women in the Ancient World, S. L. James and S. Dillon (eds). Wiley-Blackwell, Oxford, UK.

'Wudu', The Encyclopaedia of Islam (2001). In: P. J. Bearman, Th. Blanquis, C. E. Bosworth, E. van Donzel and W. P. Heinrichs (eds.), pp. 218–219.

Van der Sterren M., Rahman A., and Dennis G. R. (2012). Rainwater harvesting systems in Australia. In: Ecological water quality, K. Voudouris and D. Voutsa (eds), Published by Intech, pp. 471–496, Chapter 25.

Velenis G. (1978). Wohnviertel und Wohnungsbau in den Byzantinischen Städten, Wohnungsbau im Altertum, Discussionen zur Archäologischen Bauforschung 3, Berlin, Germany, 1978, 227–236.

Yavuz A. (2011). The baths of anatolian seljuk caravansarais. In: Bathing Culture of Anatolian Civilizations, N. Ergin (ed.). Ancient Near Eastern Studies Supl. 37, Peeters, Leuven, 77–141, Figs. 47, 59, 62, 65, 66.

Wright N. G. (2004). *Discovering Abbeys and Priories*, Shire Publ. Buckinghamshire UK, pp. 110–112.

Zarkadoulas N., Koutsoyiannis D., Mamassis N. and Papalexiou S. M. (2008). Climate, water and health in ancient Greece, European Geosciences Union General Assembly 2008, Geophysical Research Abstracts, Vol. 10, European Geosciences Union, Vienna, Austria.

Zompou-Asimi A. (1985). The Bey Hamam (Loutra Paradeisos) of Thessaloniki. In: Thessaloniki, Exhibition catalogue for the city's 2300 years, Thessaloniki, Hellas, pp. 341–361.

Du Z. Li H. and Gu T. (2007). A state of the art review on microbial fuel cells: A promising technology for wastewater treatment and bioenergy. *Biotechnology Advances*, **25**(5), 464–482.

Chapter 7

Evolution of sanitation and wastewater technologies in Iran through the centuries: Past and present

Saeid Neirizi, Seyed Ali, Seyed Navid Mahmoudnian, and Jalal Jooshesh

7.1 INTRODUCTION

By development of urban settlement and consequently growing the water supply network, human-being have considered the discharge of household sewage as an essential action to maintain their living environments clean and pure. In Iran, as the civilizations living in this region have increased, numerous witnesses have found, showing of existence of systems for collection and discharge of wastewater. The *Shahre Sookhte* (Burnt City, Figure 7.1) in the province of Sistan and Baluchistan is among the ancient civilizations in ancient Iran developed along the banks of river Helmand at a distance of 65 km from the city of Zabol. The city had regular and organized systems, and the ruins point to the fact that during the third millennium BC there was an organized system of water supply and wastewater collection and discharge. At the initial stages of archeological excavation of the city, regularly arrayed streets and houses as well as terracotta water and wastewater pipes were discovered underlining the existence of urban planning in this city (Ministry of Energy, 2012).

Among other proofs of sanitation in the Iranian civilizations is the Persepolis, where earthenware pipes were used for wastewater disposal systems. The specimens available at Persepolis demonstrate that this method was applied practically unchanged from 2500 years ago until its fold-up about a century ago. For instance the officers' quarters were equipped with 6 earthenware pipes. One end of each pipe was smaller than the other. Therefore the pipes could be connected to each other by fitting the smaller end of one pipe into the larger side of the next. Furthermore the pipes were protected by brick frames laid around their openings.

The prevailing methods used until recently, in most parts of Iran can be classified into two major groups, which were determined by soil conditions and groundwater tables. In cities where the groundwater was deep and the soil had a relatively high permeability, wastewater absorbing wells provided the principal solution. This traditional method is still applied at many urban centers of Iran. However, in cities, which had higher groundwater tables or low-permeable soils, localized and short sewer line conveys domestic wastewaters away from the city.

Figure 7.1 A view of ruins of Shahr-e-Sookhteh (Burnt city) (with permission of the Cultural Heritage Organization).

Except for Esfahan and a few areas in Tehran, where modern wastewater collection and discharge systems were created and operated since about sixty years ago, the construction of modern wastewater systems in other urban areas of Iran has been put on the agenda over the last 25 to 30 years, and currently the execution of such systems continues in many cities.

On sanitation, it is a known fact that Iranian had public baths over 3000 years ago. In addition to the purpose of cleaning the body, washing and bathing are highly instructed in Islam (Ministry of Energy, 2012).

The authors of this chapter have tried to introduce the traditional technologies used in the ancient of Iran for the collection and disposal of wastewater as well as ensuring sufficient sanitation. In conclusion, it will underline the great advances and achievements in Iranian water and wastewater management during recent years.

7.2 TRADITIONAL WASTEWATER DISPOSAL SYSTEMS IN URBAN AND RURAL REGIONS OF IRAN

Since ancient time absorbing wells have been the main means of wastewater disposal in the different geographical regions of Iran, however their types and applications vary from region to region depending on different reasons. Overally, wastewater discharge in Iran can geographically be categorized as follows:

7.2.1 Coastal regions near Persian Gulf and Caspian Sea

In these regions according to the following characteristics: (a) high groundwater tables, (b) silty soil with low permeability, and (c) high bedrock at some locations, to discharge wastewater, the land was drained in such a manner that the raw wastewater was transferred directly to water courses or rivers. This method is still practiced in most cities located on the banks of the rivers or on the seacoasts such as Ahwaz, Bushehr and a number of northern cities. In regions where either a proper drainage was not possible or the city's slope did not allow such a system, septic absorbing wells were dug in the houses. These wells are shallow with a large diameter and they are used to dispose of different wastewaters. Given the region's conditions, the intervals between the discharges of these wells are generally short.

7.2.2 Central regions

This area, which covers the largest part of Iran, shows a great diversity in soil texture. Except for zones in the proximity of rivers in which the groundwater tables are low. Absorbing wells have always been the predominant method of wastewater discharge in the central regions of Iran. The wells in these regions are relatively deep and small in diameter, and the intervals between their discharges vary with the soil texture.

7.2.3 Mountains

These include the foothill regions of Alborz and Zagros Chain Mountains where the soil texture is mainly rocky with low permeability. For this reason and given to the high costs involved, digging absorbing wells in these regions is not technically and economically recommended. Therefore the only reason for absorbing wells is the temporary collection and storage of the generated wastewater (Ministry of Energy, 2013).

7.3 ABSORBING WELL (SEPTIC AND LEACHING CESSPOOL OR CESSPIT)

Absorbing wells that are dug up in most regions of Iran are quite similar in structure. Despite the enhanced technical capacity and the improved trend of their growth and development, there have been no major changes in the structure of absorbing well and the alterations have mainly been concentrated on increasing their depths and diameters. Generally a diameter of about 1 m to 1.5 m is considered for the well's opening and its walls are commonly bare to maintain the permeability of the surrounding soil and only the initial one meter of the well collar is strengthened with bricks (National Company of Water and Wastewater, 2013). The depth of the well depends on the following parameters:

(a) Soil type: the harder the soil texture the shallower the well will be.
(b) Permeability: as the soil's permeability increases, the need to dig up a well of greater depth decreases.
(c) Groundwater tables: in areas with high groundwater tables, digging up deeper wells leads to great expenses, therefore, given the quick filling up of the well due to saturated soil texture, a shallow depth is considered for it. And
(d) Population coverage: generally in complexes composed of a number of residential units, wells of greater are excavated in addition to numerous wells to be able to increase the interval between their discharges.

With the growth of urbanization and the increase in the number of floors in residential units and consequently the increase in the number of population, a larger space in form of a septic chamber is created at the bottom of the well to increase its service life and to lengthen the discharge interval. Through this action, in addition to increasing the effective absorption surface, an appropriate space is created for storing sediments and easier discharge. This also prevents to some extent sedimentation on the well wall and increases its service life. Despite this, depending on the locale's soil type and permeability, water consumption habits and the pollution load of the generated wastewater, the well gradually loses its permeability and therefore the frequency of its discharge varies between once every two to ten years (Ministry of Energy, 2013).

After discharge the sludge resulting from absorbing wells (sludge) is transported by tankers and disposed of by different means depending on the site's conditions. The overall structure of absorbing well is shown in Figure 7.2.

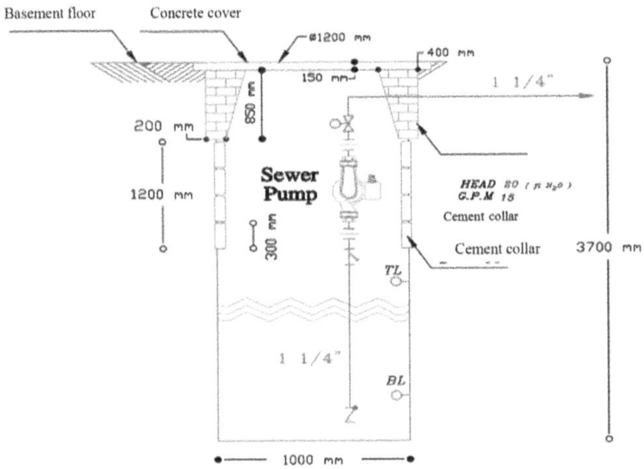

Figure 7.2 Details of absorbing well and sewer pump (with permission of the National Water and Wastewater Engineering Company).

7.3.1 The location of absorbing well within the building

Houses in Iran are generally divided in a number of categories depending on their age. Although there might be exceptions to the common rules, but most houses fit in one of the following categories:

(a) Old houses, which are generally single storied with or without a basement and possessing a yard. Most of these houses, which are over 60 years old, are devoid of bathrooms, and their toilets and kitchens are built separately in a corner of the yard. It should be mentioned that buildings with the mentioned conditions, which would still have occupants are quite rare.
(b) Single unit houses with or without basement with a life of less than 60 years, where the bathroom, kitchen and toilet are constructed inside the building. The number of these buildings in metropolis and large cities is currently very limited, but in smaller urban centers, the construction and use of these residential units is still popular.
(c) Flats and apartments, and
(d) Tall residential units and towers.

Before the creation of modern urban wastewater collection and treatment system, in almost all cases the use of absorbing wells to discharge domestic wastewater and creating a system for the limited treatment by soil was a common and normal practice. In this situation, in all buildings regardless of their age and in compliance with the religious beliefs related to segregation of grey from brown waters, more than one well would be dug up.

Based on the above and taking into account the economic and operational conditions that the shorter the length of sewer transmission line, the lower the costs and the risk of pipe blockage would be happened during the operation and consequently lessening the dissatisfaction of residents; effort would be made to construct the absorbing well at the center of wastewater sources (toilet, dish washing pipe, bathroom and rainwater outlet). According to this principle and given the mentioned categories, the location of absorbing wells can be determined as follows:

(a) Moreover than excavating a well inside the house in old buildings, a number of wells would also be built in proportion to collect and conduct wastewater from bathroom, toilet, kitchen and rain.
(b) Further than the well beneath the apartment buildings, a number of wells are excavated in the yard for collecting the rainwater and to wash the yard. And
(c) As absorbing wells cannot satisfy the needs of tall buildings and residential towers, septic tanks are built for sanitary wastewater and a few wells in the yard to collect the rainwater and to clean up the yard.

A plan for a relatively modern building, which has become quite common in recent years following by the location of basement sewer and the rainwater drainage system, is shown in Figure 7.3 (Ministry of Energy, 2013).

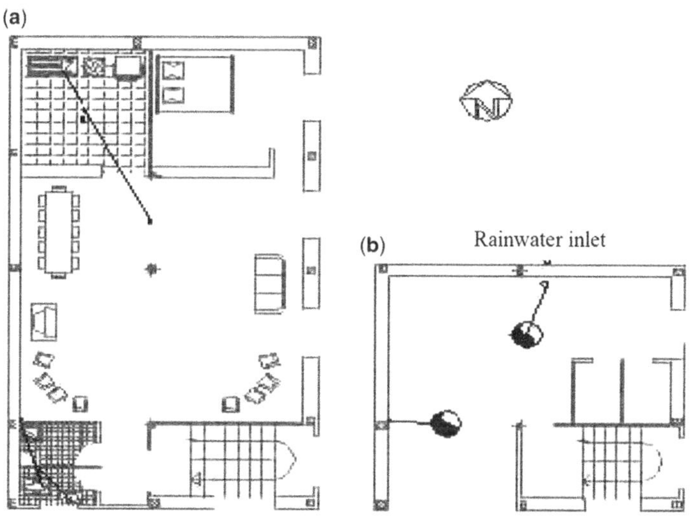

Figure 7.3 A plan for recent year's modern multi-storied building: (a) Ground floor sewer piping sitting plan and (b) Basement sewer and rainwater drain sitting plan (Scale: 1:100) (with permission of NWWEC).

7.3.2 Discharge of absorbing wells

Since ancient times emptying absorption wells was a tedious and unpleasant chore and cause of worry for many Iranians in different parts of the country. The task also required a great number of labor force.

The different methods used to empty the wells depended on the dimension and depth of the well itself and the social and natural conditions of the region. These included:

(a) Emptying the well using bucket, rope and wheel was one of the most primitive methods. The sludge would then be transported to the closest site by hand or by horse/mule drawn cart.

(b) Annual discharge by pumps and transfer by tanker.
(c) Discharge by well-emptying tankers equipped with centrifuge or screw pumps. And
(d) Currently the last method is practiced in almost all urban and rural areas of the country to empty absorbing wells.

The disposal of the sludge from absorbing wells is also realized by one of the following methods:

(a) Application as fertilizer in agriculture: In most urban areas of Iran, sludge would be transported to the agricultural lands around the towns and cities and used as fertilizer. Regarding to the health hazards of this method and the increase in public awareness, this application was gradually abandoned, and now only a small amount of sludge is sold to farmers for agriculture. The application of even this small amount should also be prevented through administrative control and applying the legislative tools.
(b) Discharge in the environment: Discharge in the environment includes methods such as discharge in the land, into the rivers or the seas. The discharge of sludge in soil is the commonest of the three and consists of spreading the sludge on the soil. In this method, tankers discharge their loads illegally in landfills around cities by spreading the sludge on the soil. In the void of laws and regulations on sludge management that would alsospecify the responsible entity, except for some rare cases, no specific action has been taken to date to address the problem in a systematic way. An example of environmentally hazardous way of discharging untreated sludge into a river located around Sari is shown in Figure 7.4.

Figure 7.4 Discharging of sludge in a river around Sari in the year 2009 (with permission of NWWEC).

(c) Discharge to urban wastewater system. In cities where the modern wastewater collection and treatment system has not been implemented, a portion of the sludge from absorbing wells from uncovered areas is generally discharged in a wastewater collection manhole either illegally or legally under the information of the waste water treatment plant's officials at the inlet of the plant to add to influent. Since joint treatment of sludge in urban wastewater plants required special measures, that had not been taken before. In cases t the sludge from absorbing wells enters the plant's influent without required pre-treatment, serious problems might occur in the operation of the plant. This happens in Kanvis, Mashhad and Bojnoord WWTPs. Following operational problems in the mentioned plants, in addition to strict legislation to stop sludge entry in the plants, the importance of a separate treatment was underlined and currently the creation of such facilities in these cities of Bojnoord, Sari and Babol figures high on the agenda (Ministry of Energy, 2012).

(d) Producing biogas from sludge: Currently the plan for producing biogas from sludge is being implemented as pilot and on a laboratory scale. Concerning the ongoing actions, this method will in near future become one important solution for sludge disposal.

(e) Sanitary disposal at waste landfill areas. Due to high humidity of sludge and risk associated with increased leachate at landfills, this method is rarely used. The Environment Protection Agency of United States (EPA) has also declared the discharge at landfills as a non-economical option. In Iran also there are no documented reports proving the use of this practice (Ministry of Energy, 2012).

7.3.3 The role of absorbing wells in treating domestic wastewater

People discharge their wastewater in absorbing wells with the aim of disposing and not treating it. However, given the regional soil texture and the depth of the well, what occurs in nature is one of the two conditions:

(a) As a septic tank the well can up to certain point and partially assist in wastewater treatment. And

(b) Soil layers around the well's walls and bottom act as a filter and in a weak form as an anaerobic unit with attached growth media, but with time and due to b locking up of the orifices, the infiltration of wastewater and nutrients as well as declining the growth and activity of microorganisms, reducing the efficiency of the system.

It should be noted that due to the current discouraged state of absorbing wells and the increased population density in cities, and consequently the filling up of wells in a relatively short period of time, one cannot consider this method as an appropriate solution in current situation (Hasanogli *et al.*, 2002).

7.3.4 Problems and consequences of absorbing wells

Absorbing wells have been used for a long time as the wastewater disposal method in Iran and in many other countries including even the United States. Undoubtedly, reliance on these wells as the sole means of disposal can entail many undesirable health, economic, social, environmental and other impacts, some of which could be potentially irreversible. The most important consequences are represented below:

7.3.4.1 Impacts on groundwater basins

Major impacts on groundwater are given as below:

(a) Groundwater resources, the main source of urban potable water, can become polluted through infiltration of wastewater from absorbing well. In regions where the groundwater tables are close to the surface and where the soil is more permeable, the infiltration and transfer of pollutions to groundwater resource will be quicker.

(b) In general, the discharge of wastewater by absorbing wells causes a rise of groundwater table. Therefore in regions where the groundwater tables are high, the risk of water infiltration in the buildings foundations and their collapse is a serious concern. The unpleasant incidents of such events occurred during recent years in some areas in south of Tehran which still haunt the minds (Ministry of Energy, 2012).

7.3.4.2 Impact on soil surrounding the absorbing wells

The major impacts are the following:

(a) Reduction of the load bearing factor of the soil beneath the foundation: In buildings the loads on the foundations are calculated in a concentrated form and then their dimension and type are determined on the basis of soil's resistance. Based on Bosinsk's studies, the closer the studied point approaches the foundation, the more the tension transferred to the point will be. If the point in soil, to which considerable tension is transferred, has a lower resistance than designed, the risk of failure is higher. It is said that the tension's bubble beneath the foundation's layer has hit the low resistance layer. This phenomenon in the ditches created in the soil to discharge wastewater, reaches its extreme, for example, the tension's bubble hits the limits of soil, which have essentially no resistance at all. Therefore, excavation of absorbing well inside the building can reduce the load bearing factor of the soil beneath the foundation. As the distance between the well and the foundation shortens, the risks of negative impacts increase. As the building ages, the negative effect of the absorbing well increases, because a larger portion of the tension bubble hits the no-resistance limits. The most undesirable occurrence is expected when the foundation is subject to tensions caused by bending moments as well. In this case, due to the severe drop of the soils shear resistance at the transfer point of the tensions, the risk of foundation's tilting increases. To reduce or eliminate the mentioned negative impacts, the absorption wells must not be constructed near the foundation of new buildings. Furthermore it is recommended that the existing buildings be strengthened by injection or by deep execution of collar at the wells' openings.

(b) Risk of severe drop in load bearing capacity of the soil over a period of time. We have time and again seen and heard about the collapse or caving in of absorbing well in a house. This is not surprising as due to the extensive use, in time the well's wall collapse and the excavated ditch gradually takes the shape of a cone, which becomes larger near the surface. As the building ages the load bearing capacity drops dramatically. In the older parts of the cities where the walls of the wells do not have proper protection, this phenomenon is quite apparent. A diagram of a well after a period of time is shown in Figure 7.5. It is obvious how such a ditch could undermine the safety of the building, particularly during an earthquake, which intensifies the collapse of the walls. In modern buildings, the well's opening is protected by a collar, and this is a relative improvement compared to the older versions.

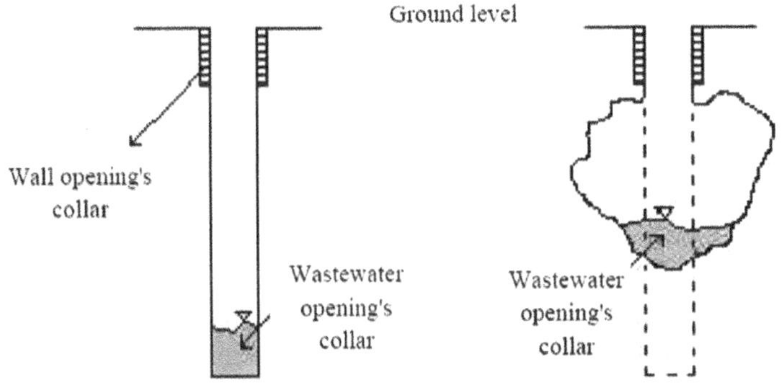

Figure 7.5 Wall collapsing over time: the figure on the right shows the well after collapsing (with permission of NWWEC).

(c) Risk of divergence of the underlying clay layers and reduced load bearing. There are special fine grain soils in nature, which are flushed away immediately in contact with water. These soils, which are mainly clayish, once exposed even to low velocity water (such as the drainage flow of wastewater in absorbing wells) are easily flushed and eroded. This is due to the weak bond between the particles of these types of clay, which is destroyed on exposure to the water and consequently the colloidal particle of the soil separate from each other and starts moving with the small force created by the slow movement of water. In practice, the digging action for wastewater absorbing wells continues until such time as the bottom of the well reached a highly permeable layer such as sand. No problem occurs at the beginning of the operation, because the percolated water flows easily. However, with the drop in permeability over time, the accumulation of water in the well can lead to the collapse of the unstable soils.

Contamination of soil surrounding absorbing wells: Contamination of soil surrounding absorbing wells is another impact of wastewater discharge by absorbing wells, and concerning the variety of wastewaters disposed, these impacts can be even quite more destructive (Hasanoqli *et al.*, 2002).

7.3.4.3 Environmental impacts

The most important environmental impacts of absorbing well are:

(a) If no provisions are made for venting the produced gases, odors propagate in the surrounding area causing disturbances for the region's residents.
(b) At the time of emptying the absorbing wells, the noise pollution caused by pumps creates nuisance for the region's residents. And
(c) The discharge of absorbing wells is generally accompanied with strong nauseating odor, which spreads over a relatively large area, causing the objection of the region's residents.

In any cases, the sludge from the absorbing wells is disposed illegally in natural environments such as rivers, sea and seasonal floodways close to cities or on the ground for agricultural consumptions. This method has the obvious special health (breeding of disease carriers and insects, spread of diseases) and environmental consequences for the different individuals in the society (Hasanoqli *et al.*, 2002).

7.3.4.4 Impact on soil fertility

The discharge of wastewater by absorbing well does not have a direct impact on soil fertility and this effect is caused by secondary reasons probably due to the application of effective factors (Hasanoqli *et al.*, 2002; Hasanoqli & Liaqat 2009; Shadkam *et al.*, 2006).

On the other hand, the diversity of wastewaters discharged through absorbing wells, the possibility of addition of toxic materials or increased concentration of heavy metals in soil is also high. Currently the following parameters caused by discharge of wastewater through absorbing wells can affect the soil fertility:

(a) Increased soil humidity
(b) Increased concentration of matters such as nitrogen and phosphor and a number of minerals available in adequate quantity or in excess in urban wastewaters.
(c) Increased porosity of soil, which facilitates the growth of some crops, and
(d) Reduced soil salinity, as mentioned by some research projects (Shadkam *et al.*, 2006).

7.3.4.5 Health impacts

Urban sewage always carries various microorganisms including the bacteria which may cause the various diseases. The entry of raw wastewater to the environment and natural water resources such as rivers and lakes causes the contamination of these water bodies with disease parameters, and through human contact with these resources, the risk of spread of infectious disease in the society increases. The main health risks of discharging wastewater through absorbing wells are related to the discharge and transfer of wastewater and include the followings:

(a) Spread of contamination through the contact of individuals with the discharge equipment in contact with wastewater.
(b) Pouring part of the wastewater on the surface of streets and the spread of pollution parameters in the air and the surrounding environment.
(c) Transfer of contamination to farmers or individuals who either use the discharged wastewater or are exposed to it, and
(d) Transfer of contamination to individuals involved in sludge processing such as production of compost from dry materials.

7.3.5 Operation and maintenance of absorbing wells

Given their simple structure and since in the past the population density and per capita wastewater generation were low, and further since the houses had large yard, the operation and maintenance of the absorbing wells were not all that important and they did not cause special problems and issues. However, today, with the increased urbanization and population density as well as the growth of the culture of apartment residence, in addition to the limitations of space for absorbing wells, the problems of their operation and maintenance have intensified due by the structural reasons. The main problems are:

(a) Filling up of the wells caused by the obstruction of orifices and infiltration pores of the liquid part of the wastewater, and the need for their constant discharge, which incur costs and cause temporary health, environmental and social problems. And
(b) The sectional or total collapse of the wells due to various reasons, which in addition to financial losses, may also cause injuries or even the loss of life.

7.3.6 Social acceptability

Given the long standing culture in Iran and lack of adequate awareness in the society, the use of absorbing wells for wastewater disposal has never been an undesirable action in public opinion. However today, with the enhanced technical know-how, level of awareness and the society's expectations, and further since the wastewater collection and discharge projects are ongoing in many urban and rural centers, the trend of using absorbing wells for discharge of wastewater is on the decline (Ministry of Energy, 2013; Molla Hosseini, 2013).

7.3.7 The overall economic impacts

The overall economic impact of using absorbing wells can be divided into the two groups of direct and indirect impacts.

(a) The direct economic impacts are: (i) the costs of excavation and construction of absorbing well, (ii) the costs of periodical discharge of the absorbing wells, and (iii) the occasionally imposed costs of well's collapse and destruction. And
(b) The indirect economic impacts are: (i) The necessary costs of preventing or removing the pollution from surface and ground waters, (ii) the necessary costs of reducing or amending the soil pollution, and (iii) the necessary cost of compiling regulations or creating a monitoring and control system.

7.4 SANITATION

There may be only a few people who would know that Iranians had public baths in their towns over 3000 years ago. Probably the first bath in history was built by Jamshid Pishdadian years ago. In addition to cleansing the body of dirt, washing the body and bathing constituted for Iranians a religious ceremony. In all the pre-Islamic religions, from sun worshippers to Zoroastrians – ablution was a prerequisite for attendance. Archeologists have dug up signs of bath in the royal palace of Persepolis and after further exploration they reached the conclusion that the construction of public baths was quite common in all the urban areas of Iran since the Achaemenids.

The people of Iran, both before Islam and after it paid great importance to cleanliness, but after the apparition of Islam this issue was further intensified. In the pre-Islamic era, the followers of Zoroaster paid attention to cleanliness and the word *Padia'o* is rooted in that era. Patio is the place for cleansing and in mosques it is referred to as *Wudukhaneh* (ablution room). The word *Garmabeh* (bath) is composed of the two words of *Garm* and *Abeh*. Here *Abeh* does not signify *Ab* (water) or in other words, *Garmabeh* does not mean warm water, but *Abeh* refers to the site of the building such as *Sardabeh* (cold house) and *Goorabeh* (cemetery or tomb). In archaic times (from about 1500 to 2000 years ago) washing did not exist in its present form. *Garmabeh* was a hothouse with a large vessel called *Abzan*. Since water was considered as sacred and they did not want to pollute it, the used water resulting from bathing was discharged on the ground or would be thrown towards the sky and it would never be allowed to soil pure water.

The heyday of baths construction in post Islamic Iran occurred during the Safavid era. Along with extensive constructions, a relatively large number of baths was built in different cities of Iran, some of which are considered as masterpieces of their kind (including the Ganjali Khan's Bath in Kerman, the Khosro Agha and the Ali Gholi Agha Baths of Esfahan and a number of others). According to Chardon, the French traveler who spent some times in Esfahan during the Safavid era, the city had 272 baths at that time.

Although after the Safavid period the Islamic architecture lost its glory, nevertheless magnificent baths were constructed during the Zandieh and Qajjar periods, the best of which were the Vakil Bath of Shiraz, constructed by Karim Khan Zand and the Ebrahim Khan Bath of Kerman dating back to the Qajjar era. The passage of time and history has offered mankind different perspectives, which can be observed in the evolution of the concept of bathing from a social and communicative activity to a completely private function such as cleaning, hygiene and religion (Molla Hosseini, 2013).

7.4.1 Comparison between the traditional Iranian baths with those of other nations Common traits

The most important traits in common between the Iranian baths and those of other nations are:

(a) The central dressing room, which in the Iranian baths is called *bineh*.
(b) Cold and hot water pools, which is the Iranian equivalent of *khazineh*.
(c) Water supply from groundwater resources or rivers. And
(d) Social–cultural gathering point.

Attention to chastity in Iranian baths has led them to be enclosed, whereas in the baths of other nations, due to neglect of this issue all spaces are overt, to the extent that the Roman Bath of Caracalla had open galleries.

Other nations had stores, libraries, gymnasium and so on in their baths, whereas these spaces were lacking in their Iranian counterparts. The shape of baths in Iran is not specific, while the baths of other nations had predetermined shapes.

7.4.2 The architecture of ancient baths in Iran

Most ancient public baths in Iran had the following sections:

Bineh or cloakroom: is the most important and most beautiful space of the public bath. It consists of a large and extensively decorated area with a large dome and a pool in the middle. All around the mid-section there are platforms or compartments for sitting, resting and getting ready to enter or leave the bath. The floor of these spaces is higher than the mid-section and their roof is generally lower. For this reason they constitute a private space overlooking it. People had to mount a few steps and wash their feet in the pools inside the platforms before entering these spaces.

Miandar: (intersection corridor) the connection between the *bineh* and the bathing space was realized by the *miandar*. This passageway, which prevented the loss of heat inside the bath, consisted of a narrow, low ceiling and sinuous corridor where usually the lavatories would also be located. The *Miandar*'s space was either in the form of a vestibule or a narrow corridor creating an indirect connection between the cloakroom, the bath, the lavatories and the depilation room. In addition to acting as a passageway, it served as insulator for heat and humidity, as well as a cordon between the different parts.

Garmkhane: (Hothouse, bath) was a smaller and simpler space than the *bineh*, and had corners for sitting and washing. The *khazineh* (pool) and more private spaces for bathing are located next to the *garmkhaneh*. *Garmkhaneh* itself has a number of small and large pools.

Khazineh: is a pool like space with warm water measuring 12–25 square meters depending on the dimensions of the bath. After passing the *garmkhaneh* (the bath floor) and going up a few steps, customers had to bend down to enter the *khazineh* through a small entrance.

Depilation room: this room was used to clean the body, apply henna or cupping.

Other sections: of ancient baths consisted of the heating system composed of a fireplace, fuel store, cauldron, ash collection pit, chimneys and canals transferring heat beneath the bath.

7.4.3 The bath's spatial connection

The plan, spaces and the architecture of the baths were designed in such a way to meet the expectations in the best possible manner. The buildings of the ancient baths in Iran were nearly all composed of three separate sections and the heating system's installations. These sections were: (i) The entrance and its relevant sections, (ii) the cloakroom and its related spaces, and (iii) the hothouse and the relevant ancillary spaces.

To regulate humidity and temperature and to prevent the damaging impact of their sudden changes on the bathers, and further to comply with ethical and religious conditions of decency and decorum, these sections were designed in a special and relatively constant style, which apparently dates back to at least the early centuries of Islamic era.

In his book titled 'the rules of city governorship' *Ebne Okhovah* has said: *'Know that the natural effect of a bath is to get warm by its air and to get wet by its water. The first locale is cooling and wetting, the second hot and softener and the third is hot and drying'*. To reach this goal the spaces in old baths were designed in a special manner and each was separated from the others by a corridor or vestibule like area (as seen in a number of baths) to maintain the temperature and humidity compared with the adjacent space and to prevent the bathers experiencing the unpleasant sudden and extreme changes in temperature and humidity. Moreover, the cloakroom space had to be protected from outside prying eyes, and in the same vein the hothouse space from the cloakroom.

The portal in most baths was simple without a special feature. However, in some case it was relatively large and bears tile decorations, stone plinths and historical inscriptions. After passing the portal – in large baths- and descending a few steps, one would enter a domed vestibule. It leaded in turn to the main cloakroom space via steps and a curved corridor. In smaller baths lacking the entrance vestibule, the portal would connect directly to the cloakroom by a curved corridor.

The entrance system was designed to prevent the scenes inside the cloakroom even from the vestibule. The *bineh* or cloakroom's space was distinguished from the others by its surface area and decorations. It was generally octagonal in shape or square with cut-off corners. The gateways were located in the smaller sides of these octagons while the rests bore the chambers and platforms for undressing or putting on one' cloths. In addition to these, the cloakroom had ample spaces for sitting, drinking tea and smoking a hookah.

The *bineh* or cloakroom would lead to the hothouse via a corridor. The lavatories would generally be built in this passageway. The hothouse included a domed space, hot, cold and tepid water pools, private spaces and in some cases a royal niche and VIP hothouse for the city's rulers and officials.

7.4.4 The climatology of traditional baths

As bathing consisted of different stages, the traditional baths in Iran had different spaces for different purposes. In this respect the inner spaces of the bath can be divided into the following three sections: (i) Temperate and semi-humid, (ii) warm and humid, and (iii) hot and very humid.

The environment conditioning system in these baths ensured that temperature and humidity increased gradually from the entrance vestibule and cloakroom (temperate and semi-humid) to the bath floor and hothouse (warm and humid) to finally the pool (very hot and very humid). These spaces needed therefore to be independent of each other, connected only by either narrow and indirect corridors or small vestibules. This is true of all traditional Iranian baths surviving from different post Islamic eras in all the four different climatic conditions of Iran.

7.4.5 The bath's position at the construction site

Generally to reduce the heat exchange between the internal and external environment of the bath, it was built inside the ground, for example, the floor of most baths was generally a few meters lower than the pavement level, thus allowing the soil surrounding the walls to act as a heat insulator and to reduce the heat exchanged between the interior and exterior of the building. Moreover, the thick body of the building and the surrounding soil functioned like a thermal capacitor, adjusting the temperature fluctuations inside the building. The other advantage of construction inside the ground was the additional resistance against the forces of earthquake. However, due to temperate conditions of the southern coasts of the Caspian Sea and the higher groundwater tables, the bath would be dug at a shallower depth in these regions. This is also true of the southern regions of the country, as in these coasts the temperature and humidity is very high and the groundwater tables are also high in many areas.

7.4.6 Materials used in the construction of the baths

The materials used in baths were always of high endurance type, and the roofs of most large and important baths of Iran were constructed in the shapes of arches and domes. However, in hillside villages or wherever wood would be in ample supply, the baths were constructed with wooden beams covered in cob (clay and straw mortar) as it was a simpler and cheaper method of construction. In any case the main materials used in the construction of ancient baths and their applications were:

Marble: Columns, floors and edges of the pools.
Tile: Plinths, pools and platforms
Brick: Walls and ceilings

Rubble: Foundation

Pipes: Terracotta pipes tailor made for the bath's water supply system

Black ash mortar (made of lime, clay, and ash and swamp reed): *khazineh* (pool), ceiling decorations, and insulation layer of the pools

Egg white or milk: These two protein substances increased the adherence of the black ash mortar. As the age of mortar increases it becomes harder and firmer.

7.4.7 Heating system in traditional baths

The needed heat was produced outside the biological environment of the baths, by burning organic fuels inside the furnace beneath the boiler of the *khazineh* (pool) heating its water as well as by transferring smoke and the heat of fire beneath the bath's floor. The boiler room (*patoon*) led to the public pavement, from where it would be supplied with fuel. This fuel consisted mainly of dung (cow or sheep), thorns and straws or dried leaves used by the stoker to fire the furnace. The resulting heat and smoke were led to the chimneys through a system of subsurface canals called *gorbehro, koochebandi* or *zirkar*. As these canals would be covered with soot and needed to be cleared, they were usually large enough for a person to crawl in and carry out the needed repairs or cleaning.

7.4.8 Water supply

The water required in these baths was supplied from streams, Qantas or from wells. In the latter case, a well would be dug next to the bath and water would be extracted either manually or by one or two castrated bulls. This water would be collected in a pool and then transferred to the *khazineh* by terracotta pipes called *tanboosheh*. Whenever they wanted to change the *khazineh*'s water, they would first empty it, wash it and then transfer water by these terracotta pipes from the collection pool to it. Change of *Khazineh* water varied from once every few days to once a week depending on the views of the proprietor and the quality and fame of the bath. Pools at different parts of the baths maintained the humidity of each section, but it should be noted that the connecting passages were mostly indirect and enclosed. Therefore the extreme humidity of the humid environments was prevented from reaching to semi-humid spaces.

7.4.9 Light and lighting

Since the major part of the auxiliary spaces of most baths was beneath the ground level, sunlight was admitted to the bath from skylights or ceiling top windows. For the purpose convex glasses called *Goljam* would be installed on top of the dome and would be sealed using either the black ash mortar or wax. The places where these glasses were installed were called *jamkhaneh*. At night or when the light was not sufficient, the lighting in the bath would be ensured by hand held lamps. Tallow burners were also used, but it burned sheep grease and therefore it had lower light density than other lamps and it burnt with an unpleasant odor.

Light controllers. Porticos and light stops were used to control light. Porticos were spaces which had a ceiling and pillars sealed on one side to protect people from rain or sunlight. In regions with strong sunlight and high temperatures, it would allow a gentle light inside thus ensuring an indirect and reflected light. *Tabesh band* or *tavoosh band*or *aftabshekan* as these awnings were called were in fact blades of 6 to 81 cm in width reaching sometimes 5 m in height built using plaster and bamboo. Generally a sheath would be drawn on top of the door or window. It was in fact the horizontal canopy called *sarsayeh*, which controlled the entry of sunlight into the space.

7.4.10 Decoration of bath

Bineh (cloakroom) had always the most decoration. This space's domes and arches on the ceiling were masterfully and attractively decorated with frameworks and frescoes and on top of the domes beautiful paned skylights would be installed (Figure 7.6). The mid-section beneath the domes and the arched chambers were mostly covered with delicate frameworks bearing plasterworks and stuccos. The regulated lighting in *bineh* added to the display and impact of these decorations (Figure 7.7).

Figure 7.6 A typical skylight window (with permission of the Cultural Heritage Organization).

Figure 7.7 Dome and arch decoration of *Bineh* (with permission of the Cultural Heritage Organization).

The floor and plinths of the hothouse were covered with carved stones, while the upper surfaces were mostly covered white lime with decorated motifs in the Springer. The internal surfaces of a number of historical baths bear colored stuccos. In the luxurious baths, the walls would be covered with tiles up to a height of approximately 1.80 m. and the rest would be covered with lime. In other baths, the walls would be covered with lime as were the arches, with occasional decorations to enhance the looks. Contrary to the practice in mosques and religious schools, images of humans and animals were painted on the tiles in baths. Of course the arabesque and geometric designs were also used (Figure 7.7).

7.4.11 Frameworks and arches

In spaces where lighting was ensured from the ceiling, the light would enter the environment directly and would light up only a part of it. In addition to beauty, frameworks and arches were used to optimize on sunlight, as they diverted it from its path to all direction and let it inside in a diffused form. In this way a uniform and non-centralized light was spread over a larger area.

7.4.12 Hoorno

This refers to the skylight on the ceiling. As execution of the works near the apex of the dome is not possible like other parts, they would leave it open to function as a light on the ceiling. For instance in the markets most of the *hoorno* holes are left open to ensure lighting and ventilation (Figure 7.8).

Figure 7.8 A typical skylight on the ceiling (with permission of the Cultural Heritage Organization).

7.4.13 Bathing rites and customs

Previously public baths were available in all parts of Iran and the local residents would visit them at least once a week for hygienic purposes, with this difference however, that men would go to the baths before sunrise until eight o'clock in the morning, and from then on until noon and even some hours into the afternoon, they would be available to women only.

7.4.14 Customs inside the bath

One of the rites was that whenever a person entered the bath, to show respect and humility towards elders sitting in the baths and being soaped, the newcomer had to take a bucket or pitcher of warm water from

Khazineh and pour it over the head of the elder. This operation was of course repeated for each and every respected elder sitting in the bath, as the newcomer considered it a duty to pour water on the head of each and every elder in order of hierarchy. Quite often one or more of these elders were in the middle of soaping and the last thing they needed was a bucketful of water poured over their heads! Another custom was that whenever the newcomer would observe a close elderly relative, he/she had to go directly to the elder and pay respect by giving him/her a massage or insistently and forcibly take the sponge from his/her hand to soap his/her back.

7.4.15 Bathing materials

Many materials were used in the traditional Baths of Iran, including gloves, cup, mascara, decanter, sponge, henna, pitcher, tallow burner, comb, basin, pumice stone, tub, hair washing mud, soap, loin cloth, white powder, henna cup.

7.4.16 People working in public baths

Many people were involved in managing the baths' affairs including:

- ***Toontab* (furnace stoker-janitor):** the person whose task was to maintain the bath, warm the water and generally clean the bath.
- ***Dallak* (body rub person):** the person who earned a living by sponging and washing the bathers.
- ***Moshtomalchi* (masseur):** this person massaged people in the bath. To call him/her the person wanting a massage had to shout '*khosh*'.[1]
- ***Hammami* (bath owner):** the private proprietor of the bath who managed the bath's affairs.
- ***Pado* (errand person):** this person's duty included arranging the bathers' shoes and sometimes carrying the bathing materials from the public bath to visitors' houses.

In addition to the owner, a *jamedar* (cloakroom attendant), masseur and pado were present in the *Bineh*. As soon as a client entered the bath, the *pado* placed the client's shoes under the platform and spread a dry loincloth over it. After undressing, the *pado* would give the client an additional loincloth. The *Moshtomalchi* performing message for a customer is shown in Figure 7.9.

Figure 7.9 The *Moshtomalchi* performing message for a customer (with permission of the Cultural Heritage Organization).

[1] Khosh in fact is the shortened word for Khoshk meaning dry. What the person shouting this word wanted at that junction of bathing was a new loin cloth to dry himself/herself as the previous one had become wet and in the same token was announcing his or her readiness for a massage

7.4.17 The time for using public baths in the past

Previously public baths were available in all parts of Iran and the local residents would visit them at least once a week for hygienic purposes, with this difference however, that men would go to the baths before sunrise until eight o'clock in the morning, and from then on until noon and even some hours into the afternoon, they would be available to women only. Today also there are public baths available in most parts of Iran, but in all the new ones, numerous showers have taken the place of *khazineh* as it did in no way concord with hygienic principles (Molla Hosseini, 2013).

7.4.18 Ancient baths in Iran

A number of ancient baths are known in Iran. Few paradigms are follows:

Ali Gholi Agha Bath. In his period (Safavid Era) Ali Gholi Agha constructed a complex in a neighborhood of Esfahan, which included a small market, a mosque, a bath, a caravanserai, a water fountain, etc. In the past and in the adage of the local people, the Ali Gholi Agha bath was known as the twin bath, since like many old baths, it was built of two separate men and women sections, which had two separate entrances as well. The *bineh* (cloakroom) in the women section was smaller with a more feminine decoration than the men's, and they could be used at intervals or simultaneously. The large and small baths of Ali Gholi Agha, which have been registered as national heritages, in addition to possessing all the features of Iranian baths are of particular historical and artistic values. The multicolored tile work with arabesque, Chinese, human and bird and flower motifs and plasterworks of the Safavid era and the murals of the Qajjar period, which were added subsequently, are among these. The paintings in the large bath or the men's bath mostly depict hunting scenes, while in the small bath they show still life of small fruit bowls.

The Four-Season Bath of Arak. This bath consisted of the three sections of men, women and religious minorities. Like all ancient bath, it used flowing water and is deeper than street level by a few meters.

The Bath's cloakroom is decorated with multi-colored tiles and the bath's ceiling lies over 8 pillars, which have beautiful twits and patterned tiles. The name of the bath refers to the tile work paintings depicting the four seasons and installed on the four corners of the men's cloakroom. The *Four-Season Bath of Arak* is shown in Figure 7.10.

Figure 7.10 The Four-Season Bath of Arak (with permission of the Cultural Heritage Organization).

The Galladari Bath in Bandar Abbas. This bath dates back to the Qajjar period and the 13th Century Hijjara. The materials used in this bat consist of corals, local ash mortar and hand pounded plaster.

The Sheikh Bahaee Bath. The date of this bath's construction, which is a Safavid period monument designed by Sheikh Bahaee, is said to be 1065 Hijjara. The heating system in this bath is an engineering feat relying on the laws of physics and chemistry. The bath's water was heated by the 'blow and gas' system, meaning that it used the methane obtained from the wastewater of the Congregational Mosque and the oil drops from the oil pressing factory located near the bath. Using this complicated engineering system, the bath was heated by a single candle, and bears all the architectural features of the baths from Safavid era.

The Ganjalikhan bath. The Ganjalikhan Bath is one of the buildings in the Ganjalikhan complex in the city of Kerman, constructed in the year 1020 Hijjara (1611) by Ganjali Beig titled as Ganjalikhan, one of the rulers of Kerman during the reign of Shah Abbas (Figure 7.11).

Figure 7.11 The Ganjalikhan bath (with permission of the Cultural Heritage Organization).

The historical bath of ZahirolEslam (Agha Naghi). The ZahirolEslam bath is considered as one of the old baths of Ardebil. From an architectural style, it resembles the Haj Sheikh Bath. Based on the thermo-luminance tests of the baths' tiles at 5 selected points, the date of its initial construction goes back to before the Safavid period, but it was still in use until the Pahlavi era. The Bath was constructed in two completely distinguished stages at an interval of approximately 200 years.

(a) This monument had a smaller building, which is the present swimming pool. It was constructed about 650 years ago. The section that is named as the en's hothouse today was built and added to the bath about 200 years thereafter, and

(b) On the present site of the bath there was a bath or another building, which dates back to 650 years ago. The women's section was built about 400 years ago after the demolition of the previous structure and the use of its bricks, while the bath itself was completed using other bricks.

This bath was first constructed before the Safavid era at second half of the seventh century Hijjara. Great efforts were made in the Safavid era to complete, extend, repair and maintain the building.

Its age at the end of Safavid era led to its demolition and subsequent closure, but it was restored and renovated once again during the Qajjar period.

The Vakil Bath in Shiraz. The Vakil bath along with other structures of Vakil Complex is located at the center of the city of Shiraz. This bath, which is a monument of Karim Khan reign, benefitted from the best architecture and technologies of its period. The availability of a royal dais has distinguished the Vakil bath from other traditional baths of Iran. The plaster works designs on the ceiling, depicting religious or folkloric and local tales – have enhanced the attraction of this bath. The Vakil Bath in Shiraz is shown in Figure 7.12 (Molla Hosseini, 2013).

Figure 7.12 The Vakil Bath in Shiraz (with permission of the Cultural Heritage Organization).

Other famous old baths in Iran are:

- The Sanandaj Old Bath
- The Vakil Bath (Kerman)
- The Pahneh Bath (Semnan)
- The Safa Bath (Qazvin)
- The Khan Bath (Yazd)
- The fin Bath (Kashan)
- The Haj Shahbaz Khan Bath (Kermanshah)
- The Sultan Amir Ahmad Bath (Esfahan)

7.5 PRESENT SITUATION

The present situation is quite different from the past. Better management, new technologies and significant achievements as is described briefly hereby:

7.5.1 Management

The Law for Establishment of Water and Wastewater Companies was ratified by the Islamic consultative Council in the year 1990. Following the approval and notification of this Law, the responsibility for creation and operation of urban water and wastewater systems was relegated to the Ministry of Energy. In the context of the mentioned Law, the Ministry of Energy established the National Water and Wastewater Engineering Company and subsequently the urban and rural water and wastewater companies.

Currently the specialized parent company of National Water and Wastewater Company manages the water and wastewater affairs of the country with the assistance of 48 provincial and 40 rural water and wastewater companies (Ministry of Energy 2012).

7.5.2 Technologies

The following three main methods are used in execution of wastewater collection networks: (a) open trench, (b) manually cut galleries, and (c) trenchless methods (such as pipe jacking and micro tunneling).

Most wastewater collection plans in Iran are implemented by open trench method, which is the simplest and least cost procedure. It is generally applied at locations where conditions such as the soil type and material, environment, site of execution (urban and per-urban roads and paths), groundwater tables, and so on are favorable. An example of this method is shown in Figure 7.13. Where there is a risk of instability of soil, wooden shields and or sheet piles are used to prevent its collapse (Figure 7.14).

Figure 7.13 Open trench execution of wastewater line: The wastewater plan in the village of Abar from Sharood County (with permission of NWWEC).

Figure 7.14 Open trench execution of wastewater line using wooden shields and sheet piles. Bushehr wastewater plan (with permission of NWWEC).

The manually cut galleries method is applied mostly in regions where the width of alleys, the density of infrastructural installations or problems due to traffic prevent the execution of project by open digging which will result in increasing the initial costs. In this method, small shafts of 1 m × 1 m are excavated at appropriate intervals (minimum 3 m) to allow workers to enter and cut the gallery at a suitable level. The dimensions of the galleries depend on the pipe diameter, but it should allow a person to excavate in a squatting position. In this method, to prevent further difficulty of work, the maximum diameter of culvert is limited to 400 mm and 500 mm.

In addition to entry and exit for workers, the shafts serve the purposes of ventilation in the galleries, measurement of discharge of debris, access to the entry of pipes and backfilling the space above the pipes until the crown of the galleries. The shafts' location is determined in such a way to be able to use as manholes. An excavated gallery in Neishaboor wastewater plan is shown in Figure 7.15.

Figure 7.15 Execution of sewer lines by manually cut gallery. Neishaboor wastewater plan (with permission of NWWEC).

The mechanized (micro tunneling) or semi mechanized (pipe jacking) methods have to date been used mostly to pass the wastewater line from streets and roads with heavy traffic load or in cities where the groundwater table and the soil texture create serious constraints for implementation. In these methods the pipe jacking operations are conducted at distances of 60 to maximum 150 m and for pipe diameters of 160 to 2000 mm. The tunneling starts at the send shaft and ends at the receive shaft, and then by transferring the excavation equipment to the new start point the operation continues. The maximum recorded speed of pipe jacking in Iran under the best conditions (soft soil and minimum error of the system) is equal to 30 m/d.

In this method the construction and preparation of the send and receive shafts and their stabilization and consolidation against collapsing walls or the penetration of ground waters. For the purpose after the excavation of the shaft and continued pumping of infiltrating waters, the walls are retained using prefabricated concrete segments of 25 × 40 × 70 cm. The application of concrete segments in the send and receive shafts of pipe jacking operations in Rasht is shown in Figure 7.16.

Figure 7.16 The application of prefabricated concrete segments in the send and receive shafts for pipe packing. Rasht wastewater plan (with permission of NWWEC).

In the semi-mechanized method, the transfer of pipe to the shaft and its positioning on the jack are done by the staff, whereas in the mechanized method, the excavation, pipe laying and transfer of debris are undertaken automatically by the boring machine. For this reason, in most coastal areas of Iran where the ground water tables are high and where it is impossible for attending personnel to work in send shaft, the mechanized pipe jacking operation is applied (Figures 7.17 and 7.18).

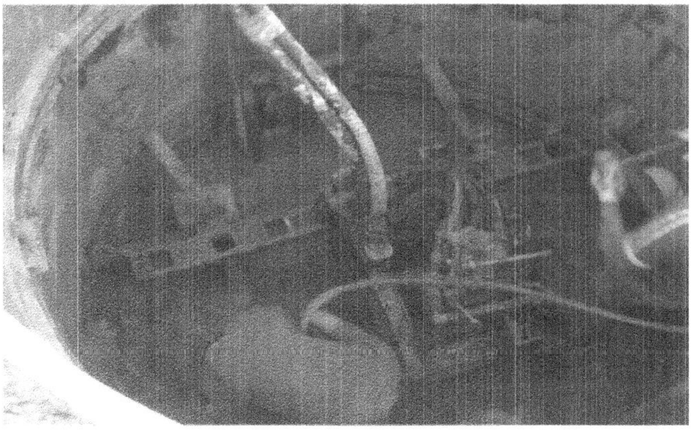

Figure 7.17 Execution of wastewater network by pipe jacking. Deilam wastewater plan (with permission of NWWEC).

Unconventional methods such as followings are generally used in the rural regions of Iran: (a) small diameter gravity sewer, (b) settled sewerage, (c) simplified sewerage, and (d) pressure sewer system. As for treatment, many plants have to date been constructed and operated using the following processes: (a) stabilization ponds, (b) aerated lagoons, and (c) activated sludge systems (Ministry of Energy 2012).

Figure 7.18 Execution of wastewater network by micro tunneling. Deilam wastewater plan (with permission of NWWEC).

7.5.3 Achievements

Passing 20 years from evolution occurred in water and wastewater management in Islamic Republic of Iran, after establishing the provincial water and wastewater companies, the authors are to introduce the achievements gained due to the evolution, as follows:

(a) **Achieving the development objectives in the water and wastewater sector.** Planning to accelerate and complete the infrastructures for water supply and distribution water as well as wastewater collection and treatment in urban and rural areas. Based on its mandate for the water and wastewater sector, the Ministry of Energy is in charge of meeting the basic demands for potable and sanitation water as well as the collection, transmission and treatment of wastewater and the reuse or the safe disposal of the treated effluent. In the context of which, it delivers these services and products to the community at large. Based on the Holy Koran's verse of 'And We created from water all things living' and the emphasis in Islam on abstention from abuse and preserving the sanctity of water, the water consumption management in Iran enjoys a special status in Iran. As Iran plans it's economic, social and cultural development plans in 5-year tranches. The status of water and wastewater indicators for year 2009 (the end of the fourth development plan) and the status in 2010 are shown in the Table 7.1.

Table 7.1 The status of water and wastewater indicators at the end of fourth development plan, 2009 and 2010.

Item	Indicator	The status at the end of fourth plan, 2009 (%)	The status in the year 2010 (%)
1	The population covered by urban water networks	98.8	100
2	The population covered by urban wastewater collection system	33	34.6
3	Rural population connected to water supply networks	74.5	77
4	Rural population having access to safe potable water	99.5	100

(b) **Financial facilities from the national sources and the international development institutes.** To implement the water and wastewater plans, in addition to the funds allocated in the budget from the national resources, the use of other financial resources, including the following examples, was considered by the sector:
   (i) *The World Bank.* The Water and wastewater sector has the largest World Bank Portfolio in Iran for related projects seven big cities including Tehran, which are recognized by the World Bank as the biggest projects in the MENA region.
   (ii) *The Islamic Development Bank.* The Islamic Development Bank has been involved five urban water and wastewater projects and the rural wastewater plans. Once more the sector became the main beneficiary of IDB financing. In the year 2010, The Islamic Development Bank approved the largest amount of financing for a single project, which concerned the execution of wastewater systems in the rural areas of Iran. Following its implementation, the plan will greatly improve the ratio of rural population's access to wastewater services.
   (iii) *ECO Development Bank.* There has been a recent boost in cooperation between the Islamic Republic of Iran and the ECO Development Bank. Three wastewater projects were proposed by the Ministry of economy and Financial Affairs to the Bank. One of them has been approved and the relevant bidding process has been launched.
   (iv) *Foreign exchange Reserve Funds.* The sector has also applied for a loan from the Foreign Exchange funds for the plan to supply water from Kochery dam to the city of Qom. This project has progressed by 100%.
   (v) *Others.* The Czech Bank and Muscat Bank in Germany have financed Karaj wastewater treatment plant at a capacity of 280,000 PE through loan agreement. The project has a current progress rate of 15%.
   (vi) *Other means of financing.* Desalination plants in south part of the country have been undertaken by private sector through BOO contracts.

(c) **Progressive boost in the implementation of wastewater plans through the country.** In recent years there has been a progressive boost in the implementation of wastewater plans in Iran, and the length of the wastewater collection system in urban areas has increased from 9,978 km in year 1997 to 41,592 km in the year 2010 as shown in Figure 7.1. Moreover the population coverage of the wastewater collection systems in urban areas has increased from 5,500,000 in the year 1997 to 18,900,900 in the year 2010. Figure 7.19 denoted the progressive trend of establishing and operating the wastewater collection systems from 1997 to 2010 showing that 243 cities are covered by the wastewater collection systems and the implementation of the concerned plans has had a positive effect on improving the sanitation, the environmental conditions and sustained development of the urban area of the country. Based on the progressive trend of the construction and operation of urban wastewater treatment plants, the number of treatment plants has increased from 30 in the year 1997 to 134 treatment plants in the year 2010 as shown in Figure 7.20. Also the trend of increase in the number of wastewater treatment plants is shown in Figure 7.20.

(d) **The status of service provision and tariffs.** The law on Targeting Subsidies was approved with the objective of establishing social equity and improving consumption efficiency in all sectors. This law came at the end year 2010. The objectives of this law are:
   (i) Reducing consumptions and encouraging the efficient use of water.
   (ii) Accelerating the implementation of policies related to Principal 44 of the constitution and expanding the participation of the private sector in water.
   (iii) Reducing the financial dependence of companies in water supply chain on public budget.

(iv) Improving efficiency and reducing losses in the different stage of supply, transmission and distribution of water.

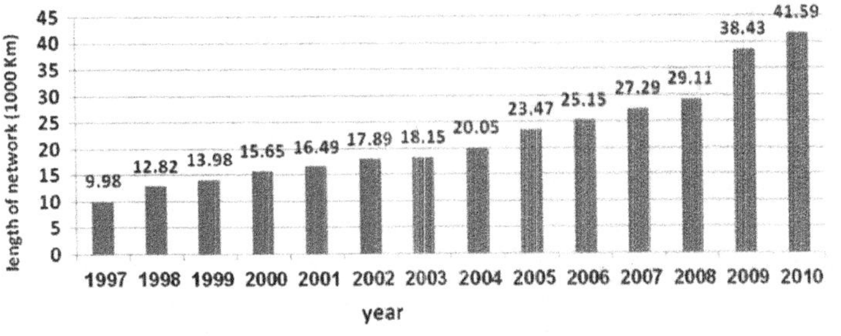

Figure 7.19 The progressive trend of the length of wastewater collection systems executed from 1997 to 2010.

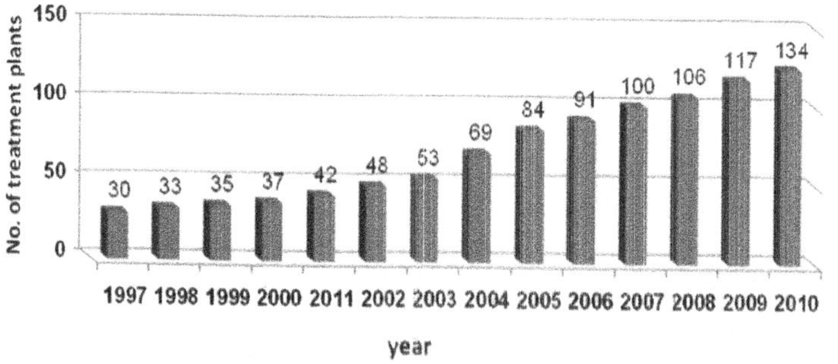

Figure 7.20 The trend of increasing the number of wastewater treatment plants.

(v) Diversifying the water supply methods (for instance desalination)

(vi) Protection of the environment and water resources, particularly in the crisis hit regions and plains, and

(vii) Financial sustainability and autonomy of companies in water supply chain.

Following the implementation of the targeted subsidies plan, the water and wastewater tariffs have been also simplified and transparent, allowing the consumers to be able to calculate their own consumption charges.

(e) **The outlook and goals of the water and wastewater sector.** Based on the outlook plan of the Islamic Republic of Iran, the objectives set for water and wastewater sector are shown in Table 7.2.

Table 7.2 The quantitative goals defined for the water and wastewater sector until 2025.

Item	Sector	Indicator	Unit	The considered goal
1	Urban water	Population coverage of the potable water networks	%	100
2	Rural Water	Access to safe potable water	%	100
3	Urban Wastewater	Population coverage of the wastewater collection systems	%	60
4	Rural wastewater	Population coverage of the wastewater collection systems	%	30

(f) **Privatization in the water and wastewater sector.** In the context of its legal mandate set in the policies of Principle 44 of the Constitution, the National Water and Wastewater Engineering Company (NWWEC) has taken steps to privatize the urban and rural water and wastewater companies by submitting its proposal to the High Council for Concessions.
The efficient management of water is among the top priorities of NWWEC, particularly during the Fifth Development Plan. Generally the programs required to reach the efficient management of urban water demand, mostly focus on the water consumption management. For this mean, particular research in the field of water and wastewater has a special status. The water and wastewater research is composed of three main axes of management, executives and customers, which form an integrated and organized system for defining the issues, planning, conducting research and implementing the results. Issues addressed by the water and wastewater research management are as follows:
    (i) Direction, management, promotion and conduction of practical research to achieve a sustainable development in water and wastewater sector.
    (ii) Compiling standards and monitoring the quality of services and products of water and wastewater industry.
    (iii) Supporting innovation in the industry active in the national water and wastewater sector. And
    (iv) Direction and execution of research projects to improve the capacity for competition, management, design and implementation of systems needed by the sector.

(g) **Women's participation in water management.** Women as the mothers, teachers and managers of household have a special effect on members of family as the smallest social unit, and for this reason their role is considered as an important topic of the international events related to management, planning and decision making on water resources. Women's participation in water management, which is quite often on voluntary and conscious basis, is crucial and undeniable in reducing the water consumption level.

Enhancing the role of women in different aspects of water management is recognized as one of the most important measures to achieve the objectives of the integrated water management particularly in the field of consumption management. The process of strengthening the participation of women who compose half of the society's population and using their potential to enhance the management of water as well as consumption have always been among the challenges of managers and executives of water sector.

Clearly as a major part of the society, the active participation of women in different aspect would greatly improve the trend of development and if their participation for any reason is not facilitated, the society would be deprived of half of its potential human resources. Women's presence at different educational sectors such as schools, cultural centers, residential complexes and their impact on information dissemination about water are among other managerial and participatory roles that can be played by women in Iran's water and wastewater companies (NWWEC, 2013).

7.6 CONCLUSIONS

As mentioned, since ancient times, the traditional system named absorbing wells have been the first and the only means of discharging domestic wastewater in urban and rural areas of Iran. They are still in use in many cities, towns and villages across the country. In towns where the level of ground water table is high, septic tanks have been also used.

In some regions of Iran, mostly in the west and northwestern parts of the country, urban areas have either been developed on rocky areas such as Khorram Abad and Kermanshah or have a high groundwater tables such as Ahwaz and Rasht, the absorbing wells system is no longer effective. Therefore the traditional sewage networks consisting of rocky canals or cement pipes were used to collect and discharge wastewater from above mentioned cities. There are no maps, technical calculations or manholes for these networks. In fact they have not been built according to sound engineering principles. Instead, these networks, which are mostly of combined types and collect surface runoffs during rainfall as well, were gradually created by municipalities or the local communities on their own initiatives, and since they were executed without proper design, map or technical calculations, they did not connect to any treatment plant and therefore, they merely conducted the raw wastewater directly to farms for reuse. Close to 70 urban centers possess these types of so called "traditional" networks.

The year 1991 became a milestone in the management of the national water and wastewater affairs. At this year, the Provincial Water and Wastewater Companies Law were ratified by the Islamic Consultative Assembly of Iran and thereafter and within the duration of 2 to 3 years, 30 water and wastewater companies were established in provinces throughout the country. This robust ratification underwent the path of development in water and wastewater sector of Iran and improved the operation and maintenance of existing systems. The establishment of these companies also meant a new wave of studies for implementation of modern wastewater collection and treatment plans in the country. Currently over 200 urban centers in the country possess modern wastewater collection and treatment systems. Iran has also launched the new experience of reusing wastewater as a substitute resource for agricultural water.

Through experiences gained from the studies and implementation of wastewater treatment and disposal plans, the country is now in possession of valuable regulations and standards related to study, design and operation of wastewater treatment and disposal plans, which could also be beneficial for other countries. The historical monuments of Iran also had a certain wastewater discharge systems, which are of interest in their own merit.

At present in most cities of Iran modern wastewater collection systems and treatment plants have been designed and are under construction or operation which has led to a better life for the people. Concerned with sanitation, public baths have been abolished and substituted by the private baths at the houses.

7.7 REFERENCES

Hasanoqli H. and Liaqat J. (2009). Variation of Deep water EC in Result of Irrigation with Raw & Treated Wastewater in Ekbatan Residential Town. Iranian Irrigation Magazine, pp. 12–20.

Hasanoqli H., Liaqat J. and Mirabzadeh M. (2002). Variation of Soil Organic Elements Content in Result of Irrigation with Domestic Sewage & Self purification. Water & Wastewater Magazine, pp. 32–39.

Ministry of Energy (2012). Achieving the Millennium Development Goals in Water and Wastewater Sector. National Water and Wastewater Engineering Company. Ministry of Energy, Tehran, Iran.

Ministry of Energy (2012). Reviewing Rural Water & Wastewater Industry. Ministry of Energy. http://www.nww.co.ir

Ministry of Energy (2013). National Water & Wastewater Engineering Co-Ministry of Energy (2013). Personal communication.

Molla Hosseini M. (2013). Culture and History. Iran's Heritage. Hamshahri online Magazine.

National Water & Wastewater Engineering Co. (2000). Technical Report of Septic Tank Design. Study Group of National Water & Wastewater Engineering Co. Tehran, Iran.

Shadkam R. Alizadeh A. and Danesh R. (2006). Considering the effect of raw & treated wastewater reuse on soil EC. In Proceeding of the 1st Environmental Engineering Conference, Tehran, Iran, pp. 112–123.

Chapter 8

Sanitation and wastewater and stormwater management in ancient Kingdom of Macedonia, Hellas

A. Kaiafa, E. Papanikolaou, V. Melfos, C. Papacharalampou, and K. Voudouris

8.1 INTRODUCTION

"Of the elements of the world, water is the best"
(Pindaros, ancient Hellene lyric poet, 5th century BC)

High-level technology and expertise for water management appeared in ancient Macedonia, Hellas, since the classical period, when settlements were classified as urban. In prehistoric Macedonia, urban hydraulic networks did not exist, because of the total absence of organised cities, in contrast to the concurrent urban centres in southern Hellas. Well organised settlements appeared in Macedonia in the 7th century BC and developed during *ca.* 6th century BC. The creation of urban centres was combined with the construction of aqueducts, water supply and wastewater systems, due to the daily requirement for drinking water and the necessity for the waste and rain water drainage for hygienic reasons. Special technical works, such as systems for collecting, transporting and storing groundwater, constructions for water supply, drainage and irrigation, and various other structures like dams, bridges, fountains, baths and latrines, were developed through the ages in several parts of Macedonia. The water availability was correlated with the climate conditions of the region, as well as with the existence of water resources (surface water, groundwater and springs). Therefore, the local geology and geomorphology controlled in many cases the development of the new cities since *ca.* 4th century BC.

The implementation of a Hippodamian poleodomic system in ancient Macedonia is attributed mainly to Philip II. His successors continued this harmonious, elegant and functional city planning. The application of Hippodamian system in the organization of the Macedonian cities in Hellenistic times, in combination with the consolidation of architectural types, undoubtedly encouraged further development of hydraulic infrastructure. The administrative, religious and commercial centres occupied separate areas of the city blocks among the private residences. In such cities, the complex water supply and drainage systems ensured comfortable living conditions to the citizens. During the Roman era, the life style and sanitation standards were advanced by the extended use of baths and toilets, therefore, urban hydraulic networks were enriched systematically.

In this chapter, special constructions related to the sanitation and wastewater and stormwater management in the most famous cities of the ancient Macedonian Kingdom, such as Pella, Aigai, Dion, Thessaloniki, Olynthus, Amphipolis and Philippoi, are examined. It is well documented that during the Hellenistic and especially the Roman period, Macedonians had a deep knowledge of the hydraulic technology and the ability to create developed and successful urban water supply and drainage systems. Urban hydraulic networks in Hellenistic and Roman Macedonia were characterized by a great technical and functional progress, planned and built to cover everyday needs. They were performed each time by a central authority, according to an existing legislation and institutional framework for water management (Kaiafa, 2008).

8.2 GENERAL CHARACTERISTICS AND WATER SUPPLY

The areas of study outlined in this chapter are situated in the northern part of Hellas, in the region of Macedonia. Moving from east to the southwestern part of Macedonia, the landmarks include Philippoi, Amphipolis, Olynthus, Pella, Dion and Aiges (Figure 8.1). The wider area has a continental climate in the mountainous areas and a semi-arid Mediterranean climate in the coastal areas, with wet winters and hot summers. The mean annual temperature ranges between 11 and 15.5°C and the mean annual rainfall varies from 465 mm to 750 mm. 65% of the total precipitation generally occurs in the wet period, which normally extends from late October through April, while summers are usually dry.

Figure 8.1 Map of central Macedonia, North Hellas, with the most significant ancient cities, discussed in this Chapter.

Human effort was first made to utilise surface waters and springs and therefore the settlements were close to them. It is pointed out that there is a numerous number of springs, torrents and rivers in Macedonia.

When the water demands were increased in connection with irrigated agriculture, the groundwater exploitation was expanded with the construction of complicated systems of qanats and dug wells.

Based on the geological features, the main aquifers for groundwater exploitation in Macedonia are developed in alluvial deposits and in the terrestrial facies of Neogene deposits (porous aquifers), as well as in carbonate rocks (karstic aquifers). Karstic aquifers in carbonate rocks (limestone, dolomites, marbles) discharge through many springs. In general, ancient people used horizontal works for water supply (tunnel, galleries, aqueduct, and qanats) as opposed to using modern vertical works (deep boreholes). During the Hellenistic (323–67 BC) and Roman period (*ca.* 67 BC–330 AD) the water supply of cities (e.g., Dion) relied on the springs and rivers. The Roman period is characterised by the construction of admirable aqueducts that carried water to the cities from a long distance (Voudouris, 2012).

A densely populated Macedonia acquired urban cities with organized formation since the *ca.* 7th and 6th century BC, associated with the systematic colonization of residents of southern Hellas and the islands. During the *ca.* 4th century BC old cities developed (Aigai) and new ones were established under the construction programs of Macedonian kings.

8.3 SANITATION AND WASTEWATER HYDRAULIC WORKS IN ANCIENT MACEDONIA

The urbanized settlements have been built according to the *Hippodamian* system, which has been launched by Philip II, in Macedonia, and preceded by Alexander III and his successors by the founding of the cities (Philippoi, Thessaloniki). This urban planning was based on a grid of parallel and perpendicular roads, with rectangular blocks and public buildings initially defined in positions that ensured functionality and defence. Its implementation increased the need of water supply and drainage, as well as these ensured on one hand comfortable living and health of residents, such as the further development of the organized cities. This need led to the construction of aqueducts for transporting water from distant natural sources. Clay circular tubes were mainly used for the transportation of water which either followed the topographic contour lines on continuous underground grooves or underground tunnels.

'Inverted siphons' had been constructed by engineers during the Hellenistic period, to convey safely the water across valleys. This technique took advantage of the differences in elevation between two points in order to create more pressure at one point, causing the water flow. The technique was applied in cases where the water pipes had to cross soil deepening such as streams and ravines while avoiding the rather longer and more expensive route of S-shaped water pipes. Roman authors such as Vitruvius and Frontinus described the various ways of water intake and water transportation (Karadedos, 1990b). Romans, taking advantage of the theoretical background of the Hellenic engineers, enriched the art of water with new technological developments and monumental constructions.

Hellenistic wastewater collection and transport systems are admired for their innovation. The Hellenes had built complex systems of rock-cut channels, clay or lead pipes and sewers constructed by stones, for water collection. Wastewater and stormwater management systems evolved slowly, and began primarily as a means to drain marshes and storm runoff, showing the care of the population on public sanitation and hygiene. The cities were equipped with advanced storm, drainage and sewer systems, planned simultaneously with the development of the roads and the construction of the buildings. The central sewer systems were closely linked with the urban plan, organized beneath the horizontal and perpendicular streets. They were constructed by stone built flat-roof channels. All central sewers of the city streets were accepting storm and wastes from pipelines or covered built conduits, which were developed under the floors of public and private buildings. Hydraulic installations were also planned for the building complexes outside the walls of the cities.

In many cases, sanitation systems appeared to be organic elements of the public or private buildings. Open channels were running around the buildings and collected the rain water. Waste flushed from the latrines flew through central channels into sewage systems and then into nearby rivers or streams.

8.3.1 Aigai (Vergina)

Aigai, the former capital of the Macedonian Kingdom, was articulated in horizontal terraces on the northern feet of mount Pieria. In spite of the fact that archaeological evidence related to the hydraulic infrastructure of the city is not enough, the necessary quantities of water were coming from the nearby springfull area of Agios Nikolaos, trapped in a closed earthen pipeline with unknown route, since at least *ca.* mid-4th century BC (Faklaris, 1996). It was running through the walls, initially in the acropolis area, and then it was branched lower to the Palace and the city. Parts of the central water supply pipe have been found, with cleaning openings scattered irregularly at the surface, probably sealed with wooden caps. Elsewhere, the earthen pipeline was protected by flat stones carved on the upper surface, shaped like an upside-down U, arranged on a line.

There is evidence that along with the city water network, an organized and functional drainage infrastructure, such as underfloor clay pipes or rockcut channels of local porous stone, was also constructed showing the care of the authorities for public cleanliness and hygiene. Equipped with all the comforts of that time, the Palace of *ca.* 4th century BC had an excellent sewerage along with the water supply system, which carried fresh water from the mountain springs. It was a square complex with a central courtyard surrounded by porticoes, with a smaller similar building next to the main, made in the end of the *ca.* 3rd century BC (Andronikos *et al.*, 1961). Sanitation systems appeared to be organic elements of the buildings at that time. In the square colonnaded courtyard an open channel was running around all four sides, collecting the rain water that was falling from the pitched roof of the perimetric stoa.

Essentially it was a continuous groove with a semi-cylindrical cross section of stones placed in a row, carefully carved on the upper surface (Drougou, 1999). A similar groove was also formed in the smaller, later yard, which was leading the wastes out of the Palace through a drain. Moreover, in three rooms at the western part of the main building, small holes in the middle of the floors were enabling percolation of dirty waters into the soil (Figure 8.2a). Similar holes have been found in houses of the Hellenistic settlement on the hill of Agios Panteleimon in Florina (Kaiafa, 2008). Wastewater systems were also found at the sanctuary of Eukleia and at the public building of the NW part of the Aigai (Saatsoglou-Paliadeli, 1996).

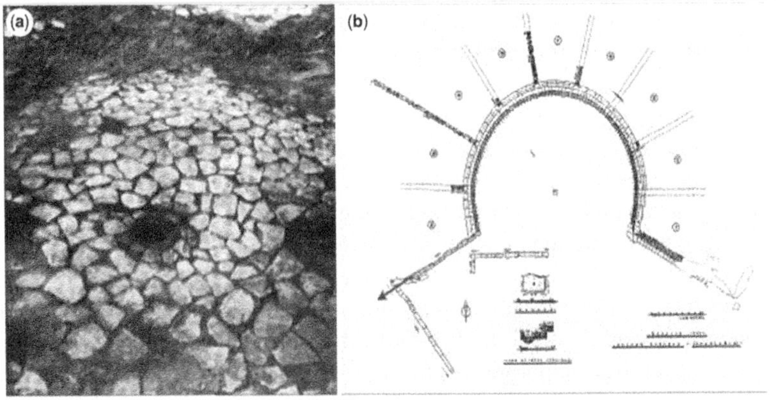

Figure 8.2 Aigai: (a) Palace, drainage holes on the floor and (b) Theatre, drainage canal (Kaiafa, 2008).

Finally, at the city Theatre there was a sewerage channel, around the orchestra, in front of the first row of the stone seats, in order to ensure the removal of stormwater (Figure 8.2b). That was a semicircular uncovered canal constructed by stones in a row carved on their surface. It was linked with a clay drainage pipe located in the western edge of the cave, which was running under the main road of the city connecting the Theatre with the sanctuary of Eukleia in a lower terrace (Drougou, 1999).

8.3.2 Pella

Pella is located in central Macedonia, approximately 40 km from Thessaloniki and is a part of the Thessaloniki plain, the largest deltaic area of Hellas (Ghilardi *et al.*, 2008) formed by the alluvial deposits of four large rivers. In the *ca.* 5th century BC, part of this area was covered by the sea and consequently Pella, the capital of ancient Macedonia Kingdom, was virtually a coastal city. By the first century BC the alluvial deposits in combination with the tectonic up-lifting of the area, encircled an expanse of the sea off the port of Pella, creating the Loudias lagoon, which later was completely cut off from the sea and became a lake (Giannitsa Lake).

Water supply and sanitation in Pella was an integral part of its urban design, planned simultaneously with the development of roads and the construction of buildings. The city, with the regular blocks and perpendicular routes, was supplied by a dense distribution network of teracotta piping beneath the roads equipped with stoneblocks or earthen jars. In addition to the long distance water supply lines, the water system included cisterns and fountains scattered around the city, as well as a large number of wells, which secured the presence of water in a more direct way.

Coexisting with the water supply, drainage facilities were essential elements in every building in Pella, public or private, connected to drains (Figures 8.3 and 8.4). The central sewer system was closely linked with the urban plan. The development of the orthogonal city planning greatly facilitated the design of integrated under street drainage. It was composed by stone built flat-roof channel under the roads, destined for the drainage of rain water and sewage. They ended up to the extensive swamps south of the city, taking advantage of the city gradient. All central sewers of the city streets were accepting storm and waste water from pipelines or conduits, which were developed under the floors of public and private buildings. Moreover, smaller lead pipes were serving sanitation purposes. These were piercing the walls of houses or public buildings, draining out dirty water from the floors cleaning. Such a pipe has been found in the House of Helen driving the waste water directly to the adjacent street (Akamatis, 1990). Stone grooves petal shape canals were preferred for the collection and removal of rainwater from the roofs. These were running along the inner side of each pillar of the courtyard. Rock cut channels have been found both in the Palace complex, private houses and the central Agora, which was equipped with a remarkable supply and sanitation system. Along with the increased number of wells, particularly at the eastern portico and less in the southern and western Stoas, water supply infrastructure of the Agora complex was strictly connected with the distribution network. Although the central yard still remains unexplored, it is speculated that there have been more than one fountains fed by the urban water supply system for covering every day needs.

The drainage infrastructure of the Agora was taking advantage of natural territorial inclination towards south and west. It was directly connected with the urban drainage network. This can be proved since a deep stone built flat covered drain from the horizontal road situated at the north of the Agora was passing through the complex. This invisible drain was crossing underneath the northeast entrance of the market, with an intense southern flow (Figure 8.4a).

The most basic element of the drainage system though is the uncovered groove around the square, made of rectangular stones in a row carved in the form of an inverted U. At intervals there were settling basins. This system was common to drain surface water from public buildings and courts (Figure 8.4b).

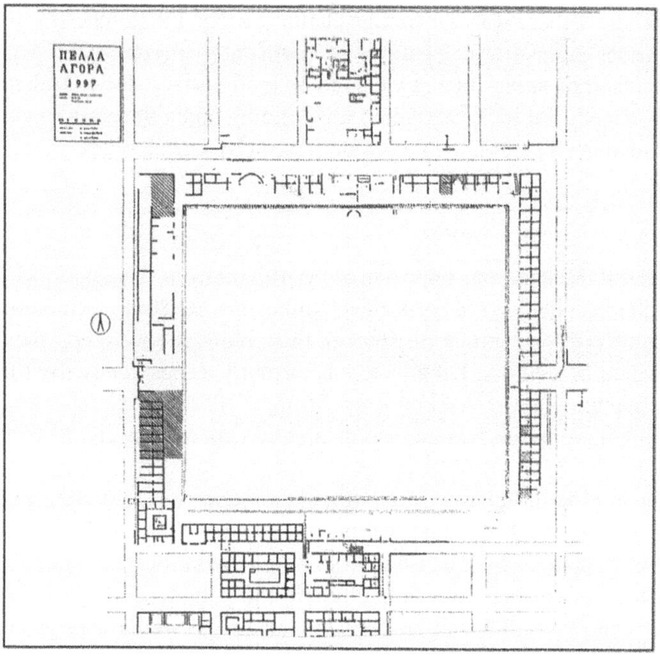

Figure 8.3 Pella: Agora, hydraulic installations for water supply and drainage system (Kaiafa, 2008).

Figure 8.4 Pella: (a) drain running through the stairs in the NE corner and (b) Agora, open stone gutter.

Similar arrangements appear to have been used in the Agora courtyards of Thasos, Thessaloniki and Philippoi. In Pella the collected water from the gutter of the Agora was channelled into another pipe at a lower level which was discharging eventually to the central sewer under the southern horizontal road. The sewer system consisted also of a significant number of surface or underground, built or carved channels, all well planned. The drainage network, due to the territorial inclination was discharging in a reinforced stone built sewer at the southern portico of the Agora, which was enhanced, improved through the centuries and finally expanded southward in the late *ca.* 2nd century BC, so as the Agora buildings were safe from rain waters (Akamatis, 2005) (Figures 8.5a, b, c, and d).

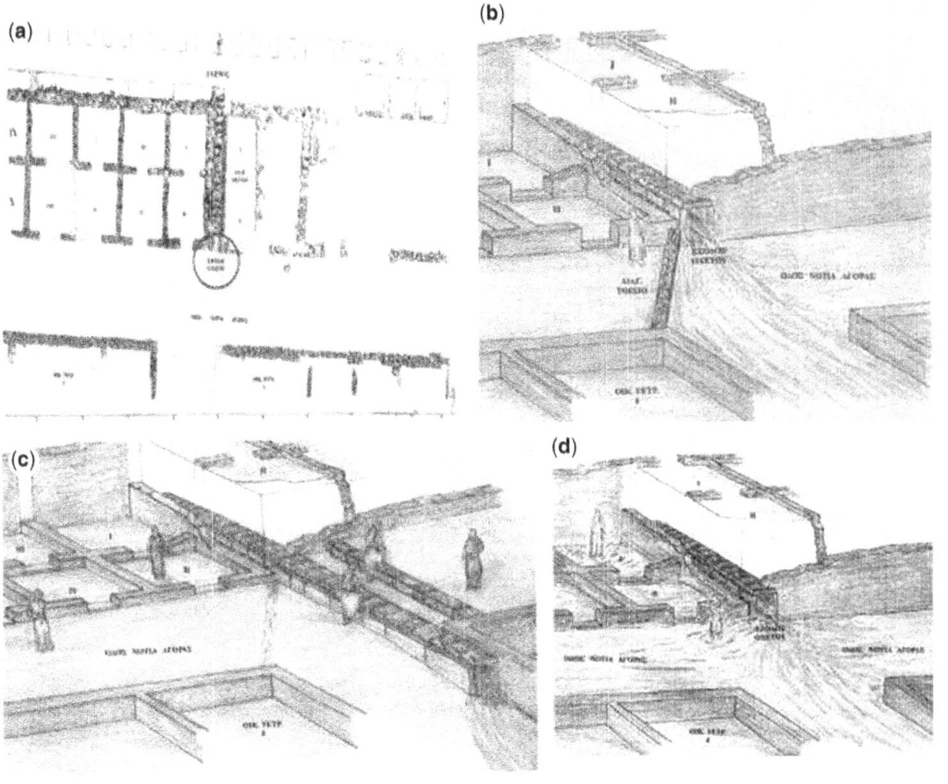

Figure 8.5 Pella: Agora, vertical sewer (a, d, c, and d) in southern portico improvements (Kaiafa, 2008).

8.3.3 Dion

Dion is situated in the southeastern part of Macedonia and it was one of the most important religious centres of ancient Hellenes, as well as during the Roman time. The city of Dion was developed in a smooth plain with no natural protection, next on the eastern foothills of mount Olympos, the highest in Hellas. Many streams were flourishing through the city, directed towards the sea. The most important is Varyfas stream which was flowing along the east walls (Figure 8.6). In the area there were also water springs and sometimes floods were threatening the entire city, which was equipped with drainage systems since Hellenistic Era.

Archaeological evidence certifies major floods that covered part of the city during the *ca.* 3rd century AD. The tumors of the stadium and the Theatre out of the city walls, were creating a protection zone from waters which were flowing to the east. This means that basically these huge embankments were hydraulic engineering projects, constructed to control the city floods. Long digged trenches were constructed along the western and southern city walls for water control and for defence purposes. The trench parallel to the south fortification was discharging water directly to the deep Vafyras stream. It is likely that a similar trench should have protected the north side of the walls, carrying flood waters to the Vafyras stream (Karadedos, 1990a; Karadedos, 2000).

The importance of water in Dion was essential, as shown by the archaeological findings. These findings certify that water, beyond its use for the needs of everyday life, was of particular importance for hygiene, medicine, decoration of buildings or gardens (Stefanidou-Tiveriou, 1998). Moreover, it had a major role in religion. At the same time there were many baths, thermals, impluviums, fountains, latrines. Beyond these, the holy water font/ basin/ sprinkler and the perceived cult of Asclepius and members of his family in the building of great thermae (baths), confirm the different uses of water (Karadedos, 1988).

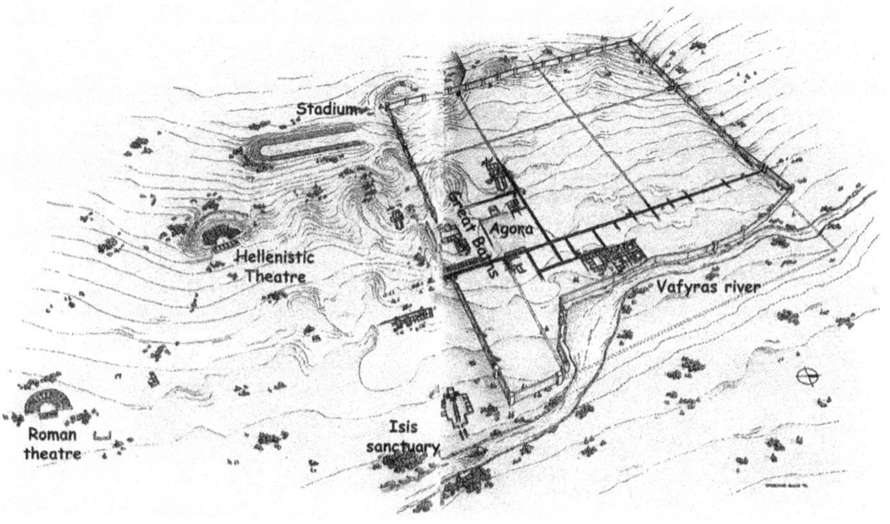

Figure 8.6 Topographic map of the wider area of Dion (Kaiafa, 2008).

Urban water supply systems and installations of Dion is a typical example of how the hydraulic networks of a Roman city were developed, as they have been described by Vitruvius and Frontinus (Karadedos, 1988, 1990b). Water was collected from local sources, at the foothills of Mount Olympos and was flowing by gravity through an underground enclosed vaulted stone built conduit. At the last parts of this conduit the water flew on an arcade ending the main distribution tank (castellum) located at the west side of the city wall. From this tank water flew to the city via three secondary castellums. Pipes of clay or lead were used to branch the supply to individual customers, public fountains and public baths. The well organized network also consisted of smaller cisterns and water-towers at the central part of the city, from which smaller pipes, usually made of lead, were placed under the sidewalks and streets and were serving the various customers. A similar water distribution system, consisting of clay or lead pipes, was also constructed in many other cities of Macedonia in Roman time.

In addition to the long distance water supply pipeline, the water system of Dion included cisterns and especially wells, with circular ground plan, coated with bricks, terracotta rings or stone lined. Apart from the water supply, the city was also equipped with an advanced storm, drainage and sewer system, planned simultaneously with the development of roads and the construction of buildings, which became more complex due to the growing needs created by the presence of baths and latrines. Hydraulic installations were also planned for the building complexes outside the walls, such as the open channel in the periphery of the Theatre orchestra, for the disposal of the rainwater collected from the seats above (Figure 8.7a). The rainwater drainage groove in the Stadion, holy cisterns in the sanctuary of Isis or the hydraulic infrastructure in the sanctuary of Demeter, with the flushing systems of the toilets connected to storm drainage conduit.

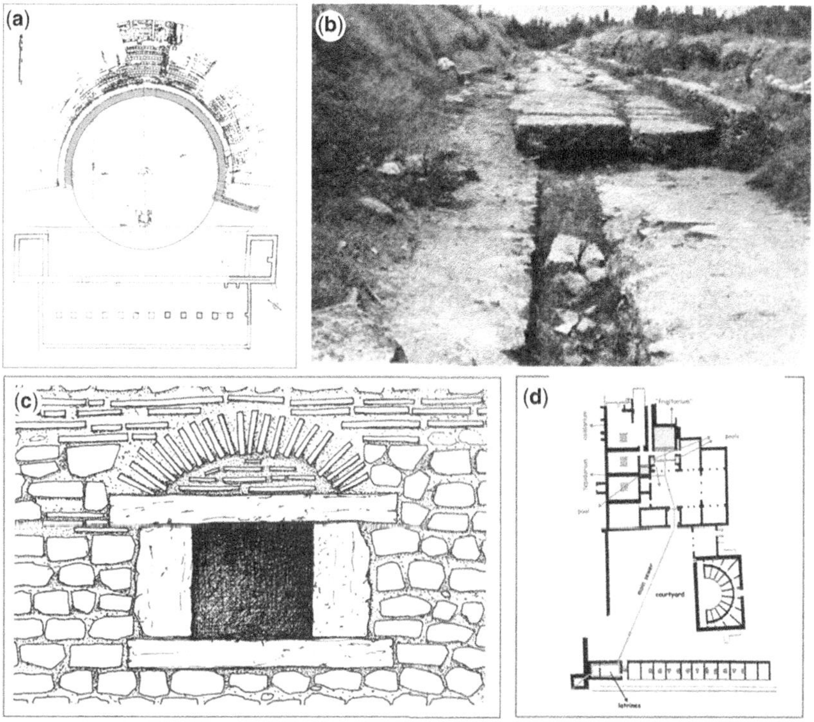

Figure 8.7 Dion: (a) drain in the Theatre (Kaiafa, 2008), (b) drain under the road (photograph by A. Kaiafa), (c) drainage opening in rampart (with permission of A. Kaiafa), and (d) Great Baths, hydraulic infrastructure for water supply and drainage (Modified from A. Kaiafa, 2008).

The urban drainage systems were organized by earthenware or lead pipes, visible rockcut tunnels, underfloor stone or brick built channels, and drains under the roads (Figure 8.7b). Both the households and public buildings had a very sophisticated system to collect and dispose the waste waters. In every case, the drainage network was functioning in conjunction with water supply systems. Part of the drainage plan was the gradients of the roadways toward the centre, which were more intense at the crossroads (Karadedos, 1985; Pantermalis, 1989; Karadedos, 2000). This solution was definitely a supplement to the well organized system of underground drains, which were covered by the slabs of the road surfaces. Rainwater, flowing on the surface was conducted into the drains through the small slots between the roadslabs. The sewers were driven in the Vafyras stream along the east side of the city, through rectangular drainage openings in the fortifications (Figure 8.7c).

Advanced knowledge and expertise of water management indicates the well organized and functional hydraulic network of the Great Baths, located in the south side of Dion (Figure 8.7d). The urban distribution network was feeding a central water tank in the north part of the impressive architectural complex, composed of bathrooms, odeum, latrines and shops, all around a court (Pantermalis, 1989; Karadedos 2000). The hydraulic infrastructure was simple in concept but complicated in its construction. There was provision for the waste disposal of the pools at the main bathroom building for the rainwater drainage from the roofs and for the waters of the floors. The wastes were channelling into clay and lead pipes, built stone conduits, brick-built drains, vertical or horizontal, inside the walls or under the floors. They were all discharging in a vaulted covered drain with manholes, which was developing beneath the central court of the complex, delivering wastes out of the city walls. Before its outflow, water was running under the latrines in the southeast corner of the complex. The sewage of the latrines was using this central duct, a technical solution that was also used in the latrines of Philippoi and Thessaloniki.

8.3.4 Olynthus

Olynthus was built on tow flat-topped hills, in a fertile plain near the neck of peninsula of Kassandra in Chalkidiki, next to the ancient Sandanos river. The Archaic city was built under a provincially urban planning extended throughout the entire hill. The city was rebuilt during the *ca.* 5th century BC on the much larger north hill, following the Hippodamian urban planning. The excavations revealed that the city was brilliantly designed with wide streets and had a remarkable water supply and sanitation system witnessed the high standard of living of the inhabitants (www.culture.gr). There were few natural wells or springs in the area and most residents relied on some kind of public water supply. An underground aqueduct by closed earthenware pressured pipes, with cleaning and venting holes on their surfaces was found coming southward for six miles from the springs on the adjacent mountain of Cholomontas.

In the last part of this conduit, pipes were laid in a vaulted tunnel with cleaning manholes. They were supplying a public fountain at the edge of the city, including a cistern with permanent taps on the front side, from where the water was flowing into a shallow basin (Robinson & Graham, 1938). Intense use of this installation is evidenced by the pottery fragments found around for water transfering. Only a few houses had their own tanks to collect the rainwater from the roof.

Apart from the long distance water supply line, many remains of sanitary facilities had been widely applied in many of the houses of the Classical city. Approximately one third of the houses had a separate bathroom, where in some cases clay bathtubs were placed on a cement floor with drainholes for driving dirty waters out. In many cases waste waters were discharging from the court by inclined the surface to one point. From there, they were simply conducted away from the houses along the centre of the streets, either directly passed through the walls or by means of stone drains. Frequently terracotta pipes from the centre of the floors were channelling the domestic sewage to a drainage path between the two rows of houses on each block, which acted as open gutters (Robinson, 1946) (Figures 8.8a and b). However, there was almost no effort made to control waste water that ended up from the pipes to the streets, which were generally were not paved, except for some areas with pebbles (Protopsalti & Athanasiou, 1997).

Figure 8.8 Olynthus: (a) terracotta drainage pipeline and (b) stone-built drain (with permission of A. Kaiafa).

8.3.5 Thessaloniki

The city of Thessaloniki was founded in 315 BC, on the east coast of Thermaikos Gulf covering a geographic area that enabled sea transportation. The choice of this location was based on the large quantities of drinking water in the area. The street plan of the Hellenistic city was based on the Hippodamean system with perpendicular intersecting streets which defined large blocks. Regular blocks and perpendicular routes were also the characteristic of the planning of the Roman city, based on Hellenistc urban articulation. Over the centuries the city has acquired three aqueducts derived from three different points, the springs of Chortiatis Mountain, of Retziki and of Lebet (Tamiolakis, 1985; Kaiafa, 2008; Papacharalampou *et al.*, 2012). The presence of three well organized water transfer projects depended on the evolution of the city, the architectural development and the population increase, so as to cover everyday water demands. Every source was combined with water distribution networks, developed all over the city, supplying individual consumers and public buildings. Urban hydraulic installations were constructed systematically due to the presence and the spreading of baths in everyday life, mainly during Roman time.

In addition to water distribution systems, Thessaloniki was also equipped with elaborate storm drainage and sewer systems under roadways, more effective and more complicated as the time went by, organized beneath the horizontal and vertical streets. Especially, during the *ca.* 2nd century AD when the political and economic situation of the city allowed much more architectural development and urban beautification, water supply and drainage systems played a major role expressed with various hydraulic installations. Moreover, the lifestyle and sanitation standards were advanced by the extended use of baths and toilets. The systems included earthenware or lead pipes, visible rockcut tunnels (Figure 8.9), underfloor stone or brick built channels, sometimes complicated, which were transferring storm and wastes out of city, into the sea, by discharging them with natural flow in invisible vaulted drains. They were built of stone, with brick floors and arches, straight sided in section, developing beneath the horizontal and vertical streets, usually in their central axis, conducting rain and waste waters to the sea. The internal width ranged between 0.60 and 1.40 m, and their height from 0.80 to 1.20 m. The urban sewerage network was always facilitated by the natural southwards inclination (Kaiafa, 2008, 2010) (Figures 8.10a, b).

Figure 8.9 Thessaloniki: Agora, stone gutter with basin (with permission of A. Kaiafa).

Figure 8.10 Thessaloniki: vaulted drains under the roads (a, b) (with permission of A. Kaiafa).

The well-organized drainage network of the city, coexisting with a well designed water supply system that was branching throughout the urban grid, is recognized among others in hydraulic infrastructures of the Palace of Galerius, or in the Agora complex both characterized by durability, sustainability and functionality.

8.3.6 Amphipolis

Amphipolis was a strategic transportation centre founded by the Athenians and built on the east bank of the Strymon river and the west feet of Mount Pangaion. The city had two long distance aqueducts from the hills, one from NNE and the other from SSE (Mee & Spawforth, 2002), which ensured uninterrupted water supply. The northern aqueduct was constructed by the Athenians in the middle of *ca.* 5th century BC, was gradually cancelled by the southern one, which was built in Roman period. In both aqueducts water was transported by closed terracotta pipes buried in the soil or protected in underground tunnels carved into the rock, not to be exposed. Both were following the topographic contour lines.

Even though archaeological findings of the urban network of the ancient site are not many, as the area is occupied by the modern city, many urban hydraulic installations have been found, such as opened rock cut channels, clay and lead, supply or drainage, pipes. However a high degree of expertise forms the impressive sewerage project of the Classical period at the north side of the fortification (Figure 8.11), where the bank of the river was close enough, so as to be very dangerous for the walls to collapse under the pressure of the inner or outer floodwaters (Lazaridi, 1990). Instead of the usual small sized drain holes, there was an innovation by the addition of vertical trapezoidal slits in the rampart, of 1.65–1.90 m high, with the narrow width side of 0.22 m outwardly. Through them the large quantities of storm water gathered there, due to the soil inclination, and could easily flew out of the city. The system was also ensuring the safety of the walls in case of river overflowing. In the Hellenistic period, a similar system was also adapted, consisting of drainage vertical grilles through the walls, rectangular in shape. It was protecting the foundations, without allowing the passage of an intruder.

Figure 8.11 Amphipolis: Sewerage systems: (a) in the Classical period and (b) in the Hellenistic period.

The Gymnasium of the city was a monumental complex, located next to a stream. Sanitation and drainage infrastructure was as important as the water supply. Supply depended on central distribution water system consisted of branched clay pipes and fed the palaestra, the altar and the cistern of the NW corner. Sanitation system, enriched through the years, included channels of various types. It was based on a large drain running covered with stoneslabs, which was crossing the courtyard and pouring out directly in the nearby stream (Figure 8.12a). It was receiving the wastes of the bathtube rooms of palaestra via two ceramic pipelines joining in a Y shape, running under a monumental staircase (Figure 8.12b). The drainage network also consisted of openings piercing supportive walls, lead pipes or rock cut channels, evacuating the complex of dirty waters. Stone built slab covered drains were also delivering stormwater from Xistos building to the main sewer, protecting it from overflows (Figure 8.12c).

Figure 8.12 Amphipolis: (a) Gymnasium hydraulic infrastructure, (b) large main drain, sewerage pipeline in bathtube rooms of the Palaestra, and (c) drain from Xistos (with permission of A. Kaiafa).

8.3.7 Philippoi

The ancient city of Philippoi was built next to the marshes at the southeast plain of modern Drama, at the feet of mount Lekani. The area around was rich in water sources since ancient times, and for this reason, the first name of the settlement was Krinides (Krini means fountain in ancient Hellenic). However, due to the poor water quality of the area, the city gained water mainly from the springs about 8 km northwest, through an aqueduct barrel vault channel, plastered with water proof limemortar, with cleaning manholes (Gounaris, 2004; Iliadis, 2010).

Apart from the well organized supply system, carrying water to houses and public buildings, attention was paid to sanitation and drainage system (Figure 8.13). An urban network integrating both stormwater drainage and sewerage under all the streets of the city was constructed (Figure 8.14a). Indeed, drains under side streets were feeding the larger collectors below the three main avenues Egnatia, Commercial and Diagonal. Through them storm and wastes were carried out of the city walls ensuring public health. The width of these collectors ranged from 0.60 m to 1.00 m and their height from 0.90 to 1.70 m.

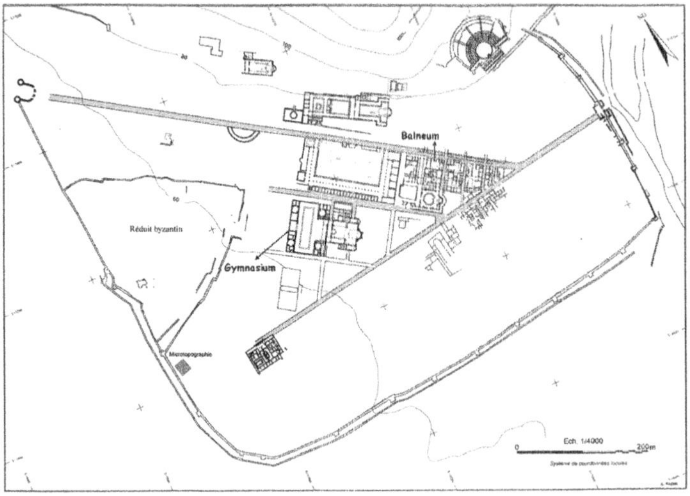

Figure 8.13 Philippoi: urban planning, sanitation and drainage system.

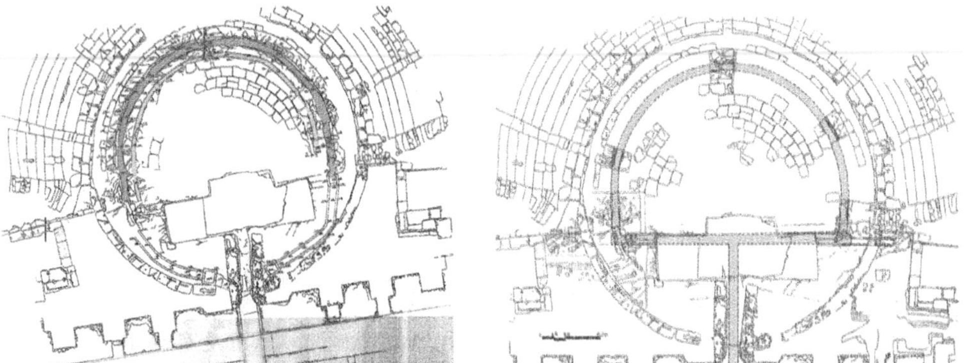

Figure 8.14 Philippoi: Theatre, storm drains (a) Hellenistic and Roman drainage channels (b) drainage systems of Roman times (Kaiafa, 2008).

Organized hydraulic installations were found in the Agora complex, which collected the excess water from the fountains in pipelines, and the rainwater in open stone gutters, leading them to the central collector of Market avenue. The city had also multi-seat public latrines in Balaneio and Palaestra, with flushing systems connected to storm drainage conduits. A very sophisticated sewerage system totally connected to the urban network was also applied at the Theatre of the city (Koukouli-Chrisanthaki & Karadedos, 2000). During Hellenistic time, an uncovered rock cut channel on the basis of the Koilon had trenches in both ends. From them two conduits were derived, joint in front of the Stage on the central axis of the Theatre. Storm waters collected there were pouring out in the urban slab covered drain. Similar arrangement appeared to have been used also in the Roman period, with few technical variations (Figure 8.14b).

8.4 CONCLUSIONS

The presented material from the main cities in ancient Macedonia proves the high level of technology and expertise in the management of water, which is characterized by both simple and more complicated structures, sometimes magnificent, dealing with the collection, conveyance, supply and use of water, the drainage of storm water and control and disposal of wastes.

The collected data, show that ancient Macedonians were definitely aware of the importance of clean water and the availability of water in large quantities in their area. They knew very well the influence of water on public health and how important was to evaluate the quality of the potable water before its consumption. This is also obvious in ancient Hellenic literature. Rivers and streams of Macedonia were imaged on coins and vases as human figures, they were mentioned in Hellenic myths or they were worshiped as gods in temples.

In the Classical, Hellinistic and Roman cities of Macedonia water supply and sanitation were adjusted to the local culture, geomorphology, topography, hydrological and climatic conditions. Finally, all the hydraulic constructions, being a significant factor of living standard, definitely ensured comfortable living conditions for Macedonian citizens.

8.5 REFERENCES

Akamatis I. (1990). Agora of Pella. The Archaeological Work in Macedonia and Thrace. Vol. 4, School of History and Archaeology, Aristotle University, Thessaloniki (in Hellenic), pp. 143–154.

Akamatis I. (2005). Pella. Annals of Investigations, Research Committee of the Aristotle University of Thessaloniki, Issue 9, pp. 4–5.

Andronikos M., Makaronas C., Moutsopoulos N. and Bakalakis G. (1961). The Palace of Vergina. Encyclopedia of ancient History. Wiley on-line library.

Drougou S. (1999). The new tomb of Heuzey. The Archaeological Work in Macedonia and Thrace. Vol. 13, School of History and Archaeology, Aristotle University, Thessaloniki (in Hellenic), pp. 535–540.

Faklaris P. (1996). Vergina: The fortified precinct and the acropolis. The Archaeological Work in Macedonia and Thrace. Vol. 10, School of History and Archaeology, Aristotle University, Thessaloniki (in Hellenic), pp. 69–78.

Ghilardi M., Fouache E., Queyrel F., Syrides G., Vouvalidis K., Kunesch S., Styllas M. and Stiros S. (2008). Human occupation and geomorphological evolution of the Thessaloniki Plain (Greece) since mid Holocene. *Journal of Archaeological Science*, **35**, 111–125.

Gounaris G. (2004). Philippoi. University Studio Press, Thessaloniki, p. 62 (in Hellenic).

Iliadis I. (2010). Philippoi. The water of the ancient city. *Archaeology*, **116**, 56–62.

Kaiafa A. (2008). Water and Sanitation systems in the Hellenistic and Roman periods in Macedonia. In Doctoral Dissertation, Dept. of Architecture, Aristotle University (in Hellenic).

Kaiafa A. (2010). Hellenistic and Roman Thessaloniki: issues of water infrastructure. *Thessalonikeon Polis*, **8**, 23–36 (in Hellenic).

Karadedos G. (1985). Water Supply and Sanitation in Ancient Dion. Archaeologists talk about Pieria, pp. 33–41 (in Hellenic).

Karadedos G. (1988). The wells of ancient Dion. *The Archaeological Work in Macedonia and Thrace*, **2**, 161–171. (in Hellenic).

Karadedos G. (1990a). The aqueduct and the central cistern of ancient Dion. *The Archaeological Work in Macedonia and Thrace*, **4**, 217–229 (in Hellenic).

Karadedos G. (1990b). Hydraulic expertise and technology of the Ancestors. *Scientific Annals of the Faculty of Architecture*, **XII**, 143–172 (in Hellenic).

Karadedos G. (2000). Hydraulic expertise and technology – The case study of Dion. In Water Relations, Water as a Source of Life in Antiquity, University Studio Press, Thessaloniki, pp. 51–68 (in Hellenic).

Koukouli-Chrisanthaki C. and Karadedos G. (2000). Excavations in the Theatre at Philippoi. *The Archaeological Work in Macedonia and Thrace*, **13**, 69–86. (in Hellenic).

Lazaridi K. (1990). The Gymnasium of Amphipolis, city and country in ancient Macedonia and Thrace, in memory of Lazaridis. In Proceedings of Archaeological Conference, Kavala, May 9–11, 1986, pp. 241–273 (in Hellenic).

Mee C. and Spawforth A. (2002). Archaeological Guide of Central, Western, Northern Greece. Leadercom.

Olynthus, Hellenic Ministry of Culture (n.d) www.culture.gr, http://odysseus.culture.gr/h/3/eh352.jsp?obj_id=2481 (Last accessed April 2013).

Pantermalis D. (1989). The Hellenistic theater at Dion. *Archaeology*, **33**, pp. 23–26 (in Hellenic).

Papacharalampou Ch., Melfos V. and Voudouris K. (2012). Water supply and related constructions since antiquity in Retziki (Pefka) of Thessaloniki, Northern Greece. In Proceedings of Third Specialized Conference on Water and Wastewater Technologies in ancient civilizations, March 22–24, Istanbul, Turkey, pp. 154–161.

Protopsalti, S. and Athanasiou, F. (1997). Ancient Olynthos: The Excavations and the works of rehabilitation and promotion of the site. *Archaeology*, **63**, 73–78 (in Hellenic).

Robinson D. M. (1946). Excavations at Olynthus, Part XII, Domestic and Public architecture, 199f. pp. 624–628.

Robinson D. M. and Graham, J. W. (1938). Excavations at Olynthus, Part VIII, The Hellenic House, a Study of the Houses Found at Olynthus with a Detailed Account of those Excavated in 1931–1934. Johns Hopkins University, Johns Hopkins Press, Baltimore.

Saatsoglou-Paliadeli Ch. (1996). The sanctuary of Eukleia at Vergina. *The Archaeological Work in Macedonia and Thrace*, **10,** 55–68, (in Hellenic).

Stefanidou-Tiveriou Th. (1998). Fortification, Excavation of Ancient Dion, 1, Thessaloniki (in Hellenic).

Tamiolakis J. (1985). The history of water supply in Thessaloniki. University Studio Press, 164 p. (in Hellenic).

Voudouris K. (2012). Diachronic evolution of water supply in eastern Mediterranean. In: "Evolution of water supply through the Millennia" A. N. Angelakis, L. W. Mays, D. Koutsoyiannis and N. Mamassis (eds), Chapter 4, pp. 77–89.

Chapter 9

The history of the development of urban sanitation and wastewater technologies in Cyprus

Panayiota Azina and Nicholas Kathijotes

9.1 PROLEGOMENA

Throughout the years and since antiquity, Cyprus has developed maintaining its own civilization, and assimilating various influences to which it has been exposed. These influences can not only be detected in buildings and temples but also in the drainage systems of these buildings found in archaeological sites. Ancient drainage systems have been discovered by various archaeological missions at Paphos, Kourion, Kition, Salamis, and other locations of Cyprus. Since ancient times, Cypriots have disposed of their excreta and other wastes on the site where wastes were generated. The first disposal technologies were probably developed in response to a need for some kind of management scheme to minimize the aesthetic and hygiene problems, associated with improper waste disposal and general well-being. It appears that something akin to a crude latrine was firstly used to minimize the offensive impression created by the indiscriminate discharge of human excreta on the ground. The pit latrine or simple privy alleviated the nuisance and sufficed until 1950, when the installation of house to house water supply for all towns and villages started. From 1950, the necessity for the disposal of liquid wastes came from the development of the water-carriage waste system. The first disposal system was a cesspool or soak-away. Septic tanks – absorption pits or ditch systems are still used in most villages in Cyprus. These decentralized systems do exhibit many economic and technical benefits. In the towns of Cyprus today, the households are connected to main sewerage systems that convey sewage to treatment plants with most of the produced treated effluent offered to farmers for agricultural use.

9.2 PHYSICAL SETTING

9.2.1 Location

Cyprus is the third largest island in the Mediterranean after Sicily and Sardinia with an area of 9251 km^2. It is located at the north-east corner of the Mediterranean basin at latitude of 34°33′ to 34°34′ North and longitude 32°16′–32°37′ East, 300 km north of Egypt, 105 km west of Syria, and 75 km south of Turkey.

Hellas lies 380 km to the north-west (Rhodes – Karpathos). Cyprus strategic position at the crossroad between three continents Europe, Asia and Africa has affected its development throughout the centuries. A map of Cyprus is shown in Figure 9.1.

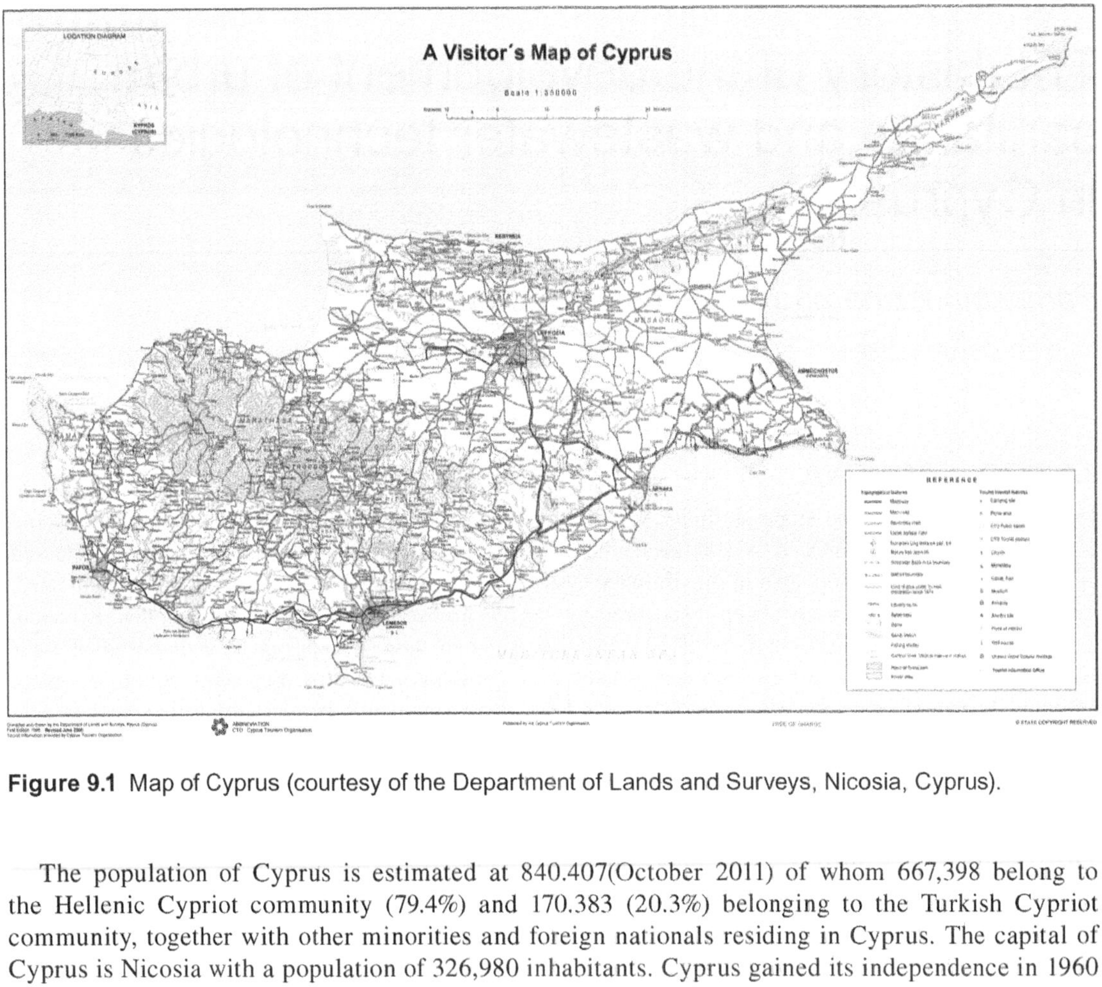

Figure 9.1 Map of Cyprus (courtesy of the Department of Lands and Surveys, Nicosia, Cyprus).

The population of Cyprus is estimated at 840.407(October 2011) of whom 667,398 belong to the Hellenic Cypriot community (79.4%) and 170.383 (20.3%) belonging to the Turkish Cypriot community, together with other minorities and foreign nationals residing in Cyprus. The capital of Cyprus is Nicosia with a population of 326,980 inhabitants. Cyprus gained its independence in 1960 after the Zurich-London Treaty. On July 20, 1974 Turkish troops invaded Cyprus occupying 36.2% of the island, and besides many UN resolutions calling Turkey to withdraw its troops, the situation as of today remains unchanged. On 1st May 2004 Cyprus joined the European Union together with nine other countries.

9.2.2 Climatic conditions

Cyprus has a Mediterranean climate with hot, dry summers from mid-May to mid-September and mild, rainy winters from November to mid-March which are separated by short autumn and spring seasons of rapid change in weather conditions. The central Troodos massif rising up to 1951 m and the long narrow Kyrenia mountain range with peaks of about 1000 m, have a significant role in the meteorology of Cyprus. The predominantly clear skies and sunshine create seasonal and daily differences between the coastal and interior of the island. Cyprus daylight varies from 9.8 hours in December to 14.5 hours in June.

In summer the island is mainly under the influence of a shallow trough of low pressure extending from the great continental depression centred over southwest Asia. Summer is a season with high temperatures and almost cloudless skies. Rainfall is negligible but isolated thunderstorms sometimes occur especially on the mountains. Winter in Cyprus is affected from a track of fairly frequent small depressions which cross the Mediterranean Sea from west to east between the continental anticyclone of Eurasia and the generally low pressure belt of North America. These depressions give periods of disturbed weather usually lasting from one to three days and produce most of the annual precipitation; the average rainfall occurs from December to January and is about 60% of the annual cumulative precipitation (Hadjioannou, 1997).

9.3 HYDROLOGY

9.3.1 Precipitation

The average annual total precipitation is 450 mm at the south western windward slopes but rises to nearly 1100 mm at the top of the central massif. On the leeward slopes amounts decrease steadily northwards and eastwards ranging between 300 and 350 mm in the central plain and the flat south-eastern parts of the island. Pentadaktylos range along the north of the island produces a relatively small increase of rainfall to about 500 mm along its ridge of about 1000 m. Rivers in Cyprus are seasonal and flow usually following heavy rain.

Rainfall in the warmer months does not contribute to water resources and agriculture. The small amounts which fall are rapidly absorbed by the very dry soil and evaporated in high temperatures and low humidity. Autumn and winter rainfall on which agriculture and water supply generally depend is somewhat variable. The average rainfall for the year as a whole is about 480 mm (covers the period 1951–1980). Statistical analysis of rainfall in Cyprus reveals a decreasing trend of rainfall amounts in the last 30 years (Hadjioannou, 1997).

Studies of the climatic conditions in the last 2000 years showed that drier hydro climatic conditions existed between AD 1400 and the twentieth century, a wetter phase during the period AD 950–1350, and a marked dry-to-wet climatic transition between 400 and 600 AD (Lionello, 2012). Other studies in the Mediterranean region during the Holocene period revealed varying climatic conditions during the last 5000 years. Despite these varying conditions there was never an abundance of water resources in the Hellenic cities of significant cultural development like Athens, Knossos, and Cyclades islands (Angelakis *et al.*, 2005; Antoniou *et al.*, 2012). These conditions apply for Cyprus as well.

9.3.2 Air temperatures

Cyprus has a hot summer and mild winter but this generalization must be modified by consideration of altitude, which lowers temperatures by about 5°C per 1000 m and of marine influences which give cooler summers and warmer winters near most of the coastline and especially on the west coast. The seasonal difference between mid-summer and mid-winter temperatures is quite large at 18°C inland and about 14°C on the coasts. Differences between day maximum and night minimum temperatures are also quite large especially inland in the summer (Hadjioannou, 1997).

9.3.3 Winds

Over the eastern Mediterranean generally surface winds are mostly westerly or south-westerly in winter and north-westerly or northerly in the summer. Usually of light or moderate strength, they rarely reach gale force. Over the island of Cyprus however winds are quite variable in direction and strength. Differences of temperature between sea and land which are built up daily in predominant periods of clear skies in the summer cause considerable sea and land breezes, whilst these are most marked near the coast they regularly penetrate far inland in the summer reaching the capital, Nicosia, and often bringing a welcome reduction of temperature and also an increase in humidity (Hadjioannou, 1997).

9.4 DEVELOPMENT OF SANITATION AND WASTEWATER THROUGH THE MELLENIA

9.4.1 Prehistoric Times

Neolithic Age (ca. 8500–3900 BC). The first Cypriot settlers arrived on the island sometime in the 9th or 10th millennium BC. They traveled in primitive boats, probably originating from Anatolia to the north of the island, and from the Syrian shore to the east. They brought with them a wide variety of mainland plants and animals, including goats, pigs, deer, foxes, dog, wheat, and barley. The daily life of the people was spent in farming, hunting, and animal husbandry.

Settlements developed near the north and south coast of the island but also very close to springs or rivers for easy access to water. Chirokoitia, Kalavasos, Lemba, Parekklisia, Kissonerga are some of the first settlement sites. Water wells excavated in Kissonerga – the earliest in the world so far discovered – demonstrate that the inhabitants lived in an organised community and were able to supply their settlements with water when surface water was not available (Karageorghis, 2012). These first settlers probably disposed their excreta and other wastes on the site where wastes were generated.

Chalcolithic Age (ca. 3900–2500 BC). The Chalcolithic Age was the transitional period between the Stone Age and Bronze Age. Although shorter in period it is known for the population growth, the development of sophisticated art, advanced religious beliefs and symbolisms and the beginning of some relations with neighboring countries. The last phase of the Chalcolithic period was marked by the introduction of small metal objects made of natural copper (Karageorghis, 2012). Despite of the population growth excavations did not reveal any findings that could indicate how those people perceived sanitation and dealt with wastewater management.

Bronze Age (ca. 2500–1050 BC). During the Bronze Age Cyprus witnessed a rapid transformation of technology and economy as well as population growth. Trading flourished with Near East, Egypt and the Aegean. Mycenaeans arrive on the island and settle spreading the Hellenic language, religion and customs. The first city-kingdoms of Paphos, Salamis, Kition and Kourion were established during this period. Population growth increased the demand for water supply but also for wastewater management. Excavations have revealed that the use of stone cut conduits for the supply of large quantities of fresh water and the removal of wastewater from settlements started this period. In Alassa, Limassol excavations have exposed conduits carved in stone for the collection and storage of storm water. In the densely populated city of Engomi near Salamina with great water needs, wells did not have the capacity to supply enough water for all irrigation and domestic water supply needs, so people developed a highly sophisticated rainwater collection system. Storm water was collected in every house through terracotta conduits, similar to those used in Amathus (Figure 9.2a) and was stored in cisterns in the house yard thus operating a rainwater harvesting process (Griva, 2008).The detailed section of this terracotta pipe is shown in Figure 9.2b.

Figure 9.2 (a) Round terracotta pipe installed under the floor possibly carrying storm-water and (b) Terracotta pipe with inspection aperture from Amathus (with permission of P. Azina).

Geometric Period (ca. 1050–750 BC). The eighth century BC is marked by an increase of wealth in Cyprus. It was a period of high prosperity as trading with the east and west was rising. Excavations of this time period did not reveal any major differences regarding the sanitation and wastewater management compared to the Bronze Age.

9.4.2 Historical Times

Archaic and Classical Period (ca. 630–323 BC). The era of prosperity continue, but the island falls prey to several conquerors. Cyprus Kingdoms become successively tributary to Assyria, Egypt and Persia (CTO, 2001). King Evagoras of Salamis (who ruled from 411–374 BC) unifies Cyprus and makes the island one of the leading political and cultural centers of the Hellenic world. In 333–325 BC the city-kingdoms of Cyprus welcome Alexander the Great, King of Macedonia, and Cyprus became part of his empire.

The water resources technologies were further developed especially since water demand is increasing due to the growing cities. In Amathus for example a very important finding was the underground Nymphaeum, an ancient tunnel at least 120 m long with a water spring (Kambanellas, 2012). After the destruction of the Nymphaeum by the earthquakes of 15BC and 76/77 AD, a reservoir and a fountain were constructed. Further excavations in Amathus have revealed a well preserved stone conduit (pipeline sculptured in stone) which was most probably used for the supply of freshwater from the water reservoir (Figure 9.3). It consists of carved stone pieces of variable length but consistent internal and external diameter. The variable length is probably a result of the stone pieces they had available to carve the conduits. The gradient is steady and relative to the ground elevation. Square stone pieces were present to support the conduits at given distances of about 2 m preventing settlement (Figure 9.3b). Apertures also exist on every piece of pipe covered by a flat stone and sealed with lime mortar for access in case of blockage or when ventilation was required.

Excavations in Amathus also revealed an under floor storm drainage system (Figure 9.4). This is a square stone channel that runs along the western side of the Agora, a map of which is shown in Figure 9.5.

Figure 9.3 (a) Stone conduit (sculptured stone water pipeline), with equally spaced small apertures used as inspection holes and (b) Detailed stone conduit connection point found in Amathus (with permission of P. Azina).

Figure 9.4 (a) Stone channel collecting storm water and (b) Under-floor drainage ditch (with permission of P. Azina).

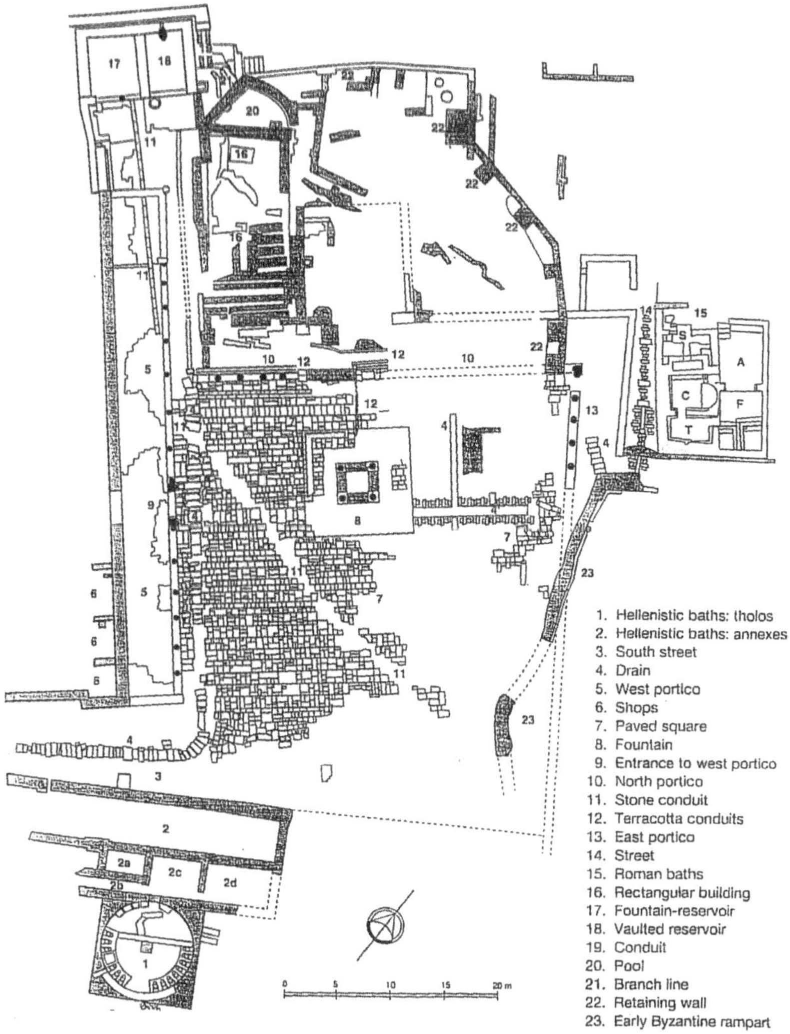

Figure 9.5 Plan of the Agora in Amathus (courtesy: Department of Antiquities, Nicosia, Cyprus).

Hellenistic Period (ca. 323–67 BC). Following the death of Alexander the Great, the Ptolemaic rule brought peace and prosperity to Cyprus. The island was unified; the Ptolemies established an efficient administration on the island dealing with various matters, such as economic, political, and cultural. Paphos became the capital in *ca.* 1st century BC. Salamis and Amathus also retained their economic importance. Culturally Cyprus was considered as part of the Hellenic world. The high officials of the Ptolemaic administration embellished Paphos with various public buildings (Karageorghis, 2012).

In Amathus excavations of the agora in the 1980s have revealed a public bath building (balaneion) consisting of an enclosed circular space and adjacent rooms and hallways (Figure 9.6). A hall realized as a roofed palaestra (exercise room) or as a dromos (race course) is present. Patrons could come to this area for exercise and then proceed to the circular bathing area. In the center of the bathing area one can observe the foundations for the furnace and of the hot water reservoir (Figure 9.6a). Two groups of eight little baths (*pueloi*), which are narrow and lacking drain holes, are aligned against the east and west walls in the southern area (Figure 9.6b). Here servants (parachytai) with buckets could rinse the patrons with cold or warm water. This circular room was partially remodeled in the second century BC when five of the little baths were replaced by larger plunge baths (Figure 9.7) in which the user could be submerged (Aupert, 2000). As no drains were found in the *balaneion*, it can be assumed that grey water was removed by the servants possibly disposing it to the large drain system which runs westwards under the street collecting storm water and the runoff from the Nymphaeum and reservoir.

Figure 9.6 Hellenistic Baths in Amathus: (a) view and (b) detailed view of the *'Pueloi'* (little baths) inside the baths (with permission of P. Azina).

Figure 9.7 Plunge baths in Hellenistic Baths in Amathus (with permission of P. Azina).

Roman Period (ca. 67 BC–330 AD). The Ptolemaic rule was weakened by internal problems, and consequently was not able to face the rising power of Rome which in 58 BC annexed the island to the Roman Empire (Karageorghis, 2012). The Roman rule was ruthless exploiting the wealth of the island in terms of metals, timber and agricultural products. The influence of the Romans is evident in many cultural and architectural developments. In the capital city of Paphos, as well as in other provincial towns, spacious villas with polychrome mosaic floors and wall paintings, theatres, stadia, baths, public roads linking the various cities and others, were found.

In Roman times sacred values were associated with the water element, originating from its importance to human life (Monteleone, 2008). Sewers and water pipes were not though invented by the Romans; they existed in other Eastern Civilizations as old as the Minoans (Angelakis *et al.*, 2005; Koutsoyiannis *et al.*, 2008). Romans have perfected them and applied this knowledge into major infrastructural works to serve all citizens (Lofrano & Brown, 2010).

The House of Theseus in Paphos is dated back to the late Hellenistic period but was built on ruins of earlier houses probably belonging to the second half of the second century AD. In the course of years the House of Theseus grew up to become the largest residential building in Cyprus. Even though not entirely excavated, it now has over one hundred rooms, corridors and passages (Figure 9.8).

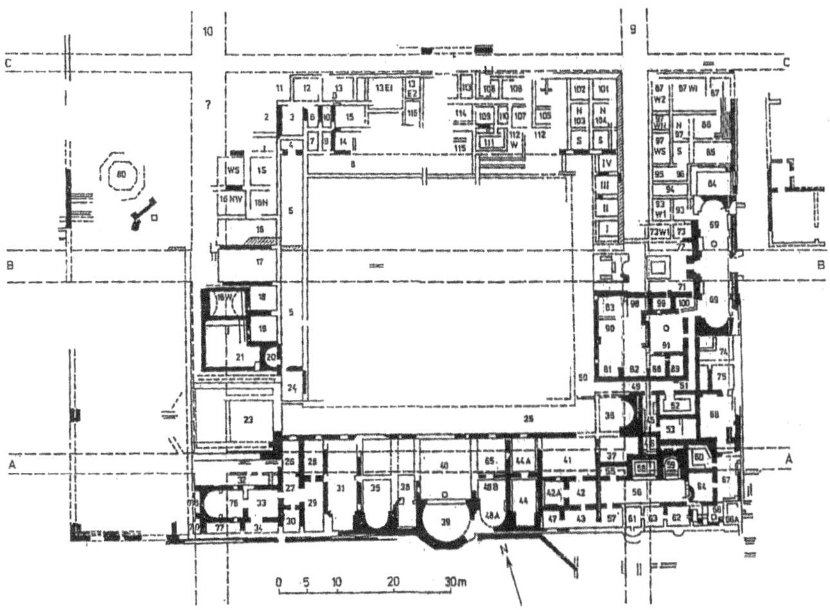

Figure 9.8 House of Theseus plan in Paphos (courtesy: Department of Antiquities, Nicosia, Cyprus).

The bath of this huge villa is inserted within the original plan. It was build upon a dissymmetrical plan in a relatively small thermal complex. It was nevertheless well organized, comfortable, appearing as a luxurious construction. A large frigidarium (cold pool area) contained two big and one small basin for cold water. The spacious warm part, composed of a *tepidarium* (warm bathroom), *sudatorium* (vaulted sweating-room) and *caldarium* (room with a hot plunge bath) was provided with three warm basins. Hot water was distributed by means of lead pipes and cold water by terracotta pipes. An ingenious system of drains allowed for the diverting of used water for flushing latrines (Figure 9.9). This shows that the Romans realized that they could recycle wastewater from spas and baths by using it in flushing latrines before discharge into sewers (Lofrano & Brown, 2010).

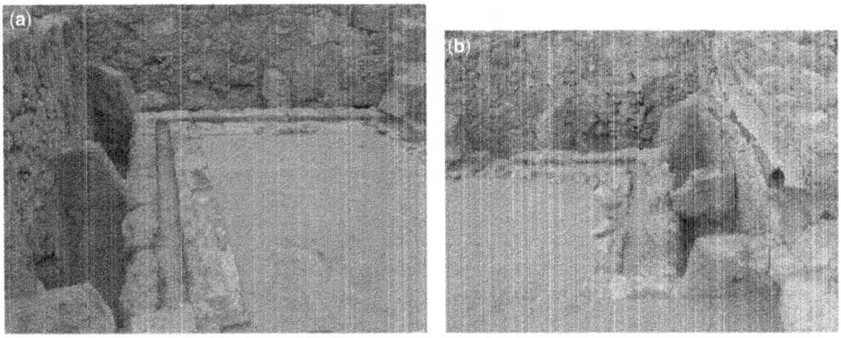

Figure 9.9 (a) Latrines in the house of Theseus and (b) Detailed view of water supply through a terracotta pipe in the wall (with permission of P. Azina).

The latrines found in the House of Theseus are very similar to the ones found at different archaeological sites throughout the Roman Empire. They could usually accommodate 12–14 persons (Figure 9.9). Although only the vertical dividing stone walls are present, when comparing it to other similar findings we can assume that there was also a stone plate covering the void space between the floor and the seat. The vertical support stones are approximately 1.2 m apart. A terracotta pipe in the wall (Figure 9.9b) is likely a provision of fresh water supply. As previously mentioned, grey water for cleaning and flushing was provided through drains from the adjacent baths and pools shown in Figure 9.10.

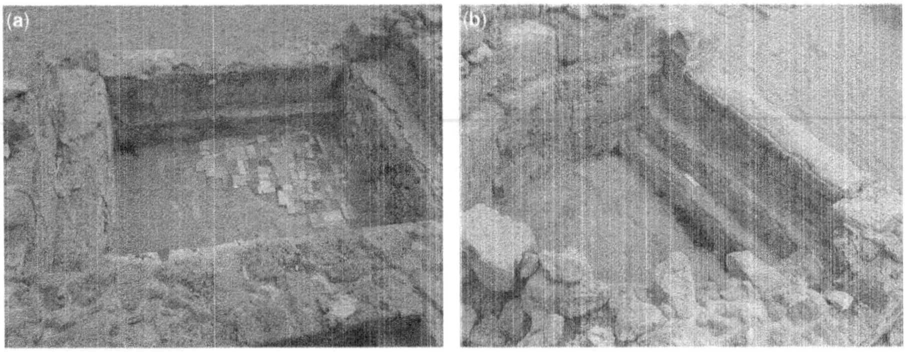

Figure 9.10 Different shapes of pools at the House of Theseus (with permission of P. Azina).

In some parts the building was probably provided with an upper floor (Karageorghis, 1985). This is supported by the fact that a vertical terracotta pipe was found which is shown in Figure 9.11, and was perhaps receiving and conveying storm water from the roof for storage and eventual use. Another option is that this pipe could have been connected to sanitary fixtures of the upper floor in order to carry and dispose wastewater.

Figure 9.11 House of Theseus bath or reservoir. A vertical terracotta drain pipe is visible, possibly supplying water from an upper floor (with permission of P. Azina).

The house of Aion, located next to the house of Theseus, was also excavated. Inside this house which obviously belonged to a wealthy citizen, one can enjoy a wonderful mosaic floor design. A floor drainage provision is available for cleaning and drainage purposes (Figure 9.12a). Just outside this room one can see the well-organized drainage system (Figure 9.12b). Findings then at the House of Theseus, as well as in other sites throughout the island show that storm water and sewerage drainage systems were carefully planned.

Figure 9.12 Drain within the House of Aion: (a) Internal drain and (b) external outlet of drain (with permission of P. Azina).

Between the House of Aion and the Villa of Theseus excavations have revealed a large sewer (width at top: 0.70–0.75 m, and at bottom 0.38 m) cut in the rock to the depth of 1.00–1.20 m built up with blocks and covered by stone slabs (Figure 9.13). The construction of this sewer is dated back to the founding of the city and discharged sewage at an outlet at the port (Mlynarczyk, 1990). Under the road and on the opposite side water supply terracotta pipes were also present. Terracotta pipes found in many locations were probably used for water supply as well as for wastewater drainage. These pipes had a length between 0.30 and 0.50 m and an inner diameter between 0.14 and 0.20 m. These pipes are 'wheel-made', interlocking, with uniform spigot and socket ends formed on molds to secure uniformity. The joints were also sealed with mortar (Haut & Viviers, 2012). In some cases a small inspection aperture was available.

Figure 9.13 Combined sewer drain (left), and water supply terracotta conduit (right) (with permission of P. Azina).

In Kourion (about 50 km east of Paphos), the remains of the House of Eustolios have been found. This House is rebuilt on the ruins of an earlier palatial private residence which was completely destroyed by the earthquakes of the *ca.* late 4th century AD. The building complex consisting of more than thirty rooms and a bathing establishment the house was used until *ca.* mid 7th century AD when it was again destroyed during the Arab raids and finally abandoned (Christou, 1996). The bathing unit as all the typical baths of that period combined bathing and sanitary facilities with relaxation and recreation units is shown in Figure 9.14. Gray water from the baths was used to clean the latrines (Figure 9.15a). The large quantity of water needed was stored in cisterns (Figure 9.15b). Rain water probably supplied these cisterns too. Stone clad rain channels probably used for rainwater harvesting are present on site (Figure 9.16). One can notice that the channel width is about twice the channel depth indicating that the people of the period had some practical knowledge of basic hydraulics. These sections are the most efficient in open channel flow. It is also evident that people of that period had the expertise and knowledge as well as the environmental consciousness to develop mixed drainage and sewerage systems, something that can be seen in today's modern societies.

Figure 9.14 Water storage tank, possibly bath basin (with permission of P. Azina).

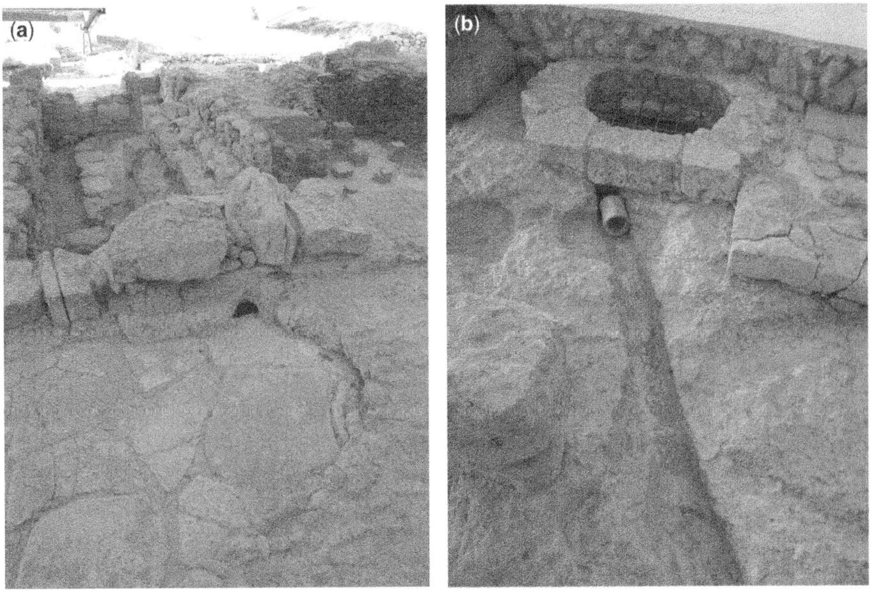

Figure 9.15 Drains from House of Eustolios in Kourion: (a) Drain system and (b) Stone cut channel/ terracotta pipe and reservoir (with permission of P. Azina).

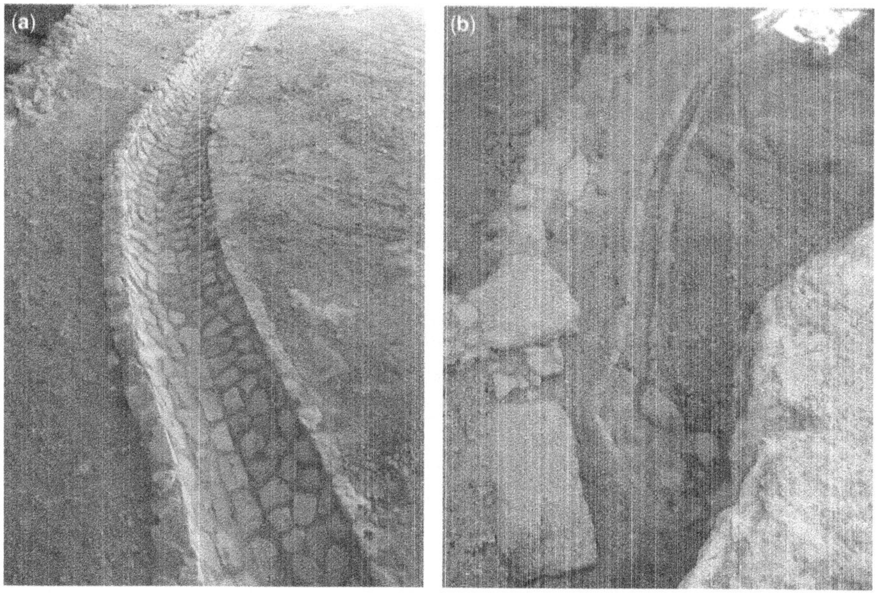

Figure 9.16 Stormwater drain and sewer in Kourion: (a) drain channel and (b) sewer (with permission of P. Azina).

9.4.3 Medieval times

Byzantine Period (ca. 330–1190 AD). The cities of Cyprus were destroyed by two successive earthquakes marking the end of the Roman Period. Following the division of the Roman Empire, Cyprus became part of the Byzantine Empire with Constantinople as the capital. This period is also marked by the Arab raids which forced the inhabitants to move towards areas less visible from the sea. They became poorer and started a new way of living, settling around sources of water like a shallow well in a lowland area or a spring on the mountains. They used terracotta pots or wineskins (storage containers made out of animal skin) for the transportation of water from the source to their houses (Kambanellas, 2012).

Frankish (Lusignian) period (1191–1489 AD) and venetian period (1489–1571 AD). The Frankish and the Venetians took over the land and organized the state on feudal principles. Agriculture was an important part of the economy during that period. Long aqueducts were constructed during this period and used for agricultural irrigation. No significant changes are observed in the domestic water supply or wastewater management.

9.4.4 Modern times (1571–1960 AD)

The Ottoman Period (1571–1878 AD). Cyprus became part of the Ottoman Empire in 1571. During this period the Turks had the control over the water resources of the island. Various chains-of-wells were constructed for good quality domestic water supply (Kambanellas, 2012). Ottoman baths (Hammams) were also constructed throughout the island. Water was important in the Koran and although Turks have made progress in the provision of both domestic and irrigation water supply, no significant progress was made in wastewater management technologies. The pit latrine or simple privy were probably still used in the towns and villages.

The British Period (1878–1960 AD). During this period Cyprus was a British colony. The British administration until the end of the second world gave no significant consideration for the development and provision of potable domestic water. It was later during the 1950s that the first villages house to house water supply was implemented (Kambanellas, 2012). Even during this period no significant changes would occur regarding wastewater management.

9.4.5 Present time (Republic of Cyprus) (1960–today)

The 20th century witnessed a European revolution in wastewater management, environmental science and societal views towards pollution. Governments began to implement wastewater treatment and wastewater treatment facilities which were constructed in major European cities (Lofrano & Brown, 2010).

In Cyprus following the independence of the island in 1960 the government made significant efforts for the supply of water to all villages and towns. Since 1967 all towns and villages of Cyprus have been serviced with a piped water supply. With urbanization cities had to construct septic systems or holding tanks and cesspools or soak-aways in buildings fitted with modern sanitary fittings and plumbing to discharge domestic sewage. Septic tanks – absorption pits or ditch systems are currently used in most villages in Cyprus. With this way there is decentralization which has many benefits like economies of scale and so on.

Today in all cities of Cyprus, the majority of households are connected with main to sewerage systems. Due to the high clay content, the soil was not capable of absorbing the seepage from the growing number of septic tanks and soak-aways. The sludge from the septic tanks that often needed to be collected was usually discharged to unauthorized drying beds creating health and nuisance problems, a current problem still existing in developing countries (Kathijotes, 2012). This led to the need for centralized wastewater systems. In 1968 a study was undertaken on the sewerage and drainage system of Nicosia with the first construction of a sewerage and storm water drainage system in Cyprus. The construction began in 1972 but the Turkish invasion halted the completion of the system which was finally completed in 1979 (Sewerage Board of Nicosia, 2013; Charalambous *et al.*, 2011). The system was upgraded in 1995 and in 2000 in order to accommodate increasing demand. According to European Directive UWWTD 91/271/EEC, Cyprus has to establish wastewater infrastructure for all the agglomerations with population equivalent (p.e.) over 2000, in order to protect the water bodies and the environment in general.

The Water Development Department (WDD) is currently (2014) providing about 55 Mm^3/year of tertiary treated sewage water for both landscape and agricultural irrigation. About 15% of treated effluent is also used for aquifer recharge (Charalambous *et al.*, 2011), often necessary to protect the island from sea-water intrusion (Kathijotes & Panayiotou, 2013). Treated water from the sewerage treatment plant of Paphos is mainly used for aquifer recharge (Yiannakou & Larkou, 2008).

The need for water conservation has lead to the development of new technologies for domestic application like the high efficiency flushing toilets. These toilets use significantly less water than conventional flushing toilets. This system uses the pressure from the water supply line to provide the energy needed to complete the flush. The Water Development Department of Cyprus offers subsidies for water saving measures like for the development of a borehole for garden irrigation and toilet flushing together with installations of grey water recycling systems. These measures are very important aiming at creating environmental consciousness to people in addition to water resources conservation.

9.5 EPILOGUE

Cyprus throughout its history has experienced multiple periods of severe water shortage due to climatic conditions. Reuse of water was and is now absolutely necessary in order to maximize the utilization of the available resources. Civilizations as early as the Minoan in Crete have realized the importance of wastewater management not only for sanitary purposes but also for reuse as a mean of conservation of water resources. Mediterranean areas like Cyprus will continue to be challenged in the area of water supply not only due to population growth but also to the changing climatic conditions, decrease in precipitation and rise of temperatures. It is imperative as a result to be able to reuse our water. This chapter shows that since antiquity early Cypriots have always harvested rain water and reused grey water for cleaning latrines and other sewage fixtures. This shows the way for sustainable water resources management that cannot only lead to conservation of our precious resource but also to the protection of surface waters from pollution.

9.6 REFERENCES

Angelakis A. N., Koutsoyiannis D. and Tchobanoglous G. (2005). Urban wastewater and storm water technologies in ancient Greece. *Water Res.*, **39**, 210–220.

Antoniou G., Kathijotes N. and Angelakis A. N. (2012). Historical Development of the Technologies on Rainwater Harvesting in the Hellenic Civilization. In The 3rd IWA-Rain Water Harvesting And Management Conference & Exhibition. Goseong, Gyeongnam, May 20–24, 2012, Republic of South Korea.

Aupert P. (2000). Guide to Amathus. The Bank of Cyprus Cultural Foundation in Collaboration with the Department of Antiquities. Nicosia, Cyprus.

Charalambous K., Bruggeman A. and Lange M. A. (2011). Policies for Improving Water Security, the Case of Cyprus'. Seventh Framework Programme, Climate Change, Hydro-conflicts and Human Security, Nicosia, Cyprus.

Christou D. (1996). Kourion. Its Monuments and Local Museum. Filokipros, Nicosia, Cyprus.

Cyprus Tourism Organization (CTO). (2001). Cyprus. Cyprus Tourism Organization, Nicosia, Cyprus. Elsevier Publications (book).

Griva E. (2008). Powered Database Ancient Hydraulic Works. Interdisciplinary Interdepartmental Postgraduate Studies (IIPS) 'Environment and Development, NTUA, Athens, Hellas (in Hellenic).

Hadjioannou L. (1997). The Climate of Cyprus. Meteorological Service, Nicosia, Cyprus.

Haut B. and Viviers D. (2012). Water supply in the middle east during Roman and Byzantine periods. In: Evolution of Water Supply throughout the Millennia, A. N. Angelakis *et al.* (eds), IWA publishing, London, UK, pp. 318–350.

Kambanellas C. A. (2012). Historical development of water supply in Cyprus. In: Evolution of Water Supply throughout the Millennia, A. N. Angelakis *et al.* (eds). IWA publishing, London, UK, pp. 301–317.

Karageorghis V. (2012). Ancient Cyprus. 9000 Years of Culture. Ministry of Foreign Affairs, Nicosia, Cyprus.

Kathijotes N. (2012). Wastewater management in developing counties: nutrient input control in coastal cities. *Elsevier Procedia Behavioral Sciences*, **42**, 259–263.

Kathijotes N. and Panayiotou Ch. (2013). Wastewater reuse for irrigation and sea water intrusion: evaluation of salinity effects on soils in Cyprus. *Journal of Water Reuse and Desalination*, **3**(4), 392–401.

Koutsoyiannis D., Zarkadoulas N., Angelakis A. N. and Tchobanoglous G. (2008). Urban water management in ancient Greece: legacies and lessons. *ASCE, Journal of Water Resources Planning and Management*, **134**, 45–54.

Lionello P. (ed.). (2012). 'The Climate of the Mediterranean Region-From Past to the Future'. Elsevier, London, UK and New York, USA.

Lofrano G. and Brown J. (2010). Waste water management through the ages: history of mankind. *Science of the Total Environment*, **408**, 5254–5264.

Mlynarczyk J. (1990). 'Nea Paphos III'. Centre of Mediterranean Archaeology of the Polish Academy of Sciences and Warsaw University, Warsaw, Poland.

Monteleone M. C. (2008). Evolution of the relation between human society and urban water from ancient roman times to modern urbanization. *Water Science & Technology Water Supply*, **8**(5), 551–556.

Sewerage Board of Nicosia. http://www.sbn.org.cy/cgibin/hweb?-A=1&-V=about (accessed 27 November 2013).

Yiannakou-Larcou A. (2008). Reuse of treated effluent in Cyprus. 48th ECCE Meeting. Water Development Department, Republic of Cyprus Ministry of Agriculture, Natural Resources and Environment, Larnaca, Cyprus, 17–18 October 2008.

Chapter 10

The history of sanitation and wastewater management in Portugal

Santino Eugénio Di Berardino

10.1 HISTORICAL CONTEXT

The civilization of Iberian Peninsula was influenced by invasion of others ancient cultures. Early in the first millennium BC several waves of Celts invaded Portugal from central Europe, settled in the North mixing with the natives, shared a common ancestry and formed different ethnic groups, with many tribes. In the southern and central parts of the country the native ancient peninsular people, the 'Lusitanos', received a major influence from Eastern Mediterranean peoples, such as the Phoenician, the Greek, the Byzantine and the Carthaginian. The Phoenicians founded semi-permanent coastal trading posts, later replaced by the Carthaginians.

The Romans stayed for 700 years in the peninsula dominating almost all the territory, playing a major role in the civilization of the Iberians. The governance and culture of Rome induced a profound and lasting influence on the language, religion, architecture, philosophy, law and administration of the territory under their control. They left a considerable patrimony. A witness demonstrating the relevant engineering knowhow: bridges, roads, buildings, water supply networks, and so on, many of them still operating under the original design in the Iberian Peninsula. With the fall of the Roman Empire, much of the scientific knowledge was displaced by scientists fleeing to the Arab world, especially Persia, characterized by a fusion of classical cultures, Barbarian and Christian teachings, centralized in Constantinople, while in Europe this knowledge was supplanted by a culture of superstition, creating the now called the Dark Ages (*ca.* 500–1000 AD).

With the decline of the Western Roman Empire, from 476 AD, began the medieval period, which survived for nearly a millennium, merging into the renaissance and the Age of discovery. The Iberian Peninsula was firstly occupied by Germanic peoples, the Visigoths at south and the Suebi in the north. In 711 the Arab subjugated the majority of Iberian Peninsula, except the North Asturias where Christians remained. In 722 the christians military force leaded by Pelagius won the Muslim Moors in the Cavadonga battle, assuring the independence of the Kingdom of Asturias for more than five hundred years. From this small reign begun the recover (Reconquista) of the Iberian peninsula territories under Arabic dominion, which created Portugal.

Much of the scientific knowledge existing Europe was suppress by barbarian invasions. The centre of culture was displaced by scientists fleeing to the Arabic world, especially Persia, resulting in a fusion of classical culture and Christianity (Barbarian and Christian teachings), centralized in Constantinople (James O. Richard, 2001). In Europe this knowledge was supplanted by a culture of superstition, creating the now called the Dark Ages (*ca.* 500–1500 AD).

The barbarians that succeed the Roman did not leave engineering heritage works of historical interest and value. But the Arabic dominion demonstrated relevant influence, appreciable in sophisticate architecture, rich of decorative elements and coating and finishing techniques, hydraulic, mechanical and civil engineering (Donald R. Hill & Ahmad y. Al-hassan, 2013). The Arabs transmitted to future generations the use of brick, plastering and tiling.

The Medieval Islam was a prosperous and dynamic civilization assisted by relevant Arabic engineering technology within the field of irrigation and water supply. The civil engineering technology dominated building of Dams, used to impound and divert irrigation water, bridges to cross canals and surveying techniques to align and level canals. Mechanical engineering developed water-raising machines, an integral part of hydraulic engineering schemes for transporting water to the fields and to urban communities. They developed also machineries for mill grinding, moved by water flows, which still continues in time. The history of engineering in Islam is a very wide subject indeed, which left as heritage, great works of irrigation, process of water extraction and water mills, masterpieces of Arabian Peninsula engineering.

A copy of a very popular hydraulic wheel placed in Tomar (Portugal), which was used for milling and providing power for industrial operations, originating from Arab heritage is represented in Figure 10.1. In the same Figure is represented the photo of a Noria at Hama, Syria (from Donald R. Hill & Ahmad y. Al-hassan, 2013).

Figure 10.1 A copy of Arab Style Hydraulic wheel (Noria) in Tomar-Portugal (left) and a Noria at Hama (Siria).

In the beginning of the eighth century an Iberian-Christian movement started, also known as Christian-Conquest, aiming at recovering the lands lost during the Muslim invasion. In 868 the County of Portugal was formed in the Iberian Peninsula as part of the Kingdom of Galicia and later integrated into the Kingdom of León.

The history of Portugal as a European nation dates back to the middle Ages, when the County of Portugal 'Dom' Henrique became a fief of the Kingdom of León in 1095, and encompassed the territory from the river Minho to Coimbra. In 1128 'Dom' Afonso Henriques succeeded his father in the government of the County of Portugal and, after a major victory against the Moors at the Battle of Ourique in 1139, became independent and founded the Kingdom of Portugal, the most ancient kingdom of Europe, recognized in 1143 in the Treaty of Zamora. Afonso Henriques and his successors were engaged in an independence war, expanding the territory through expeditions to the centre and South, under Arabic dominion.

The evolution of the borders during the conquest is shown in Figure 10.2. The strategy outlined in the court of King Afonso Henriques to ensure the kingdom autonomy, was centered in gaining recognition and support of the Pope. The common interest between the church and the new King was evident, as the conquest was carried-out against the infidels Arabs.

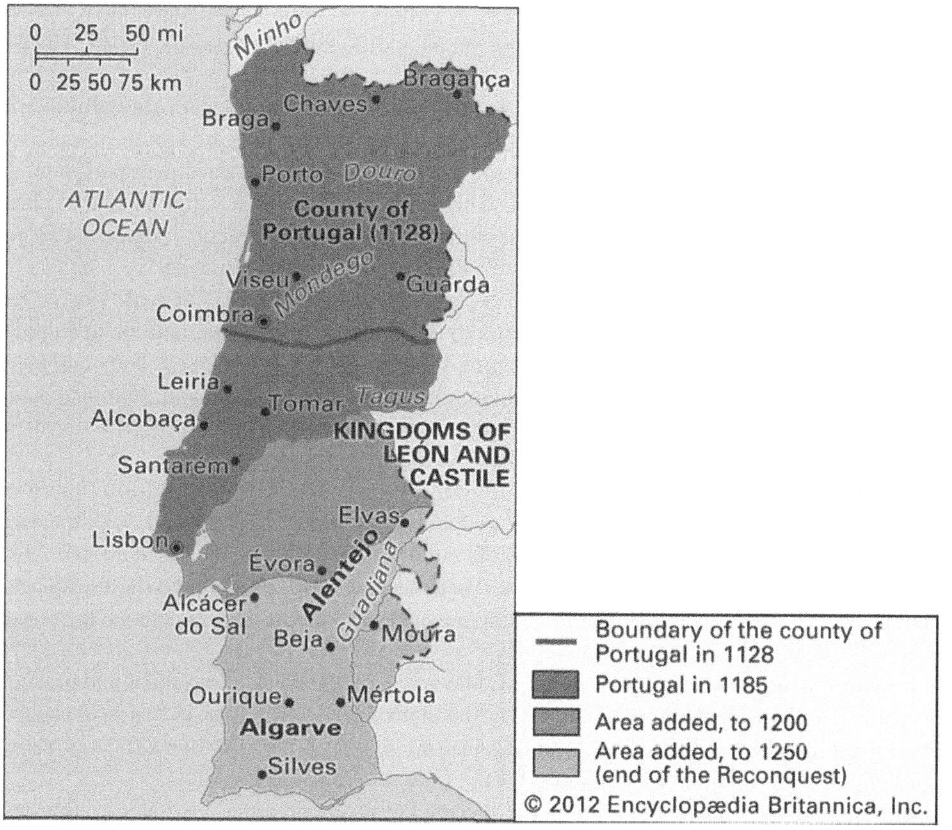

Figure 10.2 The evolution of the borders of Portugal during the conquest, from 1128 to 1250 AD (*Source*: http://www.britannica.com/EBchecked/media/3168/Portugal-in-the-12th-and-13th-centuries).

Afonso Henriques achieved the influence of the most important religious figure of his time: S. Bernard of Clairvaux and the Cistercians. He was politically and religiously a key player, as mentor of Innocent II of the Templars and preacher of the Second Crusade, displaying an essential role on the legitimacy of the new kingdom.

Afonso Henriques continued his policy of expansion of the Kingdom to the south, winning the line of the Tagus, with the conquest of Santarem and Lisbon in 1147 AD, with the help of the Crusaders. In order to ensure the control of strategic positions that would guarantee the defense of those territories, the king sought aid from other forces, particularly the Cistercian Monks and the military religious orders, the Templars. On behalf of the Catholic religion, they formed a monastic-military army.

Besides making donations to the Crusaders, who wanted to stay in Portugal, he offered to the Cistercian monks domains with a dimension and unusual importance in the region of Alcobaça (1153), crucial for their settlement in the Iberian Peninsula. In 1159 the Order of the Templars was transferred to the Castle of Tomar, along the river Zêzere, being assigned to ensure the defense of Santarem and Lisbon and protect these cities against Muslim incursions coming from east.

After these achievements, military activity slowed down and it was time to change policy and strategy, to find people, organize territory and promote population setting. Only an effective population settlement and prosperity could induce the desired stability in the conquered area. The Cistercians represented the most appropriate organization for this job, fulfilling yet an additional task: to normalize the Christian rite and remove Islamic religion.

Thereafter his son D. Sancho I devoted almost all the years of his reign to the administration of territories, promoting municipalism, a suitable politic and administrative option to fix the population in the new territories and ensure its governance and control. The key objectives of the erection of municipalities governed by operating rules and protected by city walls were integrated in a policy of spatial planning of territory for the establishment of a social structure of the people. The Cistercians religious order provided a considerable support to the development of the economy, predominantly agrarian.

The Cistercian religious orders also contributed decisively to the culture and spirituality of the Kingdom of Portugal in the vast areas of occupied land, providing religious, humanistic and technical education. They disseminated intensive farming irrigated by innovative hydraulic techniques. They employed great discipline and methodologies for consolidating the territory: afforestation of land, regularization of the rivers banks and streams, planning of hydraulic piping for water supply, sanitation works, construction of underground conduits, diversion of flows, hydraulic wheels for energy production.

Victorious battles allowed the complete removal of invaders from all the territory, after more than 800 years of governor. The current borders were stabilized later in 1249 by D. Afonso III, with the reconquest of the south (the Algarve) from the Moors taking the denomination of King of Portugal and Algarve. In 1290 AD the King D. Diniz, in the regained entire Portuguese territory proclaimed the Portuguese as the official language of his kingdom and created the first university in Lisbon. This concluded the construction of the reign of Portugal, the first European country.

In that time the Europe was still living in the Middle Ages, on the basis of the spirit and precepts of St. Benedict's Rule, practicing agricultural economy, as the main activity. The Iberian Peninsula keep out this Middle Ages from its history due to the Arab occupation, progressively shrank away from the territory under their control, creating the right condition for the growth of a new kingdom.

The constitution of an independent reign in the Iberian Peninsula provided favorable conditions for the consolidation of an organized society, requiring a set of laws and regulations acting in many fields: juridical, administrative, technical, social and so on, making Portugal quite advanced, as a nation. Obviously, the public works represented a way to consolidate the reign and show the capacity of the government. For these reasons, sanitary and hydraulic structures played an important role, which contributes to the politic of stabilization of the reign since the beginning of nationality.

During the reign of D. Afonso Henriques and after his death, until the XV century, several prestigious works of sanitary/hydraulic structures were inaugurated such as: bridges, hospitals, monasteries, hospitals, lazarettos, fonts, aqueducts, water cisterns, spa, sewage pipes and so on, spread all over the occupied territory, providing examples of brilliant technological capacity (Correia, 1954). The best national and foreign engineering technicians were engaged in undertaking the design and construction of these works, creating remarkable examples of the powerful new kingdom.

10.2 MIDDLE AGES SANITATION IN PORTUGAL

Many historical reports described the middle ages as the times with poor water, sanitation and hygiene control, generating epidemic diseases and causing the substantial decrease of the European population. Nevertheless, the medieval city was frequently labeled as a hell toilet, providing a set of conditions favorable to disease propagation: unpaved and dirty roads, chaotic urbanization, low illumination and air ventilation in houses, garbage accumulation and contamination of the water supply sources, and so on.

Yet this opinion is not completely valid for Portugal. Available studies (Pereira, 2005) report that the Portuguese mediaeval cities were healthy planned, protecting against propagation of epidemic diseases, during a long time.

The iconography of Portuguese medieval cities, eventually influenced by the Cistercian and Templars culture, shows well planned arrangements, with compact city center provided with paved roads, accommodating the nobility. Around the center were situated spacious and distinct surroundings areas, endowed with plentiful green spaces, and not crowded with people.

The city center of the Portuguese municipalities was not a structure built from the spontaneous growth of rural nucleus. It was a planned area demarcating the separation between the nobility and the gentry, outlining the social status. This living area was exclusively restricted to the dominating class, dictating the local economy and governing the rural population. These medieval cities even when accommodated higher population number inside the medieval walls, were protected from disease propagation by the surrounding intramural spaces and rural areas, during more than a century. The contagion did not take place in the entire middle age period, starting in a later time, in the transition period from the low middle ages and acting until the modern ages (1400–1600 AD), provoked by the migration and excessive concentration of country side population inside the city Walls.

These medieval cities even when high population density living inside the medieval walls were surrounded by intramural land reserves and surrounding rural areas, protect them disease propagation during more than a century.

The contagion did not take place in the entire middle age period, starting in a later time, in the transition period from the low middle ages and acting until the modern ages (1400–1600 AD), provoked by the migration and excessive concentration of country side population inside the city Walls. The relevant period in the evolution of sanitation History is shown in Figure 10.3.

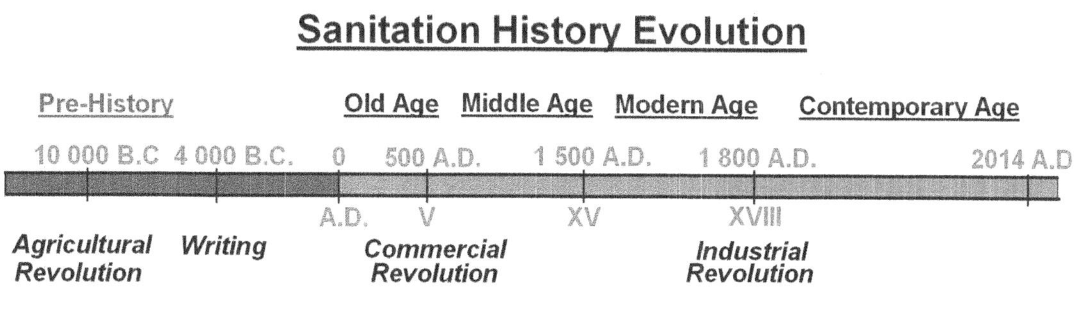

Figure 10.3 Periods of evolution of sanitation History.

This migration was motivated by symbolic, economic and defense reasons. Since the late middle ages, the transfer of rich farmers desirous of living in the city provoked a great citizen growth and demand for renting houses. The Portuguese nobility looked to this population as an opportunity to enhance incomes from rental homes. This process of concentrating people inside the medieval towns also depended on safety awareness. The compact group of houses with a common wall was seen as defensive protection block against enemy attacks, an additional barrier fortifying ground defense value. Both medieval military practice and the legislation looked favorably to this option of the tenants, promoting its implementation and provoking uncontrolled increase of population density, the main cause of the urban crowding.

The medieval urbanization in Portugal was a process, carried-out by rural people, accustomed to practicing an economy of self-sufficiency and food processing daily activities, which moved living in the town. This countryside medieval people, when residing in the fields, currently used their own and livestock excrements as fertilizer, mixed with forest residues, composted or stabilized during storage, having an immediately and recognizable utility value. Their wastes did not contain contaminants harmful to health and, perhaps, were not smelly. Likewise, other forms of waste, if any, were produced and discarded with a speed that allowed the regeneration by the action of the environment, without causing accumulation.

In the period from the late middle Ages to the Modern Age, a good portion of this productive waste management process stayed in the countryside, bringing benefits. However, when an important number of peasants established living inside the city walls, keeping consumption habits proper to the dominant rural countryside economy, the natural cycle of waste reuse was interrupted.

This new citizens continued to play in his small townhouse activities like food preparation for immediate consumption (cleaning, peeling or grinding) or for preservation (drying, salting or smoking), using for this purpose urban backyards, balconies or terraces, spaces indispensable to carry-out everyday healthy life in the medieval urban environment. These various factors lead to crowding of the city and made the corresponding urban area insufficient for the needs, taking place the appropriation of the streets with wastes. The urban condition was not enough to accommodate the total recycling process practiced in the peasant economy and could not manage digestion of such remnants into fertilizers, the solution applied in the countryside environment, establishing the kingdom of putrid (Pereira, 2005).

The city was the great inventor of nauseating smells. The plagues appeared in the medieval city simultaneously with the settlement of rotting, leading residents to establish an interconnection of cause and effect between them. The putrid organic urban waste was identified as the main cause of illness of the inhabitants. Portuguese cities were faced with relevant sanitary problems, unless they were not overpopulated inside the walls (Pereira, 2005).

Thus, the king D. John I (João) understood the causes as he said in a speech in Evora in 1392: 'Because of dirt and manure, rotten and nasty things and fumes that dumped in many places regrowth damage and pain to the bodies'. The wastes and residues set-up in the cities created putridity. Hence the city invented the bad smell. As a consequence plagues occurred in many places around the country. The monasteries and others isolated structures maintained a high sanitation level and were saved from the disease.

The XIV century was cauterized in Europe by big pestilences that in many localities killed more than half of the population. This debacle caused a profound renewal of the social structures and distribution of wealth and largely changed the economy, marking the end of the Middle Age. These epidemics event increased the awareness on urban sanitation and to leave the unhealthy products out of the urban space. It motivated the establishment and application of standardized procedures for waste disposal control and practical measures aiming at improving the sanitary level in the cities. A complete scientific procedure for disease control and sanitary improvement was developed. The set of measures and practices recommended by the Crown and by the municipal authorities were based on perception of odors, used as pestilence indicators, followed by implementation of prophylactic measures, namely: isolation, evacuation and cleaning of houses. Thereafter, a treatment processes consisting of aeration, disinfection with vinegar and odor control with perfumes to purify de pestilent air, was applied. The cadavers were buried in cemeteries.

In Portugal the epidemic disease struck all the cities, even the ones who adopted the measures prescribed by the authorities to eradicate pests and improve the health. In the second half of the fourteenth century, outbreaks of bubonic plague led to severe depopulation. The economy was highly localized in a few cities and the migration from the field provoked the abandonment of agriculture and the rise in unemployment in the villages.

The sea offered alternatives, a way to escape from pestilent cities, and most population settled in coast for fishing and trade. So the diseases occurred in this century became the driving force to stimulate Portuguese people to go out in the ocean, to navigate far away, thus contributing to discovery of new territories and to the evolution of the world.

Unfortunately, the great epidemics sprayed also in the great cities of the Portuguese empire (Lisbon XV century, Goa XVI century, and Salvador XVII century), reinforcing the concept that living in cities involves many risks. In the fifteenth century, the causes for disease were not clear. The authorities attempting to limit the scourge with emergency measures reinforced the notion that living in the city facilitates transmission, recommending isolation. As a consequence, in the seventeenth century, the wealthy families having second homes outside the urban city got the habit to live in their recreational farms, receiving considerable benefits in terms of health, savings and wellness.

The escape was a way to stand-up to unsanitary cities. Authorities clearly acquired the notion of the dangers coming from putrefaction and odours in the atmosphere, but were not capable to identify and recognize the cause and the way of transmission. At the same time, it was necessary to introduce concrete measures to protect population. In the perspective of the cities administration the best option was the application of measure for preventing diseases epidemics, rather than applying emergency practices to halt propagation. This option requires drafting and enacting legislation, defining and regulating preventive actions capable of overcoming the problem, assigning a precise role to each involved actors.

10.3 SANITATION IN RECENT TIMES: EVACUATION, DRAINAGE AND TREATMENT

As a result of occurrences of outbreaks of pests, the municipalities developed legislation aimed specifically at improving sanitation of the cities based on olfactory criteria, which led to great consequences with regard to life style in the city.

The new legislation specifying preventive measures and new rules on living in the city was promulgated. The legal texts prepared by the municipal authority established and specified that health was a fundamental component in the construction of urbanity, and the concept of 'Urbanization' became synonymous to 'Civilization'. According to this criterion, the responsibility for removing out the garbage heaps, for preservation of water supply sources and for unclogging pipes was attributed to the local authorities. The citizens or even entire communities were called to contribute and collaborate in the process, being obliged to respect the rules of urban conduct adequately prepared by the sanitary and administrative authorities.

The habits of body washing or cleaning the environment in which they lived became indicators of social class to assign status. The condition of city dweller imposed a serious attitude with regards to the disposal of excrement and other waste, being the citizen the responsible for its production and release. The cities were not rigged, physically or institutionally, to deal with excrement and garbage, and the first step was to establish the link between the waste and its producers, calling and obliging them to prevent the accumulation or participate in the treatment and disposal. The construction of urbanity involved as first step the control of waste by the respective producers, and anyone who disposes his excretions in streets and public areas was made to serve a penalty.

In the fifteenth-century the municipality of Lisbon evaluated the problem of waste dumping, recording all the dumps and heaps disseminated throughout the city, exhaling pestilential air. Thereafter established and indicated the sites for authorized waste disposal as well as regulated the procedures to comply. Nevertheless, a widespread disregard and non compliance for these regulations occurred. The municipal institutions did not yet assumed the role of providing public services for cleanliness of the city and transportation of waste to the sites assigned for its release. These tasks were regarded as an obligation of private residents. In turn, the citizens considered these tasks as a burden, which many seek to escape. In addition, the few attempts to create municipal services for garbage collection did not advance in this century, in small-medium communities. In contrast, in the larger cities, the transport of waste was controlled by a multifaceted market providing small urban services. Along with chargers firewood or water, there were slaves passing by houses to collect garbage.

In the seventeenth-century in Lisbon 'a thousand black women were walking around town with canasta cleaning up the city'. They used a specific container for waste collection called '*Calhandra*' (Isabel Castro Henriques, 2011). Due to the incomes from the discoveries, the city was rich in everything, also in trash, having generated another 'profession': the seeker of urban residues (Pereira, 2005).

Relatively to the drainage and evacuation of sewage and polluted slurries in Portugal, the historical elements reporting the status of advance and the current problems also date back to the fifteenth century, after the occurrence of the plagues. The widespread paving of streets in European and Portuguese cities was provided with drainage channels, for removing rain water and avoid flooding. The construction of a separate sewer system was expensive and complex to implement, and not recognized as an absolute necessity, once the scientific knowledge did not prove that the sanitary sewage was one of the main cause of pestilence and vehicle of contamination. In this context the rainwater drainage channel was the available evacuation system, thus transporting undesirable dregs of the streets towards the rivers and lakes, produced odors and, in addition, became dangerously polluted for water supply.

In that period, the excessive population concentration in the cities acquired the habit to dispose all kinds of manures and filth in the rain water drainage pipes, quickly transporting 'pestilent waters' everywhere and spreading the disease without any possible control. King D. João II played an active role to control the disease propagation, issuing proper measures and providing active cleaning actions.

The modern age, between the sixteenth and mid-eighteenth century, was the period which revealed high sanitary problems and, also, the higher growth of public health. In the course of these centuries industrial development and the consequent concentration of population in big cities increased the problem of wastewater drainage from the urban areas. The water supply and street cleaning did not follow the rate of the urban expansion. The drainage of domestic wastewater by the existing storm water networks was authorized, as a practical and urgent measure. But this procedure greatly aggravated the risk of transmission of waterborne diseases due to poor conditions of service of the networking.

While the proliferation of industries and its wastewaters aggravated environmental pollution of water resources, serious epidemics, especially cholera and typhoid fever, transmitted by water contaminated with domestic sewage, appeared in Europe, decimating population. Mortality was exacerbated by scarce food quality and poor living and labor conditions of the working class.

The remarkable population growth in Lisbon until the earthquake of 1755 intensified the episode of 'fearsome floods' and the health problems associated with hygiene and cleanliness of the city (Prazeres, 1954). Under the city streets were placed stone channels for the drainage of storm water which, later, also drained domestic waste water, acting as unitary collectors. The living conditions in the city are described as terrible and only the early nineteenth century started improvements in life standards.

The 1755 earthquake and the consequent tsunami destroyed a great portion of the city started an era of progress in the capital, marked by the rebuilding of the city and the organization and construction of a modern and methodical drainage system, based on collectors arranged in a grid, connected to the Tagus estuary (Matos, 2006). In the eighteenth century the beginning of sanitation science started, based on the evolution of medical beliefs and, even more, on the progress of chemistry, enabling to overcome the empirical medieval and Renaissance approach and systematization, provided new tools to shape and plan proper policy sanitation, based on a science of health in place of episodic performances of previous centuries.

In the Portuguese world, the recently created sanitary science produced more diagnostic than unhealthy remedial actions (Pereira, 2005) and did not expanded greatly in the current practice. A generation of doctors and naturalists trained at the University of Coimbra begin to circulate in the colonies, as holders of scientific knowledge existing at the time.

In the eighteenth century water supply from abundant external catchments, to reinforce the withdrawal from wells was the main concern in Lisbon and Rio de Janeiro, led to the construction of two relevant aqueducts, granting safe and controlled water to populations. In the second half of the nineteenth century the evolution of sanitation in Portugal and its colonies, in terms of approach, technologies and solutions, were inspired on the best practices implemented worldwide by the British, French, German or U.S. technologies. The adopted measures effectively attenuate the outbreak of epidemics, but did not solve entirely the problem. The sanitary control and hydraulic works did not cover all the territory, once involved large investment in water supply, drainage and treatment facilities, making pestilence reappear.

As result of the Lisbon epidemic cholera outbreak in 1856, Bernardino Gomes, by order of the Royal Academy of Sciences, elaborated a report of the situation in the Portuguese capital, recommending the absolute need of installation of separate drainage systems, like those existing in other European cities such as Paris, London or Brussels. In 1884 Ressano Garcia prepared drainage plans for Lisbon inspired by the principles of the hygienist Edwin Chadwick, largely responsible for alerting and advocating the need for planning and construction of urban drainage (Matos *et al.*, 2009). The adopted solutions were the right response to large cities sanitary problems, requiring separate drainage, large-scale hydraulic works on collection, pumping systems, and wastewater treatment plants.

The historical evolution of sanitation in Portugal was carried-out by competent sanitary and hydraulic engineers, spreading all over the world (Portuguese colonies), relevant technologies and solutions. In the twentieth century was started and gradually implemented in all the cities, the separate collection of sewage and rainwater, followed by efficient wastewater treatment plants (WWTP), allowing to fulfill the standards imposed by the environmental legislation and ensure the safe disposal of the treated effluent.

The treatment of the sewage generates huge quantities of sludge, by creating problems with its treatment and disposal, is today regarded as the most relevant issue and challenge under resolution by sanitary engineering. Sanitation in Portugal today ensures fulfillment of high level standards. Water supply serves almost 100% of population, guarantying sanitary controlled high quality freshwater. Currently, more than 95% of the wastewater produced is treated and adequately dispose (Rea, 2012).

Moreover, the adopted solutions are based on the best technologies developed in industrialized countries which make use of relevant energy and natural resources. They are expensive in capital and running costs as well as unsustainable for less economically developed countries, which have not achieved a significant degree of industrialization relative to their populations, and have, in most cases, a medium to low standard of living.

There is a strong correlation between low income and high population growth, which makes it difficult to provide adequate sanitation level worldwide and guarantee safe and sustainable use of resources, using these expensive and sophisticated technologies. Hence, a novel approach is suggested today with regards to the sustainable sanitation concept, based on source separation of the used waters, according to their characteristics, onsite treatment and the recovery and reuse of the valuable components: water, nitrogen, phosphate, organic matter, and so on. This can avoid long distance transportation and, consequently, high investment in hydraulic infrastructures.

Furthermore, this approach is rather complex to implement in existing large cities, but can be proposed for new cities and houses especially in high population growing countries. This can also be a way to make compatible sanitation and planet resources, allowing health accommodation of more population. New innovative solution are proposed as the reuse of sewage in water supply, a solution already under implementation in some countries with limited amount of freshwater supply (Singapore), shortening the water cycle. There was a time in Portugal when the concept of sustainable use of sewage and water, was applied, anticipating the future. Some remarkable examples since the twelfth century until the seventeenth century are the witness of feasibility of implementing sustainability concepts and practices.

10.4 EXAMPLE OF SUSTAINABLE SANITATION IN PORTUGAL THE CHRIST CONVENT

10.4.1 Introduction

During the reconquest of the Portuguese territory under Arab control, Dom Afonso Henriques and its successors continued their expeditions to the South, making donations to the Crusaders, the Knights Templar and the Cistercian monks staying in Portugal who received large domains to colonize, from the recently recovered territory. The Templars were, thereafter, in charge of the defense of Santarem and Lisbon, in order to avoid the attacks of the Muslims, now coming from the east. Supported by this strategic defense of the recovered territories and politic conversion of the habitants, the king could continue to run the independence campaigns towards the South.

The construction of the castle of Tomar as a fortified enclosure, is a fine example of military architecture, started in 1160 by the Portuguese Templar Knights. This monumental building is another important piece of the defense strategy. The construction stands-up relevant features as the tower in the citadel, the bailey fortress with ramp defense stones and the Rotunda of the Templar (Figure 10.4).

Figure 10.4 Tomar Templars Castle in Tomar.

After the dissolution of the Templars by decree of Pope Clement V in 1312, the Portuguese King D. Dinis founded the new Order of Knights of Christ, to defend the country, and gave them the privileges and goods of the Templars militia. For this purpose he built on 14 March 1319 a monastery, under Cistercian custody, annexed to the castle, with the apostolic approval of Pope John XXII. The Knights of Christ moved their headquarters to the castle of Tomar, in 1357, expanding and adapting the original polygonal Oratory of the Templars (the so-called Charola, built in 1190) to their needs. Details of the convent and its evolution were described by Virgolino (2012).

The high-lying plant of Christ Monastery was built on the west side of the castle of Tomar. The magnificent complex, as it is observable today, with eight cloisters of different ages and styles, is strongly influenced by the major construction projects under King D. João III reign, corresponding to the needs and functions of a community of monastic knights.

The Convent of Christ occupies major monumental complexes in space and time of the peninsular and European architecture. It covers a vast area of forty-five acres, five of which correspond to the Templar Castle and the Convent of Christ, and the rest around the old convent, now known as 'Mata dos Sete Montes'. The six centuries that took the construction involved the largest investments and most important works in Portugal, a financial effort that has not slowed down, even during the regency by the kingdom of Portugal of the Spanish King Felipe II.

With the abolition of the religious orders in 1834, parts of the monastery became a military hospital, others were used for other purposes and the remaining parts of the building were publicly auctioned.

Despite its unfinished majesty, the damage caused by the earthquake of 1755, the devastation of the Napoleonic troops (1809–1811), the Royal Monastery of the Order of Jesus Christ is one of the outstanding witnesses of national monastic architecture of the sixteenth Century, through its dominant and unique monumentality, which is of great cultural and aesthetic importance in its inconstancy architectural development. The monastery was declared a national monument in 1907 and added to the list of World Heritage Sites by UNESCO in 1983 (Virgolino *et al.*, 2005).

The convent of Christ in Tomar is not just a very important historical remain with artistic architecture relevance. It is also a witness of an excellent water supply and sanitation technology and a model of sustainability and excellent water and waste management. It was provided with a sophisticate system of water supply and wastewater management which covered all the duties and need of the religious community. At that time, the convent was surrounded by forestry and agricultural territories, providing food and wood, making the convent self-sufficient, as well as allowing recycling of all the generated wastes. Thus, this is a nice example of full sustainability.

A medieval monastery, to operate under appropriate hygienic and sanitary conditions, required a solid and technically efficient hydraulic infrastructure, providing the withdrawal of groundwater, its gravity transport, storage and distribution in the interior of the monastery, collection of rainwater and the evacuation, treatment and disposal of sewage and other used effluents. The monks needed running water supply systems for cooking, drinking, for their hygiene (e.g., ablutions, shaving, and tonsure). They also need water for irrigation of gardens, orchards and agricultural activities, to get self-sustainability.

In terms of sanitation the latrines were the dispositive of evacuation of human excreta, which was treated in a wastewater treatment plant and joined to forest residues for composting and creation of a fertiliser. The supported model adopted in the Monastery of Tomar, shows the important contribution to the knowledge of the Cistercians in sanitation technology.

10.5 WATER SUPPLY TO THE CASTLE AND THE MONASTERY

10.5.1 Rain water supply

The hill on which the castle and the monastery were built is flanked by two deep valleys which not only represent a strategic point for military occupation but also a major problem for water supply. As highlighted in 1590 by brother Jerónimo Román, the monastery, due to its high altitude position, did not have wells or other natural sources of water. Therefore, since creation of the castle and convent, the rain water collected from roofs, porches and cloister paved areas and stored in cisterns, having enough capacity to cover the needs of the community, provided water supply. A system of cisterns and tanks collecting water from roofs and open areas were strategically built. Two of them, well known, are located within the walls of the simple, fortified medieval village. The most important is in the churchyard which has a depth of 18, 50 m, while the other is located near the Almedina gate and has a depth of about 10 m.

Regarding the tanks, it is likely that at least three units were built originally in the medieval period: in the courtyard of the castle, next to the large tower at the southwest corner of the wall and the fountain cistern of the pharmacist-court. With modifications and extensions of the castle to be the seat of the Order of Christ, the respective cisterns were built in the laundry and Graveyard cloisters. Nonetheless, the cemetery is the oldest cloister provided with a cistern. Although, built in the sixteenth Century its water was allowed to feed the sinks of the adjacent sacristy.

10.5.2 Water supply of the monastery by an aqueduct

The Aqueduct of Christ convent is a monumental work, whose grandeur has been highlighted by several authors since immemorial times, being considered 'probably the grandest and most elegant aqueduct of the country'. It was commissioned by Felipe II of Spain to the architect Filipe Terzi and was intended to make the convent and its land totally autonomous in water supply.

Its primary function was to provide water to the monastery for daily drinking and food preparation, draining high quality ground water from natural springs in the mountains. The channeling of the aqueduct by gravity over a distance of more than 5, 8 km consists entirely of hewn stone troughs. To overcome the valleys, it was sometimes necessary to built beautiful arcades, whose dimensions correspond to the winding valleys.

The aqueduct reaches the convent building by an arch structure built into the south wall (Figure 10.5), and ends in a water reservoir constructed in 1617 in the southern part of the dormitory. At the end of the aqueduct has a small strainer before the water flows into the water tank (capacity of 4 m^3).

Figure 10.5 Convent aqueduct in Tomar.

The channel of the aqueduct leads to a desilting pool into the building. The system is provided with a storage tank, accumulating the water amount from the basin of the aqueduct in low consumption period, and capable to reinforce the water flow in peak demand. Nevertheless, a lock enabled the control of water flow in the basin. While, the other side of the tank was provided with a lock regulating the amount of water in the channel. The evacuation was carried out in a stream outside the building. Probably, the building and the control device were built in 1600–1613 as reported by written sources. Two independent clay pipes integrated in pillars ran down to supply water, to the cistern of latrines in the cloister of the 'Necessárias' and to feed another tank, probably used for several purposes such as watering animals, washing vegetable and fruits, and fish farming.

10.5.3 Storage and distribution of water

The rainwater from roofs and the other open paved spaces, as well the high quality water from mountain underground water used essentially for drinking and food preparation was conducted to reservoirs and to a supply system, distributing it to the faucet of the kitchen, lavatories and other potable water fountains. The monastery was provided with several separate reservoirs, having various end-use: drinking and cooking, garden irrigation, fountain feeding, washings, sewage draining, latrine flushing. Practically almost all the cloisters are provided with a cistern placed in the centre (Cloister of 'Lavagem', of cemetery, of 'Corvos', of 'Micha' e of 'Necessárias') or under one of the lateral buildings Cloister of 'Hospedaria'. The plant of the main core of the convent is shown in Figure 10.6 (adapted from http://tomar.com.sapo.pt/convento_planta.html). The cloisters provided with cistern are in green color.

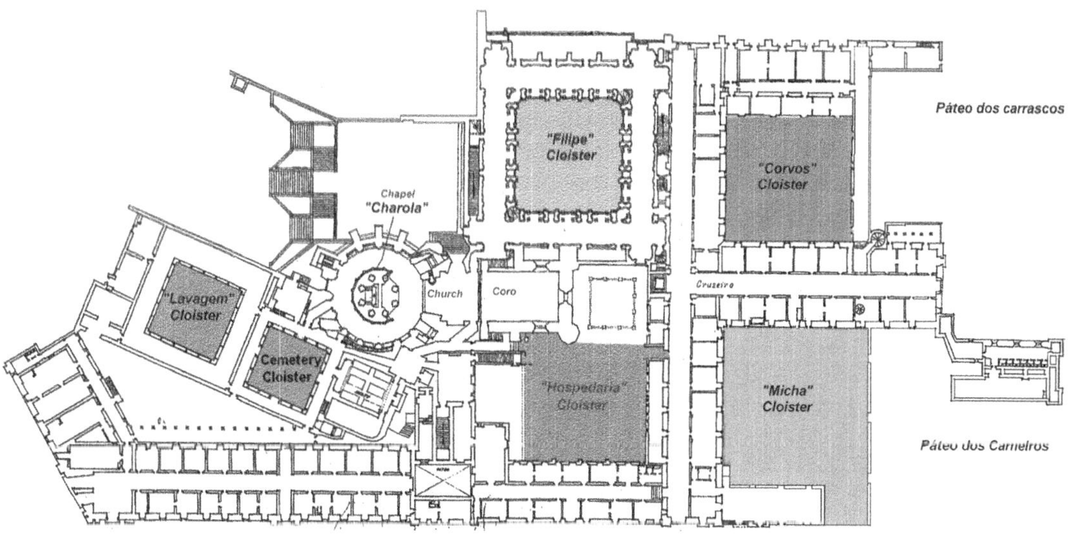

Figure 10.6 Convent plant and cloisters (adapted from: http://tomar.com.sapo.pt/convento_planta.html.).

The storage and hydraulic system was completely independent according to its use. All the potable water system was isolated from sewage and the other waters (rain water), completely preventing any possible contamination.

As the water availability was not abundant, before the aqueduct entered in service, the system reused efficiently the collected rain water, using various reservoirs storing different water quality and interconnected between them, facilitating the water management procedures. Large water tanks were built, of which it is possible to emphasize the cistern of the 'Micha' cloister (Grand cistern) and the Raven cloister (605 m^3 volume), and a cistern close to the 'guest' cloister. In the center of the great cloister (or D. João III cloister) a fountain existed. It is very likely that the water from this fountain was originally drawn for the basin of the refectory.

The cistern of 'Necessarias' cloister, with a capacity of 240 m^3, was used to supply water to the latrines using two channels. The cistern was supplied by rainwater and low polluted used waters. Inside the cistern can be seen on the south wall, a vertical clay drain pipe which is assumed to be a supply pipe from the aqueduct. Special care was dedicated to the design of the inflows and of the edge of the roofs.

The water supply of the Great Cistern ('Micha' cloister), was amplified by water from the aqueduct, through the distribution system of the monastery and receives rainwater collected in the neighboring Santa Barbara cloister. Also, the water of the central fountain of the D. João III cloister, if necessary, could be discharged into the cisternae of the Raven and 'Micha' cloister as well as into the cistern in the southeast corner of the guest cloister. The latter are anyway cisterns fed by rainwater collected in the cloister.

10.5.4 Runoff and drainage

In the Christ convent in terms of its drainage and sewage treatment, two separate distinguishable systems have been identified (Jorge and Mascarenhas, 1999). One is the simple connection of wastes which drives the wastewater from the cuisine, the olive oil room and the oven room to the outside of the convent. The size of the manholes and collectors where the sewers were collected have remarkable dimension, draining considerable amount of wastewater from these three convent spaces. There is no evidence that these sewers can have a possible connection with the integrated drainage and sewage treatment of the second system. This low contaminated wastewater ends in the courtyard and was used for irrigation of gardens and agriculture fields (Jorge and Mascarenhas, 1999).

The sanitary contaminated and polluted sewage from latrine was drained by the second system to a wastewater treatment and disposal system, displaying a more complex lay-out. It is an integrated system with important derivations and connections between them, designated to meet precise functions. The system ends in the 'Bloco das Necessárias', where there are the latrines and the drainage and convergence of the sewage and its subsequent treatment. Although, details of the complex drainage system were described by Jorge and Mascarenhas (1999). The localisation of the main pipe systems were indicated using precise stone signals and marks (Figure 10.7), thus making it easy to proceed for repairs or interventions.

Figure 10.7 Marks indicating pipe localization inside the monastery.

10.6 SEWAGE DRAINAGE AND TREATMENT

Sanitation of domestic sewage was carefully planned. In the west side of the convent, in a position clearly individualized relative to the body formed by the remaining cloisters, there is a sixteenth century building called the 'Bloco das Necessarias', rectangular with three floors above ground and one floor in basement, represented in Figure 10.8. In this specific sector of the convent the latrine and personal hygiene facilities, were placed in the upper floors of the building.

Figure 10.8 The 'Bloco das Necessarias' in Tomar containing latrines and the WWTP.

Although separate, once it was intended to accommodate latrines, this 'Bloco das Necessarias' was well connected to the rest of the convent. At the ground floor level, it was accessible through the small cloister of the same name. At the level of the two upper floors, the allowable physical connection between the spaces of the convent and the 'Bloco das Necessarias' is through small corridors, one on each floor, which allow direct access from the dormitories to the area of latrines. Special care has been dedicated to the sanitary odours control of the monastery. These corridors are 'S' shaped, provided with ventilation windows placed strategically in the curves, to prevent and eliminate any possible odour propagation inside the living area of the monstery.

The wastewater treatment plant (WWTP) was placed in the underground of this block of latrines, which is wide and elongated, with inside dimensions of 12.6 × 9.0 m. The evacuation seats of latrines discharge directly over the treatment tank the excrement, where it is provided the necessary treatment. The treatment tank on which the excrement fell is divided into two compartments, one for each tile of latrines, separated by a central wall, provided with two large venting channels, constituting the aeration system (Figure 10.9).

The wastes which fell down into the tank were retained for a period of about 12 months (Figure 10.10). The liquid level increased during storage until it reached its maximum eight (1 m approximately) and completely filled the two compartments. The cleaning and sludge removal from the tank was carried-out by flushing abundant water volume, using the amount of poor quality water joined by the complex hydraulic system and stored in the reservoir of 'necessarias', placed close to WWTP.

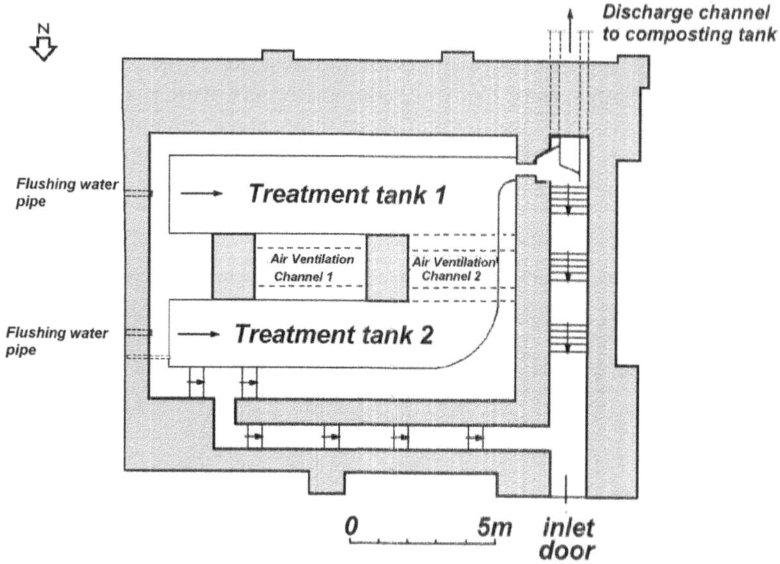

Figure 10.9 Wastewater treatment System Plant of Tomar monastery (from Virgolino, 1999).

Figure 10.10 The wastewater Treatment Plant in Tomar monastery.

During storage, the settled suspended solids underwent a process of hydrolysis and degradation using biological processes. The upper water layer was in contact with air renovated by the natural aeration system of the building. Presumably the biological activity becomes aerobic in surface layer and anaerobic in the bottom. Once accumulated in this space in the basement, sewage went sliding to the west, transported by a stone made tunnel of the sixteenth century. The Latrine drain was a channel at the output of the building provided with a lock, making possible to adjust the supply for the amount of water for hygiene and sanitation needs. The disposal channel was opened as a ditch, easily accessible, and is located at the foot of the gallery. Nevertheless, it measured 0.51 m high and 0.48 m wide. It was surmounted by a gallery, which is 1.95 m high and 1.12 m wide, consisting of masonry of hewn stone, and is covered with horizontal slabs.

Moreover, the tunnel can be accessed from outside the so-called 'cloaca', opened in the basement of the block of 'necessarias' which runs through the courtyard of the Marauders heading south, passing under the arches of the aqueduct, and always continuing northwards south with a slight rising inflection, and finishes in the 'Mata dos Sete Montes'. The flushed waste was sent to a nitrification system or other device, inexistent today, where vegetable residues were joined to the waste, forming a fertiliser to be used for the agriculture and forest fertilisation, contributing to enrich the soil of its boundary.

The ventilation system of the treatment chamber is an important tool to control hydrogen sulphide buildup, being toxic at certain level of concentration in air. The ventilation probably also provided air to the upper liquid layer in the treatment tank, maintaining possibly an aerobic zone, useful to prevent smell, to control hydrogen sulphide buildup and perhaps to improve stabilisation. Therefore, it is possible to conclude that the monks had a high level of knowledge in terms of water management, security of supply, air ventilation, treatment technology and waste disposal.

10.7 CONCLUDING REMARKS

The history of Portugal dates back to the Middle Age, the year 1139, when the prince Afonso Henriques proclaimed himself King of the most ancient kingdom of Europe. The independence started from a County in the north of Portugal and expanded to the south, conquering territories controlled by the infidel Arabs, thus creating common motivations with the catholic church at that time of crusades. The church recognised the new kingdom, provided religious and military support through Cistercian order and Templar Knights, becoming a vital support to the expansion strategy. Due to its contribute, the Cistercian religious order staying in Portugal received large domains to colonize, contributing decisively to the culture and spirituality of the Kingdom of Portugal and to the settlement and development of the vast areas of occupied land.

The medieval Cistercian monasteries displayed a high level of knowledge and domain in several fields: hygiene and sanitary procedures, hydraulic infrastructures, energy production, irrigation and agriculture. These monasteries were excellent example of self-sustainable system. From the beginning of the reign until the fifteenth century, several prestigious works of sanitary/hydraulic structures were spread all over the country, providing examples of brilliant activity. The convent of Christ in Tomar is not just a very important historical remain with artistic architecture relevance; it is also a witness of an excellent water supply and sanitation technology, a model of waste and water management and sustainability. It is provided with one of the most ancient wastewater treatment system.

The heritage of Cistercian culture influenced the planning and approach of Portuguese medieval age cities, which were constituted by a defined city centre surrounded by abundant reserve area, protecting population from pestilence until the end of fourteenth century. As a consequence the Middle ages in Portugal has not been as dark as in the rest of Europe, where sanitation in urban areas virtually did not show significant progress until the seventeenth century.

But from the second half of the fourteenth century pestilence also attacked Portuguese cities with outbreaks of bubonic plague, which led to severe depopulation: the economy was highly localized in a few cities and the migration from the field led to the abandonment of agriculture and the rise in unemployment in the villages. The sea offered alternatives, a way to escape from pestilent cities, and most population set in coast for fishing and trade. The diseases became also a driving force to stimulate Portuguese people to go out in the ocean, to navigate far away, thus contributing to discovery of new territories and to evolution of the world. Nevertheless, the disease is also a good navigator and the great epidemics sprayed also in the Portuguese empire (Lisbon fifteenth century, Goa sixteenth century, and Salvador seventeenth Century), a globalization phenomenon.

Unfortunately, the sanitary sapience of the Cistercian order did not spread enough outside, to attenuate pestilences. Thus, remains as great heritage the Monastery of Christ in Tomar and its convent garden with innovative aspects, testifying the experience and technical expertise of its builders in the field of hydraulic and sanitary engineering in the sixteenth and seventeenth century. Therefore, this knowledge is a valuable testimony, in part, still intact and in use, unique in the history of the medieval and modern hydraulic monasteries in Portugal.

Acknowledgements

The author would like to acknowledge the Portuguese Ministry of Culture and Patrimony (Direção-Geral do Património Cultural – DGPC) and address special thanks for the received support to:

(a) Director of Convent of Christ Dr Ana Carvalho Dias, for authorizing and approving visits to the restricted access areas and for providing valuable references.
(b) Arquitect Álvaro Barbosa for the precious support during the technical visit and the interesting explanations.
(c) Prof. Virgolino Jorge for providing the excellent texts and drawings concerning the Convent of Christ water and sanitary facilities.
(d) Prof. Maria Paula Marques Valente for the support historical background.

10.8 REFERENCES

Agencia Portuguesa so Ambiente-APA.I.P (2012). REA 2012 PORTUGAL. Relatório do Estado do Ambiente. http://sniamb.apambiente.pt/docs/REA/rea2012.pdf

Agnelo P. (1954) História, evolução e técnica geral de saneamento. Curso de Aperfeiçoamento de engenharia sanitária para Engenheiros Municipais (History, evolution and technology of sanitation. Training course on Sanitary Engineering for municipal Engineers), Lisbon, Vol. 1. pp. 230–246.

Anon (n.d) http://tomar.com.sapo.pt/convento_planta.html (accessed 30 April 2014).

Correia Fernando da Silva (1954). A engenharia e a saúde pública em Portugal. Curso de Aperfeiçoamento de engenharia sanitária para Engenheiros Municipais (The engineering and public Health in Portugal. Training course on Sanitary Engineering for municipal Engineers), Lisbon, Vol. 1, pp. 9–5.

Encyclopaedia Britannica (n.d) Portugal: Portugal, 12th and 13th centuries. http://www.britannica.com/EBchecked/media/3168/Portugal-in-the-12th-and-13th-centuries (accessed 30 April 2014).

Henriques I. C. (2011). Os africanos em Portugal – história e memória | séculos XV–XXI (The africans in Portugal – history and memory | Centuries XV–XXI). Edited by «Comité Português do Projecto Unesco «A Rota do Escravo». 1ª edition November 2011.

Hill D. R. and Al-hassan A. y. History of Science and Technology in Islam. 3rd edn, http://www.history-science-technology.com/articles/articles%2011.htm (accessed 30 December 2013).

Jorge V. F. (2012). Os Cistercienses e a Água. Revista Portuguesa de História, Tomo XLIII.

Jorge V. F. and Mascarenhas J. M. (1999). Die Wasserversorgung des Christus-Klosters in Tomar (Portugal). Wasser, Lebensquelle und Bedeutungsträger S. pp. 44–57.

Matos, José de Saldanha (2003). Aspectos Históricos a Actuais da Evolução da Drenagem de Águas Residuais em Meio Urbano (Historical and actual aspects of evolution of wastewater drainage in urban area). Engenharia Civil-UM N. 16.

Richard J. O. (2001) The Fusion of Classical Culture and Christianity: Antagonism, Adjustment and Synthesis (200–500). Pdf file. http://www.gordonstate.edu/PT_Faculty/j_richards/lectures/Hist1121FusionGrecoRomanChristian.PDF

Roberto de Mello P. M. (2005). Some aspects of the sanitary issue of the cities in Portugal and Portuguese colonies. *TOPOI*, **6**(10), 99–142.

Chapter 11

From Volubilis to Fez: Water, witnessed a transfer of an International Heritage

H. Benqlilou, S. Bensaid, and M. El Faiz

11.1 INTRODUCTION

For their territorial expansion, the Romans were content to reap the hydraulic techniques of their time and train engineers. The exchange and development of hydraulic expertise is made within the vast territory stretching from Roman Britain via North Africa to Arabia. Water and wastewater projects in the city of Volubilis, located at north of West Africa comprised of aqueducts bringing water from the river to supply several fountains, public baths, toilets bathrooms and mills through a distribution system coupled with a system of sewage and drainage of similar consolidation. Several houses were provided with water supply and a sewage disposal system. The Romans left the mark of their hydraulic engineering, focused on domestic water in the Volubilis in North Africa.

The water was indeed a central place in the cities and homes that were built around an atrium in the middle of which is a basin to collect rain. The remains of cities of the Roman Empire are therefore a legacy in terms of expertise in the integrated management of water resources and wastewater and storm water as well.

The end of Rome did not mean an interruption in the flow of science and hydraulic expertise. Retransmission of this hydraulic development recorded in manuscripts has not been able to upgrade and build a rich heritage to the coming of Islam. The Muslim expansion reached the Iberian Peninsula and the Great Maghreb allowed assimilating all hydraulic techniques developed in Mesopotamia and the science of collecting groundwater by completing by transition in Egypt. Inheritance and mixing all this knowledge has to exercise great influence around the Mediterranean, including water mills, the water clocks and other hydraulic motrices such as sewers and drains.

The ancestral sewage system of the city of Fez is studied to illustrate the diversity of this knowledge on wastewater management in the Mediterranean region. The ancestral traditional latrines of Fez are located near the main and antic mosques. These traditional latrines are equipped with bathrooms to allow the faithful to their ablutions before prayer and visitors to the city of Fez to use as a public latrine. These traditional latrines are connected to traditional pipes for sewage disposal.

This Chapter analyzes the similarity between systems collecting wastewater in the both ancient Moroccan cities: Volubilis founded by Romans and Fes founded by Muslims at the end of the eighth century. This analysis will highlight the influence of the two cultures on the evolution of wastewater systems in the Mediterranean region.

These systems which influence in the past water and wastewater technologies development around the Mediterranean region can serve as a basis for the development of drinking water and sanitation for small low income communities due to the absence of excessive costs related to basic investments and operating costs. Similarly, for small low income communities, participative management approach based on the Al Fachtali model can be used for pricing models.

11.2 VOLUBILIS HYDRAULIC SYSTEM

The archaeological Roman city of Volubilis is located in North Africa in Morocco (Figure 11.1). The remains Carthaginian occupation of this city dates back to the *ca.* 3rd century BC. The city was then part of the Mauretanien Kingdom at the 2nd century BC, before being occupied by the Roman Empire in AD 40 and founded by the Idrisid dynasty between 789 and 808 (Panetier, 2002).

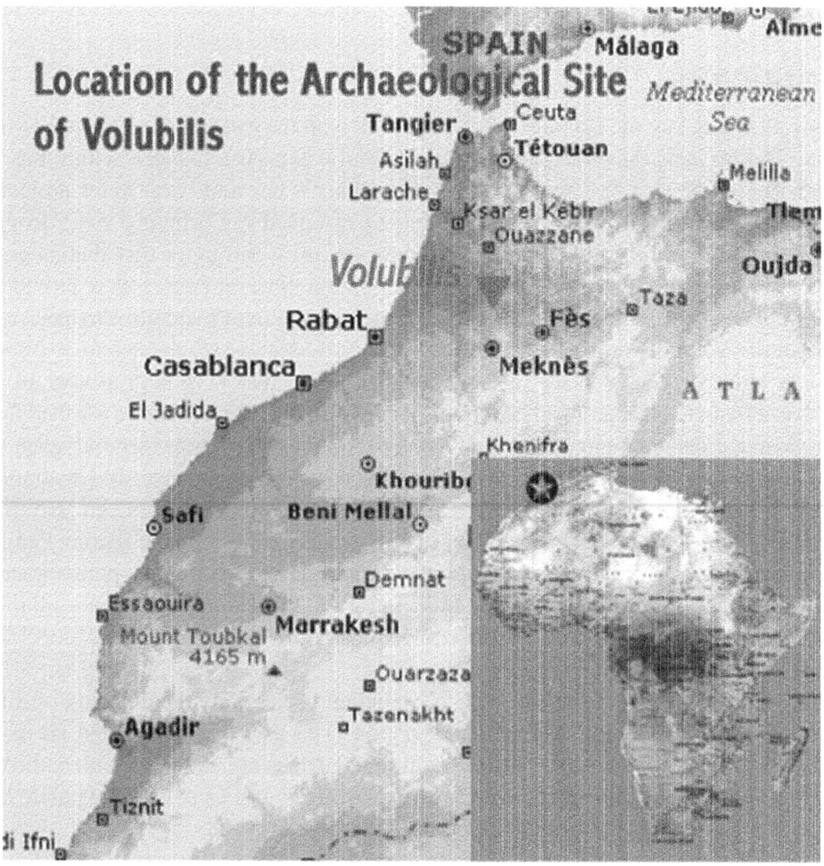

Figure 11.1 Location of the archaeological site of Volubilis (African World heritage Sites: http://www.africanworldheritagesites.org/cultural-places/frontiers-of-the-roman-empire/volubilis.html).

Volubilis is planned according to the same architectural Roman cities with a central role for water in its future development. An integrated hydraulic system allows the water supply and the wastewater disposal of the city. Water sources located at foothill of the mountains (Jebel Zerhoun) upstream of the city of Volubilis are transported by gravity and are stored in cisterns to supply houses, thermal baths, fountains, water basins and latrines of the Roman city Volubilis. Particular wells were also used to supply some individual homes. Then, at the end of the 1st century AD an aqueduct was built to supply water from a source located one kilometer at the east of the city, to ensure the increase of water needs related to urban development of the Volubilis city according to the Roman style of life. The aqueduct buried on the most part of it course emerged in the first secondary decumano, was built masonry without arches (Figure 11.2). A second water connection made from lead supplied the city from another source.

Figure 11.2 Volubilis Aqueduct (with permission of H. Benqlilou).

The water captured from the different sources is then stored in circular cisterns that functions as storage tank for regulating the water flow and gravity distribution for various private and public uses. The distribution is ensured by lead pipes (Figure 11.3) consist of 2.25 m to 2.50 m (Panetier, 2002).

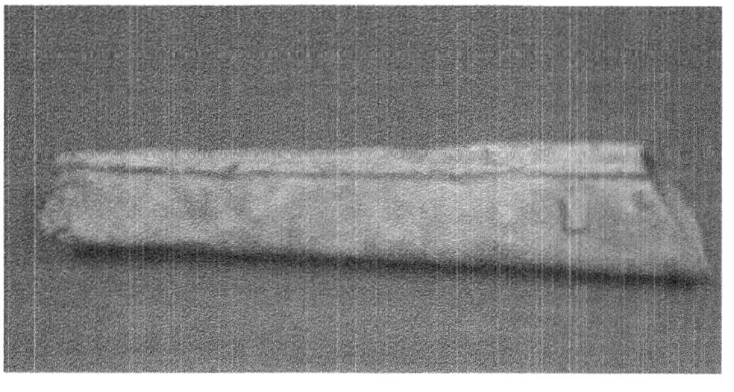

Figure 11.3 Lead pipes (with permission of M. Atki).
Source: Conservation of the archaeological site of Volubilis.

The sewage and rain water as well as overflow water from basins and fountains are evacuated in a central canal located in the main street of the city called Decumanus Maximus (Figure 11.4). The wastewater canal has a width of 60 cm, and a depth of 1.2 m. The canal was built in stonework and covered by slabs made of limestone (Panetier, 2002). The limestone slabs are slightly spaced to recover the trickling rainwater (Figure 11.4). The canal is also constituted by septic tank covered by a limestone plates with a rose in their center (Figure 11.5). The wastewater canal crossed by a network of drains collects wastewater from homes, fountains, thermal bath and latrines before being discharged into the river Khoumane. A secondary wastewater canal and Fertassa river allowed also evacuating the wastewater from the nearest houses as well as thermal bath (Figure 11.6).

Figure 11.4 Main wastewater and stormwater sewer (with permission of H. Benqlilou).

Figure 11.5 Manhole cover with a rose in their center (with permission of H. Benqlilou).

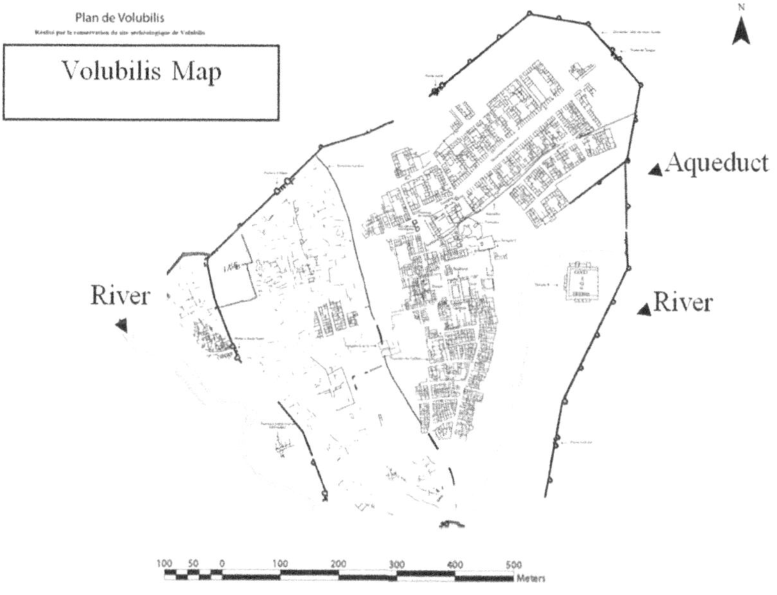

Figure 11.6 Map illustrating the Volubilis aqueduct and the two rivers playing the role of ultimate collector of sewer network (Conservation of the archaeological site of Volubilis, with permission of M. Atki).

11.2.1 Public fountains

Some public fountains, provided with storage chambers are used to filter water and to ensure water distribution of the secondary networks. These fountains were built on two levels to allow the drinking water supply and livestock watering (Figure 11.7). Fountains are also equipped for washing clothes. The overflows of the public fountains ensure flushing the sewer system and facilitate the gravity flow of wastewater to their disposal at the river (Figure 11.8).

Figure 11.7 Public fountain in two floors (with permission of H. Benqlilou).

Figure 11.8 Fountain fitted out for washing clothes (with permission of H. Benqlilou).

11.2.2 Volubilis houses

Water is at the heart of architectural plans of Volubilis houses built under square or rectangular shape with atriums and peristyles with fountains in different shapes (rectangular, circular, . . .) placed in the middle and are supplied by lead pipes from the aqueduct. The emptying and overflow water from basins are flushed down the drain. The houses also have a system for rainwater recovery 'Impluvium' (Panetier, 2002).

The houses have kitchens with basins; the excess water of these basins is discharged to the secluded subjacent latrines before evacuating into the sewer system which join the river downstream at a lower level than the city. Some homes also have private baths. The hydraulic system of the Palace of Gordian provides an overview of the water supply system (Figure 11.9).

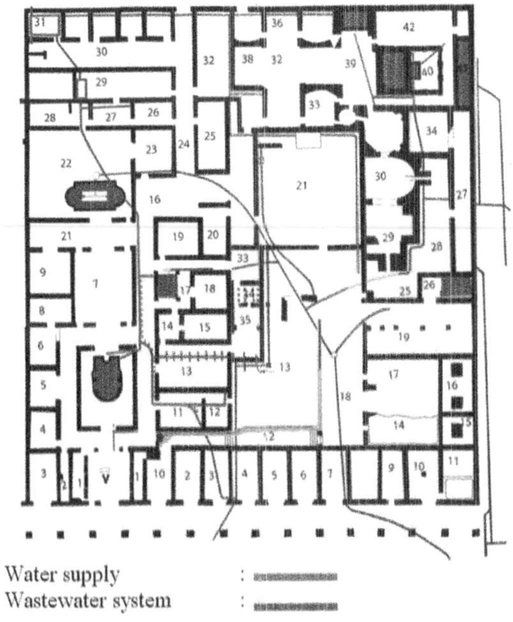

Figure 11.9 Water supply system of Gordien Palace (Thouvenot, 1958) (with permission of M. Atki).

11.2.3 Thermal baths

The city of Volubilis has four public thermal baths: thermal baths of the Northern Capitol, the House with Cistern and those called Gallien (Figure 11.10). Some homes have private thermal baths. The thermal baths has a typical classical Roman architectural. Thermal baths have three areas: cold, warm and hot rooms with basins and pools.

Figure 11.10 Thermal baths called Gallien (with permission of H. Benqlilou).

11.2.4 Latrines

Public latrines in a U shape around a central fountain is equipped with a drain to evacuate waste water. Latrines are located at the lowest part of the city with respect to the fountains and public thermal baths (Figure 11.11) levels.

Figure 11.11 Public latrines with a drain to evacuate waste water (with permission of H. Benqlilou).

Volubilis joins cities built at BC Roman Empire marked by the emergence of effective sanitation. The sanitation system of Volubilis is similar to the storm sewers covered with large slabs of stone described

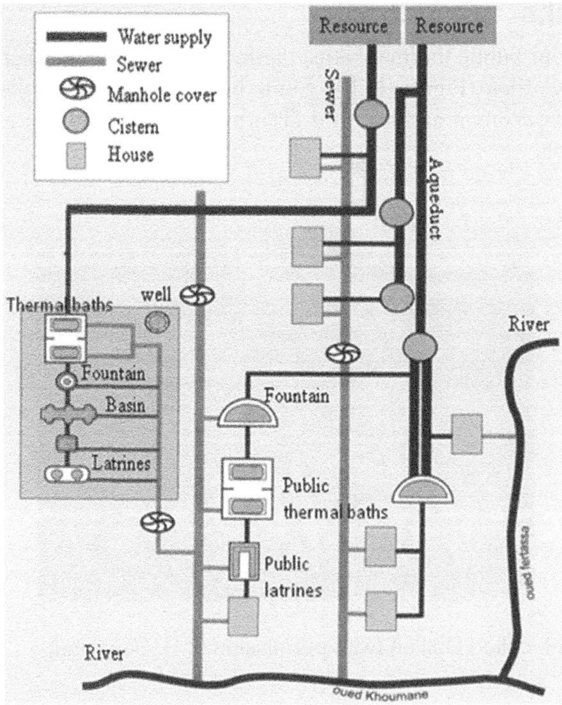

Figure 11.12 Volubilis Hydraulic System (Panetier, 2002).

in the Tharros Sardinia city, by G. Pesce and C. Poirieux (Bonnin, 1984). The continuous spillage of the Volubilis fountains (Figure 11.12) contributed to facilitate sewerage effectively, developed in the Roman cities before development of the toilet flushes (Bonnin, 1984). At Volubilis the waste water discharges at the river serving as the final collector is similar to the Tiber river collecting waste water from Rome (Bonnin, 1984).

11.3 MANAGEMENT AND MAINTENANCE OF THE VOLUBILIS HYDRAULIC SYSTEM

The management and maintenance of the Volubilis hydraulic system were provided by municipal institutions on the basis of special taxes in the same Manner of organization in the different cities built in the Roman Empire (Bonnin, 1984). The decline of the Roman Empire at Volubilis at 285 was marked by the abandonment of the aqueduct and the development of the city along the river in the south near the Khoumane river. The city was then the cradle of the first Moroccan Muslim dynasty founded by Idriss 1st at 789 AD how transferred his capital to Fez. Islamic thermal baths have been discovered in the southern district of Volubilis along the river and a wide aqueduct was constructed by Idriss 1 (Figure 11.13).

Volubilis was therefore a major milestone between the Carthaginian, Maurétanienne, Roman and Muslim civilizations in terms of transfer and exchange of hydraulic knowledge. The city witnessed a transfer of an International water Heritage.

Figure 11.13 Aqueduct constructed by Idriss 1 (with permission of M. Atki).
Source: Conservation of the archaeological site of Volubilis.

11.4 FES HYDRAULIC SYSTEM

The site selection for Fez foundation was done on the basis of abundance of water resources (springs and rivers), the same as the major cities around the world, from the oldest to present time. On its fertile land and availability of water resources, the first Moroccan Muslim dynasty was installed initially in Volubilis. For its expansion, the rights to water and land use has been purchased from Moroccan native population and non by requisition as in the past (El Faïz 2005). Fez was constructed on the both banks of the river 'Oued Fez' provides an ingenious system of water channels draining springs and rivers by gravity.

The complex appearance of the hydraulic system of Fez. A network whose oldest parts date back to the Almoravid period (eleventh century). The layout of the Oued Fez with its multiple branches and branches, its equipment and the wider ramifications is well known today thanks to studies by academics and experts from international organizations (UNESCO, World Bank, ...). The hydraulic network Fez is constituted by three main components: A primary network formed by the major diversions of Oued Fez in fact akin to rivers (also known as wadis in local terminology). At this level, the channels are divided into two distinct categories: those drinking water (al-ma lahlou or al-ma 'al-Tahir) and those of the wastewater (al-ma al-Harr); a secondary network consists of pipes pottery (kawâdis) who liaise between the main channels and buildings and facilities served, and finally, a tertiary network of pipes of different diameters and assuming the dual role of the drinking water supply and sanitation to the most basic levels of hydraulic mesh (houses, mills, religious buildings, public baths, etc.). Especially this level is relevant when managing the network analysis because it is directly related the needs of water users. Two elements characterize the service network: the famous device of Ma'ida (dispatchers) that collect and distribute water to drinking water pipes, and the grids (the chabkat-Qadus) for episodic duct cleaning and waste disposal that may clog.

Water is distributed between the different districts and for various purposes by intermediate water splitters appropriately dimensioned (Figure 11.14). The water splitters allow managing the principle of overflow and delays for the first proprietor to retain their priority. Water channels present a height and width between 1 and 2 m that depend on the topographic levels (Madani, 2003).

Figure 11.14 Main water splitter located in Bab Boujloud, one of the city gates (with permission of F. Serghini, ADER–FEZ).

A small tank called 'Ma'ida' allows different fonctions (storage, decompression and distribution) is placed through the water supply system within the city and even in the buildings (e.g., houses, schools, and mosques). The Ma'ida presentes a size of about 0.45 m in diameter and 0.60 m in deep and has various shapes (Figure 11.15). The Ma'ida system is designed to supply a group of houses or buildings, or other resources where new pipelines leave. The size of these distributors is based on the number of houses to be supplied. The amount of water allocated to each home is measured and calculated fingers with a clipboard called Debha (groove).

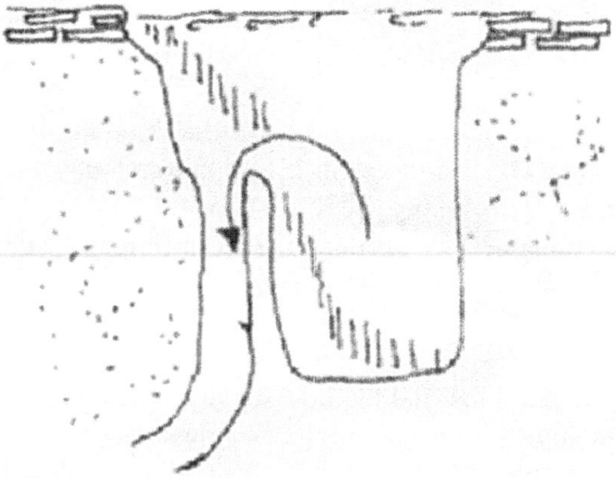

Figure 11.15 Longitudinal section of the traditional system 'Ma'ida' (with permission of. F. Serghini, ADER–FEZ). *Source*: ADER-Fès.

A distribution network consisting of many small pipes of different diameters (from 0.12 to 0.80 m) called 'kadus' provides water supply of the city for various needs: household, religious and crafts places (Madani, 2013). The pipes are made of clay allowing easy fitting to realize the numerous branches (Figure 11.16). To facilitate the drainage of wastewater, the excess water from the overflow of fountains and basins supplied by clean water channels are discharged into the sewerage system. The wastewater canal network (70 km length) is connected to rivers downstream of the city. Sewers could be considered as a continuation of the river 'Oued Fez' (Madani, 2003) (Figure 11.17).

Figure 11.16 traditionnel pipes (with permission of F. Serghini, ADER–FEZ).
Source: ADER-Fès.

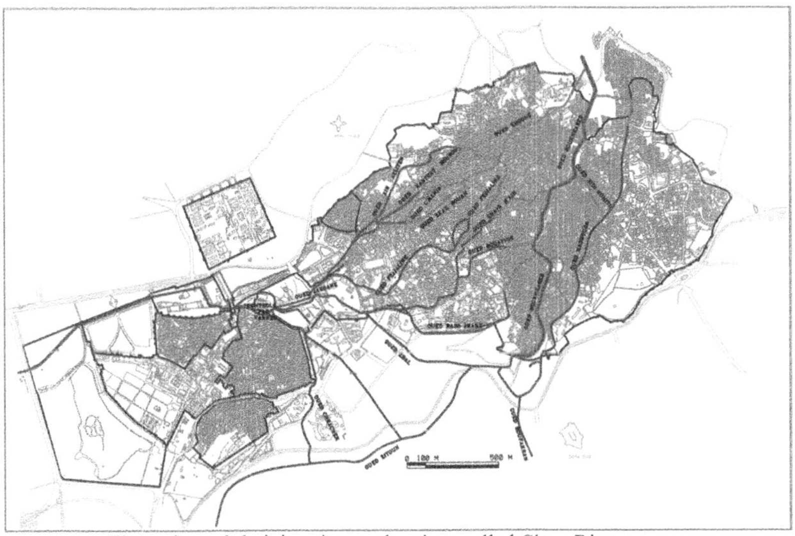

: Water channel draining river and springs called Clean River
: Wastewater canal called Wastewater River

Figure 11.17 Map of localization of clean and wastewater sewers of Fès (with permission of F. Serghini, ADER–FEZ).
Source: ADER-Fès.

Wastewaters Canals are located lower level than those intended for drinking water and mosques. These underground canals built with local materials (e.g., bricks, sand, and lime) cross homes, places and mosques and are discovered in different places (Figure 11.18). At the level of individual buildings and houses a system made up of small septic tank 'Slûkiyya' with a rectangular 0.40 m wide and about 0.60 m long (Figure 11.19).

Figure 11.18 Traditional sewer under houses (with permission of H. Benqlilou).

Figure 11.19 Small septic tank 'Slûkiyya' (with permission of H. Benqlilou).

The channels of clean water and wastewater canals are closely linked and branched into houses and the city so that the water is used for several purposes depending on its quality before being discharged into the sewage canals which are connected to rivers.

11.4.1 Mosques and Islamic schools

For purification (ablution), in the Mosques and the Islamic schools, a wall fountain and a central basin are mainly supplied by pure water channels drained from water springs (Figure 11.20). These water channels also supplied rectangular basins or fountains around which are arranged latrines available for students, residents and visitors of mosques and Islamic schools (Figure 11.21).

Figure 11.20 Water channel at the Islamic school Al Bû-nâniyya (with permission of H. Benqlilou).

Figure 11.21 Latrines around a fountain at Al Bu-nâniyya Koranic school (with permission of H. Benqlilou).

In 2013 until today, the water sources continue to fuel tanks, basins and fountain wall (Figure 11.22) of several 'mosques' and Koranic schools. The flow of water fountains and basins is done continuously for purification (ablution), for watering gardens and to help make flushing water into the sewer system (Figure 11.23).

Figure 11.22 Basin of spring resurgence in the Mosque 'Ayn Azlîtan'. Latrines are supplied by the basin (with permission of H. Benqlilou).

Figure 11.23 Wall fountain for ablution supplied by source 'Ayn Azlîtan' (with permission of H. Benqlilou).

In the basin supplied by source 'Ayn Azlîtan' fish species are observed (Figure 11.24), which is an indicator of high water quality and quantity of the source (74 m^3 flow given by Madani, 2003), a witness of traditional water system durability.

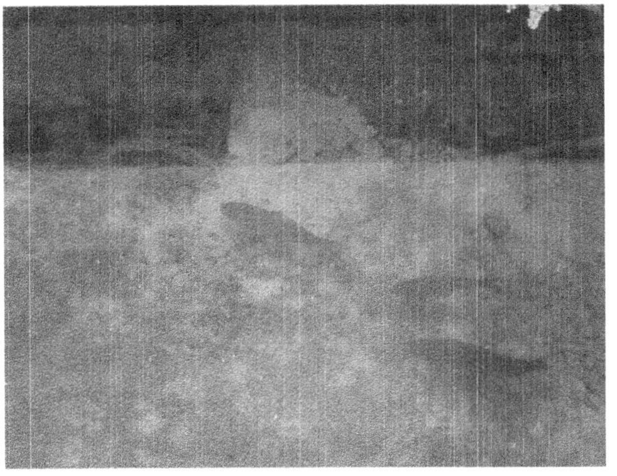

Figure 11.24 Fishes observed in 'Mosque Ayn Azlîtan' basin (with permission of H. Benqlilou).

11.4.2 Fez houses

Water is once again at the heart of traditional Fez houses built with a square or a rectangular shape with open atrium with a fountain at the middle and gardens goshawks supplied by water springs or rivers. Water is stored in a sort of small storage tank 'ma'dda' for water decompression and distribution. The drain and overflow fountains are evacuated to kitchens and latrines located behind the house. The evacuation of water overflows fountains and pools and all water already used for different needs at the level of sewage pipes to improve the efficiency of the wastewater system of the city.

11.4.3 Public latrines

Public toilets for visitors to the city are supplied by water from springs and rivers and are arranged around a fountain generally rectangular. The water flow is continuously and the slope of the ground allows that the water excess is discharged to the latrines to ensure the wastewater (Figure 11.25).

Figure 11.25 Public toilets for visitors (with permission of H. Benqlilou).

11.4.4 Handicraft

At Fez, the tanneries are among the main water consumers of craft trades. For economic issues, water sources are used primarily due to their sustainability compared with those from rivers. The water flows continuously for washing leathers. Releases of the factory are discharged into the sewerage network continuously.

11.4.5 Bathhouse

Public baths consist of three main parts: cold temperature in the first part followed by a warm room and at the end a very hot room. A wall fountain is fed by two water taps hot and cold (Madani, 2003).

11.5 THE ISSUE OF MAINTENANCE OF THE FEZ HYDRAULIC SYSTEM: THE AL-FACHTALI MODEL

The secret of the extension of the water network of Fez and its longevity lies in two essential conditions, namely: maintenance and repair. Both types of work, whose execution is the responsibility of the corporation qwadsiyya (fountain), were still funded by contributions from water users. These were responsible for the maintenance of pipes from their homes and those who came to the general collector of the district. It was broad cooperation at the neighborhood level and all the ramifications of the Oued Fes to ensure sustainable management of the network.

Due to the significant urbanization of the city of Fez during Saadian dynasty, marked by the creation of new districts and the proliferation of structural equipment (Existence of 84,000 houses, 366 water mills, 600 fountains and ablution rooms, 700 mosques, etc. (Mezzine, 1986)). This urbanization implies a densification of the hydraulic system and maintenance accompanying measures. The city of Fes was also marked by the earthquake (1033 years AH (1624)), causing damage to homes and their supply network. These are probably the reasons of developments of The Al-Fachtali model to address the issue of determining quotas maintenance of Fez hydraulic system which is the greatest innovation in hydraulics in the seventeenth century. The author divides his model into three aspects:

(a) solution according to the rules of jurisprudence (Fiqh);
(b) mathematical problem solving; and
(c) general considerations of the cities development.

11.5.1 The legal constraints of the model

Regarding the first aspect, Al-Fachtali recalls the teaching of Judge Muhammad Al-Yafranî Miknâsî (1436–1512) (El Faïz, 2002). The magistrate Fez relates the opinions of Ibn Rushd (d. 1112) and Asbagh (d. 886) who specify that the cleaning and maintenance of pipelines is performed according to the shares and rights and not according to the number of people. This is the first constraint.

The second major legal constraint is that when a line is clogged after having served a user which is upstream, it is not obliged to cooperate with the following user. Because it cannot make a profit from this work, or suffer harm by renouncing it, since it is possible to water in both cases. The same constraint operates when it comes to perform maintenance.

Al-Miknâsî noted that this opinion of Ibn Rushd is only valid if the user uses the upstream part water completely without having to return to the channel. If this is the case, it must contribute to the cost of cleaning with its immediate neighbor and so on until the water is returned to the branch of the great river. Based on these two fundamental constraints, Al-Fachtali was engaged in the construction of its model recalling the mathematician and has no specific expertise in architecture.

11.5.2 The mathematical formulation of the model

The author begins by telling us that the distances which Ibn Ghazi talks are those between the mother source (ra's al-'ayn) to each of the distributors for supplying homes with water. Dispatchers that do not directly serve this purpose and are created either to collect water or to increase their speed, are not considered in the calculations. Only count dispatchers serving users. The principle applied here is that the user pays.

Suppose, writes Al-Fachtali, a bottleneck occurs in any part of the pipeline, the distance to remember is calculated relative to the dispatcher just after the location of the obstruction and not before (Applying the second stress in the model). If this is the case, we first calculate the number of spans distance from the source to dispatcher used first, second distance from the source to dispatcher used secondly, the third distance from the source to look used third, the fourth distance, and then the fifth and so on until the last distributor actually used.

Then, the author adds, the sum of distances spans are made and reduces the result of common factors. The divider can be a composite number or integer. Then, multiply the quotient by the maintenance capital (sum of repair costs) and divide them result by the common divisor. The resulting figure is the financial contribution of the calculated distance.

As an example, facilitate understanding the Al-Fachtali model. Suppose a pipeline that supplies three dispatchers. The distance is spans 10 relative to the first splitter, 15 spans (span (chibr) is approximately 0.231 m) from the second and 20 spans from the third. Assume also that the failure occurs before the first splitter and we have spent on the repair costs 90 Oukiyya (the common currency to Saadian time were Mitqâl (equivalent dinar gold or 3,548 g) and ouqiyya (equivalent dirham silver or 3.3105 gr) (Mezzine, 1986). If the sum of the three spans distances we do, we found 45 that spans divided into divisors of 9 and 5. Disposing the first assembly in a column. Knowing that the sum of the spans of the three distances is divisible by 5, it reduces distance to each denominator in accordance with the recommendation of Ibn Ghazi.

Thereby obtaining the first distance to the figure 2, the second 3 and a third 4. The sum of these numbers gives us 9 which is the common divisor (al-Imam, the master). We put all these results in a second column, capital maintenance in a third and a fourth column dividers. If you wish to calculate the distance to each corresponding costs, the ratio of the first distance (2) is multiplied by the capital maintenance (90). This gives you 180 divided by the common factor (9), which gives us 20, that is, the amount to pay repair costs by the first distance. The author applies the same method to calculate the maintenance costs for the second and third distance. He finally gives us the calculations, which are summarized in Table 11.1.

Table 11.1 Caclulated maintainance costs at diferent distances (spans).

Distance (spans)	Dividers 5 et 9	Maintenance costs 90
10	02	20
15	03	30
20	04	40
45	09	90

The second step in the construction of al-fachtali model is to consider for each distributor a number of houses to be served with their respective shares in the water rights, suppose the first exchange serves housing 6, the first holding units of water 1/9th, 1/8th second, third 1/6th, 1/5th the fourth, the fifth fourth and sixth third. One puts all these fractions in a first column, and a second column has the least common denominators of these fractions, following the recommendations of ibn ghazi, the least common denominator of all units is 360. This gives 40 shares for the first house, 45 for second, 60 for third, 72 for fourth, 90 for fifth and sixth 120. The sum of the shares is equal to 427 is divisible by 7 and 61. We put all these parts in a second column, the contribution to maintenance costs (20 in this case) in a third column and dividers shares in a fourth column. Then we multiply each home its share of contribution to maintenance costs (20 in the case of first division) and dividing the result by the dividers reported in the fourth column (Table 11.2).

Table 11.2 Calculations for first distance (10 spans) and dispatcher n°1.

Fractional units	Units 427	Maintenance costs 20	Dividers	
			61	07
1/9	40	01	53	–
1/8	45	02	06	04
1/6	60	02	49	03
1/5	72	03	22	05
1/4	90	04	13	01
1/3	120	05	37	06

These calculations (Table 11.2) gives for the first house a financial contribution of ouqiyya and 53 shares of 61, for the second house two ouqiyya, 6/61ème ouqiyya and 4/7ème from 61, for the third house two oukiyya, 49/61ème oukiyya and 3/7ème from 61 and for the fourth house three oukiyya, 22/61ème oukiyya and 5/7ème from 61, for the fifth house four oukiyya, 13/61ème oukiyya and 1/7th share of 61, and the sixth home five oukiyya, 37/61ème oukiyya and 6/7ème from 61. Al-Fachtali apply the same method to calculate the contributions of the houses within the second and third distance (Table 11.3).

Table 11.3 Calculations for second distance (15 spans) dispatcher n°2.

Fractional	Units 5624	Maintenance costs 30	Dividers		
			37	19	08
1/11	630	03	13	06	04
1/10	693	03	25	14	06
1/9	770	04	03	18	04
1/7	990	05	10	07	04
1/6	1155	06	05	18	01
1/5	1386	07	14	10	04

The results presented in Table 11.3 indicate for the first house a financial contribution: 3 oukiyya, 13/37ème oukiyya, 6/19ème from 37 and 4/8th part of the 19th from the 37th of oukiyya share, for the second 3 oukiyya house, part of the 25/37ème oukiyya, 14/19ème from 37 6/8ème and part of the 19th from the 37th of oukiyya share, for the third house 4 oukiyya, 3/37ème share of 'oukiyya, 18th/19th from 37 and 4/8th share of 19 shares of the 37th part of the oukiyya and so on. The calculations second distance (20 spans) and dispatcher n°3 are given in Table 11.4.

Table 11.4 Calculations for second distance (20 spans) and dispatcher n°3.

Fractions	The units 1026	Corresponding 513	Maintenance fees 40	Dividers		
				19	09	03
1/8	108	54	04	04	00	00
1/7	120	60	04	12	08	00
1/6	140	70	05	08	06	01
1/5	168	84	06	10	04	00
1/4	210	105	08	03	05	00
1/3	280	140	10	17	03	02

These calculations (Table 11.4) gives, for the first house a financial contribution and 4 oukiyya 4/19ème from the oukiyya, the second house 4 oukiyya, 12/19ème from the oukiyya and 8/9ème from 19, for the third oukiyya house 5, part of the 8/19ème oukiyya, 6/9ème from 19 and 1/3 of the ninth part of the 19 shares oukiyya and so on.

11.5.3 The terminological aspects of the problem

Al-Fachtali reserve the third part of his treatise to discuss issues of terminology. It segregates the seguia, exposed pipe for irrigation of gardens and orchards Qadus, buried pipe, used to supply homes with drinking water. It also tackles the system measures used. Assuming chibr (span), which is a half cubit Ibn Luyun, the author was able to show, based on his own experience, how this value could vary from one person to another. In addition to the definitions of terms used, al-Fachtali arrives to explain the formulas used, referring to what is treated in detail to mathematics and known in his time calculation.

11.5.4 Scope and contributions of the mathematical model of al-Fachtali

The Al-Fachtali mathematical model is, as we have seen, both a model of water distribution and calculation of quotas network maintenance. It thus addresses the fundamental aspects of the distribution and financing are the keys to understanding the urban water since the eleventh century in Fez. All these elements show that the issue of maintenance and maintenance of the hydraulic system was not a new concern. However, if the legal and statutory standards of arbitration were known, we cannot settle yet a management tool of the pipe network on a scientific basis.

It is al-Fachtali return that deserves to meet this intellectual challenge at a time of great dynastic instability in Morocco and weakening of urban solidarity. The al-Fachtali model was both realistic and pragmatic in that it took into account the legal provisions as well as the actual operation of infrastructure.

Just remember the statement of constraints and the choice of examples to convince his empirical orientation. Another feature characterizes this model: this is the philosophy that underlies it and which aspires to achieve the ideals of justice and equity in distribution of water and calculating the financial contribution of users. In addition to the principle of user pays, al-Fachtali, pushed his arithmetic of water through its boundaries. Its distributive genius expressed itself perfectly into the architecture of a hydraulic model which splits an infinitesimal degree contributions and quotas and seems to have been designed with mathematical precision, at a time when conflicts between users Urban Water Fes became more acute.

What is also striking in the model developed by al-Fachtali is its universality. It can be applied in all Arab-Islamic cities with a similar Fez network and operating according to the same legal constraints. We know for example that Damascus and Aleppo distribution network maintenance expenses was in proportion to the distance and consumption. The measurements were made in finger (Usbu) and chibr (span). When the amount of water supplied, it is not measured in m or second liter, but share the flow channel at the point considered. However, no text has been preserved for us the attempt to model the distribution of water shares and financial contributions. We can also add that where the system Ma'ida was common even in the cities supplied by sources other than those from wadis (khettaras or qanat), it became possible to apply arithmetic al-Fichtâlî.

The hydraulic system of Fez is an exemplary network of its size, its age and complexity. If the various branches and branches of this system are now known, it remained to illuminate the much finer issue of articulation between the technical structure and social organization responsible for the management and ensure its sustainability. We had so far only very vague notions of network operation to the most basic, most notably at the supply houses drinking water and determining the role played by the user community levels. It is this gray area, marked by the interweaving of socio-technical, al-Fichtâlî tried to explore for us, bringing its skills as a mathematician, physician and magistrate specializing in science the determination of quotas.

The al-Fachtali model, developed during the first half of the seventeenth century, also demonstrates that the reign of the Saadian dynasty in Morocco was not only a great moment in the development of hydraulic engineering and agricultural industries and sugar. He had, a few years before dying, a glimmer of genius who helped build a successful management tool, capable of regulating the hydraulic network of Fez. In addition, the model was designed at a level of generality as its parameters could be applied to similar supply systems in other Islamic cities.

11.6 SIMILARITIES BETWEEN FES AND VOLUBILIS HYDRAULIC SYSTEMS

The two cities of Fes and Volubilis have strong similarities in their hydraulic systems. Both cities were developed to be gravity supplied by springs and rivers located upstream in the foothills. Water channels draining river and spring were supplied houses, public fountains, and public and private thermal baths, latrines with an integrated ressources management approach.

Overflows and drain fountains and basins as well as the evacuation of rainwater canals in sewage allow for regular flushing into the sewer system which allows wastewater discharge in the natural environment at the river downstream to the lowest level of pollution.

The water system supplying Fes and Volubilis houses is also designed focusing on a multi-use of the same water depending on its quality. The water fountains are used for drinking and overflows are routed to the kitchen and then to the latrine before being discharged into the sewage system. The water is discharged into the sewer once the level of pollution has reached its maximum. Public latrines in the two cities are arranged around a central fountain allowing water continuously flowing to discharge wastewater. A great similarity between the Fes public baths and Volubilis thermal bath is observed either for the heating system or as for the architectural and the temperature of the different rooms (Madani, 2003).

11.7 EVOLUTION OF THE HYDRAULIC SYSTEM OF FES AND VOLUBILIS

The decline of the Roman Empire led to the abandonment of the Volubilis complex and ingenious water system because of the failure of the collection of taxes by the public administration. At Fez, population growth, as well as the failure of social cohesion, contribute that customary law 'urf' is no longer applied to maintain and repair the traditional hydraulic systems. However, 20% of the sewerage network continues to function after several rehabilitation programs.

Currently two wastewater canals are being built along the river that has been used in the past to collect wastewater from both sides of the city of Fez (Figure 11.26). The new wastewater canals allow the evacuation of overflow rain into the river used in the past to dispose wastewater from the medieval Fes city.

Figure 11.26 Two wastewater canals are being built along the river 'Oued Fez' (with permission of H. Benqlilou).

Fez and Volubilis traditional hydraulic systems, witnessed a transfer of an International Heritage. Volubilis Roman and medieval Fez Muslim cities were built and developed by integrating sustainable cycle of available water resources: springs and rivers upstream in the foothills by a gravity water supply. Water channels that carried the water from springs and rivers followed the backbone of arteries and streets of the two cities through the soles in different buildings without the use of lifting devices. Wastewater canals and rivers used to dispose of wastewater after a rational use of the same water for different needs even within the same dwelling and within the city.

Today we talk about integrated management of water resources in the past master plans for the cities and architectural plans for houses were thought in an integrated water management approach. If the cities of Volubilis and Fez are classified by UNESCO as World Heritage Sites, the hydraulic heritages of these two cities mirror of water Roman either Arab-Muslim civilizations. These water transfer heritage is an important resource in terms of development of cities and territories by considering water as a central element to allow drawing several solutions and answers to the current problems of water raised through the Mediterranean and the world to deal with the scarcity of water is increasing as a result of climate change.

In the both cities the clean water resources bring human and economic development and the wastewater is evacuated to a river on the basis of a branched and interconnected water canals called clean and dirty rivers. Clean and dirty branching Fez channels rivers remind by their complexities and similarities arteries and veins of body blood circulation. The rivers and springs bring clean water (source of life) and reject dirty water to natural outfalls within the same water natural gravity system. The Fez hydraulic heritage is a witnessed that the water was and still at the heart of human, cities and civilizations development.

11.8 DISCUSSION AND CONCLUSIONS

The comparison between the hydraulic system of Volubilis, an ancient Roman city, and the water system of the medieval Fez Muslim cities, shows the evolution of the public water supply systems from ancient to modern times. The two cities of Fes and Volubilis have strong similarities in their hydraulic systems. Both cities were developed to be gravity supplied by springs and rivers located upstream in the foothills. Water channels that carried the water from springs and rivers followed the backbone of arteries and streets of the two cities through the soles in different buildings without the use of lifting devices. Wastewater canals and rivers used to dispose of wastewater after a rational use of the same water for different needs even within the same dwelling and within the city. Water channels draining river and spring were supplied houses, public fountains, and public and private thermal baths, latrines with an integrated resources management approach. The two cities Fez and Volubilis present an ingenious water supply and sanitation systems characterized by:

(a) A direct and continuous intake of water upstream (springs and rivers) without using pumping systems.
(b) The distribution is made according to the piezometric level of resources recurring to a dispatching system in the different neighborhoods and buildings.
(c) A small storage is just used at fountains.
(d) The water excess is evacuated by flushing at the sewerage network, continuously.
(e) Water is used for several purposes before being discharged into the sewage system, and only when it reaches its maximum level of pollution.
(f) The continuous flow of water in fountains, toilets and basins, ensures a continuous drainage of wastewater which limits the maintenance and management works of the sewerage network.
(g) Participative management using defined quotas calculated on the basis of the distances of drinking water supply and sewage and not on the number of users ensures rigorous maintenance of water and sanitation systems.

Fez and Volubilis traditional hydraulic systems, witnessed a transfer of an International Heritage. If the cities of Volubilis and Fez are classified by UNESCO as Word Heritage Sites, the hydraulic heritages of these two cities mirror of water Roman either Arab-Muslim civilizations. These systems which influence in the past water and wastewater technologies development around the Mediterranean region can serve as a basis for the development of drinking water and sanitation for small low income communities due to the absence of excessive costs related to basic investments and operating costs. Similarly, for small low income communities, participative management approach based on the Al Fachtali model can be used for pricing models.

These systems similar to veins of the human body, canals bring clean water rich in oxygen and discharge dirty water sometimes at the same drainage system, reflect ultimately the integration of the city to the water resources in a sustainable and integrated approach in terms of exploitation of clean water and wastewater discharge without using pumping systems.

Acknowledgments

For their great contributions to this Chapter, we are thanking the following:

(a) Serghini F., Directeur Général ADER–FES, Agency for De-densification and Rehabilitation of the Medina of Fez.
(b) Chafi A. ADER–FES, Agency for De-densification and Rehabilitation of the Medina of Fez.
(c) M. Mustapha Atki, Conservation of the archaeological site of Volubilis.
(d) LaouimirA., National Office of Electricity and Potable Water ONEE, Meknes.
(e) Lahlou M., National Office of Electricity and Potable Water ONEE, Fez.
(f) Amarti Riffi S., Autonome Intercommunale water and electricity distribution RADEEF, Fez.
(g) Moumni A., Autonome Intercommunale water and electricity distribution RADEEF, Fez
(h) Faouzi R., Autonome Intercommunale water and electricity distribution RADEEF, Fez.

11.9 REFERENCES

Bonnin J. (1984). L'eau dans l'antiquité, l'hydraulique avant notre ère (The water in ancient, hydraulic BCE). Eyrolles, Paris, France.

El Faïz M. (2002). L'apport des mécaniciens arabes à l'évolution de l'hydraulique médiévale (The Contribution of Arab engineers in the development of medieval hydraulic). La Houille Blanche, France.

El Faïz M. (2005). Les maîtres de l'eau, (The masters of water). History of the Arab Hydraulics, Actes Sud, France.

Madani T. (2003). L'eau dans le monde musulman médiéval (The water in the medieval muslim world). Université Lyon II, Auguste et louis Lumière, Lyon, France.

Maps, pictures: Conservation of the archaeological site of Volubilis, Walili, Morocco.

Maps, pictures: ADER- FES: Agency for De-densification and Rehabilitation of the Medina of Fez.

Mezzine M. (1986). Fâs wa-bâdiyyatuhâ (Fès et sa campagne à l'époque saâdienne, 1549–1637) (Fes and its countryside at the Saadian period). I, Rabat, 1986, p. 257.

Mezzine M. (1986). Fâs wa-bâdiyyatuhâ (Fès et sa campagne à l'époque saâdienne, 1549–1637) (Fes and its countryside at the Saadian period). I, Rabat, 1986, p. 458.

Panetier J. L. and Limane H. (2002). Volubilis une cité du Maroc antique (Volubilis an ancient city of Morocco). Maisonneuve et Larose, Paris, France.

Thouvenot R. (1958). Maisons de Volubilis: le Palais dit de Gordien et la Maison à la mosaïque de Vénus (The gordian palace and the house of venus mosaic). Antiquities Department of Morocco (ed.), Service des antiquités du Maroc, Rabat, Morocco.

UNESCO, website (archaeological World Heritage). http://whc.unesco.org/en/list/836 and http://whc.unesco.org/en/list/170 (accessed 5 May 2014).

Chapter 12

Sanitation and wastewater technologies in ancient Roman cities

Giovanni De Feo, Sabino De Gisi, and Meisha Hunter

12.1 INTRODUCTION

The history of sanitation can be traced to the beginning of human existence. Evidence of the oldest known wastewater drainage has been dated to the Neolithic Age (approximately 6500 BC) in El Kowm (modern day Syria). Early Mesopotamian cities featured networks of wastewater and stormwater drainage. During the early Bronze Age, Mohenjo-Daro (located in the Indus Valley of modern day Pakistan) included both water supply and effluent disposal systems. During the Bronze Age, wastewater management was practiced in several Minoan palaces and settlements (modern day Crete). Recognizing the significance of water in terms of public health, the Greeks organized baths, toilets, and sewerage and drainage systems. The Romans designed and built systematic engineering works for the management of sanitation and wastewater (De Feo *et al.*, 2013).

The principal aim of this Chapter is to perform a review of the principal aspects of sanitation and wastewater technologies in three ancient Roman cities: Pompeii, Herculaneum and Ostia. Consequently, the Chapter is organized into three principal sections corresponding with each city, each with two sub-sections: (1) toilets and cesspits, and (2) drainage and sewerage systems.

A cesspit is a hole in the ground for collecting the urine, feces and solid waste that flows from a toilet. Toilets during the Roman era can be divided into two groups, namely, public and private (Vuorinen *et al.*, 2010; De Feo *et al.*, 2010). According to Hobson (2009) and Jensen *et al.* (2011), ancient Roman toilets can be further defined as either public multi-seat toilets accessible in public areas (*foricae*), or private toilets with one, two, or more seats.

Drainage systems were used for the disposal of surplus water, and were found both in cities and in the country. In cities, drainage systems carried rainfall and overflow from fountains and bathrooms. In the country, drainage systems functioned to prevent flooding in the fields. Sewerage systems, on the other hand, served to convey domestic wastewater, and were specific to urban environments, where high population densities made sewerage systems necessary (Hodge, 2002; De Feo *et al.*, 2011).

The city of Rome was famous for its sewers. The most celebrated sewer in Rome was the *Cloaca Maxima*, constructed beneath the Forum, whose effluent was discharged into the Tiber River (Asplund Ingemark, 2008). Tradition ascribes the *Cloaca's* construction to Tarquinius Priscus, king of Rome

616–578 B.C. The *Cloaca Maxima* (4.2 m high, 3.2 m wide) was covered by stone vaulting, while its floor was paved with basalt. It served to remove wastewater and rainwater, and also functioned as swamp drainage. The *Cloaca Maxima* drain into the Tiber still exists in Rome, but is partially concealed by the modern *Lungotevere* Embankment (Hodge, 2002; De Feo *et al.*, 2010).

The wastewater facilities discovered by archaeological excavations in the cities of Pompeii, Herculaneum, and Ostia evidence sophistication in design and construction that are not uniformly present in modern wastewater infrastructure and merit review by modern engineers (NSSW, 2009).

12.2 SANITATION AND WASTEWATER TECHNOLOGIES IN POMPEII

Pompeii, approximately four fifths of whose urban area has been excavated, is arguably the most well-known ancient Roman archaeological site in the world. The eruption of Vesuvius in 79 AD buried the city under layers of ash and small pieces of lava (*lapillus*) 6–7 meters thick. Most of the inhabitants, who fled from their homes, were killed on the coast. Those who sought shelter in the basements of homes, died of asphyxiation: the molds of their bodies in agony, obtained by pouring liquid plaster into the cavities left by the bodies in the layers of ash, testify to this tragedy. Visiting the excavations offers a unique opportunity for modern-day viewers to appreciate public and private life in ancient Pompeii, including residences, both luxurious and humble, as well as commercial shops, taverns, and bakeries, etc. Since many furnishings are intact, intimate aspects of the lives of the ancient Romans are revealed, including those activities associated with personal hygiene.

The *Forum* was the heart of Pompeii's economic, commercial and political activities. All of the city's principal public buildings were situated in this vast space (38 m × 142 m), surrounded by a double colonnade and adorned with statues of bronze and marble: the Basilica where justice was administered, the offices of *duumviri*, the *decuriones* and *ediles*, who ruled the city, the *Comitium* for the vote of the judges, the Temple of Jupiter, the Temple of Apollo and many other sacred buildings.

The *Forum* was also the locus of trade. In the *macellum*, a covered market built during the imperial age, Pompeians sold their crops, as well as meat and vegetables. The Building of Eumachia, priestess of Venus, the home of the corporation of *Fullones* (launderers, dyers and manufacturers of cloths), was intended for sale at auction of wool, clothing and other apparel items. Along the *Via dell'Abbondanza* were the *thermopolia* (a commercial establishment which sold ready-to-eat food) such as that of Asellina, for the pouring of hot and cold drinks, the *caupone* (taverns and restaurants), and *fullonicae* (dry cleaners) such as that of Stephanus. In addition to the bakeries, the sale of *garum* (sauce made of dried fish, moray eels, tuna and mackerel) occurred throughout the city in the artisan workshops of iron, ceramics, and goldsmiths.

Bathing complexes (*thermae*) were constructed in the most higly trafficked parts of the cuty, such as the *Forum*'s, Central and Stabian Baths. These buildings, divided into male and female sections, included the *caldarium* (hot bath), *tepidarium* (warm bath), and the *firigidarium* (cold bath). The care of the body was completed in gyms (*palaestra*) where Romans practiced sports activities (e.g., the grand imperial era palace of the Big Gymnasium). Theatrical performances were hosted in the Large Theatre (200–150 BC), with a capacity of 5000 spectators, or the Small Theatre (post 80 BC). The Large Theatre featured classical works, comedies or tragedies, while the Small Theater featured musical performances and mime.

12.2.1 Toilets and cesspits

Arguably, it is difficult for modern day visitors to physically identify ancient toilets in the archaeological site of Pompeii. The limited public awareness of toilets at Pompeii is underscored by the fact that archaeologists who have performed excavations on the site, such as Giuseppe Fiorelli, have provided only brief descriptions and analyses (Jansen *et al.*, 2011). In Roman towns, public toilets (*foricae*) were a commonplace (Hobson, 2009). Physical evidence indicates the city of Pompeii featured a large number of toilets and cesspits, both domestic and public.

The majority of domestic toilets in Pompeii drained directly into cesspits, which typically were constructed beneath the sidewalks of the streets. Some latrines were located within residential interiors, often in the kitchen. Cesspits associated with latrines were also located on the interiors of Pompeian homes. In some houses, the depth of cesspits equals that of a large amphora (*dolium);* in others, the cesspits measure up to 10 m deep. When full, the portable receptacles would have been removed by *stercorarii* (slaves) and replaced by another vessel. For cleaning, water was either poured down the toilet or males urinated directly onto the tiles (Hobson, 2009). Disposal of domestic human waste with water was a late development in Pompeii: some residences connected to the piped water supply in the post-Augustan period did install flushing toilets (Kolowski-Ostrow, 1996).

Pompeian toilets generally had sloping floors in front of them, and they often consisted of wooden seats resting on stone supports, which surrounded the entrance to the cesspit or drain (Jansen, 2000; Flohr, 2011). Some *foricae* were built in the *Forum*, the gymnasium and some of the baths after the earthquake in 62 AD (Keenan, 2005). The remains of the multi-seat latrine in the northwest corner of the *Forum* are shown in Figure 12.1. The protruding stonework around the edges of the latrine would have had wooden seats resting on them. Beneath the seats, a flow of water flushed away waste toward the drain.

Figure 12.1 Remains of the multi-seat latrine in the *Forum* of Pompeii (the protruding stonework around the edges would have had wooden seats resting on them).

Shops and bars had toilets as well. In some areas of Pompeii, there are a number of single rooms opening directly onto the street that have been identified as latrines (Hobson, 2009).

12.2.2 Drainage and sewerage systems

In addition to toilets and cesspits, Pompeii featured a scattered network of sewers. Rainwater and wastewater were mainly disposed of along the streets (Jansen, 2000) that would have also contained some human waste (Keenan, 2005).

At the time of Vesuvius' eruption, Pompeii's street drains existed only in the vicinity of the *Forum*. The streets were a sort of open channel conveying water from public fountains, rainwater and sewage. The overflow from the cisterns (via outlet pipes to the street) and the road-side fountains and towers of the piped distribution system would have assisted in removing both human and animal waste from the streets (Keenan, 2005). Therefore, as shown in Figure 12.2, in the streets featured raised sidewalks (50–60 cm high) with stepping stones (*pondera*) at the intersections to enable pedestrians to cross from one side to the other without stepping down (Hodge, 2002; De Feo *et al.*, 2010).

At the downhill gates on the south and west portions of the city, the *Portae Nocerua*, *Stabia*, *Nola* and *Sarno*, gutters in the thresholds allowed the flow of untreated sewage and stormwater into the Sarnus river and thence into the Bay of Naples (Koga, 1991; Koloski-Ostrow, 1996; Keenan, 2005). The *Via Stabia* was the major artery in this drainage system. There were a few underground drains, mainly in the area of the Civic *Forum* and the old city. Some of these underground drains consisted of earlier street beds that had subsequently been reused after the street level was raised (arguably due to earthquakes that occurred prior to the volcanic eruption). Many of these drains received wastewater from depressions within the town (Koga, 1991; Keenan, 2005).

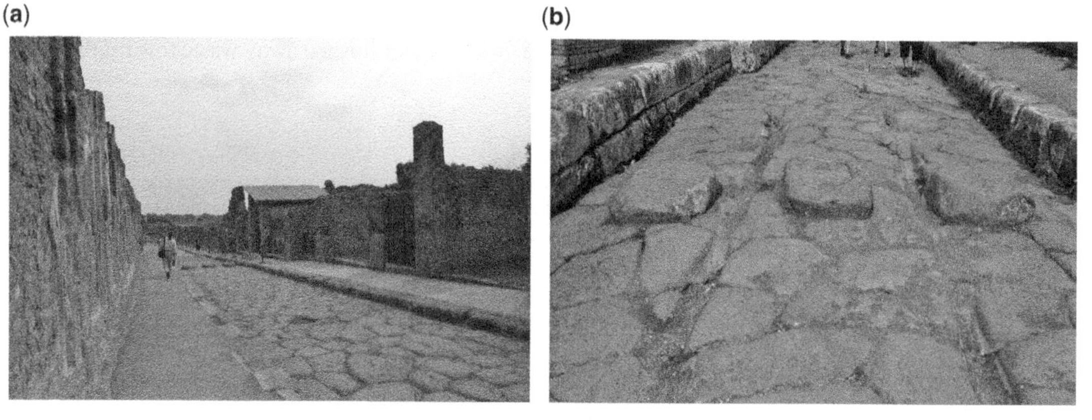

Figure 12.2 Stepping stones (*pondera*) in Pompeii: (a) overall view of *Via dell'Abbondanza*; (b) with a detail of *pondera*.

Notably, Pompeii's sub-grade drainage system has not been subjected to widespread archaeological investigation and analysis. Rome had an extensive drainage system, as did some of the *coloniae* (an autonomous community comprised of Roman citizens, located in a territory conquered by Rome). Arguably, when Pompeii became a *colonia*, the capital investment necessary to finance a system-wide retrofit the city's drainage was not available and the gradient slope of the streets permitted their use for drainage. However evidence of substantial drainage systems has been found in some premises such as in the fullers' workshops (*fullonica*) (Koga, 1991; Keenan, 2005). For instance, De Feo and De Gisi (2013) describe the drainage system of the *fullonica* of Stephanus. The basins of the *fullonica* feature a drainage channel that extends all the way through the house and terminates at the *Via dell'Abbondanza*. The channel bends slightly to the east, so the actual outlet is in front of the neighboring property: this is precisely at the point where the *Abbondanza* starts to descend toward the east so the water was drained off not through the *Via Stabiana* but through the *Via dell'Abbondanza* (De Feo & De Gisi, 2013).

12.3 SANITATION AND WASTEWATER TECHNOLOGIES IN HERCULANEUM

The 24 August 79 AD eruption of Vesuvius devastated numerous Roman cities, including Pompeii and Herculaneum. Unlike Pompeii, buried by layers of ash and *lapillus*, Herculaneum was buried under a layer of mud and lava up to 25 meters thick. The sealing action of the mud has preserved everything: wood, cloth and food have undergone a slow transformation, remaining unchanged in their wrapping, almost petrified. In 1709 the Prince of Elboeuf, during the excavation of a well in one of his villas, exposed structures of the ancient theater of Herculaneum. In 1738, King Charles of Bourbon ordered the official commencement of archaeological excavations at Herculaneum. The most sensational discovery was the *Villa dei Papiri*, from which bronze and marble sculptures (now in the National Archaeological Museum of Naples) and a papyrus library consisting of more than 1800 philosophical texts (now in the National Library of Naples) were extracted. In 1927, excavations of houses and public buildings had extended as far north as the *Forum*, as far east as the *Palestra*, and as far south as the Suburban Baths (*Terme suburbane*). At Herculaneum wealthy Romans enjoyed their holidays, as evidenced by the sumptuous villas on the sea. The streets, paved with Vesuvian lava or limestone, outline the characteristicly Roman system of *insulae* (apartment building that housed most of the urban citizen population).

With the aim of supporting the Italian State in preserving the archaeological site of Herculaneum, the Herculaneum Conservation Project (HCP) was established by David W. Packard, president of the Packard Humanities Institute, to assist the local heritage authority (Soprintendenza Speciale per i Beni Archeologici di Napoli e Pompei). The HCP was established as part of the international community's response to the extraordinary challenges of conservation faced by the Vesuvian sites, which arguably cannot be shouldered by the national heritage authority alone (Wallace-Hadrill, 2007).

12.3.1 Toilets and cesspits

Although only a few kilometers from Pompeii, the town of Herculaneum is different in many ways, including the distribution of toilets and the limited number of cesspits (Jansen, 2000; Hobson, 2009). The principal reason for the limited presence of cesspits is that the ground in Herculaneum is too rocky for draining away urine and feces (NOFSR, 1970). Toilets are more frequently visible in association with commercial or public uses, and are less frequently seen in residences. The only multi-seat latrine supplied the men's baths (*thermae*) (Hobson, 2009), as shown in Figure 12.3.

Pipes connected to upper storey latrines do exist in Herculaneum. However, in contrast to Pompeii where almost all toilets drain into cesspits, detailed inspection of those in Herculaneum appear to drain into sewers (Hobson, 2009).

In terms of location, toilets occurred in public baths and commercial spaces such as workshops and taverns (*taberna*). The Gemmarius workshop (Ins. Or. II, 10) is one example. The plan of the shop included three main areas: the kitchen with an attached toilet, a cubicle and a back office connected by means of a long, narrow corridor with the kitchen (Marchetti, 2009). The *Taberna vasaria* in the *Cardine* IV (north-south oriented road) is another example. Accessible from the *Decumanus Inferior* (east-west oriented road), wedged between the housing and commercial sectors of the *Thermopolium*, the tavern was equipped with a latrine in the southwest (Marchetti, 2009).

Toilets were located in domestic, as well as commercial, buildings. For instance, the House of Neptune and Amphitrite (*Cardine* V, 7) is equipped with an upper floor, partly preserved and visible in cross-section along the street, cantilevering over the sidewalk. The plan resulted in an articulated entrance flanked by kitchen, toilet and closet, leading into the atrium (Marchetti, 2009). To the rear of the House of the Grand Portal, in a space under the stairs, is a single-seat domestic toilet. Two pedestals define where the wooden seat was laid, and there is a stone foot-rest in front. What might initially appear to be a water tank is in fact the flushing floor that runs beneath the foot-rest (Hobson, 2009).

(a) (b) (c) (d)

Figure 12.3 Multi-seat latrine in the men's baths, Herculaneum: (a) general view; (b) detail of the channels for clean water (on the left) and wastewater (on the right); (c) detail of the clean water channel; (d) detail of the wastewater channel.

In the House of the Mosaic Atrium (*Cardine* IV, 1–2) there is a latrine with a preserved hydraulic mortar coating on the walls, located on the south side of the small room. The floor surface in *cocciopesto* or *opus signinum* (tiles broken up into very small pieces, mixed with mortar, and then crushed) slopes toward the exterior that was intended to accommodate rapid outside flow for wastewater. The central area of the floor in *cocciopesto* retains clear traces of consumption related to the use of the plan. The latrine may have served as a 'flushing toilet' which would have necessitated a connection to the public sewerage system (Guidobaldi *et al.*, 2006).

12.3.2 Drainage and sewerage systems

Herculaneum had a systematic network of sewers where rainwater and wastewater were directed away from the town by means of paved streets (Jansen, 2000). As shown in Figures 12.4 and 12.5, the sewage system was constructed under the street along the perpendicular thoroughfares (*Cardines* III, IV, and V) intersected by the *Decumanus Superior* and *Inferior*.

A sewer (0.60 cm wide and 1.05 m high) was built beneath the pavement of *Cardine* III, along its entire length. Arguably, this sewer collected water from the *Forum* and the *thermae*. Much of the water came from the *Decumanus Inferior* as well as from houses that opened onto *Cardine* III. A channel that delivered water into the sewer has been discovered in the House of the Skeleton; it was also equipped with an inspection sump (see Figure 12.6). The absence of discharge holes in the sidewalk supports the premise that the majority of residences opening onto *Cardine* III were served by this sewer. Therefore, it is possible to argue that water was discharged directly from the buildings onto the street (Camardo, 2005).

Figure 12.4 Views of Herculaneum: (a) plan; (b) aerial view annotated to indicate direction of wastewater and rainwater flows; (c) aerial view annotated to indicate numbered photographs with orientations, as shown on Figure 12.5.

The presence of several discharge holes along the sidewalk of *Cardine* IV prompted some archeologists to initially hypothesize this road was not served by a sewer along its entire length. In fact, conduit outlet holes are located along the sidewalk of the street where each outlet hole corresponds with each house (Camardo, 2005). Notably, the presence of a sewer under *Cardine* IV was excluded by Amedeo Maiuri (Maiuri, 1958), director of the excavations at Herculaneum from 1927.

Figure 12.5 Views of Herculaneum, according to Figure 12.4: (a) photograph from point of view (POW) n. 1; (b) photograph from POW n. 2; (c) photograph from POW n. 3 (*Cardine* V); (d) photograph from POW n. 4 (*Decumanus Superios*); (e) photograph from POW n. 5 (*Cardine* IV); (f) photograph from POW n. 6 (*Decumanus Inferior*); (g) photograph from POW n. 7 (*Cardine* IV); (h) photograph from POW n. 8 (*Cardine* III).

Physical evidence supports the thesis that a sewer was constructed under *Cardine* IV. Inside the House of the Mosaic Atrium (*Cardine* IV, 1–2) a discharge channel of the sewer was found. This channel exits the house and passes under the sidewalk, where near its edge another inspection hole is located. Arguably, this discharge was constructed under the pavement of *Cardine* IV which at this location, features the intersection of two paving types. Here, large basalt paving units meet small pebbles of the ramp leading to the city gate located at the end of *Cardine* IV.

Figure 12.6 Herculaneum, House of the Skeleton (*Insula* III, 3, *Cardine* III inf.), inspection sump of the conduit used to convey wastewater into the sewer.

Additional physical evidence supports the thesis that a sewer ran under the street. At the House of the Mosaic Atrium, the presence of a discharge hole in the floor (a short distance from that of the latrine) may have been channeled towards the sewer, where an inspection hole was situated at the edge of the sidewalk. On behalf of the HCP, a team led by Eng. Massari illustrated that outside the city gate of *Cardine* IV and near the *Sacellum* B (Urban Area Sacra) there is a still-active vertical shaft that is arguably in contact with a sewerage branch located under the street. At the meeting point of the shaft and the street is a gutter with a sliding surface comprised of flat tiles. The presence of a sewer under the street had also been hypothesized due to the existence of a discharge hole located under the fourth arch of the *fornices* (boat shelters, warehouses) of the Sacred Area, located to the right of *Cardine* IV.

Other data have been obtained with a geoelectric and GPR investigation performed by archaeologists along *Cardine* IV. These explorations commenced outside the city gate and extended to the intersection of *Cardine* IV and the *Decumanus Inferior.* The results of both surveys support the presence of a sewer conduit of a size similar to that under *Cardine* III. Finally, the existence of discharge holes in the sidewalk under each *domus* supports their connectivity with a sophisticated system of wastewater and rainwater discharge that may have differentiated between rainwater (channeled through the sidewalks on the street), and wastewater (discharged into the sewer below street level) (Guidobaldi *et al.*, 2006).

The main sewer (80.30 m in length; 3.60 m high; 80 cm wide) runs parallel to *Cardine* V, under the buildings of the *Insula Orientalis* II. It was discovered and excavated by Amedeo Maiuri in 1949. The HCP has endeavored to empty the sewer of more than five decades of accumulated debris, and to complete the excavation of the main conduit (approximately 20 m). A series of minor branches of the sewer will also be investigated; these separate at right angles from the *Cardine* V toward the *palaestra* (gymnasium).

Wastewater from public streets, latrines and downspouts of houses and shops of the *insula* were discharged into the main sewer by means of a series of smaller channels. The channel is formed of masonry walls made of *opus incertum* (irregular masterwork) or coarse cross-linked, typical of building techniques in Herculaneum. The sewer's vaulted roof was built by means of cement casting; some traces of the original wooden formwork are still visible. The bottom of the channel is of *opus cementicium* (a hydraulic-setting cement with many material qualities similar to modern Portland cement) and the corners were rounded where the channel intersected with other branches so as to maximize flow amidst decreasing pressure.

The modernity of the ancient sewer system of Herculaneum is demonstrated by the presence of pipes for the discharge of rainwater in the walls of the buildings of the *Insula Orientalis* II. These pipes were connected to the sewerage system and built directly into the walls of the front of the block. Analogously, the modernity of the system is witnessed by the constant disposition of the latrines of each shop discharging into the sewer of the *Insula Orientalis* II (Camardo, 2005).

The HCP has been responsible for clearing large sections of the Roman sewer system of organic waste deposits; these deposits were also mapped prior to removal. The opportunity to study the city 'from below' also allowed a whole series of building phases of the *domus* above to be confirmed, simply by studying the position, the dimension and the construction technique of their waste outlets. The stratigraphic excavation of deposits made up of organic remains and kitchen waste in the *Insula Orientalis* II sewer has yielded important information about the ancient diet of Herculaneum's population. The clearing of the sewer has also returned this infrastructure to its purpose built use, thereby permitting the channeling of modern water from the eastern portion of the site (Camardo, 2007).

Between 2008 and 2009, the HCP performed sewer restoration efforts, either to collect and drain water (*Cardine* III; *Decumanus Inferior*) or to accommodate new conduits (*Cardine* V). The sewer under the *Decumanus Inferior* was emptied in the service corridor of the *Forum* baths. In the sewer under the southern sidewalk of the *Decumanus Inferior*, materials which had fallen into the *specus* (sewers' channel) after an initial cleaning performed during the excavations of Maiuri (October, 1932) were removed. The survey permitted archaeologists to assess the size and path of the sewer in terms of potential reuse for the collection of rainwater. The sewer, which maintains a gradient slope from east to west, flows into the sewer of *Cardine* III, and ends about 6 m before intersecting with *Cardine* IV.

In the final portion of the sewer, archaeologists excavated a wall against which the channel ends. At the terminus of the sewer, the discharge for latrines at no. 4 and no. 9 *Insula* III were excavated. The channel was still partially filled with organic debris and fragments of pottery. Samples of organic deposits, recovered from some fragments of common pottery, fire and bricks, were analyzed. These samples had been dumped into the channel from the first floors of the building. A pipeline of terracotta conveyed the discharges of kitchens and latrines at the upper floors into the sewer below (SSBANP, 2009). The channel that discharged wastewater (or rainwater) onto the street of the House of the Mosaic Atrium (*Cardine* IV) and the secondary water tower (*castellum aquae secondarium*) with the inscription forbidding the discharge of feces, respectively is illustrated in last two Figures of the section (12.7 and 12.8). Also Figure 12.7 illustrates that water was not discharged perpendicularly to the orientation of the street but instead was sloped to maximize space efficiency and facilitate flow into the street. The edict prohibits disposal of feces near the *castellum secondarium* of the aqueduct. Offenders would have been punished with a fine of a *denarius* (small silver coin first minted about 211 BC during the Second Punic War), if they were free men, or whipped if they were slaves.

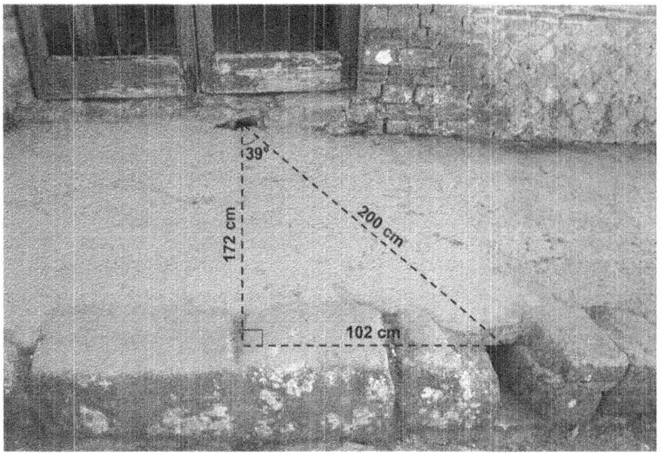

Figure 12.7 Channel to discharge wastewater (or rainwater) onto the street of the House of the Mosaic Atrium (*Cardine* IV, 1–2).

Figure 12.8 Herculaneum, *castellum aquae secondarium* featuring an inscription forbidding the discharge of feces.

12.4 SANITATION AND WASTEWATER TECHNOLOGIES IN OSTIA

Ostia was founded in the *ca*. 4th century B.C. near the mouth (*ostium*) of the River Tiber. Originally a fortified citadel (*castrum*) controlling water-borne access to Rome, Ostia grew over five centuries to become an important city. The first construction campaign of the city's perimeter walls were exceeded in the *ca.* 1st century B.C. by a new tier of walls, complete with turrets and three gates: *Porta Romana*, *Porta Marina* and *Porta Laurentina*. The main road, the *Decumanus*, is paved with basalt paving slabs and measures approximately 9 m wide and 2 km long. The city experienced a long period of prosperity from the *ca.* 1st to the 3rd centuries AD, which was followed by a phase of crisis and recession in the late empire. The city was abandoned between the *ca.* 9th and 10th centuries AD, when the surviving population moved to Gregoriopolis, a mediaeval village which had sprung up to the east of the city.

12.4.1 Toilets and cesspits

Due to the height of the groundwater, no cesspits existed in Ostia (Jansen, 2000). If the use of cesspits had been implemented, the water would have seeped through the piled up stones of the cesspit, mixing with the waste, and raising both water and wastewater to the surface. Therefore, a sewerage system was constructed (NOFSR, 1970).

A diverse typology of latrines exists in Ostia, including: V.II.4–5 (e.g., region V, block II, buildings 4–5), where a two-seater is located on the ground floor below the staircase in the House of the Porch; V.II.6–7, where a ten to fifteen seat latrine is located in front of the entrance to the Baths of the Philosopher; V.II.8, where a lavish one-seater is located below the staircase in the House of Fortuna Annonaria; and V.II.13, where a one-seater is situated below the staircase at the south-east corner of the House of the Well (Boersma, 1996; Hobson, 2009). Of the single domestic latrines observed in Ostia, five are situated under staircases (*subscalaria*) (I.VI.1; I.VII.1; I.XI.2; I.XII.2; I.XII.3). The seats are made from slabs of white marble (possibly salvaged), as are the walls, the small fountain and the pavement. The latrines date to the 2nd century AD, and are contemporary with the building.

There are a considerable number of multi-seat latrines in Ostia, which suggests that these were a primary option. The availability of multi-seat latrines also applied in Rome where, in the *ca.* 4th century AD, 114 *foricae* have been reported (Hobson, 2009).

A toilet – possibly a two-seater – is located in the Tavern of Fortunatus (II.VI.1). This toilet has a large water tank next to it with a feeder pipe and an exit pipe to permit the tiles in the latrine to be flushed. It is possible that the water from such a large tank also served other needs within the tavern. The latrine in the School of Trajan has four seats, with marble behind, some of which is a reconstruction (Hobson, 2009). The most famous multi-seat latrine in Ostia, situated near the *Forum*. is illustrated in Figures 12.9, 12.10, and 12.11. The doorways indicate that there may have been revolving doors (Hobson, 2009), as shown in Figure 12.11h. Another large multi-seat latrine is located in the Barracks of the Vigiles (II.V.1–2) (Hobson, 2009).

12.4.2 Drainage and sewerage systems

Ostia had a complete and uniform sewer system (Jansen, 2000). The sewer in Ostia has remained intact because it has always remained underground. Due to the dark and damp conditions, the sewer has attracted a large population of toads (NOFSR, 1970).

A sewer with a Cappuchin and barrel vault is situated under the sidewalk of the *Via Ostiense*. The vault of the sewer is located under the street that runs in front of the Theater, outside the entrance, 1.55 m below street level. The sewer (1.05 m high; 0.60 m wide) extends from north to south, in the direction of the street. The Capuchin vault consists of large terracotta shingles. At almost 5 m from the entrance of the Theater, to the south, the sewer is divided into two branches: one branch descends to the south-west and another descends to the east. The latter sewer is broader and deeper: its walls are clad in brick, with an arched vault above and large tile shingle paving below. Both elbows of the sewer have round-headed arches, with the presence of marble fragments (Paschetto, 1912).

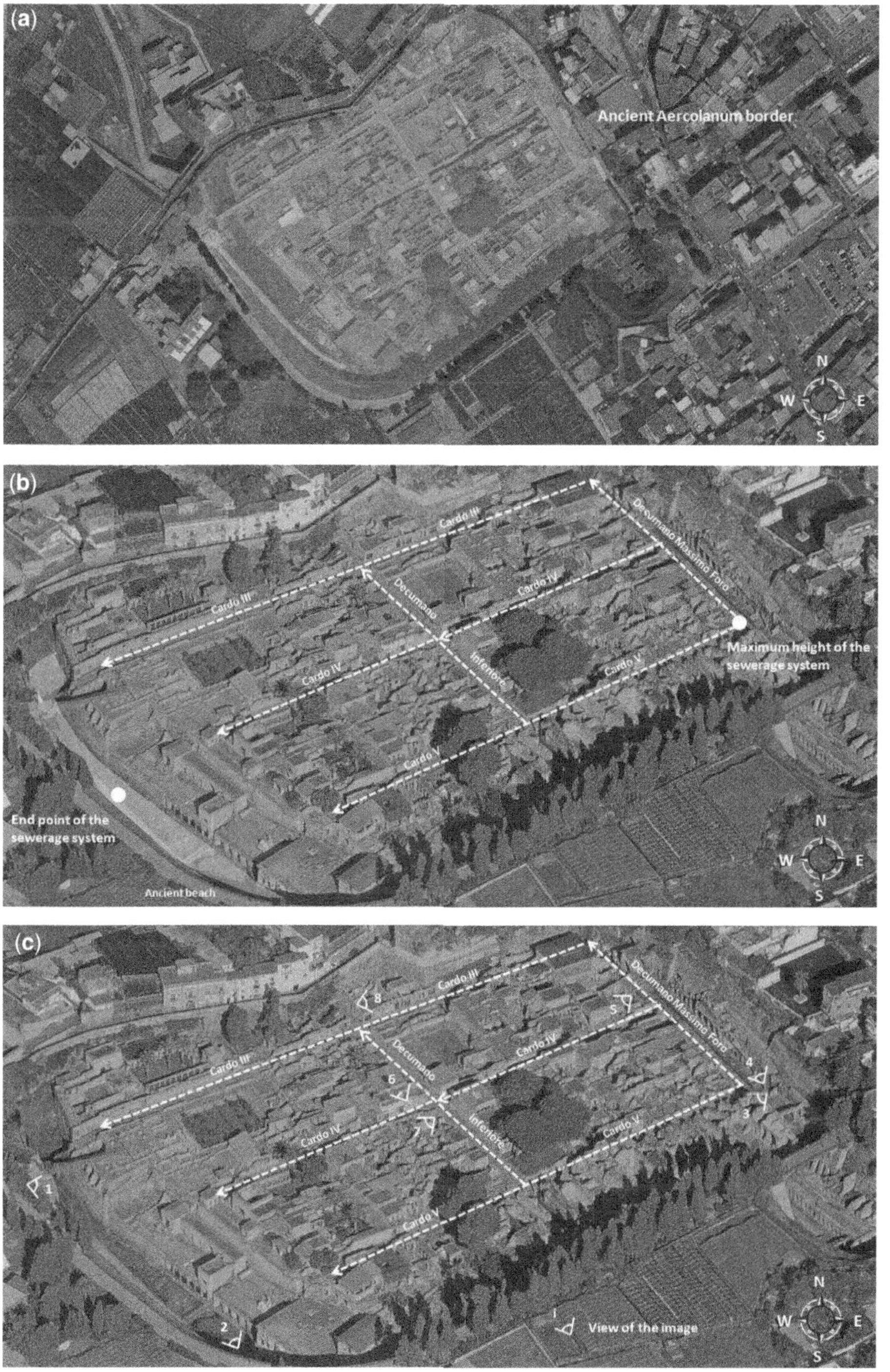

Figure 12.9 Ostia: (a) plan; (b) and (c) aerial views of the public multi-seat latrine in Ostia near the *Forum*.

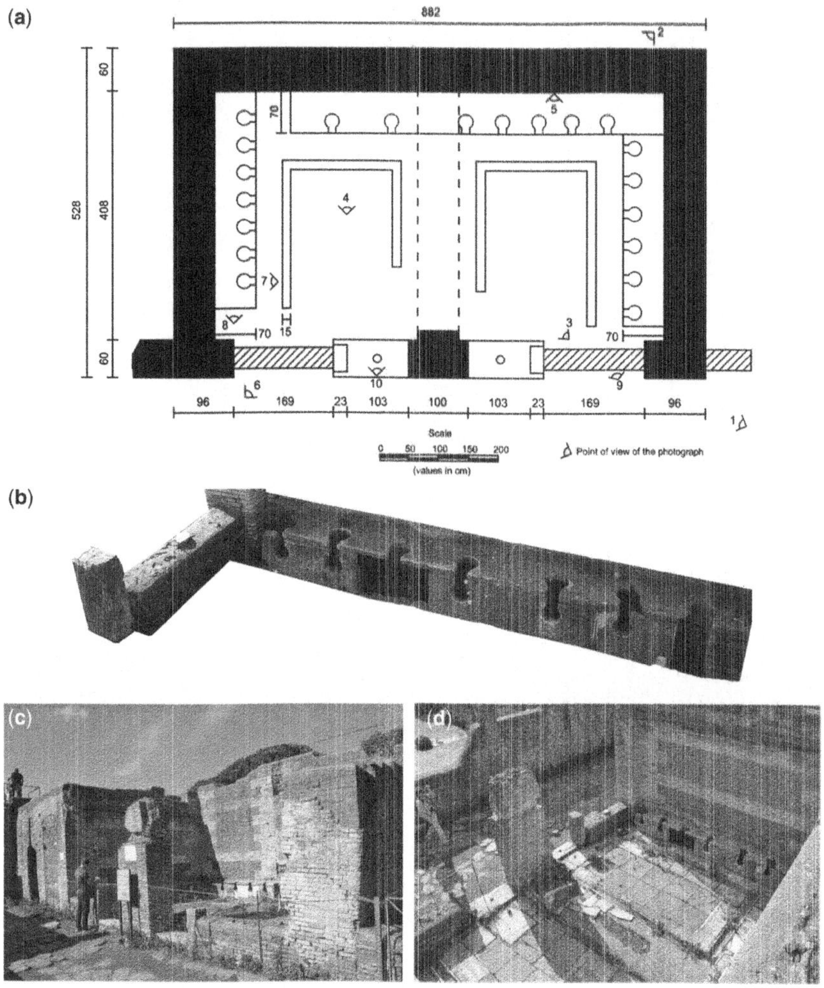

Figure 12.10 Views of the public multi-seat latrine in Ostia near the *Forum*: (a) plan; (b) axonometric view; (c) photograph from point of view (POW) n. 1; (d) photograph from POW n. 2.

Another sewer crosses the square of *Scholae* (guild buildings), behind the Theater, passing beneath the Temple that stood in the center of the square. The walls are of *opus reticulatum* (diamond-shaped bricks of tuff placed around a core of *opus caementicium*) and brick. The sewer features brick walls, and is covered either with a round barrel arch or with a Capuchin vault with large shingles. Behind the Temple, archaeologists excavated a sewer covered with eight amphorae that could represent a later restoration. In front of the Temple, a branch of the sewer proceeds south and another branch is located behind the Temple in the same direction. Further west, another sewer branch proceeds north (Paschetto, 1912).

A large sewer passes under the middle of the street running between the Mill building and the great Temple. It functioned to channel wastewater collected from the little sewers exiting from the *domus* on both sides of the street. One of these secondary lateral sewers measures 0.81 × 0.51 m and is built with a

Figure 12.11 Views of the public multi-seat latrine in Ostia near the *Forum* (points of views indicated in Fig. 12.10): (a) photograph from point of view (POW) n. 3; (b) photograph from POW n. 4; (c) photograph from POW n. 5; (d) photograph from POW n. 6; (e) photograph from POW n. 7; (f) photograph from POW n. 8; (g) photograph from POW n. 9; (h) photograph from POW n. 10.

Capuchin vault and brick walls. It is sloped toward the central sewer, which is lower. Finally, another large and well-built sewer is located along the street of the Vigiles (Paschetto, 1912).

12.5 CONCLUSIONS

The principal aim of this Chapter has been to review the core aspects of sanitation and wastewater technologies in the ancient Roman cities of Pompeii, Herculaneum and Ostia. Due to the absence of knowledge about bacteria and viruses (they have been discovered only in the *ca.* 17th and 19th century, respectively), ancient Roman cities were primarily concerned with alleviating visual and olfactory conditions caused by human waste rather than addressing public health issues. From this preceding synoptic discussion, the following conclusions can be made:

(a) Pompeii was characterized by a diffuse presence of toilets and cesspits. The majority of domestic toilets drained directly into cesspits, which were typically located under the sidewalks of the streets. Some latrines were located deep within the houses, often in the kitchen, and their cesspits were located within the houses as well.
(b) Pompeii had a scattered network of sewers. Rainwater and wastewater were primarily disposed of along the streets. Consequently, the streets would have contained some quantities of human waste.
(c) Herculaneum had far fewer cesspits than Pompeii. The ground is too rocky for draining away urine and feces. Few toilets are visible in the houses and the only multi-seat latrine supplied the men's baths.
(d) Herculaneum had a more systematic network of sewers than Pompeii. Rainwater and wastewater were directed away from the town by means of paved streets. The sewage system was constructed under the street along the *Cardines* (III, IV, and V) intersected by the *Decumanus Superior* and *Inferior.*
(e) In Herculaneum, the existence of discharge holes in the sidewalk under each *domus* of *Cardine* III suggests the presence of a sophisticated system of wastewater and rainwater discharge. The sewer system may have differentiated between rainwater (channeled through the sidewalks on the street) and wastewater (discharged into the sewer below street level).
(f) In Ostia there were no cesspits because the groundwater is too high for them.
(g) Ostia had a complete and uniform sewerage system. The system has been well preserved because it has always remained underground.
(h) The sanitation and wastewater facilities excavated in Herculaneum, Pompeii, and Ostia can be appreciated by modern engineers for their technologically sophisticated design and construction.

12.6 REFERENCES

Asplund Ingemark C. (2008). The Octopus in the Sewers: An Ancient Legend Analogue. *J. Folklore Res,* **45**(2), 145–170.

Boersma J. (1996). Private latrines in Ostia: a case study. *Babesch Bull. Ant. Bes,* **71**, 151–160.

Camardo D. (2005). L'acqua: fontane e fogne. *Forma Urbis,* **X**(3), 24–26 (in Italian).

Camardo D. (2007). Archaeology and conservation at Herculaneum: from the Maiuri campaign to the Herculaneum Conservation Project, *Conservation and Management of Archaeological Sites,* **8**(4), 205–214.

De Feo G., Mays L. W. and Angelakis A. N. (2011). Water and Wastewater Management Technologies in the Ancient Greek and Roman Civilizations. In: Treatise on Water Science, P. Wilderer (ed.), vol. 4, 3–22, Oxford: Academic Press.

De Feo G., Laureano P., Drusiani R. and Angelakis A. N. (2010). Water and wastewater management technologies through the centuries, *Water Science and Technology,* **10**(3), 337–349.

De Feo G., Antoniou G. P., Mays L., Dragoni W., Fardin H. F., El-Gohary F., Laureano P., Kanetaki E. I., Zheng X. Y. and Angelakis A. N. (2013). Historical Development of Waste Water Management. In: Handbook of Engineering Hydrology, S. Eslamian (ed.), Taylor & Francis Group, 6000 Broken Sound Parkway NW, Boca Raton, Florida 33487, USA, Vol. 3, Ch. 9, pp. 163–217.

De Feo G. and De Gisi S. (2013). Water and wastewater management in the treatment process of a Roman fullonica, *Water Science and Technology*, **13**(3), 599–605.

Flohr M. (2011). Cleaning the Laundries III. Report of the 2008 campaign, *Fasti Online Documents & Research*, **214**, 1–13.

Guidobaldi M. P., Basile, F., Camardo D. and Tommasino E. (2006). Indagini archeologiche nella casa dell'Atrio a Mosaico di Ercolano (IV,2; 1), *Fasti Online Documents & Research*, 1–13, Available on-line at http://www.fastionline.org/docs/FOLDER-it-2006-61.pdf (in Italian) (accessed 7 May 2014).

Hobson B. (2009). Latrinae et foricae. Toilest in the Roman World. Duckworth, London.

Hodge A. T. (2002). Roman aqueducts & water supply, 2nd edn. Gerald Duckworth & Co. Ltd, London, UK.

Jansen G. (2000). Systems for the Disposal of Waste and Excreta in Roman Cities. The Situation in Pompeii, Herculaneum and Ostia, Sordes Urbis. In: Monografias de la Escuela Espanola de la Historia y Arqueologia en Roma (Monographs of the Spanish School of History and Archaeology in Rome), X. Dupre (ed.), 37–49, Escuela Española de Historia y Arqueología en Roma, Roma, Italy.

Jansen G. C. M., Koloski-Ostrow, A. O. and Moormann, E. M. (2011). Introduction, In: G. C. M. Jansen, A. O. Koloski-Ostrow, and M. Moormann (eds), Roman Toilets. Peeters Publishers, Leuven, Belgium.

Keenan D. (2005) Water supply and wastewater disposal in Pompeii: an overview, *Ancient history: resources for teachers*, **34**(2), 149–158.

Koga M. (1991). The surface drainage system of Pompeii. *Opuscula Pompeiana* **1**, 57–72.

Kolowski-Ostrow A. O. (1996). Finding Social Meaning in the Public Latrines of Pompeii'. In: N. de Haan. and G. C. M. Jansen (eds), Cura Aquarum in Campania: Proceedings of the Ninth International Congress on the History of Water Management and Hydraulic Engineering in the Mediterranean region, Pompeii, 1–8 October 1994 Leiden, 79–86,

Maiuri A. (1958). Ercolano. I nuovi scavi (1927–1958), Roma, Italy (in Italian).

Marchetti C. M. (2009). Il culto domestico ad Ercolano: le evidenze archeologiche. Tesi di Specializzazione in Archeologia e Storia dell'Arte Romana (The domestic cult at Herculaneum: archaeological evidence. Postgraduate Thesis in Archaeology and History of Roman Art), Università del Salento, Scuola di Specializzazione in Beni archeologici 'Dinu Adamesteanu', Lecce, Italy (in Italian).

NOFSR (Netherlands Organization For Scientific Research) (1970). The Romans Preferred Small-Scale Solutions to Aqueducts and Sewers. ScienceDaily. Retrieved May 5, 2013, from http://www.sciencedaily.com/releases/2002/06/020627010620.htm (accessed 7 May 2014).

NSSW (Nlingenieurs Sewer Systems Workgroup) (2009). Sewer Systems Module for Higher Professional Education, KIVI Niria, NLIngenieurs, RIONED Foundation, available on-line at http://www.kiviniria.nl/docenten/rioned_hbo_reader_2009_engelse_versie.pdf (accessed 7 May 2014).

Paschetto L. (1912). Ostia, Colonia Romana (Ostia, Roman Colony), Atti Pont. Acc. Rom. Arck. s. II, vol. X, Roma, Italy (in Italian).

SSBANP (2009). Soprintendenza Speciale per i Beni Archeologici di Napoli e Pompei: (Superintendence for Archaeological Heritage of Naples and Pompeii). The Packard Humanities Institute: The British School at Rome, COMITATO SCIENTIFICO 2009, Rapporti di aggiornamento 2008–2009 Per l'incontro del 16 luglio 2009 ad Ercolano (SCIENTIFIC COMMITTEE 2009, Update reports 2008–2009 For the meeting of 16 July 2009 in Herculaneum) (in Italian).

Vuorinen H. S. (2010). Water, toilets and public health in the roman era. *Water Science and Technology*, **10**(3), 411–415.

Vuorinen H. S., Juuti P. S. and Katko T. S. (2007) History of water and health from ancient civilizations to modern times. *Water Science and Technology*, **7**(1), 49–57.

Wallace-Hadrill A. (2007). The Herculaneum Conservation Project: Introduction, *Conservation and Management of Archaeological Sites*, **8**, 187–190.

Chapter 13

The sanitary system in ancient Roman civilization: An insight on Tunisia

Olfa Mahjoub and Mohamed Thameur Chaibi

13.1 INTRODUCTION

13.1.1 A historical glimpse on Roman Tunisia (*Africa Proconsularis*)

Tunisia is situated in North Africa. It is bound on the north and on the east by the Mediterranean Sea. Tunisia has taken advantage from this strategic location by being on the forefront of several occupations. It has been and the battle field of several civilizations as well. After the Phoenicians settlement in Carthage and its foundation in 814 BC, the Greek invasion took place in 310 BC and resulted in the destruction of a great part of the city. Years later, the Punic wars between the Phoenicians and the Romans started in 264 BC. Around 149 B.C, Carthage was blockaded by Scipion Emilien for three years, a period which ended in 146 BC by the destruction of the city and its defeat by the Roman Republic. Carthage was burnt, its soil tilled and spread with salt and be declared damned city (Slim *et al.*, 2010). After the Third Punic War (149–146 BC), part of the current Tunisian territory became a Roman domain and was the first colony in Africa incorporated under the Roman Empire. The land which is today Tunisia was known as '*Provincia Africa*' or '*Africa Proconsularis*' (Graham, 1902).

After its destruction, Carthage was rebuilt by the emperor Augustus and re-founded as a Roman city, and named as the capital of *Africa Proconsularis*. During the Roman civilization, Carthage became the metropolis of Africa and one of the wealthiest cities of the world (Graham, 1902). This area of the previous Phoenician territory, about the third of Tunisia nowadays, was delimited by ditches starting from Thabarca (Tabarka) to Thaenae (Tina, close to Sfax). The colony included most of today's Tunisia and Tripolitania in today's Libya. As it goes along, the Roman territory in North Africa expanded to the Maghreb by merging Numidia, Mauritania Caesarenis (Algeria nowadays) and Mauretania Tingitana (Morocco nowadays) (Mahjoubi, 2002) and to east to Cyrenaica (part of Libya) (Figure 13.1). According to Graham (1902), the coast under the control of Carthage expanded from the West of Cyene in Greece to the Atlantic with a seaboard of more than 2000 miles (more than 3200 km) was. The area has also included the Nile delta in addition to some Egyptian oases (Murphey, 1951).

Water, as the elixir of civilizations, has played a key role in the development of the Roman Empire. Water management was of prime importance for potable supply, on one hand, and for the economic wealth as a whole, on the other hand. In Tunisia, water used to be abundant in the north. Water from aquifers and springs was drawn to be stored in cisterns located in basements. Water channelling was also achieved through aqueducts and underground channels that reached the plumbed private houses and the public premises (baths and latrines). To discharge used water (waste water) from water facilities like baths and latrines and houses, the Romans used engineered (sewerage network) and simple solutions (septic tanks). Romans were deemed to be advanced on their time in dealing with hydraulics and sanitation. Despite the progress shown in transforming the urban landscape of Tunisia for water supply and the importance the Roman gave to hygiene, they were estimated to lack the global concept of sanitation.

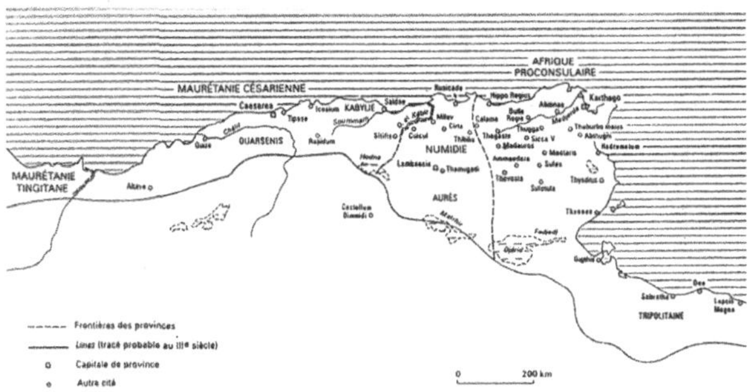

Figure 13.1 African provinces in the early 3rd century BC and extension of the Roman Empire in North Africa (Mahjoubi, 2011).

Around last of the 1st century AD, the Roman Empire expended more and the inland cities like *Thuburbo Minus, Thuburbo Majus, Uthina,* Dougga *(Thugga), Bulla Regia, Mactaris, Thysdrus,* Sbeitla *(Sufetula),* and so on. gained in prosperity thanks to their agricultural activities (Mahjoubi, 2002). All these cities are still holding vestiges of the Roman civilization. The Roman Empire has known the peak of its prosperity in the 2nd century and Tunisia counted more than 200 cities, some of them endowed with autonomy like *Bulla Regia.*

During the 3rd century AD, the decline of the Roman Empire started and expanded over about three centuries, till the end of the 5th century AD. Water management issues have been mentioned among the main theories put forward to explain the accelerated and the progressive fall of the Roman Empire and the beginning of it decline in western part as well (Middle Ages). Raised concerns were related chiefly to mismanagement, overexploitation and water pollution of water resources. Even after the Vandals invasion and the defeat of Rome in 410 AD, Carthage was considered as the metropolis of the Empire where some 'leaders' took refuge; it was called the African Rome. The Carthaginians conserved their luxury despite the hard circumstances few after the 4th century, after Rome conquest (Slim *et al.*, 2010)

13.1.2 Chapter structure and scope

This chapter presents the sanitary system in the Roman Tunisia from the post-Punic period to the early 3rd century, for example, the beginning of the Roman Empire dislocation and the start of the Byzantine invasion. This chapter is made up of two main parts: (a) a narrative part based on literature review on the Roman Tunisia, and (b) a richly illustrated part organized around examples of famous ancient Roman cities like Carthage, *Bulla Regia*, *Thuburbo Majus*, *Thysdrus*, and *Dougga.*

Through its different sections, the Chapter starts by presenting briefly the water supply system as the source of clean water. Afterwards, it examines the sanitary systems developed in Tunisia during the Roman civilization, taking into consideration the urban planning and the socio-cultural factors. Then it gives an overview on wastewater collection in public and private premises by focusing on baths and latrines. Wastewater treatment and reuse are presented as well, and finally the health implications of the sanitary are detailed with emphasis on water born diseases. It is worth underlining the paucity of references on the sanitary system in Tunisian cities, its planning, management and maintenance. Most of the available literature was focused on clean water supply rather than waste water discharge and recycling. The authors are aware of the lack of figures on water consumption and wastewater discharge of the population in Roman Tunisia.

Before exploring the ways the Romans in Tunisia managed their waste waters within the sanitary system starting from collection till disposal, it is of prime importance to have an overview on the water resources exploited for water supply, the techniques they used to channel and store it, and finally the ultimate domestic uses (drinking, bathing, garden watering, cleaning, etc.). The following section is a non exhaustive review on methods of urban water supply in Roman Tunisia.

13.2 WATER RESOURCES AND SUPPLY

Water availability is critical for the development of civilizations. Nevertheless, the Roman civilization has developed in the Mediterranean Basin where water was rather not abundant in some regions (Mays *et al.*, 2012). It is to say that supplying the population with water was one of the Romans' concerns in *Africa Proconsularis*. Alike the other North African Roman cities, water-supply in Tunisia was conditioned by drought periods. To counterbalance the uneven occurrence of drought, Roman used multiple water supply sources and adapted them to their physical environment by: digging out springs, remote springs and wells in shallow and deep aquifers; building cisterns to collect rain water from terraces; bringing groundwater in by aqueducts; constructing storage reservoirs, and so on. (Wilson, 1997; Mahjoubi, 2004b). In large Roman cities, water was delivered by using aqueducts. Romans were known, between 50 and 300 A.D to have built hundreds of aqueducts all over their Empire, especially in the Mediterranean Basin (Passchier *et al.*, 2011). Aqueducts were large water supply systems; one of the prestigious Roman heritages (Chanson, 2002). They are long subterranean canals, following topographic contours lines. Though they were simple, they used to have their own complexity by their adaptation to slope to have enough elevation for water flow. In their adaptation to the relief, some included arcades and bridges, but almost were built at or below the natural ground level (Chanson, 2002). Aqueducts represent a great achievement: construction of aqueducts was costly, laborious and time consuming that many of them were used for centuries. The aqueduct of Carthage was kept in service until the invasion of Vandals in 439 A.D (Figueiredo *et al.*, 2001). Some sections are still standing near Tunis (Tardieu, 1986; Foil *et al.*, 1993; Chanson, 2002).

Aqueducts were used everywhere in the Roman Empire. They were built primarily to supply water for public health and sanitary needs in towns and cities. They provided large volumes of water that were chiefly used for drinking, bathing and supplying fountains. Therefore they did not constitute the original drinking water supply, and were not only used to supply private houses, but also to supply other needs such as irrigation in rural area, public fountains, public baths, large baths or *thermae* (Aicher, 1995), toilets and latrines; they helped out as well with the fight against fires since houses were easily flammable at that time (Chanson, 2002). The overflow was used to flush through the sewers (Wilson, 1994)

Romans were lacking the techniques for the elevation of water therefore they searched for highly located springs to assure flowing. Indeed, Romans systems were relying on the principle of gravity flow (Tardieu, 1986; Foil *et al.*, 1993). Following this rule, the first major aqueducts were built to supply waters to the city of Rome. Later on, others were built in other North African cities. To supply the capital Carthage with fresh water, the Zaghouan aqueduct was built (rebuilt) during the time of the Emperor Hadrian to supply the city with fresh water (Moussa, 2000; Figueiredo *et al.*, 2001; Mays *et al.*, 2012). A *nymphaeum* (water temple) still in evidence in Zaghouan marked the beginning of the aqueduct. The complex described as 'a jewel of structural and archaeology' (Figueiredo *et al.*, 2001) comprises three water captions: Zaghouan at an altitude of 289 m (at 2 km of the actual Zaghouan city), Aïn Djougar at 360 m and Aïn Djour (Ferchiou, 1999; Mahjoubi, 2004b). This aqueduct is the most spectacular work ever realized in the early 2nd century AD. It was one of the longest aqueducts built anywhere in the Roman Empire. The aqueduct stretches for some 132 km, channelling water at a daily flow estimated at 25,000 m^3 from Zaghouan Mountain (*Mons Zeugitanus*) and Djougar *(Zucchara)* to Carthage. In Carthage, between 15 and 24 large cisterns known as Citerns of Maalga (La Malaga), were located on the north-west side of the Byrsa Hill from where the water was distributed to rebuild the city. The capacity of the cisterns was the largest in the Empire, estimated at 51,000 m^3 (Hurst *et al.*, 1999). The main goal of bringing water to Carthage was not for land irrigation or crops growing but the well-being of the inhabitants. This was a proof that water was not scarce in that region and that floods were frequently occurring (Peyras, 2006). In contrast, some have reported exceptional drought periods during the 1st century. From 123 to 128 AD the water resources have dried up obliging the Roman to fetch for new resources to supply the capital, Carthage (Moussa, 2000; Figueiredo *et al.*, 2001).

Being at the vanguard of engineering and pioneers in hydraulic technology, the Romans were completely aware of and concerned by the water management at local and at regional levels. Starting from the source and before reaching the city, water passes a distribution system composed of several basins with specific functions (Figure 13.2). The networks were composed of series of reservoirs, clarifiers, and settling basins. Reservoirs were constructed at regular distances to facilitate drainage and fixing pipes (Foil *et al.*, 1993). The main reservoir or storage basin is built to regulate the flow that comes for the water sources. Sometimes it has the function of settling tank to allow removal of pollution from the sources. It is connected to the city reservoir, also called castella (*castellum divisorium*). It is the distribution basin that leads the water to the customers through lead or clay pipes. Secondary castellas may exist before delivery to houses and baths. For example at Carthage, after reaching the basin of 70,000 m^3, the water coming from Zaghouan is diverted to Malaga and Bordj Djedid citerns. In the case of Tebourba (*Tuburbo minus*), the distribution basin had a completely different design with three outlets (Germain De Montozan, 1907).

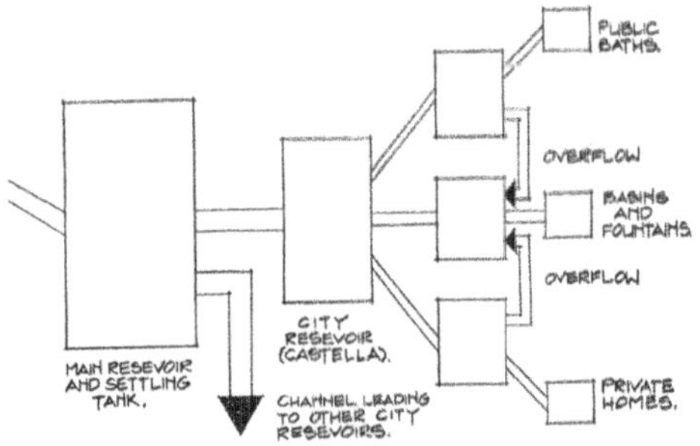

Figure 13.2 Example of a water distribution system (Tardieu, 1986).

Romans used to give high importance to the quality of the water they use either for drinking purposes or regular bathing; water of lower quality was used for secondary uses such as irrigation, cleaning, flushing o latrines, and so on. This has led them to built aqueducts and thermal stations close to thermal water sources. Aqueducts had the utility to provide water for public baths, fountains and latrines, and to flush sewers (Wilson, 1994). Aqueducts were also supplying water for numerous public fountains in the Roman cities which were supplied for the wellbeing of the dwellers (Graham, 1902). During the Roman civilization, the construction of aqueducts has flourished supplying the cities all along several kilometers with water. Aqueducts were even constructed to bring water into the villas of the richest people. In fact, for social considerations, only wealthy people used to have access to running water and were connected to aqueducts through channels. The rest of the population had to access water only through public constructions: public fountains or public baths (Malissard, 2002).

In urban area, and due to occurrence of draught periods, water of the roofs was as far as possible channelled into cisterns (Graham, 1902). It is reported that Romans used to take benefit from rain water through harvesting rather than relying on groundwater. In the Northern cities of Tunisia benefiting from the humid climate, in spite of the water abundance, Romans did their best to collect excess water by installing several devices including reservoirs, cisterns, fountains, and so on. In Carthage, reservoirs and cisterns were used extensively to regulate the water storage and distribution (Crow, 2007; Chanson, 2008). In arid and semi-arid regions of Tunisia where the landscape was a real constraint, water was provided through wells (Shaw, 1984), otherwise through cisterns built on-site (Leone, 2012). Unlikely, the Central and Southern parts were suffering water scarcity and shortage during the year which promoted water economy and waste limitation. *Sufetula* (Sbeitla), was provided with water through springs which water is carried through an aqueduct (Moussa, 2000). In villages lacking access to water, river floods were collected and directed to tanks (Moussa, 2000). In *Hadrumetum* (Sousse) water coming from wells was rather brackish and stored into tanks. It was receiving drainage in Wadi Kharroub via an underground pipeline carrying150 m^3/day (Moussa, 2000). For water supply, some regions of the Roman Tunisia were relying on rain water collected in cisterns. Large cisterns of various shapes, but almost rectangular, were built in each house to collect rain water from roofs and terraces. To increase the collected volume of water and to assure regular water supply, Romans found more spectacular solutions and built subterranean cisterns to store water abstracted from deep aquifers. It is the case of Rougga (*Bararus*) vaulted cisterns (Slim *et al.*, 2010) and *Thysdrus* (El Jem) where numerous cisterns were excavated. At *Thysdrus*, water resources are scare compared to the cities in the north. To store water, the population used tanks. For water supply, the city was relying exclusively on rain water collection and storage in cisterns built in the courtyards or nearby the gardens. The excess of water flowing from the series of cisterns was used for the irrigation of the gardens or poured in the street. The archaeological site shows that the 'House of Lucius Verus' for instance, is the only house known to be supplied with running water in the city (Wilson, 1994). It is also provided with several cisterns built in the nearby (Figure 13.3). In the 'House of Africa' a central part was used to be filled with water for cooling during the hot weather (Figure 13.4). Because of water scarcity, this case was rare: the second one existing in the city (Institut National du Patrimoine, 2002).

In some region water was not only scarce but also of low quality. Since water wells was brackish in *Thysdrus*, Romans fetched water far from the city at 13 km and used an underground aqueduct at 15 m deep to carry a water rich in sulphates (Moussa, 2000). The brackish water withdrawn from deep aquifers (35 m deep) was improper for potable uses, so it was mainly used for cleaning purposes (Moussa, 2000; Institut National du Patrimoine, 2002).

In more extreme climatic conditions, other strategies were adopted to prevent waste of water. In the Sahara and sub-Saharan region for instance, a system of dams was developed to control the water that goes to oasis (Mahjoubi & Slim). In view of the low water availability and the high demand from north to south, water scarcity and drought experienced in some regions has constrained the inhabitants in Tunisia to reduce their water consumption associated to luxurious behaviours like having private baths. Since then, public baths and latrines have developed all over the territory. In addition and according to some authors, a correlation seems to exist between the provision of the cities with water through aqueducts supply and construction of public baths and latrines during the Roman period, not only in Roman Tunisia, but all over the Roman Empire. Supplying public baths with water was a matter of euergetism (Mattingly & Hitchner, 1995) because aqueducts construction was prohibitively expensive in some cities. Indeed, the elite benefactors had the tendency to be generous and sometimes help the state in building aqueducts, supplying cities with water, constructing public latrines, and so on. All this was to keep the distance between the wealthiest people and the other citizens. Well-being of the society, social harmony and political stability were dependent, to a certain extent, on the charity of the local elite (Zuiderhoek, 2009).

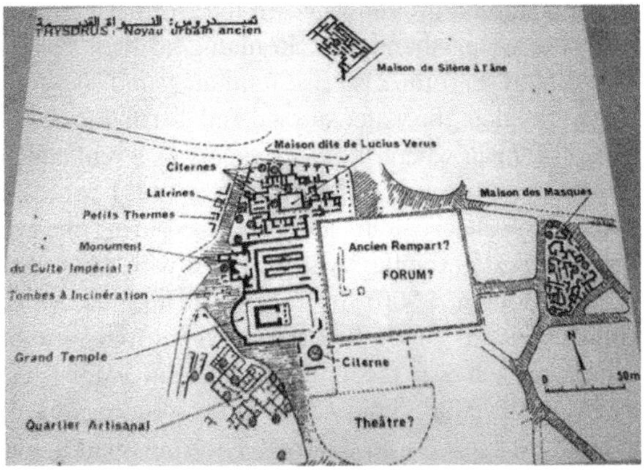

Figure 13.3 Outlay of the ancient urban center at Thysdrus (El Jem) showing the distribution of the cisterns (grey dots) (with permission of O. Mahjoub).

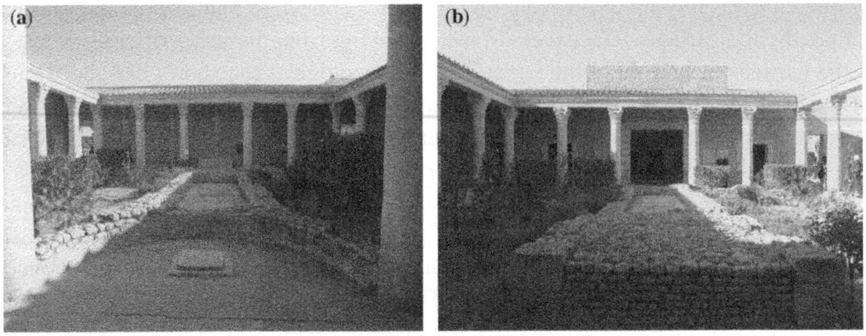

Figure 13.4 Front (a) and back (b) views of the main courtyard where water used to flow in the central part in the 'House of Africa' at *Thysdrus* (El Jem) (with permission of O. Mahjoub).

Given the multiple water resources the Romans used for domestic uses and wellbeing, one should imagine the huge volume of waste water disposed off from buildings and the techniques used to collect this water, treat it, and convey it to its final point of discharge or reuse it.

13.3 SANITARY SYSTEM IN ROMAN TUNISIA

'Sanitation', the word as known nowadays and world widely used by international organizations is associated with different relatively innovative concepts like 'ecological sanitation', 'on-site sanitation', 'decentralized sanitations', 'integrated sanitation' mostly adopted for population lacking access to bath and toilet facilities in less developed regions. According to the contemporary terminology, the Roman sanitation system could be qualified as sustainable and decentralised, even if in its detailed design it is not classified as such by international organization (WHO/UNICEF) (Bond et *al.*, 2013). In order to refer to one specific meaning, the definition of the dictionary could be the most suitable. 'Sanitation' is defined as 'conditions relating to public health, especially the provision of clean drinking water and adequate sewage disposal' (Oxford University Press, 2012). Nowadays, sanitation has become more related to collection, treatment, and disposal of human liquid and solid wastes. Hence, the sanitary system encompasses all the connections and piping within private and public buildings used to convey liquid wastes and discharge them (Nadkarni, 2004).

In comparison with the extensive research work that has been done and the informations made available on the sanitary system of large urbanized Roman cities like Rome and Pompeii, few materials was made accessible on the sanitary system in the Roman cities of Tunisia among which famous ones like Carthage and *Bulla Regia*. Most studies carried out on water were facilitated by excavations and almost were oriented toward the study of water supply systems and the distribution and management of 'clean' water within regions, cities, quarters, and villas rather than the management of waste water within the sanitary system (Rakob, 1979; Wilson, 1997; Ferchiou, 1999; Figueiredo *et al.*, 2001; Malissard, 2002; Mahjoubi, 2004a; Chanson, 2008; Leone, 2012; Mays *et al.*, 2012).

The sanitation system encompasses collection, transport, treatment and disposal/reuse of liquid and solid wastes produced by humans' domestic and industrial activities, as well as runoff and storm water. This section and the followings (13.4 and 13.5) are dedicated to the description of these steps. Wastewater discharged by craft workshops and 'industrial' activity

13.3.1 Sanitary system and urban planning

In the Roman cities, since water is continuously flowing in public and plumbed private buildings, it has to be disposed off in two cases: when it is in excess to avoid floods, and when it is polluted by use to prevent contamination of fresh resources. The situation is that the overflow of water ran out to the roadway, and this was true for waste water including that from the latrines which ended up in the same place. Hence, the first sewers that the Romans have constructed to collect used/polluted water were simple canals dogged in the roadway (Malissard, 2002).

One of the earliest sewage systems and the best-known sanitation artefacts constructed during the Roman civilization was the *Cloaca Maxima*, constructed in the city of Rome to collect drainage water and remove wastes. These sewers were open drains designed initially to carry stormwater. The system in Rome was copied in all provincial towns and cities of the Roman Empire, not only in public premises but even down to villas of wealthy people that could afford the plumbing (Malissard, 2002). However, since excrement and household rubbish were thrown together into the streets, they were flushed away with the flowing rain water and the overflow of water into the municipal drain and thence to the river (Mattingly & Hitchner, 1995; Malissard, 2002).

Based on historical literature and excavations, Roman urban planning was focused mainly on determining the location of the city as described by Vitruvius, the most famous Roman urban engineer and architect (Morgan, 1914). Accordingly, three types of public buildings existed: building of defensive measures (towers, gates), buildings of civic and religious nature (temples) that highlights the achievements of the

Roman culture and signified their way of life, and finally public utilities (baths, markets, amphitheatres) (Morgan, 1914; Leight, 2008). Consequently, the set up of a sewer system in urban areas depended largely on the city establishment and its urban planning. Urban management and development of the cities on the Tunisian territory have not seen the light with the Roman civilization but they dates back in time to more than 7 centuries when the Phoenicians civilization set up their urban planning. This was discovered during the 70's thanks to the excavation of Carthage and Kerkouane (Slim *et al.*, 2010). Several evidences show the early stages of urbanization in Punic cities like *Utica* (Utique), *Carthago* (Carthage), and *Hippo Diarrhytus* (Bizerte), most of them located on the littoral. To have a clear idea of the urban management in Tunisia during the Roman civilization, one had to wait till the beginning of the excavation of cities (Hurst *et al.*, 1999). At Carthage, ruins have shown that Roman cadastre has just followed the organization of the Punic metropolis (Slim *et al.*, 2010). The rectangular layout, already adopted by the Phoenicians, was followed (Mahjoubi, 2004b). African cities during the Roman civilization were built according to very stringent principles. It is a sacred ritual act that takes into consideration the master plan and the instructions of the authority. The *decumanus maximus*, the main axis, traced according to the east-west orientation, and the *cardo*, the perpendicular axis (north-south) delimited a grid according to which the city was built (Slim *et al.*, 2010). Thus, buildings were constructed according to an orthogonal design (Mahjoubi, 2004b) where almost streets run parallel then meet in the centre of the city. The figures below (Figure 13.5) depict two examples from Carthage and *Bulla Regia* showing the orthogonal design. In or close to the latter, one can find a place called *Forum*, surrounded by administrative buildings, shops and markets, baths, basilicas, temples, and so on (Mahjoubi, 2002). This urban plan dictates exactly the way the sewer system has to follow and several examples can be seen in the Roman Tunisia cities like *Carthage, Bulla Regia, Thuburbo Majus* and many others. As a common plan for their roads, the Romans used to start by digging two parallel channels as sewers where the waste water can circulate then they start constructing the road. The latter is made of several layers of sand and mortar, that could reach 1.5 m thick, covered with flat stones made to resist over time. The road should be bulging to allow the drainage of rainwater into the canals.

Figure 13.5 Views on the market place in *Bulla Regia* (left) and House of the *Cryptoporticus* at Carthage showing the orthogonal design followed in urban planning within cities and quarters (with permission of O. Mahjoub)

Africa Proconsularis, like few of the North African and Eastern Mediterranean areas, was among the most urbanized areas after Rome since cities and have showed an exceptional development during the Roman civilization (Vuorinen *et al.*, 2007). Till 100 BC, the city of Carthage was in construction witnessing demolition and levelling of the Punic structures. Street pavement has to wait till 200 BC. Before laying stone slabs the streets in the quarter of Magon were just repaired with fresh clay till the second half of the second century when they were paved for the first time (Hurst *et al.*, 1999). It was discovered that within its boundaries, the Roman Africa encompassed more than 200 cities built in replacement of Numidian and Phoenician ones (Mahjoubi, 2002; Slim *et al.*, 2010). After excavations, only about 20 of them are still showing monuments like baths, latrines, temples, and so on. The most famous are *Carthage, Thuburbo Majus, Bulla Regia*, and so on, which relics are described hereafter.

Despite the high level of urbanization in Tunisia, the rural communities were still important and scattered far away from the urban centres, living in the farms or gathered in hamlets. The centre-east and south-east regions of *Africa Proconsularis* were not so highly populated with 1000 to 3000 inhabitants (Slim *et al.*, 2010) and this had influenced in a large part the management of wastes (liquid and solid) and the rate of connection to the sewage system. This difference between rural and urban population reflects in a certain way their social status which was the main factor influencing access to sanitation facilities.

13.3.2 Access to sanitation and social considerations

Access to running water and sanitation in the Roman Tunisia is comparable to all the other cities of the Roman Empire. It was ruled by several considerations, not only the type of settlements (urban vs. rural) but also the social ranking as well as the property/ownership of the space (public vs. private), which seems to be closely correlated. Wealthy people used to donate money or finance building of public facilities, like construction of baths (and latrines) and gymnasium, and so on, to their cities receiving in counterpart a statue or an inscription that perpetuates and gives honour to their generosity over time (Kalinowski, 1996); a concept called euergetism (Zuiderhoek, 2009).

Usually, cities have public baths in each of their quarters. These are of modest importance compared to the central baths. The latter are larger, imposing and lavishly decorated. In many of the Tunisian roman cities, we can distinguish between two types of baths: winter baths and summers baths. These are built separately and the latter are larger than the former allowing cooling and much more activities. Ruins of winter and summer baths are found in cities like *Thuburbo Maius, Mactar and Sbeitla*, and *Bulla Regia* (Slim *et al.*, 2010). Romans used to spend a lot of money in cleaning themselves and their bodies by taking baths in public premises. Public baths were open to all categories of people and gender for free or through paying a modest charge. This ritual was considered as a collective ceremony rather than a real care of their hygiene (Malissard, 2002).

Romans used to take baths at the end of the day and bathing was a social phenomenon after which they used to enjoy delicacies. Baths were dedicated also for culture, sport, shopping, and so on as they included museums, libraries, gymnasiums, stores, and so on, (Malissard, 2002). The link between access to clean water and sanitation system in private building is generally closely related to the social class to which one belongs. The Roman society was extremely divided with a social hierarchy based almost exclusively on wealth (Antonio, 1979). Within the urbanized cities, access of dwellers to sanitation services depended also in a large part on their social status. As for baths, they were of prime importance in the Roman's social lives. So important that at the exception of luxurious villas, almost private houses was lacking baths and latrines. When they existed in private villas, latrines typically lie near the *atrium* or kitchen of the villa (Slim *et al.*, 2010). Again, if there were no toilet facilities, chamber pots were typically used, and their contents were dumped periodically into cesspits.

The Roman population was difficult to assess. The informations available estimated at 300 000 the inhabitants of large cities like Carthage during the most prosperous period. The less populated cities accounted 10,000 inhabitants (Slim *et al.*, 2010). Between the 2nd and the 3rd century, the total population in Roman Tunisia was estimated at 2,500,000 inhabitants with a great majority of Berber. The wealthiest were a minority composed of Italians descendents of Roman colons (Slim *et al.*, 2010). As underlined in several archaeological research works, access to running water was inextricably related to wealth and power. In Italy for instance, decuriones (leading members and then powerful political figures) were honoured by receiving a private donation of water, and the volume of water depended on the social status as well: the higher the status the higher the amount of water delivered, provided they proofed the existence of private bath (Wilson, 1997; Dessales, 2006). This was also interesting to notice in North African Roman cities. Excavation in the rich quarters of *Bulla Regia,* which owes its name to its status of autonomous Royal Residence, showed that houses were equipped with private bath and latrine in addition to the existence of one or several cisterns to store water coming from pipes, rainwater collected from the roof for domestic uses, and wells used for drinking purposes (Wilson, 1994).

Based on social classes and subsequent access to water, only apartments situated in basement were connected to water pipes and latrines (Magnabal, 2004). These apartments can have private latrines connected to the sewerage network where waste water is discharged. When latrines are built close to the roadside, they could be open to the public (semi-public) offering to the owner a complement of salary. The richest used also to have their private heated baths (Magnabal, 2004). To pride their wealth and make their houses visible to visitors and guests, Romans acquired private connection to running water; obviously the primary objective was not the access to a better quality drinking water but for supplying their private pools and gardens. Fortunately, water of lower quality (non-potable) was used for this purpose. It was noticed that at El Jem (*Thysdrus*) only one house can display this social showiness, the House of *Lucius Verus*. It is an uncommon luxurious house known to have running water used among other purposed to water the garden (Wilson, 1994). While wealthy Romans used to afford to hook up their homes to the sewer system, the less well to do had to rely on public bathrooms, which were a vast improvement over anything built before. In fact, by the 4th century AD over 140 of these public facilities were available (Magnabal, 2004).

Whereas the wealthy citizens attended during daylight hours, for social considerations, people from lower social classes had access to different systems of sanitation (Malissard, 2002) and the less fortunate people followed at the end of the day or in the evening when the admission to public baths is cheaper (Malissard, 2002). Poor and disabled, but also mean people did merely have access to public sanitation. Instead, they used to use chamber pots which they empty either in containers or in cesspools located in the nearby. For the laziest, a common practice was throwing out their wastes from windows directly into the street, which was often causing the anger of passers-by (Malissard, 2002). Poor and people of low-status were living in upper floors of buildings or renting individual rooms, others in slums, and some used to sleep in the public baths, especially in winter time (Scobie, 1986). Poor used to live in abject poverty in the most unpleasant and unsafe environment. Slaves, considered as the Roman underclass represented about 25% of Italy's population and 10% of that in the provinces (Antonio, 1979). All this population was constrained to go to common latrines reserved for the whole building or simply to public latrines (Magnabal, 2004).

Given these social disparities that Roman used to deepen and the distribution of social classes, collection of wastewater from public and public premises was not done systematically and the same way.

In the following section structures used by Romans to collect used water and wastes thrown or discharged by the dwellers in public and private premises are presented.

13.4 CONNECTION TO SEWERS AND WASTEWATER COLLECTION

It is estimated that the first Roman sewers were built between 800 and 735 BC. The Roman sewers network was as complex as the modern existing nowadays with the difference that they connect mainly central areas. The sewer system was not intended to directly receive excrement and other wastes because wastes were usually thrown directly into the streets and water was used to flush the wastes daily for cleaning, then the water runs into underground drains (Foil *et al.*, 1993). This shows that Romans used to have combined sewer system where runoff (rainwater and excess water) and sewage are collected and flushed all together (Scobie, 1986). Drains and sewers were aimed to as to transfer waste and surface runoff to discharge it out of the city. In the case of coastal cities, direct discharge of the wastes used to occur directly into the sea. The case of the Antonine Baths at Carthage is the best example existing in Tunisia with three drains (Wilson, 1997). In 200 BC a vaulted drain was constructed in the Magon quarter in Carthage (Hurst *et al.*, 1999). In *Bulla Regia*, the system of sewers seems to end downstream into the river according to sayings of autochthones. During excavations that took place in the 80's, water conduits and drains were discovered in *Bulla Regia*.

A general overview of the sewer system leads to the more or less detailed description carried out by Wilson (1997). The urban drains can be classified as first-second, third and fourth order drains corresponding to their connections the ones to the others. In fact, the connection starts from the first and second drains which are connected to individual buildings of small size. These lead to wider collector of large size that carries the water and discharge it into the river. The first and the second order drains consisted in canals under the houses, thereby they were not apparent, while the third and the fourth are dogged in the street to collect the water coming from the first and second drains. This is a simple description to show that urban drains may channel not only runoff, dirty water from cleaning, but also flushes public and private latrines and baths. Despite they were deemed as the most technologically advanced and the most elaborated, the Roman sewers were ignoring basic sanitary principles. Sanitary systems allowed wastes to be discharged directly to rivers via open pipes, as it was the case in Rome where the waste were discharged into the Tiber river (Malissard, 2002). Urban runoff was mixed with water discharged from baths, latrines, fountains, and other premises, and water pipes, drains and toilets all go together and drains joined together with sewers to carry off rain water and sewage. This made it necessary to have large open canals along the streets (Farnsworth Gray, 1940; Hansen, 2007). It is therefore interesting to examine how public buildings and private houses were connected.

13.4.1 Public premises

Romans achieved a remarkable level of progress in providing the population in the different cities with facilities like public baths and latrines. However, some shortcomings were well identified later. Despite the construction of large sewer system network, households were still lacking individual connections and the inadequacy of the sanitary systems was constraining dwellers to spend much of their time outside their apartments mainly in latrines and baths (Hansen, 2007). A part from the markets, the shops, the public places and forum, public baths and latrines are by far the largest and most visited public premises.

13.4.1.1 Public baths

The first roman baths were built around 312 BC, about 4 centuries after the construction of the sewer system. The process of development of baths during the Roman period has been discussed by archeologists. Almost agreed that the 2nd and the 3rd centuries were crucial periods for the evolution of these buildings (Fagan, 2001). Roman baths were very famous till the empire collapsed in the 6th century AD (Malissard, 2002). Baths are the main public area that can gather the largest number of people and into which large amount of water enter and comes out. In Tunisia, each city used to have its public baths, more or less modest, with some others considered as public monuments like the Antonine baths in Carthage or the Memmian baths in *Bulla Regia*. Public baths were essentials for hygiene especially that houses were lacking private baths and even latrines (Mahjoubi, 2004b). Ancient Roman baths formed an integral part of public and private life. Some of the largest complexes were only rendered possible by conveyance of huge quantities of water channelled continuously and regularly through aqueducts into the cities (Graham, 1902; Mahjoubi, 2004b).

From an architectural point of view, Roman baths are likely to follow a regular, almost symmetrical design described to be copied all over the Roman cities with rooms and halls coming off the central axis. During his passage, the bather can successively go through the *Apodyterium* (changing room), *Frigidarium* (cold water pool), *Natatio* (open air pool), *Tepidarium* (heated covered pools with warm water), *Calidarium* (pool with hot water) and *Laconica* or *Sudatorium* (hot vapour room, the equivalent of a Turkish bath). The succession of rooms and the presence of a common heated swimming pool (*solia* or *alevi*) were features of Roman baths and evidences for their identification (Fagan, 2001). The baths comprised also other rooms dedicated to culture, sport and leisure (music concert, conferences, libraries gymnasiums (*palaestrae*), game rooms, etc.). The exterior would be surrounded by gardens and parks for taking walks or simply resting in the shade (Milani-Santarpia, 2011). The succession of rooms supplied with cool and hot water implies the circulation of water inside the bath: water coming in and out the building according to a certain path and at a given rate that avoids mixing of clean and used water, and assures renewal of water for bathers in-between.

13.4.1.2 Latrines

'Latrines' originates from the contraction of the word *lavatrina*. The word in itself is of recent use, since only the 1st century BC. This implies and confirms that the Roman system of latrines is the oldest one known in history. Latrines formed part of the developments of public spaces and were considered valuable institutions where traders used to negotiate business (Malissard, 2002). Latrines design and structure have been fully described in the historical literature (Germain De Montozan, 1907; Amulree, 1973; Tardieu, 1986; Wilson, 1994; Mahjoubi, 2004b; Vuorinen *et al.*, 2007). In the particular case of North African cities, the latrines seats were usually cut into stone (Wilson, 1994) and some are equipped with armrests like the latrines of the Antonine Baths. Some rare ones were cut into marble like the vestiges found in *Bulla Regia*.

The communal latrines are constant consumers of water. Water circulating underneath is initially supplied by the aqueducts and used for flushing and carrying the excrement away, and clean the sewers (Soifer, 1999; Malissard, 2002). A flow of water circulated under the seats at the bottom of the peripheral ditch. There was often provision for washing the hands after use in a basin fed from an adjoining fountain usually juxtaposing the wall. Fountain provided water for ablutions and allowed a flow of water to run into the gutters of the pavestone (Graham, 1902). Latrines often communicated with the sewers and were flushed with water on a 'trough closet' system. A fine example of this existed in North African cities in Tunisia (Amulree, 1973). In some excavated ruins, the facilities had seats cut into marble ledged on three or four sides, with no walls separating the seats.

Ever since flushing latrines required a lot of water, many examples of Tunisian Roman cities have the latrines located within the baths building (Antonine Baths in Carthage, Memmian Baths at *Bulla Regia*, and Summer Baths at *Thuburo Maius*). Either continuously or regularly practiced, it is not well known how often the latrines were flushed. Within the baths, the frequency seems to be related to the water circulation for hygienic purposes and for easy maintenance (Wilson, 1994). Wastewater flushed from the latrines flowed through a central channel into the main sewage system and then into a nearby river or stream (Farnsworth Gray, 1940; Hansen, 2007).

In view of the high visiting frequency and the long time the Romans used to spend in communal latrines, a small charge was made, but some latrines were completely for free. Sanitary conveniences were regarded as a source of revenue and the idea of putting a tax upon urine originated with Vespasian at a time when the imperial treasury was running dry and the emperor was interested in raising money by this means (Amulree, 1973). This made the famous saying about the sewer '*pecunia non olet*' (money does not smell) (Graham, 1902).

13.4.1.3 Other premises

Large public premises like amphitheatres, markets, public places, gardens, theatres, and so on, receive huge qualities of water from the rainfall, cleaning, and so on that should be disposed off to avoid floods inside the buildings, and water stagnation in the street. These are connected directly to the tertiary and quaternary drain as described by Wilson (1997).

Public premises that have flourished during the Roman time are the urinals. Romans were known to have pioneered the men's public urinal, which was referred to as the 'pissoir' or 'Vespasienne' in reference to the Emperor Vespasian. Either simply located outside in the open air or with a minimal degree of privacy, these devices collected the urine to be sold to fullers. This service of collecting urine was charged by the authorities, namely the Emperor, to fatten his coffers. Fullers used to have workshops called *fullonicae* or dyeing shops. Numbers of them were found in Rome (Fowler, 2007; Hansen, 2007). In Roman Tunisia, purple dyeing workshops were developed in cities like Carthage, Kerkouane, Utique, and so on (Slim *et al.*, 2010). However, at our knowledge, none have been evidenced through the excavations in Tunisia and described from the point of view of wastewater collection. It was described, elsewhere, that in workshops of fullers, the used water used to be collected in drains underneath the workshop and discharged into the drain in the street (De Feo & De Gisi, 2013). The latter have depicted the fluxes of water and wastewater, collection of effluents and recycling within the *fullonica* in Pompei.

Wilson (1997) during his study on the water management in North African cities under the Roman Empire has pointed out that wastewater reuse could not be excluded; on the contrary it should have been the rule. Nonetheless, there are rather no investigations on the status of wastewater reuse in North Africa, which is not the case for larger Roman cities like Rome. The Romans recycled also public bath waste water by using it as part of the flow that flushed the latrines. Cesspits existed in public baths and were emptied at regular intervals. Their contents removed from the city and used as manure on the fields (Amulree, 1973).

13.4.2 Private houses

As highlighted previously, only wealthy people benefiting of a certain social status used to have private connections to running water. Consequently, private latrines and baths were considered signs of luxury. Connection to the sewer system was not an obligation for the private houses because the owner is expected to bear for all the expenses (Scobie, 1986). Several have been found in the sumptuous House of the Hunt in *Bulla Regia*. Rich residences are usually sited on hill slopes with a beautiful landscape as it can be seen in *Bulla Regia* with a view on the theatre, or on the sea side as it is the case of Carthage. Boutiques and shops are usually located on the ground floor. The first floor is composed of rooms, gardens, basins, and pools, lavishly decorated with marble and mosaics. It was suggested by some authors that private latrines and baths were installed only to keep the social distance between rich and poor (Thébert, 1987; Slim *et al.*, 2010).

In villas, the primary bathrooms were sometimes used also as latrines (Malissard, 2002). In such buildings, water runs from reservoir located on the roof and are used only for cold or warm shower, when heated by the sun. Otherwise water usually should be heated before being poured into the bathtub (Malissard, 2002). For rich people, private villa owners may have their own baths constructed exactly in the same design as the public ones, with small cold and hot pools. The hot pool may be constructed in the basement of the house to keep the heat and a small hole in the roof is used for lighting (Graham, 1902). Since in the warm and hot room were heated by a system of *hypocaust* a complicated system based on underground heating, literally meaning 'heat from below, analogically private houses have their baths built on the backing furnace (Milani-Santarpia, 2011). Actually, the private baths seem have influenced the design of the public ones with the linear sequence of rooms. Early private baths were located near the kitchen to benefit from the heat source, the backing furnace (Fagan, 2001).

The existence of private bath and latrine could not be considered the unique signs of wealth because even some rich houses did not have latrines and their inhabitants were rather relying on the use of chamber pots (Malissard, 2002; Hansen, 2007). In private households the latrines, which were little more than cesspits, were in general located near kitchens to reuse the water (Hansen, 2007). Some authors highlighted that a combination of kitchen/latrine existed with the advantage of dumping the 'grey water' and wastes directly into the cesspit. The lavatory which is usually located beside or inside the kitchen had the disadvantage of favoring stench and cockroaches (Scobie, 1986). Ancient Roman sewer designers developed cesspits. The latter were likely the ancestors of the nowadays septic tanks. Sewage first flowed into a large stone or concrete tank where wastes were kept to settle to the bottom (Uy, 2007). For maintenance, they had to be cleaned out by hand (Hansen, 2007). Once cleaned, wastes were used as fertilizers or just dumped in the street. Since Romans had to pay for connection to the sewer system, they used to prefer cesspits. However, these facilities need regular maintenance operated by slaves. Discharge of liquid and solid wastes into the sewers cause the sludge to accumulate which can generate toxic gases like hydrogen sulphide. The latter in contact with air give sulphuric acid that can cause corrosion of the sewer walls and can expose cleaners to death (Scobie, 1986).

After collection all sewage is channeled into sewers. The following section will describe how the sanitary system allowed treatment of the sewage collected from public and private buildings.

13.5 WASTEWATER TREATMENT AND REUSE

Despite the rudimentary structure of the sewerage system, the management of wastewater was deemed to be innovative during the Roman time and Roman were considered among the most advanced in this field (Farnsworth Gray, 1940). However it is difficult to imagine what could be the equivalent of our current treatment plants where wastewater could be collected, treated to a certain level, then released in the receiving environment or reused.

Since water was running continuously within the public premises and some of the private houses and in regards of the structure of the sewerage network, the flow of water can be considered as a natural treatment in itself. Nevertheless, the treatment efficiency will depend on the climatic conditions, the contact time between wastewater and the flushed solid wastes, and the frequency at which the sewers and cesspools are cleaned. The wastewater treatment during the Roman times has some similarities, to a certain extent, with the relatively new concepts developed for population lacking access to sanitation or improved sanitation. The wastewater collection, treatment for reuse during the Roman era could be qualified of what is called nowadays 'decentralized sanitation' or 'on-site sanitation'. However, it was lacking the principles of integration and prevention that should be prevailing on treatment and disposal (Huibers *et al.*, 2010). Because of the absence of sewer system, the lack of waste treatment of any kind, the environmental impact of human activity was high. These practices had high environmental consequences on dwellers in urbanized area and around cities (Malissard, 2002).

In private buildings having cesspools, similar processes could have been used in the cities throughout the whole Empire. Sewage first flowed into a large stone or concrete tank where heavy solids settled to the bottom and lighter particles floated to the top. The partially cleaned liquid flowed into a nearby body of water.

Collected sludge was either used as fertilizer or simply buried (Uy, 2007). According to Scobie (1986), some agricultural writers referred to the use of human excreta instead of animal waste to be used as fertilizers.

Another way to recycle their liquid and solid sewage products was the use of the supernatant obtained after the decantation of the urine. The supernatant is collected in cisterns for cleaning purposes. Indeed, the fullers discovered the cleaning properties of the ammonia contained in urine (Hansen, 2007) and were using it for laundry as a bleaching agent or dye for their clothing. Not only human, but also animal urine was used to clean clothes and to dye them after applying some receipts. Roman fullers used to soak animal leather in urine in order to remove the hair fibers. Urine was also used to prepare textile for dying (Scobie, 1986). Human urine is collected from public latrines in the vicinity and in other cities as well as from reservoirs they put in the streets.

In regards of the common practices of reusing wastes and recycled water, exposure to infectious agents is not excluded, which may imply serious health issues.

13.6 ROMAN SANITARY SYSTEM AND PUBLIC HEALTH

The relation between health and sanitation has been of particular concern since the antiquity. As stated recently by the Director General of WHO 'Sanitation is the cornerstone of public health' (WHO, 2008). When the sanitation facilities are inadequate or inaccessible, human health, especially children can be threatened (Drewko, 2007). Lack of hygiene and failings of the sanitary system may endanger human health and end in death. Diseases causal agents were not well known at the Roman time as well as their associated symptoms and exposure to insalubrities would have been lethal (Scobie, 1986). However, in latrines, the evacuation of liquid and solid wastes by the flowing water was considered a highly hygienic procedure (Haut & Viviers, 2012).

Given the state of the sanitation system in Roman North African cities, one should expect exposure to the same health risks and threats for the inhabitants in Tunisia as with some regional differences. Scobie (1986) has described how the sanitary system has affected public health in large Roman cities. He highlighted the deficiencies in waste water and waste management and associated it to the high registered mortality rate. In fact, some historians have blamed the Roman practices, such as the use of cesspits, latrines and their location close to the kitchen, and made them responsible of the exposure of humans to pathogens and the spread of intestinal diseases thus causing mortality (Scobie, 1986). It is well know nowadays that diarrheic diseases are the most common ones associated with contact with wastewater and excreta and unsanitary excreta disposal and absence of nearby sources of safe water (Jiménez *et al.*, 2010). In deed among the most spread diseases by the unhygienic conditions in the public and private latrines during the Roman time one can cite diarrhoea, dysentery, typhoid fever, and various intestinal symptoms due to intestinal parasites (protozoan and helminths). The care that the authorities of Imperial Rome conferred on water supplies and sewage disposal should have helped to prevent, or at least to reduce, epidemics of typhoid fever and dysentery. Typhus fever may have been prevented as well by the Roman fondness for bathing (Amulree, 1973). Two main water related diseases largely spread around the Mediterranean countries were managed: malaria and bilharzias (Vuorinen *et al.*, 2007).

Life-time is related to the occurrence of lethal diseases. Historians have estimated life expectancy for the whole Empire being between 20 and 30 years (Kron, 2012). In Tunisia, life expectancy was beyond this range, but sill it was very low evaluated at 30–35 years. Mortality used to be high, estimated roughly at 45%, as being highly affected by epidermis. This has made the demographic balance quite precarious and encouraged the population to procreate and have many children (Mahjoubi, 2004b). Based on records on the Roman world, most registered deaths was associated with infectious diseases (gastro-intestinal infections). However, some authors do not agree on this theory and stated that Roman's water quality management was relatively sound by selecting the appropriate tools to mitigate or avoid contamination (Kron, 2012). The Roman system should not be blamed because it was praised for providing basic facilities with running water, and stringent water quality standards and regulations were set to avoid contamination of water supply and the use of unsafe water sources (Scobie, 1986).

During the Roman time, infection rates and mortality were deemed to have seasonal trend: death rate increased in the late summer to early fall. At the opposite, in late Roman Carthage and in Alexandria (Egypt), the mortality was aseasonal with rather regional trend (Scheidel, 2010a). This observation was associated with the time when malaria, which appears to have been endemic and used to spread, exacerbate other diseases (Scheidel, 2010a, b). Several examples and significant records from the empire territory from north to south, including Italy and the Iberian peninsula, and from east to west, including the territory from Mauretania Caesariensis to Jordan suggest that death rates varied greatly over the course of the year indicating a high disease loads in these regions (Scheidel, 2010a).

Public health is known to be intimately connected to water quality and liquid and solid wastes management. Urban drainage and surface runoff in addition to the evacuation of used water and sewage is of high significance for public health (Wilson, 1997). Romans were recognized as giving high importance to water quality. Before domestic use and in order to prevent health risks, water quality was improved by the mean of physical treatment like settling, sieving, filtering and boiling (Vuorinen *et al.*, 2007). Moreover, the high quality water from springs and wells was dedicated for drinking purposes, while the poorer-quality (collected rain water, saline and brackish water) was used in the public baths and in the latrines (Foil *et al.*, 1993). In cities lacking water, cisterns were used to store water and use it for domestic uses and hygiene purposes. However, water turnover was slow and has sometimes caused the spread of diseases (Vuorinen *et al.*, 2007; Leone, 2012).

Physical wellbeing was unevenly distributed in the Roman Empire due to several considerations that make distinction between rural and urban population, rich and poor classes, dry and rainy regions, and so on, Concerning the social status, rich Romans in large cities used to live in houses built on featured hills offering a slope for gravity drainage, good aeration, and exposure to sun meanwhile poors were constrained to live down the hills exposed to threats like floods (Scobie, 1986). At the opposite of the huge amounts of water they used for their daily baths, the Romans used little water for their morning wash and for cleaning themselves (Malissard, 2002). Bowls or jugs were usually used in case the house lack running water (Wilson, 1997). Even though it was reported by the Roman ear that obliging people to respect health rules and sanitation was considered an intrusion of individual rights and in private life (Foil *et al.*, 1993).

Regardless of the correlation existing between health and nutrition and the registered fact that the north-western Europe inhabitants were on average taller than Mediterranean populations, skeletal observations and evidences were partly explained by low population density and in consequence the lower disease loads (Scheidel, 2010a). Comparably, in Tunisia, the population density was high. Only in the Medjerda valley, 150 cities were clustered in an area of 175 km by 120 km (Mahjoubi, 2004b). In some regions, the density was exceeding 100 inhabitant per square meter (Slim *et al.*, 2010). Coastal lands, cities along rivers, and farming land were more prone disease areas with potentially adverse consequences for morbidity and mortality. In addition, according to observations in Roman burial sites, among them an archaeological site at Carthage, it has been found that some skeleton exhibited cranial lesions. The incidence of these lesions called *cribra orbitalia* was comparable to Egypt but higher in central Italy. One possible explanation is the occurrence of malaria, a disease that causes anaemia (Scheidel, 2010a). However, as mentioned by researchers, it remains still unclear to what extent the negative effects of urban crowding were compensated by the water and sanitary system management like the construction of aqueducts, latrines, and sewers (Scheidel, 2010a). It is obvious in some cases that the permanent flow of water from the source to the drains would have avoided health hazard caused by water stagnation, especially during the hot season (Wilson, 1997).

In terms of quantity, in spite of the abundance of water, Roman societies were lacking means to deal with public health. Urban centres were suffering serious public health problems because of low level of waste (water) management. Hygiene was quite poor in public as well as in private toilets. In public, it was usual to see latrine either incorporated into a public bath building or to be adjacent. Public latrines were lacking privacy. On average, they could accommodate between 10 and 20 people at a time. Water running through a channel beneath the seats removed the waste, and toilet sponges' mosses on sticks were available for rinsing after defecation; these reusable sponges were a breeding-ground for bacteria (Vuorinen *et al.*, 2007). In public places, these facilities were common and hand washing was rare or quite absent. The situation was not improved in private houses where running water is lacking. Since private latrines were located near kitchens, dissemination of pathogens was rendered quite easy resulting in frequent intestinal diseases (dysentery, diarrhoea). This is to say that difference in social status between rich and poor people has also influenced their respective health (Vuorinen *et al.*, 2007).

A part from ordinary dwellers, slaves, represented mainly by the fullers, belong to the category of people with the highest rate of exposure to human wastes (urine, faeces) and wastewater including runoff. For slaves and prisoners who were assigned the chores of the maintenance of the sewers (municipal aediles), several health dangers and risks were reported such as asphyxiation, exposure to pathogens in human wastes and spread of diseases like the leptospiral jaundice (Weil's disease) transmitted by rats (Wilson, 1997). In the North African cities, the drains under the streets were covered with slabs that were removed for regular cleaning out. Although, some uncovered drains existed and have caused health hazards to the population either by falling in or by the odours (Wilson, 1997). In addition to diseases caused by pathogens and rodents, the fullers in Roman their laundries and workshops were exposed to 'natural' derived chemicals, like ammonia contained in urine, and sulphur used for clothes bleaching. This situation has caused exposure of workers to an unhealthy environment with continuous pollution emission and foul-smelling air in addition to the constant skin contact with the chemicals in water. As a result fullers ran a high risk of developing work-related strains and illnesses. In addition to cleaners, some animals were helping in getting rid of wastes such as vultures and dogs. However, these animals which were quite familiar, were also carrying several diseases (rabies, skin disease) since they were in contact with waste, in addition to contaminating public places, especially water basins that could extend to the contamination of water supply. Accumulation of rubbish and wastes in the streets resulted in the proliferation of insects like flies especially in summer. Contamination of food and water with animal and human wastes was not well managed by the authorities leading to environmental and health problems like infectious diseases (Scobie, 1986).

13.7 CONCLUSION

An insight on the sanitary system developed in cities of Tunisia (*Africa Proconsularis*) during the Roman era is presented in this chapter. It takes into consideration water circulation in baths and latrines as the main buildings requiring water supply. Examples from prosperous Tunisian Roman cities are presented with illustrations to highlight the concern of water adduction, distribution and management within those buildings.

The sanitary system in Tunisian cities followed the one existing in large Roman cities. Informations made available on the sanitary system within the ruins in Tunisia were facilitated by excavations carried out during the early 20th century but many of the Roman waste water structures are still not well investigated. In fact, few literature reviews have described the circulation of used water within public and private premises in Tunisia, in comparison with those existing in Rome or Pompeii.

Romans were deemed as the most advanced in water engineering and sanitation and this has made Tunisia among the most prosperous region of the Empire. The rigorous urban planning followed for constructing cities and streets was dictating the circulation of clean and used water. The sewer system developed for water disposal consisted in drainage canals for runoff collection that may connect villas and public baths and latrines as well. This system with its connections was considered as the most spectacular.

Cities of the Roman Tunisia were among the wealthiest of the Empire during the Roman civilization thanks to the availability of water resources in some regions and the engineered solution developed to bring water into public and private buildings. Even though, access to water for cleaning and wellbeing depended largely on the social ranking and status of wealth. Accordingly, the richest people may have plumbed villas with private latrines and baths while the more impoverished have to rely on public sanitation facilities.

Despite the relatively well managed sewage water and the maintenance of the sewerage network, basic principles of sanitation and safety were lacking thereby exposing dwellers and labourers (mainly slaves) to health problems ending to death.

The decline of the Roman Empire starting from the late 3rd century was associated with arise of water issues related to quality and quantity. The poor sanitary conditions enhanced by over population in urban cities and recrudescence of poverty (in urban and rural area) has enhanced the development of plagues and water related diseases thus lowering the life-expectancy. In addition, over abstraction of water resources and degradation of the agricultural environment have been reported.

13.8 SANITARY FACILITIES IN SELECTED ROMAN CITIES IN TUNISIA

Based on the informations presented in the first part of the Chapter, it becomes obvious that baths and latrines are the major water consumers' buildings, where water used to circulate abundantly and continuously. In this second part which is rather descriptive, illustrations are dedicated to the depiction, and to some extent, to the follow-up of the pathway of clean and polluted water in selected examples of famous Roman Tunisian cities. Even though they may seem to have the same outlay, the main characteristics of the buildings where water circulate are described in each of them.

Photos displayed in this chapter were recently taken by the author during 2009 and 2012, except some outlays found in museums.

13.8.1 Sanitation facilities in Carthage

International excavations have been carried out during the 70's and the 80's in Tunisia and the Carthage was qualified as 'one of the greatest cities of the ancient world' (Hurst *et al.*, 1999). The main Roman ruins excavated in Carthage are the Antonine Baths and the communal latrines.

13.8.1.1 Antonine Baths

Situated in Carthage, on the sea shore, the Antonine Baths is one of the most impressing roman buildings. They are the largest in North Africa, and third largest in the whole Roman world. It spreads over more than 4 ha with a series of monuments onto paved streets. The baths was built under the reign of the Emperor Hadrien and his successor the Emperor Antoninus Pius (145–162) (Mahjoubi, 2002). Excavations of the Antonine Baths took place in 1954 and 1955 and showed the damage caused by the Vandles during their invasion and the restorations done by the Byzantins. The canal that used to channel water from the Bordj Djedid cisterns was no more used, and reservoirs and other cisterns were built instead (Lézine *et al.*, 1956).

Like common large central Roman public buildings, the bath is symmetrically designed. The bath was supplied by the aqueduct of Zaghouan channelling water in underground canals to the hill of Bordj Jedid and to the Malaga cisterns (Mahjoubi, 2002). However, according to Wilson (1994), the Bordj Jedid cisterns are of 20,000 m^3 and are the ones associated with the baths. The basement, which supported the vaulted rooms of the first floor (Figure 13.6) and the huge bathrooms, is all that remain today. In its symmetry, the bath has two parts with separate entrances, one for women and one for men. The sequential arrangement from cold (north) to hot (south) is well designed. The bath encompasses a series of polygonal rooms surrounding a large *frigidarium* (rectangular shape) supported by 8 large granite columns of more than 20 m high (Figure 13.6).

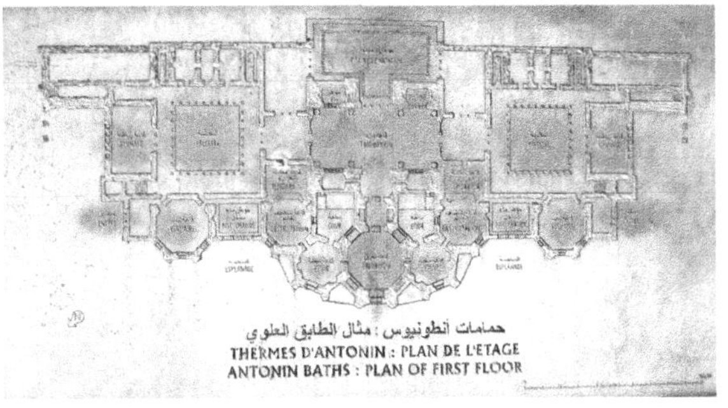

Figure 13.6 Plan of the first floor of the Antonine baths. The diagramme shows symmetrical parts dedicated separately for men and women (with permission of O. Mahjoub).

Figure 13.7 Different views of the channel between the latrines and the hypocaust. According to the land slop, part of the water used to flow by gravity toward the latrines channelled into covered or open canals (with permission of O. Mahjoub).

Figure 13.8 Views of different sections of the channel that carries out the water to/from the bath. It is clear from the photo that the canal follows the relief. The water seems to start flowing from the highest point, following the contour of the bath. The canal sections are different in shapes and depths depending on the flow: sometimes it is straight, and sometimes it is curve or sinuous (with permission of O. Mahjoub).

On the archaeological site of the Antonine Baths, one can notice the existence of a water canal that follows, to a certain extent, the contours of the remaining buildings (Figure 13.8). By following the canal, the water inlet providing the baths with water was identified. The depth and the width of the canal varied in different points indicating how the Romans were controlling the water flow. The series of photos shows various points of the channel from the identified inlet, following gravity, to the outlet close to the sea (Figure 13.7–13.10).

Figure 13.9 View of the hypocaust in basement of the Antonine baths. Canals are covered with pieces of stones. When following these canals upward the slope, a kind of cistern is found upstream. These canals may have been used to carry hot water (with permission of O. Mahjoub).

Since the baths are located on the sea shore, the runoff and waste water were directly channelled to the sea. The Antonine Baths have three drains to collect waste water from pools and latrines to discharge them into the sea (Figure 13.11).

Figure 13.10 View of 2 drains at the Antonine baths, one of them is being restored. The photos are taken toward the sea (with permission of O. Mahjoub).

Figure 13.11 Drain canals carrying the water through the bath rooms toward to the sea (with permission of O. Mahjoub).

13.8.1.2 Latrines

The latrines at the Antonine Baths archaeological park have two latrines at their entrances. Only one located on the east exists thanks to reconstruction works. Latrines are of semi-circular shape surrounded by a water circulation canal that passes under the seats. The seats parts do no more exist. All that remains is just the marble plate with the keyhole design (Figure 13.12–13.14). Latrines were continuously flushed with a water flows under the seats. Urine and human wastes were discharged in the sewerage system.

Figure 13.12 (a–b) Panoramic views of the public latrines built next to the south west portico of the Antonine baths. The latrines are hemicycle shaped with 35 m of diameter. Parts of the broken wash basin, in front of the seats, are visible in the centre. Arrows of same colour are used to indicate the same columns on both photos (with permission of O. Mahjoub).

Figure 13.13 Relics of the marble plate of the seat and marble column used to decorate the communal latrine (with permission of O. Mahjoub).

Figure 13.14 (a–b) Water that flows under the seats is collected in the sewer system visible through drains hole (with permission of O. Mahjoub).

13.8.2 Sanitation and water facilities in *Bulla Regia* (Henchir Hammam Darradji)

Bulla Regia or *Royal Bulla* is one of the residential cities that passed in 46 BC under the Roman Empire. It was prosperous and promoted to the state of municipe under Vespasian and honorary city later under Hadrian. The hydraulic installations in *Bulla Regia* were not imposing, like those of Carthage. It is apparent that the archaeological site is not well investigated yet; less than 25% of the territory is excavated till now.

Nevertheless the city has humid climate, private houses used to rely on cisterns and water reservoirs that collect rain water used for domestic purposes, other than potable supply. Figures 13.15a–c show cistern used to collect rainwater. The underground drain was used to collect overflowing water in private (e.g., House of the Hunt) and public premises (e.g., amphitheatre) (Figures 13.15–13.17).

Albeit the water supply sources are well known, the sanitation system is not well drawn yet. Few investigations and very few references exist on the final destination of the sewage water and waste. Since the city is located on a hill, the most probable final discharge point is the river downstream. The only available sayings are explanations about the sewers cleaning and maintenance (Figure 13.18 and 13.19).

Figure 13.15 (a–c) General view of the peristyle courtyard at the underground of the House of the New Hunt showing the cistern for collecting rainwater and the underground drain (indicated by the white arrow) where the overflowing water is discharge (left). Zoom in view of the cistern opening (top right) and the drain (bottom right) (with permission of O. Mahjoub).

Figure 13.16 Openings of the underground canals (white arrows) on the right corner to collect water used for cleaning at the House of Amphitrite (left) and orchestra opening drain in the amphitheatre at *Bulla Regia* (right) (with permission of O. Mahjoub).

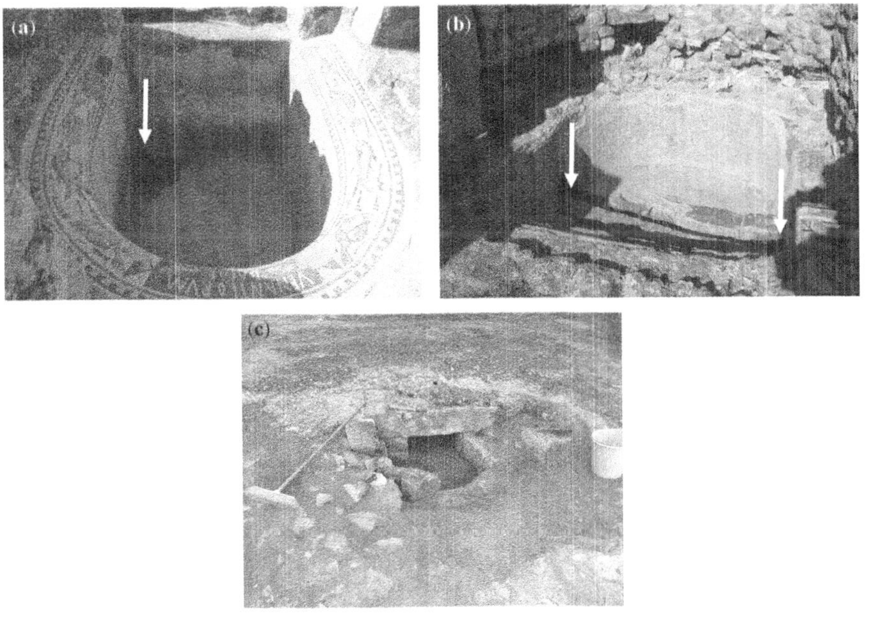

Figure 13.17 (a–c) Views of different system to discharge used water in private and public bath at *Bulla Regia*. Drains' openings in private bath at the House of Amphitrite (top left) and the canal that channel the circulating water in the bath of the House of the Hunt (top right). A drainage system for the collection of overflowing water in one of the rooms of the public bath (bottom centre) with maintenance work to remove sediments and avoid clogging of the system (with permission of O. Mahjoub).

13.8.3 Sanitation facilities in *Tuburbo Majus* (Henchir Kasbat)

Thuburbo Majus was a Berber-Carthaginian settlement long before the Romans arrived. The name was actually based on a common root in the indigenous local language, used in place names like *Thuburbo Minus* (modern Tebourba). The emperor Augustus started to build *Thuburbo Majus* in 27 BC. Called also the 'Roman Colonia Ivlia Avrelia Commoda', it is known as the fifth major Roman sites in Tunisia after *Bulla Regia, Dougga, Makthar, and Sbeitla. Thuburbo Majus* has a strategic location with access to trade routes made. *Tuburbo Majus* has several baths (Figure 13.20) for summer and winter seasons. Vestiges of latrines can be seen showing clearly the canal under the seats to flush wastes. In front of the seats, gutters where water is circulating are still visible showing inlet and outlet of the water (Figure 13.21).

13.8.4 Sanitation facilities in Thugga (Dougga)

In order to preserve hygiene and wellbeing of citizens, Roman has made of the drainage and sanitary system one their main concerns. At Thugga, vestiges of the sewer system testimony of the structures developed at that time (Figure 13.22). A visit to the archaeological site showed different design of a well conserved public latrine. As one can notice, latrines are cut into blocks of stone. The gutter in front will serve to dip the mosses the visitors of latrines used for cleaning. A washbasin is used for hand washing. At Thougga, one can see latrines consisting of 12 seats made of limestone. A gutter where water flows is meant for excrements flushing and elimination of bad smell.

Figure 13.18 A two-seater private latrines at House of the Hunt (top right) at *Bulla Regia* with a partially broken floor that let show the sewer drain underneath, and remains of public latrines (right) with the canal that brings water in front of the seats then dispose it off in the drain. The example below is vestiges of two seats latrines cut into grey-blue marble showing a part of the gutter in front in Bulla Regia (with permission of O. Mahjoub).

Figure 13.19 General view of slabs used to lay the street and cover the drainage canals and sewage network at *Bulla Regia*. The partially degraded cement between slabs let show the sewer beneath (left). Sometimes the slabs collapse or are deformed causing deformation of the street (with permission of O. Mahjoub).

Figure 13.20 A general view of the vestiges of the winter baths (Thermes d'hiver) at the archeological site of *Tuburbo Majus* (with permission of O. Mahjoub).

Figure 13.21 Relic of latrines showing the canals under the seats allowing water circulation and the canal in front of the seats with the inlet and outlet of water (with permission of O. Mahjoub).

Figure 13.22 Well conserved small communal latrines with U-shaped gutters part of the Cyclopes thermes (left) and a washbasin located in front of the seats at *Thugga* (right) (with permission of O. Mahjoub).

13.9 CONCLUSIONS

The overview on the sanitary system in Tunisia during the Roman civilization has highlighted the advanced knowledge of the Romans in water and wastewater management. Part of which is now Tunisia as one of the most important and wealthiest regions of the Roman Empire, was completely transformed to guarantee the wellbeing of the dwellers especially in private premises, by collecting and withdrawing every available drop of water. The huge efforts made to manage water resources in the cities were driven by social, cultural, and political motivations which have significantly influenced the access to sanitation facilities and the perception of personal hygiene.

Public baths and latrines were by far the largest consumers of water during the Roman time, where water used to flow continuously, therefore wastewater and wastes released in the drainage and wastewater collection system of the city used to be discharged directly into rivers and streams. Several environmental and health implications have resulted from this (mis)management of the sewage water and wastes.

The various illustrations of archaeological sites in Tunisia showing examples of sanitation facilities depicted the magnificence of the edifices and the associated sanitary system, at a time, from engineered and architectural points of view. A part from those existing in large prestigious Tunisian cities as witnessed here, few sanitary facilities are known. It comes out that, at the opposite of the large number of research addressing features of the water supply system in Tunisia during the Roman era, wastewater management and reuse options remained insufficiently investigated, at our knowledge.

13.10 REFERENCES

Aicher P. J. (1995). Guide to Aqueducts of ancient Rome. Bolchazy-Carducci Inc., Wauconda, Illinois, p. 183.

Amulree L. (1973). Hygienic conditions in ancient Rome and modern London. *Medical History*, **17**(3), 244–255.

Antonio R. J. (1979). The Contradiction of domination and production in bureaucracy: the contribution of organizational efficiency to the decline of the Roman Empire. *American Sociological Review*, **44**(6), 895–912.

Bond T., Roma E., Foxon K. M., Templeton M. R. and Buckley C. A. (2013). Ancient water and sanitation systems – applicability for the contemporary urban developing world. *Water Science & Technology*, **67**(5), 935–941.

Chanson H. (2002). Certains aspects de la conception hydrauliques des aqueducs Romains. (Some aspects on the hydraulic design of Roman aqueducts). *La Houille Blanche*, **6/7**, 43–57.

Chanson H. (2008). The hydraulics of Roman aqueducts: What do we know? Why should we learn? In: Proc. of the World Environmental and Water Resources Congress, Hawai, USA, pp. 1–16.

Crow J. (2007). Infrastructure of a great city: earth, walls and water in late antique Constantinople. In: Technology in Transition. Late Antique Archeology, E. Zanini (ed.), BRILL, Leiden, Boston, pp. 251–285.

De Feo G. and De Gisi S. (2013). Water and wastewater management in the treatment process of a Roman fullonica. *Water Science & Technology: Water Supply*, **13**(3), 599–605.

Dessales H. (2006). Le prix de l'eau dans l'habitat romain: une étude des modes de gestion à Pompéi (Water price in the Roman habitat: a study on the management at Pompei). In: Proc. of the Vers une gestion intégrée de l'eau dans l'empire romain (Towards an integrated water managment in the Roman Empire). Actes du Colloque International de l'Université Laval (Proceedings of the International Colloquium of th Laval university), Rome, pp. 55–66.

Drewko A. (2007). Resource-oriented Public Toilets in Developing Countries: Ideas, Design, Operation and Maintenance for Arba Minch, Ethiopia. Hamburg University of Technology, Hamburg, p. 148.

Fagan G. G. (2001). The genesis of the Roman public baths: recent approaches and future directions. *American Journal of Archaeology*, **105**(3), 403–426.

Farnsworth Gray H. (1940). Sewerage in ancient and mediaeval times. *Sewage Works Journal*, **12**(5), 939–946.

Ferchiou N. (1999). Les Aqueducs de Zaghouan à Carthage et leurs structures complémentaires. Note Préliminaire (Aqueducts of Zaghouan at Carthage and their complementary structures, Preliminary notes). *Africa*, **17**, 69–86.

Figueiredo M. O., Veiga J. P. and Silva T. P. (2001). Materials and reconstruction techniques at the Aqueduct of Carthage since the Roman period. In: Historical Constructions, P. Roca (ed.), University of Minho, Guimarães/ Portugal, pp. 391–400.

Foil J. L., Cerwick J. A. and White J. E. (1993). Collection systems, past and present. A historical perspective of design, operation, and mintenance. *Operation Forum Magazine*, Vol. 10.

Fowler R. (2007). The fullers of Rome. http://suite101.com/article/the-fullers-of-rome-a25153 (accessed 12 December 2012).

Germain De Montozan G. (1907). Rapport sur une mission scientifique en Italie et en Tunisie (Report of a scientific mission in Italy and Tunisia). *Nouvelles Archives des Missions Scientifiques*, **XV**(fasc. II), 71–123.

Graham A. (1902). Roman Africa. An outline of the history of the Roman occupation of the North Africa. Longmans, Green and Co., London.

Hansen R. D. (2007). Water and wastewater systems in imperial Rome. *Journal of the American Water Resources Association*, **19**(2), 263–269.

Haut B. and Viviers D. (2012). Water supply in the Middle East during Roman and Byzantine periods. In: Evolution of Water Supply Through the Milleania, N. Mamassis (ed.), IWA Publishing, London, pp. 319–350.

Huibers F., Redwood M. and Raschid-Sally L. (2010). Challenging conventional approaches to managing wastewater use in agriculture. In: Wastewater Irrigation and Health, A. Bahri (ed.), 1st, International Development Research Centre (IDRC) and the International Water Management Institute (IWMI), London, UK, pp. 287–302.

Hurst H., Ben Abdallah Z., Fulford M. G. and Henson S. (1999). The sanctuary of Tanit at Carthage in the Roman period. A re-interpretation. *Journal of Roman Archaeology Supplementary Series*, **30**, 136, Portsmouth.

Institut National du Patrimoine. (2002). La maison d'Africa à Thysdrus (El Jem) (House of Africa at Thysdrus (El Jem)). Ministère de la Culture, Tunis, p. 14.

Jiménez B., Drechsel P., Koné D., Bahri A., Raschid-Sally L. and Qadir M. (2010). Wastewater, sludge and excreta use in developing countries: an overview. In: Wastewater Irrigation and Health, A. Bahri (ed.), 1st, International Development Research Centre (IDRC) and the International Water Management Institute (IWMI), London, UK, pp. 3–28.

Kalinowski A. V. (1996). Patterns of Patronage: The Politics and Ideology of Public Building in the Eastern Roman Empire (31 BCE-600 CE). Department of Classics. University of Toronto, Toronto, p. 268.

Kron G. (2012). L'impatto della 'peste antonina' (Impact of the 'peste antonina'). In: Nutrition, Hygiene and Mortality. Setting Parameters for Roman Health and Life Expectancy Consistent with Our Comparative Evidence, E. Lo Casio (ed.), Edipuglia, Bari, pp. 193–252.

Leight M. (2008). Investigating Urban Centers in the Ancient Roman Empire: City Design Based on the Writings of Vitrivius. West Virginia University, Virginia, USA, p. 85.

Leone A. (2012). Water management in late antique North Africa: agricultural irrigation. *Water History*, **4**(1), 119–133.

Lézine A., Picard C. and Picard C. (1956) Observations sur la ruine des Thermes d'Antonin à Carthage (Observations on the ruines of the Antonin Baths at Carthage) 100ème année(N. 4), 425–430.

Magnabal G. (2004). Servitudes urbaines et réglements d'urbanisme dans l'habitat de la Rome Impériale (27 avant Jésu Christ – 192 après Jésus Christ (Urban servitude and urban regulations in habitats of the Imperial Rome). Université de Marne la Vallée, Champs sur Marne, p. 46.

Mahjoubi A. (2002). La Tunisie Romano-africaine 146 AV.J.C. – 439 AP. J.C. (The Romano-african Tunisia 146 B.C.439 A.D.). In: R. 3/AMVPPC, (ed.), 2nd edn, Ministère de la culture, Tunis, p. 17.

Mahjoubi A. (2004a). Cités antiques de Tunisie (Ancient cities in Tunisia). In: Agence de mise en valeur du patrimoine et de promotion culturelle, (ed.). Top Printing, Tunis, p. 176.

Mahjoubi A. (2004b). La Tunisie à l'époque Romaine (Tunisia at the Roman era), In: I. Zeidoun, (ed.) p. 125.

Mahjoubi A. (2011). Villes et structures urbaines de la province romaine d'Afrique (Cities and urban structures of the Roman province of Africa), 2nd edn, Centre de publication universitaire, Tunis, p. 316.

Mahjoubi A. and Slim H. Water Control During Ancient Times. http://www.semide.tn/english/contextedeleau/cuturel.pdf (accessed 13 December 2012)

Malissard A. (2002). Les Romains et l'eau : fontaines, salles de bain, thèmes, égouts, aqueducs...(Romans and water: fountains, baths, thermes, sewers, aqueducts, ...). REALIA/Les belles lettres, Paris, p. 350.

Mattingly D. J. and Hitchner R. B. (1995). Roman Africa: an archaeological review. *The Journal of Roman Studies*, **85**, 165–213.

Mays L. W., Sklivaniotis M. and Angelakis A. N. (2012). Water for human consumption through history. In: Evolution of Water Supply Through the Millennia, N. Mamassis (ed.), IWA Publishing, London, pp. 19–42.

Milani-Santarpia G. (2011). Ancient Roman Baths. http://www.mariamilani.com/ancient_rome/roman_baths.htm#roman_baths (accessed 13 December 2012)

Morgan M. H. (1914). Vitruvius, The Ten Books on Architecture, Translated by Morris Hicky Morgan. Harvard University Press, London, p. 365.

Moussa M. (2000). Drinking Water Supply Techniques. www.semide.tn/english/contextedeleau/Culture4.pdf (accessed 13 December 2012).

Murphey R. (1951). The decline of North Africa since the Roman occupation: climatic or human? *ANNALS, Association of American Geographers*, **XLI**(2), 116–132.

Nadkarni M. (2004). The Meaning of Sanitation: An Ecosystem Approach. CERNA, Centre d'économie industrielle, Ecole nationale Supérieure des Mines de Paris, Paris.

Oxford University Press (2012). Oxford Dictionaries. oxforddictionaries.com (accessed 12 December 2012)

Passchier C. W., Wiplinger G., Sürmelihindi G., Kessener P. and Güngr T. (2011). Roman aqueducts as indicators of historically active faults in the Mediterranean Basin. In: Proc. of the 2nd INQUA-IGCP-567 International Workshop on Active Tectonics, Earthquake Geology, Archaeology and Engineering, Corinth, Greece, pp. 168–189.

Peyras J. (2006). La gestion intégrée de l'eau dans l'antiquité tardive: La réception du droit Romain dans l'Afrique Mineure (Water management in the late antiquity: the Roman right installation in Minor Africa). In: Proc. of the Vers une gestion intégrée de l'eau dans l'Empire romain: Actes du colloque international, Université Laval, Rome, pp. 253–264.

Rakob F. (1979). A Carthage, l'aqueduc des thermes impérieux (At Carthage, the aqueduct of the impérial thermes). *Dossiers de l'Archéologie*, **38**(oct–nov), 34–42.

Scheidel W. (2010a). Physical Wellbeing in the Roman World. Stanford University, Princeton.

Scheidel W. (2010b). Age and health in Roman Egypt. Stanford University, Stanford.

Scobie A. (1986). Slum, sanitation, and mortality in the Roman world. *Klio*, **68**, 399–433.

Shaw B. D. (1984). Water and society in the Ancient Maghrib: technology, property and development. *Antiquités Africaines*, **20**, 121–173.

Slim H., Mahjoubi A., Belkhoja K. and Ennabli A. (2010). L'antiquité (The antiquity). In: Sud Editions, (ed.), 2nd edn, Histoire générale de la Tunisie, Vol. I, Tunis, p. 464.

Soifer S. (1999). The Evolution of the Bathroom and the Implications for Paruresis. http://paruresis.org (accessed 10 June 2012)

Tardieu G. (1986). The Roman Water System: Its Contribution to the Success and Failure of Ancient Rome. The Faculty of the Department of History. San Jose State University, USA, p. 45.

Thébert Y. (1987). A history of private life: from pagan Rome to Byzantium. In: Private Life and Domestic Architecture in Roman Africa, P. Veyne (ed.), Harvard University Press., Cambridge, Mass., and London, pp. 313–409.

Uy J. I. (2007). Dam Nation: dispatches from the water underground. In: History of Sewers Cesspools and Cholera: The Development of the Modern Sewer, L. Allen (ed.), Soft Skull Press, Brooklyn, NY, pp. 112–119.

Vuorinen H. S., Juui P. S. and Katko T. S. (2007). History of water and health from ancient civilizations to modern times. *Water Science and Technology*, **7**(1), 49–57.

WHO (2008). Poor Sanitation Threatens Public Health. http://www.who.int/mediacentre/news/releases/2008/pr08/en/index.html (accessed 10 November 2012)

Wilson A. (1997). Water Management and Usage in Roman North Africa. A Social and Technological Study. PhD thesis, University of Oxford, Oxford, United Kingdom, p. 589.

Wilson A. (1994). Running water and social status in North Africa. In: Proc. of the North Africa from Antiquity to Islam, Bristol, pp. 52–56.

Zuiderhoek A. (2009). Introducing euergetism: questions, definitions and data. In: The Politics of Munificience in the Roman Empire: Citizens, Elites, and Benefactors in Asia Minor, J. Elsner (ed.), Cambridge University Press, USA, p. 204.

Chapter 14

Revisiting the technical and social aspects of wastewater management in ancient Korea

Mooyoung Han and Mikyeong Kim

14.1 INTRODUCTION

Since the history of mankind, water and wastewater management has been the major concern of civilization, especially for agricultural based societies where people had to settle at one place for a long period. Some earlier civilizations in the world are known to disappear because of improper water and wastewater management.

Ancient Korea was known as an agricultural based society since over 7000 year earlier (Lee, 2004). The appropriate water and wastewater managements are crucial with regards to the economic and political sustainability of the nation. Korea is located in a monsoon region, experiencing summer flooding and spring drought; furthermore, the rocky and steep topography of the country poses a greater challenge in water and wastewater management. Despite of such hard natural conditions, Korean people have survived for more than 5000 year history cultivating its unique culture and tradition.

During ancient Korea, five dynasties were made which lasted for longer than 500 years: Gojoseon dynasty (2333–108 BC), Goguryeo dynasty (37–668 BC), Baekje dynasty (18–660 BC), Silla dynasty (57 BC–935 AD), and Joseon dynasty (1392–1910 AD). Especially, Seorabeol which was the capital city of Silla dynasty maintained sustainability for more than 1000 years. For a city to last for such a long time, not only the technology of water infrastructure but also the social aspect such as tradition and education had to be well established

Korea has cultivated unique techniques for both wastewater management technology and culture for agricultural maintenance, which are necessary for long-term period of settlement. Therefore, it is justified that ancient Korea had proven management and technology philosophies regarding sustainable wastewater management. Wastewater management, including domestic wastewater and human wastes, requires technical and social knowledge. The analysis of the traditional values and culture are necessary to serve as guidance for an effective and efficient resource management.

This study will present the technical and social aspects of the Korean traditional wastewater management practices utilized from traditional houses to villages. Furthermore, the comparison of the ancient and present wastewater management practices was performed to identify the knowledge that can be adapted from the past. Lessons from the past can be beneficial for addressing future challenges, to build sustainable cities, caused by climate change and/ or the limitation of resources.

14.2 WASTEWATER MANAGEMENT IN TRADITIONAL HOUSES OF ANCIENT KOREA

14.2.1 Water and wastewater flow in a traditional household

Figure 14.1 shows the water and wastewater flow diagram in a traditional house during ancient Korea. The cycle starts from the collection of water for household consumptions (e.g., drinking, cooking, laundry, washing); which can be harvested from community wells, streams or rooftops by rainwater harvesting (Han *et al.*, 2013a; Yu, 2007). A *Pit Latrine toilet* is usually built in around backyard or gate (Kim, 2002; Suh, 2012) while, the animals grown (i.e., cows, pigs, hens and dogs) were kept in the front yard. The wastewater effluents, collected from traditional toilet systems, were normally reused for various purposes. For example, special jars located near the *Pit Latrine toilet*, were kept for the purpose of collecting separated wastes (i.e., urine and faeces) (Kim, 2002; Yun *et al.*, 2011). The urine storage pot is called '*ojum-janggun*' and the faeces storage pot is called '*ddong-janggun*'; both waste types were used as fertilizers for farming (Kim, 2002; Yun *et al.*, 2011). The collected water effluents from laundry and washing activities were kept in a specific vessel called '*tteun-multong*', which can be easily reclaimed for fodder making (Exhibition, The national folk museum of Korea). Wastewater generated from traditional houses were directed and treated in a small wetland called '*water-dropwort field*', which has natural water purification functions; afterwards, the treated wastewater were used for farmland irrigations such as rice paddies (Ahn, 2009; Yu, 2007).

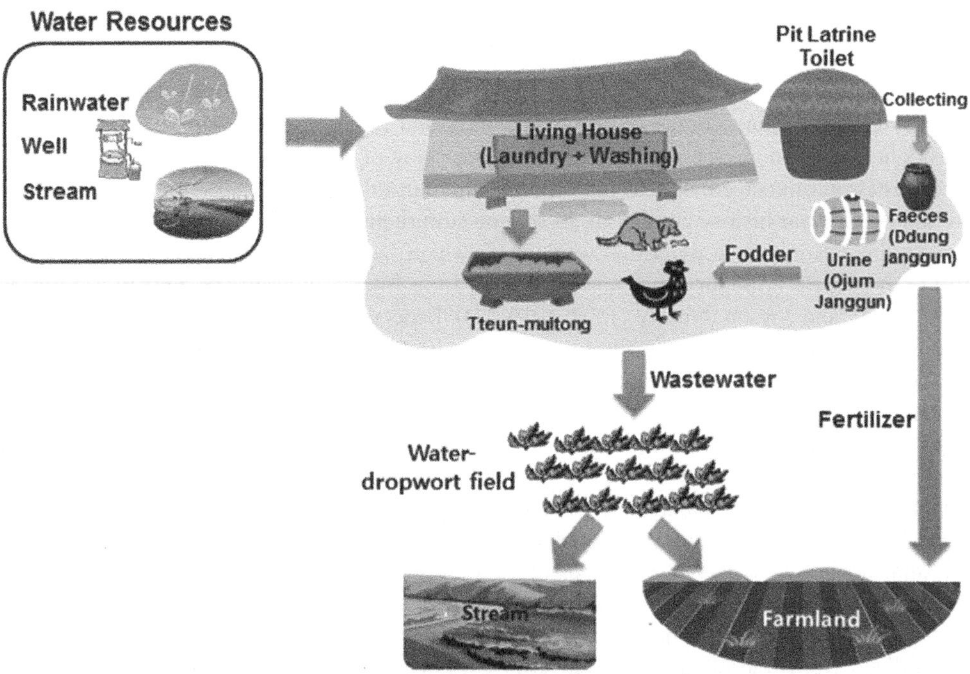

Figure 14.1 Water and wastewater flow diagram of traditional households during ancient Korea.

Therefore, the wastewater managements practiced in traditional households during ancient Korea were characterized to minimize the quantity of wastewater, through recycling and reusing; an approach which is similar to the 4R activities emphasized in the present time: Reduce, Recycle, Reuse and Revitalization.

14.2.2 Nutrient flow generated by typical traditional house

The food cycle can be further analyzed to two subcategories, which was characterized through the consideration of two different types of material transportation as shown in Figure 14.2. For the first type of food-waste cycle, the cycle starts with the person who emits food wastes through food consumption. Consequently, the food-waste was used as fodder for animal feeding. Lastly, the livestock animals serve as a food source for human consumption. For the second food-waster cycle: the excreta from human beings and animals were reclaimed to serve as agricultural fertilizers, which were matured in farmland for a fixed period of time. Lastly, the fertilizers give nourishment to the agricultural plants, which is also beneficial for grain production for human consumption.

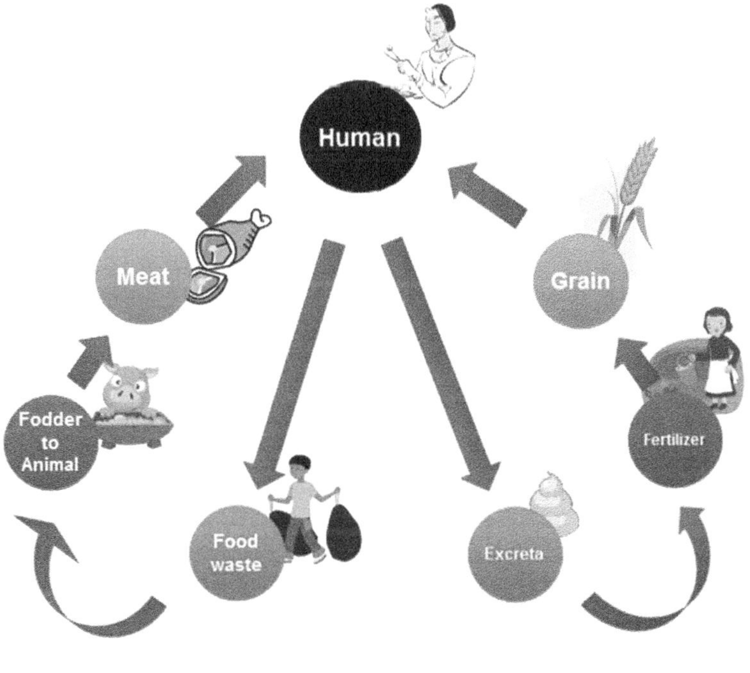

Figure 14.2 Food-waste cycle in a traditional house in ancient Korea.

Comparing the traditional food-waste cycle with the modern trend, a one-sided structure is usually observed in the present time, where in the structure can be summarized into three simultaneous steps, production, consumption and emission. However, the nutrient flow generated in the past was almost zero, due to the recycling of waste effluents resulting to minimal wastes.

14.2.3 Wastewater management system in a traditional village

Early Koreans constructed the village with having the values of environmental, economical, landscape, and social sustainability (Shin, 2006). Based on the reference of Taekriji, geography book published in middle of 18 century, Wi (1992) introduced a paradigm that when people selected their residential location they should consider nature (water inlet, field shape, mountain form, soil color, water utilization, and water discharge), economy (regional economy), community (importance of society) and environment (environment-human being relationships).

The layout of a village-scaled wastewater management practice in traditional villages is presented in Figure 14.3 (Han *et al.*, 2013a; Yu, 2007; Ahn, 2009). Traditional villages were usually located in front of mountains and next to streams or rivers. Furthermore, traditional houses are usually situated within the vicinity of a small village pond, which has self-purification functions and which serves as a primary treatment facility. Furthermore, the agricultural land such as rice paddies is usually situated beside the village pond, to reuse the recycled wastewater. The discharged wastewater from village ponds and farmlands, which has low pollution loadings, is directed towards the stream.

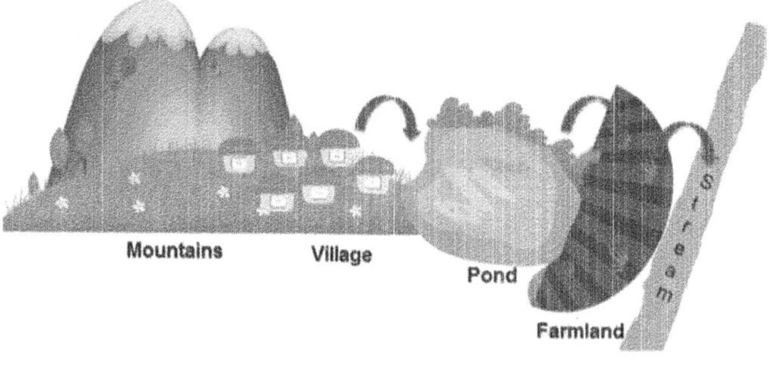

Figure 14.3 Layout of a typical wastewater management system in traditional village unit during ancient Korea.

During ancient times, the traditional villages constructed self-purifying wastewater management systems, which are small and decentralized in scale, to minimize the impacts on downstream areas.

An example of a Korean traditional village pond, which still exists, is situated in Goseong city. The traditional village pond is called as '*Jangsan* pond' which is shown in Figure 14.4. Both the village and village pond were constructed during Joseon dynasty in year 1406. The village pond was built for multipurpose uses such as wastewater purification, culturing fish, aesthetic purpose, and so on.

Figure 14.4 Image of a traditional village pond, *Jangsan* pond located in Goseong city, Korea (with permission of M. Han).

14.3 TECHNICAL ASPECTS OF TRADITIONAL WASTEWATER TREATMENT

In ancient Korea, numerous technical methods for wastewater management already existed. Management methods such as the reusing and recycling of wastewater with excreta were the examples of earlier management practices; consequently, the treated wastewater was usually discharged to nearby streams and rivers. The main purposes of earlier wastewater management practices were to mitigate the impacts on both water and the environment. The characteristics of the technical aspects of wastewater treatment and management by ancient Koreans are as follows:

14.3.1 Water dropwort fields

Water dropwort fields were usually grown in man-made puddles, located around houses and farmlands. The primary function of cultivating water dropwort was to purify the wastewater that passes through the area, which was discharged to farmlands and rivers. An example of the typical scene of water dropwort field is shown in Figure 14.5. The impurities in wastewater settled down to the bottom of the puddle through sedimentation, while the organic matter is absorbed by water dropwort; hence, the wastewater was treated resulting in clean effluent. Recent studies on the utilization of water dropwort have been carried out for wastewater treatment (Lee *et al.*, 2003; Reddy *et al.*, 1985).

Figure 14.5 Image of a typical water-dropwort field for wastewater treatment.
Source: http://blog.ohmynews.com/q9447/235984.

14.3.2 *Jetgan* toilet

During ancient Korea, traditional toilets were usually located apart from main houses. *Jetgan* is a traditional type of toilet usually used by houses situated in mountainous areas and islands (Kim, 2002). An example of a traditional toilet, known as *jetgan* is shown in Figure 14.6. A typical *jetgan* consists of two large stepping stones; and furnace ashes deposited in between the two stepping stones, which receives the excreta. After using the *jetgan*, the chaff of grains were placed and mixed together with urine, faeces and furnace ashes. The mixture was stored in a corner to contain in the odours and further used to make fertilizers.

Figure 14.6 An image of a *Jetgan* found in traditional house (Kim, 2002).

14.3.3 *Pit latrine* toilet

Pit latrine toilets, for urine and faeces receiving function, had to be installed in a certain depth and width to allow the excreta to be properly fermented (Kim, 2002). Wide storage facilities were usually constructed to provide access for reclaiming urine and faeces. Time is essential parameter in the storage and reuse of excreta. The cycle starts from the storage of excreta, followed by the separation of solid wastes with the liquid wastes. During the storage of excreta, the solid wastes floats and ferments, while the liquid wastes settles down to the bottom of the storage. The liquid wastes, stored at the bottom of the container, were used as liquid fertilizers for cultivation. However, since the produced fertilizers were consumed only during spring season, pots or puddles were also installed around the pit latrine toilet, to keep wastes for fertilizer purposes. Examples of a *pit latrine* toilet and urine jar are shown in Figure 14.7. Furthermore, during ancient Korea, numerous toilet types were installed, since each toilet type has a unique composting operation (Kim, 2002; Yun *et al.*, 2011).

Figure 14.7 *Pit latrine* toilet (left) and urine jar (right) for excreta collection and storage (Kim, 2002).

14.3.4 Separated excreta jars (*Ojum-janggun* and *Ddung-janggun*)

Early Koreans practiced the separation of excreta (e.g., faeces and urine) by having different containers for each waste. The excreta used as fertilizers functioned better when separated as compared with the use of mixed excreta (Gunter *et al.*, 2005). During the *Joseon* dynasty, urine jars were usually situated nearby rooms for easier access. The collected urine wastes were fermented to serve as agricultural fertilizers. Furthermore, the different fermentation stages of urine were made possible through the use of several urine jars, which were stored in an organized method. The farming equipments called as *ojum-janggun* and *ddung-janggun* were used to carry separated urine and faeces, respectively. Figure 14.8 shows both the typical *ojum-janggun* and *ddung-janggun* used to contain urine and faeces, respectively; which were used as farming equipments to contain agricultural fertilizers.

Figure 14.8 Examples of *Ojum-janggun* and *Ddung-janggun* as farming equipments, exhibited at Goyang exhibition hall for sanitation in Gyeonggi-Do, Korea (with permission of M. Han).

14.4 SOCIAL EDUCATION FOR WASTEWATER MANAGEMENT

Numerous proverbs regarding wastewater management (i.e., urine and faeces) have been passed down from many generations. The existing proverbs reflect the wisdom and thoughts of early Koreans. In this study, the philosophies of early Koreans on wastewater management are investigated through the analysis of several proverbs.

Some proverbs indicate that the early Koreans valued the utilization of excreta as agricultural fertilizers. For example, the proverb: '*I would rather give a bowl of rice, than a basket of excreta and ashes*' which means that excreta and ashes were considered important (Choi, 1986). Moreover, the proverb: '*Mixing fertilizer with dog's faeces improves the performance of the fertilizer*' which implies that early Koreans also treasured animal faeces (Choi, 1986).

On the other hand, proverbs regarding raw water management states that: '*You might drink poop water if your poop is in wells*' which warns the people about the danger of discharging excreta to the aquatic system without performing any treatment measures (Lim, 2002). Furthermore, the proverbs: '*If you pee in streams your penis will swell*' and '*If men and women pee in streams, they will have trouble having babies in the future*' which prohibits people from urinating in the streams (Choi, 1986; Lim, 2002).

14.5 IMPORTANCE AND CHARACTERISTICS OF TRADITIONAL WASTEWATER MANAGEMENT PRACTICES

The characteristics of traditional wastewater treatment and management in Korea are briefly discussed in this section. Furthermore, the importance of each wastewater practices is also discussed in this study. The traditional wastewater management practices during ancient Korea are as follows:

14.5.1 The separation of faeces with urine

The separation of faeces and urine, traditional wastewater management system practices in Korea, was performed through the use of 'faeces pots' and 'urine pots' (Kim, 2002; Yun *et al.*, 2011). Early Koreans used separated faeces and urine wastes as agricultural fertilizers (Kim, 2002; Yun *et al.*, 2011). The fertilizers made from separated wastes, contains more nutrients (e.g., N, P and K) as compared with the use of mixed excreta as fertilizers (Langergraber & Muelleggera, 2005). Therefore, early Koreans were aware that the practices of urine and faeces separation were beneficial for farming purposes, even with the lack of knowledge in modern science.

14.5.2 Resource recovery from wastewater

Early Koreans made efforts to recycle wastewater as a water resource and to control the wastewater discharge through reusing the water. For example, urine and faeces from a toilet were reused as a fertilizers and wastewater from a kitchen were further boiled with other ingredients, creating food for stocks (Kim, 2002; Yun *et al.*, 2011; Exhibition, The national folk museum of Korea). Furthermore, water resource management came to be realized naturally through minimizing the pollutant load rates of effluents and reducing the direct discharge into the natural aquatic system (Ahn, 2009; Yu, 2007).

14.5.3 Small-scaled and decentralized management

We could find the wisdom of the source control in wastewater management which minimize the impact of the wastewater on the natural aquatic system by managing the wastewater generated in each unit from a house to the village. Early Korean tried to mitigate pollutant load rates to the natural aquatic system by reusing and pre-treating the wastewater generated from each house and village (Ahn, 2009; Yu, 2007). The continuous decentralized wastewater management seems to be beneficial to improve and maintain the ecosystem of the aquatic system.

14.5.4 Ecological concepts in wastewater management system

Early Koreans considered the ecological concepts and principles in wastewater management. During ancient Korea, the treatment of wastewater was usually addressed through the utilization of natural resources such as water parley beds, water dropwort, and so on (Ahn, 2009; Yu, 2007). Furthermore, the utilization of ecological methods improves the production of food and fodder. The collected the water effluents from laundry and washing activities were reused for fodder making (Exhibition, The national folk museum of Korea) and there was a type of toilet called 'Tongsi' feeding excreta to pig in some region of Korea (Kim, 2002). Therefore, it is proven that early Koreans considered the ecological concepts of living harmoniously with nature.

14.5.5 Emphasis on social responsibility

Farmers in ancient Korea were in need of manpower to be able to perform farming tasks. Early Koreans were united and the village community helped one another. A culture named Du-rae is a kind of farm working system consisted of communal concepts of work and play sequences (Guh, 2009). Early Koreans disseminated information regarding wastewater management in a convenient way, through education and proverbs (Choi, 1986; Lim, 2002). Furthermore, each individual was cooperative in minimizing the amount of pollutants and hence, the pollutant discharge of the village was minimal.

14.6 LESSONS FROM THE PAST FOR THE IMPROVEMENT OF CURRENT WASTEWATER MANAGEMENT SYSTEMS

The modern wastewater management systems, currently practiced by the general society, have remarkably contributed to human development through the improvement of public health. However, this kind of technology requires abundant monetary and energy resources for managing the wastes generated by urban areas with high population densities. For example, in flushing toilets, excessive volume of water is used to dispose a small amount of excreta. Consequently, water is inefficiently used and thus wasted; generating environmental problems such as soil pollution, groundwater contamination and river pollution.

Since 1970s, modern flush type toilets have been established in Korea. However, flush type toilets consume a considerable amount of water, which results to the generation of highly concentrated wastes with nitrogen and phosphorus pollutant concentrations. Therefore, the modern wastewater management systems in Korea are considered as inefficient. Eventually, a limit will be reached by the current wastewater management system due to the lack of energy and resources sustainability.

Lessons from ancient Korea can be adapted to improve the current wastewater management systems. A schematic diagram showing the modern waste water management practices is shown in Figure 14.9. Based from the comparison from the past with the present wastewater management practices, conclusions on the improvement of the current wastewater management practices are summarized as follows:

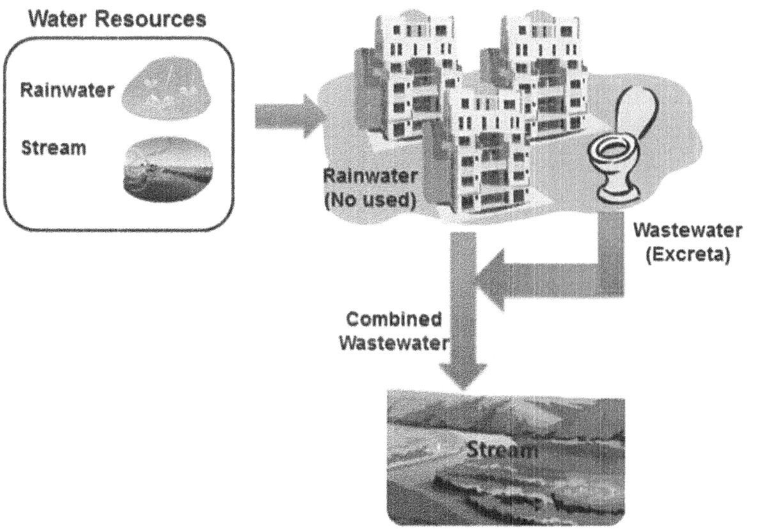

Figure 14.9 The layout of a typical wastewater management process in modern times.

First, the reusing and recycling of wastewater which were practiced in the past should be considered to serve as a paradigm for addressing the flaws in the current wastewater management system. Even, the earlier civilization was already knowledgeable about the value of water as natural resources.

Second, pollutant loads discharged in the environment should be minimized, through practicing on-site wastewater treatment methods to purify the effluents. Traditional wastewater management system separated rainwater from wastewater discharges. Moreover, early Koreans collected rainwater for domestic consumption. However, the current wastewater management system mixes rainwater; hence, the rainwater becomes stormwater runoff which was hard to handle in terms of quantity and quality.

Third, the separation of urine and faeces should be performed to minimize the impacts in the environment. Although, urine accounts to less than 1% of the total volume in wastewater, most of the nutrient loads in wastewater were generated from urine (i.e., 50% of P, and 80% of N). Ancient Korean practiced the separation of excreta, which were used as fertilizers, to minimize the pollutant loads discharged in the environment. However, the current wastewater management practice combines the excreta, which is directed to centralized treatment plants which consume a lot of monetary and energy resources, for operation.

The population continuously increases (UN, 2009) and is dependent on the agricultural food produce, which requires industrial technologies. Awareness and the efficient utilization of the available resources and technology, will results to an easier collection of nutrients (e.g., N and P) from excreta.

Although numerous toilet types (e.g., water-saving, waterless, urine and faeces separator, recovery, and composting toilets) have been recently developed, to address the issues known for flush type toilets, information dissemination to educate the society about the new solutions are not made efficiently. Education and information dissemination will aid in the improvement of the current wastewater management systems.

14.7 CONCLUSIONS

Water and wastewater management has been the major concern throughout the civilization, particularly for the citizens residing in agricultural areas, who continuously resides in the same area. Korea was not exempted from such situation; moreover, Korea must overcome the worst natural conditions caused by uneven seasonal distribution of rainfall and the rocky and steep topography of the country. The practice of wastewater management in rural areas reflects the ancient wisdom of cohabiting with the environment. Proverbs and cultural traditions may reflect the history of wastewater management, which can also provide basic philosophies of handling human waste in accordance with local environmental laws. The traditional wastewater management of ancient Korea was based on decentralized scaled operations, oriented to resource management, eco-friendly, and the consumption of minimal volume of water. It implies the presence of widespread prevalence of social responsibility amongst individuals with the society. Some traditional practices can contribute to significant implications for the current engineering and science technologies; since it also provides effective ways of involving and educating individuals in wastewater management. Therefore, the wisdom and principles found from ancient Korean wastewater management practices, can be a good lesson to the improvement of wastewater management plans in future development of cities, which are threatened by both climate change and urbanization.

The current wastewater management have made a big contribution to improve public health since the 20th century. New paradigms for wastewater management are necessary to address the new threatening problems (e.g., monetary and energy dissipation). Early Koreans employed eco-friendly and sustainable wastewater managements for on-site treatment, water reuse, urines and faeces separation, pit latrine toilets, and composting. Guided by the wisdom of early Koreans, the 4R, 'reduce, reproduce, and reuse water, revitalization' can be achieved. Therefore, issues regarding water consumption and wastewater management practices can be solved through the adaptation of good ecological principles from the past.

14.8 REFERENCES

Ahn B. (2009). Study on the Holistic Value Analysis and Efficient Use of Korean Traditional Village Bangjuk [dike]. Doctoral thesis, Korea University.

Choi K. (1986). Korea Proverb Dictionary. Munhak publishing, Seoul, Korea.

Cosgrove J. and Rijsberman F. R. (2000). World Water Vision: Making Water Everybody's Business. Earthscan, UK. http://www.worldwatercouncil.org/fileadmin/wwc/Library/WWVision/TableOfContents.pdf

Gu J. (2009). Up-to-date significance of Korean conventional agriculture. *Korea Agricultural History Association*, **8**(3), 157–206.

Han M. Y., Kim J. Y., Kim M. K. and Woen J. Y. (2013). Integrated water management system at a planned city of Hwaseong built at AD 1796 in Joseon Dynasty, Korea, Int. Water History Association Conference Proceeding, China.

Kim D. (2006). The natural environment control system of Korea traditional architecture: comparison with Korean contemporary architecture. *Building and Environment*, **41**(12), 1905–1912.

Kim K. (2002). Latrines of the East Asia. Minsokwon, Seoul.

Langergraber G. and Muelleggera E. (2005). Ecological Sanitation – a way to solve global sanitation problems? *Environmental International*, **31**(3), 433–444.

Lee H. (2004). The Significance of Korean agriculture, its value and prospect. *Korea Agricultural History Association* **3**(2), 133–148.

Lee Y. and Kim C. (2003). A studies on removal of nutrient material by using dropwort field. *Korea Journal of Environmental Health Science*, **29**(3), 16–20.

Lim D. (2002). Proverb Dictionary. Minsokwon.

Reddy K. R. and Debusk W. F. (1985). Growth characteristics of aquatic macrophytes cultured in nutrient enriched water: II. Azolla, Duckweed, and Salvinia. *Economic Botany*, **39**(2), 200–208.

Shin S. (2006). Environmental design principles of traditional settlements reading ecological code. *Planning and Policy*, **298**, 120–128.

Suh B. (2012). A study on the toilet history of civilization in Korean building, the society of air-conditioning and refrigerating engineers of Korea. In: Proceeding of the Conference in Summer, pp. 293–296.

United Nations (2009). World Population Prospects: 2008 Revision, Economic and Social Affairs, United Nations. http://www.un.org/en/development/desa/population/

Wi W. (1992). Study on the Traditional Regional Theory. Doctoral thesis, Dankook Univresity.

Yu H. (2007). The Characteristics of the Traditional Korean Village through the Water Utilization. Master thesis, Sungkyunkwan University, http://blog.ohmynews.com/q9447/235984

Yun Y., Kang B., Kim M. and Rhee S. (2011). Modern application of traditional knowledge on Korean culture of eco toilet, 'Dwit Gan'. *Korea Agricultural History Association*, **10**(1), 48–72.

Chapter 15

Drainage and sewerage systems at ancient Athens, Hellas

E. D. Chiotis and L. E. Chioti

15.1 INTRODUCTION

Urban wastewater and stormwater technologies in Hellas have already been reviewed by Angelakis et al. (2005), with a brief reference to ancient Athens. The present study focuses on Athens in particular and relies extensively on published archaeological evidence extending from the Classical to the Late Roman period[1].

The city walls shown in Figure 15.1, adapted from Theocharaki's study (2011), are a good marker of the urban area occupied by the ancient city of Athens; the figure displays also a general plan of the terrain, the natural drainage network of streams and rivers and the mountains of Parnitha (Pa), Penteli (Pe), Hymettus (Hy) and Aegalaio (Ae) surrounding the Kifissos plain.

Modern urbanization has covered up most of the natural drainage network. Despite that, the streams at elevations above fifty meters can be reliably delineated on detailed topographic maps, especially if in addition maps of the 19th century are taken into account; this combination was actually applied for the compilation of the Figure 15.1.

Torrential rains cause significant natural disaster in the Mediterranean region and Hellas in particular. The evaluation of precipitation in Athens for a long period of time (1891–2000) has revealed that extreme daily precipitation of more than 10 mm occurs statistically in 3 to 24 days per year and more than 20 mm in 0 to 11 days (Paliatsos *et al.*, 2005). Thus, torrential rainfalls represent a significant fraction of the average annual precipitation of 400 mm in Athens and this necessitates an effective drainage system. Furthermore, local geology favours surface run-off of rainwater since the broader area is covered by schists and only about 5% of the precipitation penetrates the ground. Only the tops of the hills are covered by limestones which are good water reservoirs, of limited areal extent and capacity though.

[1] The following historical periods are often mentioned below: Classical (480–323 BC, from the first Persian war to the death of Alexander the Great), Hellenistic (323–146 BC, i.e., up to the control of Hellas by the Romans), Roman (146 BC–267 AD, i.e., up to the Herulians' invasion) and Late Roman period, roughly up to late 4th/early 5th century AD.

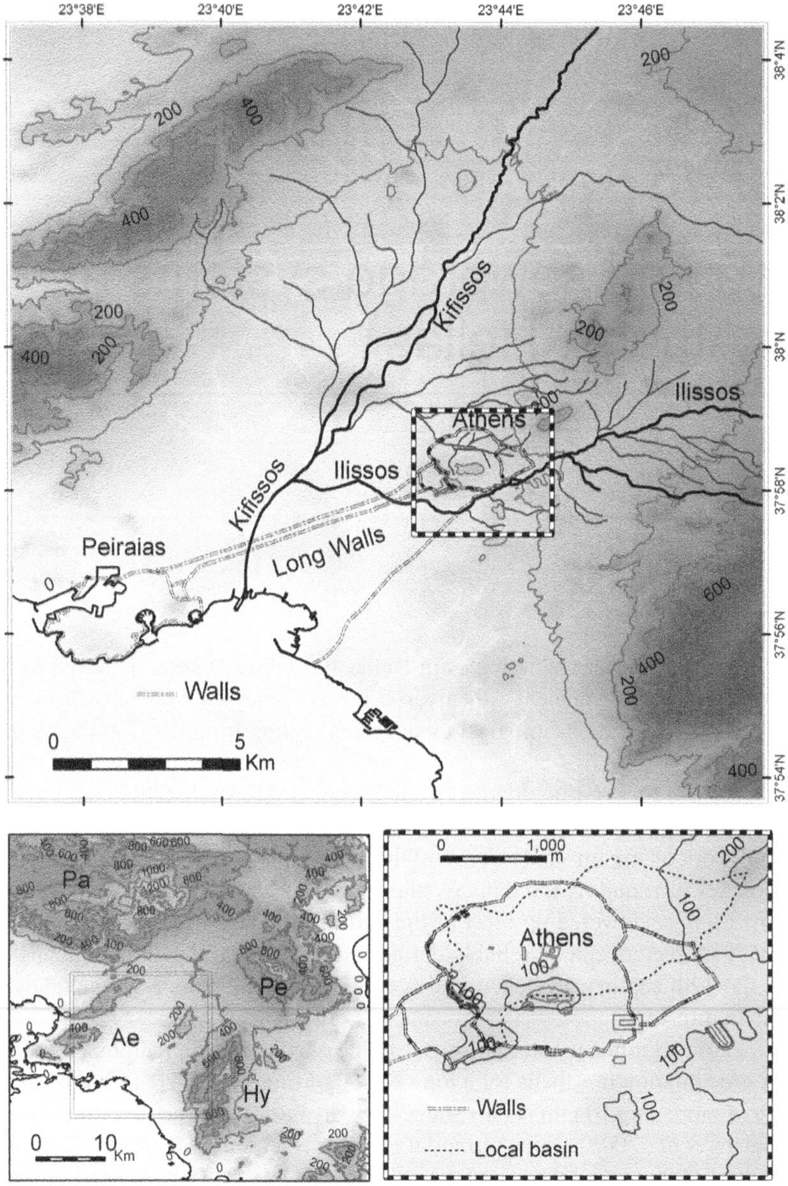

Figure 15.1 Rivers and streams surrounding the area of study (top); western Attica and the adjacent mountains (lower left); the area of study (lower right). Inset frames delineate magnified areas.

Two hydrologic domains can be distinguished in Athens, separated by the east–west trending watershed passing through the Acropolis hill: (a) the steeper southern slopes of the Acropolis hill and (b) the rather flat domain north of the Acropolis hill, which is drained by a network of streams, as shown in the Figure 15.2.

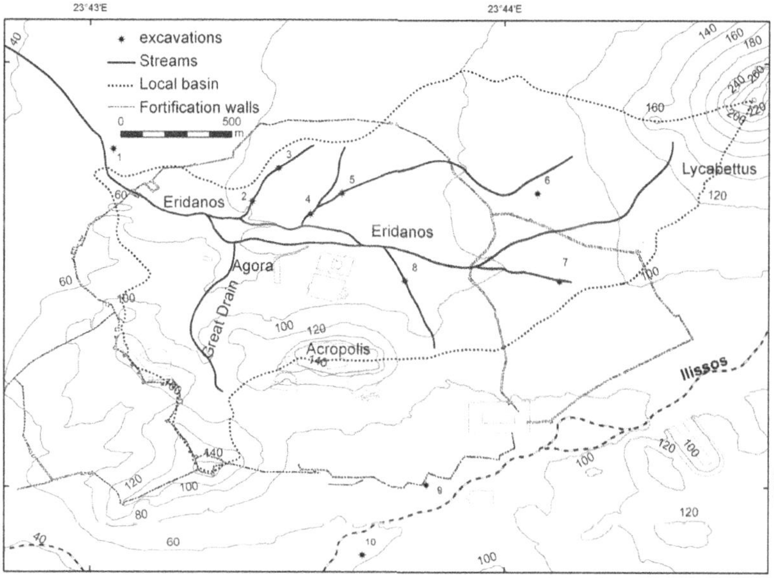

Figure 15.2 The network of streams of the local hydrologic basin of ancient Athens.

In the northern domain the streams were modified into artificial drainage channels. In the late 6th or early 5th century BC, the so called Great Drain (Figure 15.2) was the first stream transformed into an artificial drainage channel at the western side of the Agora, which was the political centre of the city. This made possible the establishment of public buildings in an area previously occupied by swamps. The Great Drain was built with heavy walls of polygonal-shaped stone blocks covered with stone plates to serve as a street. It is very well preserved and measures internally 1 × 1 m. It has served as the prototype of drainage channels that followed in the next centuries. As public buildings expanded in the Agora, additional stone-built channels were progressively constructed which discharged into the Great Drain, as described by Chiotis and Chioti (2012, Figure 16.22). The Great Drain continues to drain the Agora today. The images 2000.03.0008, 1997.04.0260 and 1997.04.0289 at http://www.agathe.gr of ASCSA (American School of the Classical Studies at Athens) are representative pictures of the Great Drain. In the southern domain Underground pipelines, usually along the streets, were used as sewers, due to the lack of natural runoff courses. Drainage control is aggravated by the steep dip of the terrain and the dense occupation of the area during the Classical-Roman times. In addition, the Ilissos valley was dotted with sanctuaries and public buildings; therefore restrictions were implemented to prevent pollution of the river and the adjacent sanctuaries from tanning, as implied from the inscription IG I³ 257 (Karouzos, 1923).

An additional major feature of the city's drainage system was the trench surrounding the circuit walls.

15.2 TYPES OF DRAINAGE AND SEWERAGE STRUCTURES: PIPELINES, CHANNELS AND GALLERIES

The same structures were used both for wastewater and rainwater at Athens during the studied period. Various types of terracotta pipes were used depending on the particular application (image/1997.09.0382 ASCSA). Special terracotta pipes (image/1997.04.0152 ASCSA) were used in the Peisistratean aqueduct late in the 6th century BC; they were tightly connected with lime mortar. A hole closed by a lid was available for the placement of the mortar and perhaps for cleaning the calcite deposits. The pipes were placed in trenches with the hole closer to the end of the previously laid pipe section (image/2004.01.1571, ASCSA; Lang, 1968).

Rectangular sections were commonly applied either to transfer waste from the buildings into larger pipes or to serve as gutters for the flow of rainwater along the streets. When placed in the ground they were covered with terracotta plates or tiles originally used for other applications, such as tiles for the lining of wells (image/2004.01.1091 ASCSA); occasionally, they were covered with pieces of U-shaped tiles. The U-shaped tiles were extensively used in couples for the construction of elliptical pipelines for drainage or water supply. The pipes were composed of two symmetrically placed parts 0.45 × 0.45 m, up to 5 cm thick (image/1997.09.0397 ASCSA); manholes were also constructed at intervals (image/1997.09.0393 ASCSA). U-shaped tiles were used in the 4th century BC in the Acharnian aqueduct which was designed for the water supply from the springs at the foothills of Parnitha Mountain to the city of Athens (Vanderpool, 1965).

Elliptical pipes laid in trenches became very common in underground drainage lines along the streets in the Hellenistic and the Roman times; amphorae were usually placed on top to reduce, as believed, the overburden. Burying amphorae layers in drainage ditches was a technique applied also in the 1st century BC in Rome in land improvement projects (Wilson, 2000b, p. 316). It is therefore suggested that amphorae, commonly placed in Athens on top of U-shaped pipe lines in trenches, would also contribute to the drainage of rainwater from the surface of the streets, as the voids between the amphorae provide a horizontal way of flow of easier circulation.

Underground channels built of bricks were developed in Roman times, particularly in great buildings, like the Hadrian's Library, to drain the rainwater from large open spaces (Tigginaga, 2008). Channels made of cast concrete were also introduced in the Roman times.

In nearly flat areas or in swamps drainage was accomplished by means of dipping underground galleries carved in soft rock; a section of a 33 m long drainage gallery was studied by Axioti (2009) in an excavation at the centre of Piraeus, about 100 meters NW of the ancient theatre. The shallow gallery, trending NW-SE, was uncovered beneath the axis of a street heading towards the harbour of Zea; it was dug in the soft rock, 0.55 m wide and 1.10 m high. It is emphasized that drainage galleries were conveniently accommodated in the rectangular Hippodamean urban plan of Piraeus. Rainwater was collected in the galleries through a system of shafts and trenches. The cisterns which accumulated and stored the rainwater from the roofs of the adjacent houses were dug in the yards, the location of which was generally anticipated in the city-plan (Axioti, 2009).

A similar drainage gallery, 1.70 m high and 0.50 m wide, was excavated over a length of 65 m at the Kerameikos cemetery, beneath a mass burial, located in the middle of a swampy area (Baziotopoulou-Valavani and Tsirigoti-Drakotou, 2000).

It is summarized that the commonest principal structures of drainage and sewerage can be classified into three main types: (a) pipelines, most commonly elliptical, buried in trenches along streets (b) stone masonry channels along streams or streets and (c) unlined galleries hewn in soft bedrock. The close connection of the drainage networks with the streets is emphasized.

As Costaki (2006) notes, water supply and drainage systems installed beneath the streets was rather elaborate, ranging from great drains and aqueducts in public areas to drain pipes piercing walls of individual houses and emptying into street drains.

Water pipes and sewers in parallel lines along streets are described below at the excavation for the Acropolis Museum (Section 15.3.2.2). A unique case of drainage and water flow in opposite directions is described by Camp (2007) along the north-south street west of the Painted Stoa, at the Agora's NW corner, where water in pressure pipes was directed uphill and waste in elliptical tile pipes southwards emptying into the so called Eridanos River by gravity flow.

The function of roads not only as conduits for people, goods, and wheeled traffic but also for water and waste management is perfectly demonstrated at a Roman road at ancient Corinth recently studied by Palinkas and Herbst (2011). It is impressive that the road in the Panayia Field of the Roman forum at Corinth 'contains 31 successive lines of pipes and drains that represent solutions to water supply and sewage management within the city' over a period of six centuries.

15.3 DRAINAGE OF REPRESENTATIVE AREAS

Drainage structures are described below in two representative areas of ancient Athens, that is, the local hydrologic basin and the southern slopes of the Acropolis hill.

15.3.1 Drainage in the hydrologic basin north of the Acropolis hill

Archaeological excavations reveal the function of streams in the drainage of the buildings and confirm at the same time the geometry of the network of streams as drawn in the Figure 15.2.

Excavations at the site No 5 (Figure 15.2) uncovered a sanctuary of Zeus and Athena along an ancient road passing over a channel (Kyparisses & Thompson, 1938). The walls of the channel were made of large blocks of limestone and conglomerate and were spanned by a carefully constructed flat vault of bricks. The top of the vault was approximately at the level of the road. The walls of the channel are dated to the Hellenistic and the brickwork to the Roman period. The channel is located along a stream (Figure 15.2, site No 5) and the vault served actually as a bridge. The excavators noted that the marked topographic irregularity, of five meters difference in elevation across the street would certainly not have been suspected from the present level surface of the region; it is expected though between the bank and the bottom of a stream. Similar bridges over streams were also uncovered across the so called Eridanos River at the Sacred Gate, NW of the Agora and at the Monastiraki square at the recent excavation for the Metro station.

A similar drain channel was excavated at the location No 4 (Alexandri, 1967a); the walls, made again of large stone blocks, are spanned by a vaulted roof of bricks; the structure is dated to the times of the Roman Emperor Hadrian (117–138 AD).

At the site No 2, an elliptical tile pipeline set below a Roman street drains towards the adjacent stream (C´ Archaeol. District, 1964); the roof of the pipe is covered with amphorae to reduce the overburden. At another excavation (Alexandri, 1973), some fifty meters N-NE from the site No 2, another channel built of brick-walls and floor of square terracotta plates conveyed rain water and domestic waste to the same stream. Furthermore, a channel of monumental size was uncovered some 50 meters NW from the site No 2 next to hypocausts for baths. It is enclosed in a rectangular structure made of Roman concrete (opus caementitium) with external dimensions 2.80 × 2.45 m (Alexandri, 1975). Waste was emptied directly into the same stream by rectangular tile pipes at the site No 3 (Stavropoullos, 1965a).

Streams of the local hydrologic basin of Athens were also excavated at the sites No 6 and No 7. It is suggested that archaeological excavations with fluvial or torrential sediments should not be correlated in space as points arranged linearly, but through the dendrogramme of the Figure 15.2.

Another interesting drainage pattern was excavated some fifty meters west of the site No 4 (Alexandri, 1967b), where the local drainage lines are arranged along the streets. This solution seems to be preferable in areas closer to a street than a stream. The channels, enclosed between the retaining walls of the streets, are dated in the beginning of the 4th century BC. They collected the waste directly from the houses by means of terra cotta pipes placed below the floor of the buildings. Houses were being erected here over a long period, from the beginning of the 4th century BC till the Late Roman times and the drainage structure was maintained in use over this period.

The artificial drainage system interlaced with the network of streams was developed continuously through the Classical to the late Roman times with remarkable technical improvements, such as the construction of bridges with vaulted roofs and drainage channels of cast concrete in the Roman and Late Roman times.

The drainage works described in this section can be classified into categories similar to those suggested by Wilson (p. 152, 2000a). The first order drainage from the buildings is connected to the second order drain along streets which in turn empties into the third order drain of the natural streams. Finally, the Eridanos River collects all streams and represents the fourth order drainage. Naturally, buildings close to a stream drain directly to it.

15.3.2 Drainage at the southern slopes of the Acropolis hill

Drainage and sewerage will be commented in two extensive excavated areas south of the Acropolis: the area 'A', south of the Roman Odeion of Herodes Atticus and the area 'B', south of the Theatre of Dionysos (Figure 15.3).

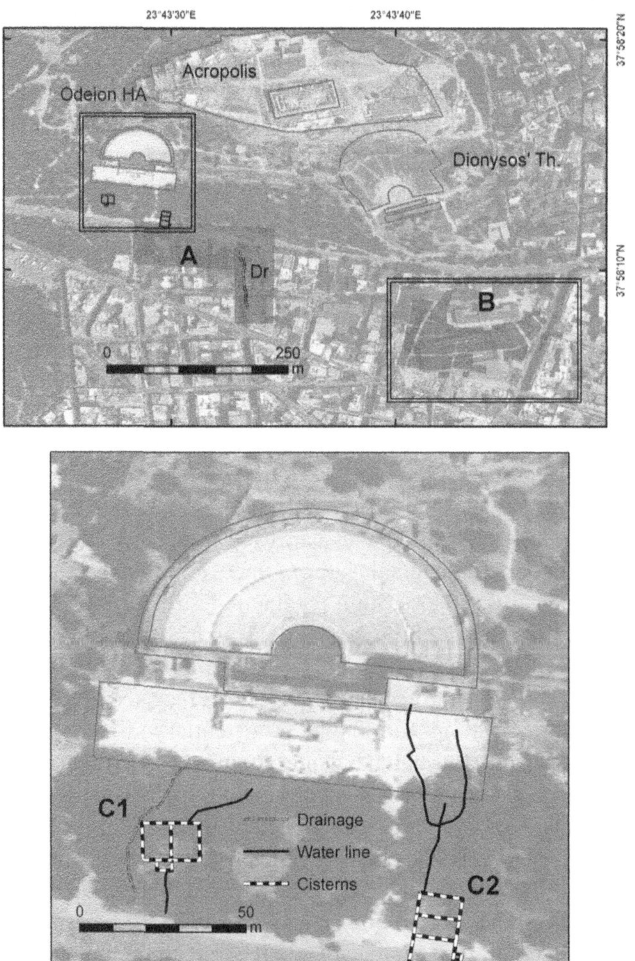

Figure 15.3 General plan of the southern slopes of the Acropolis hill projected on KTIMATOLOGIO air photo, where the areas A and B are delineated (top); the cisterns C1 and C2 south of the Odeion of Herodes Atticus are also shown (bottom).

15.3.2.1 Area south of the Odeion of Herodes Atticus

The area south of the Odeion of Herodes Atticus (area A in Figure 15.3) is of special historical importance due to the almost continuous occupation in the last millennia and the wealthy residences during the Roman period (Dontas, 1962a, 1962b). New buildings were erected on the ruins of older ones and the remains are therefore rather fragmentary; however the drainage system (Dr), shown in Figure 15.3 (top), has been preserved in good condition, as it was constructed in the ground.

The oldest recognizable drainage network is dated to the middle of the 2nd century BC. It was composed of a main gallery hewn in the schists with lateral branches and manholes. The gallery was later opened and terracotta pipes were used instead, a measure which would protect ground water from pollution.

Elliptical tile pipes were used in the Roman period and the drainage network was extended with more lateral branches and tile-lined manholes; baths and a private latrine are dated in this period. Majestic baths were built around 400 AD. The route of the main drainage (Dr) is shown in the Figure 15.3 (top); the pipeline was placed in a trench in schists and extended over a length of 80 meters at least.

This drainage pattern became the standard practice in the Roman times. A central sewer dipping downhill was made of couples of U-shaped tiles and collected the waste from lateral sub-horizontal branches. Tile-lined manholes provided access for cleaning of the central drainage line.

The central line in the area A was placed below the foundation of the buildings with a slope about 10%. The drainage lines are in general arranged along streets where the terrain is gentler, dipping not more than 5%, as in the area B uncovered in the excavation for the foundation of the New Acropolis Museum.

The well preserved Roman cisterns C1 and C2 were essential for the water supply of the area at elevations above the 84 meters, since the aqueducts could not reach so high by gravity flow, as explained below. Drainage of the Odeion of Herodes Atticus was performed through a tile pipeline, the route of which is partly shown west of the cistern C1. The routes of the pipes supplying water to the cistern, adapted from Brouskari (2002), are also shown in the Figure 15.3.

15.3.2.2 Excavation for the New Acropolis Museum and the Metro Station

The excavation for the New Acropolis Museum and the adjacent Metro station (area B in the Figure 15.3) covered an area of 10,000 sq.m. and disclosed a complete part of the ancient city; continuity in the habitation of the area extended from the Late Neolithic to the Middle Byzantine times (Eleutheratou 2008).

The area emerged in the classical times as a thriving quarter of houses with yards, wells and cisterns. Urban partitioning of the area was already shaped in the last decades of the 5th century BC and was only marginally changed up to the 3rd century AD. The character of this densely occupied district changed from residential to manufacturing after Sulla's sack of Athens in 86 BC, when pottery, marble and steel workshops were developed. New and prosperous dwellings with private baths and lavatories were built in the 2nd and 3rd century AD, to suffer a new destruction during the Herulian invasion in 267 AD. Two baths existed in the Roman period and a third was added in the Late Roman times

The drainage network during the 2nd and 3rd centuries AD is shown in the Figure 15.4, adapted and simplified from Eleutheratou (2008). The triangular sector delimited between the main streets was occupied by houses with wells for drinking water in the yards and lavatories arranged close to the main street. A drainage network made of elliptical terracotta pipes existed along the streets which collected both rainwater and domestic waste, whereas the waste of the lavatories was accumulated in pits along the streets.

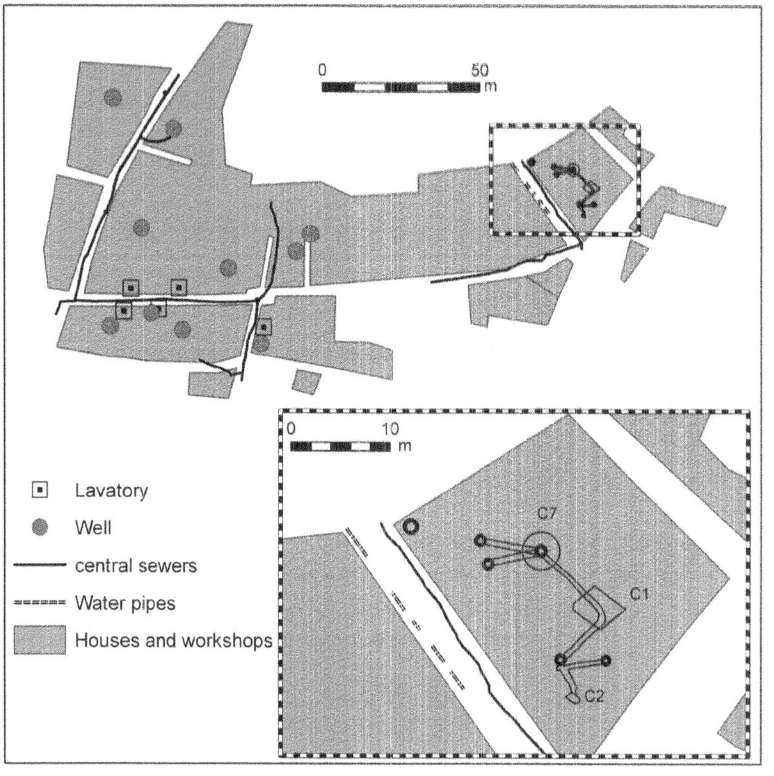

Figure 15.4 Hydraulic features of the excavation for the New Acropolis Museum, adapted from Eleutheratou (2008): the excavation area, 'B', is delineated in the Figure 15.3 (top).

Water pipes are also shown in the above figure which are dated to the second half of the 5th century BC and are considered to belong to an extension of the older Peisistratean aqueduct (Eleutheratou, 2006).

15.4 UNDERGROUND STRUCTURES OF DOUBTFUL INTERPRETATION

Underground works for aqueducts and sewers are often of similar construction and as a result they are occasionally misinterpreted. Some of them deserve further investigation and are therefore summarized below.

A case of misinterpretation is believed to be that at the site No 8 in the Figure 15.2. A surface channel hewn in the bedrock has been interpreted as the Hadrianic Aqueduct (Stavropoullos 1965b). The channel was spanned with marble plates; there are no built or lined walls and the site of the excavation falls along a local stream of the basin of Athens; it should therefore be reinterpreted as a typical drain channel.

A significant structure of disputed interpretation is the gallery excavated in the National Garden close to the Valerian Wall; it was built in a shallow trench in the Athenian Schists. Its outline measured 2.05 m wide and 1–1.50 m high, the walls were 0.55 to 0.60 m thick, the roof was vaulted and the floor was paved with stones; it was inclined towards the city and was ascribed to the Hadrianic Aqueduct (Spathari & Chatzioti, 1983). Another section of the same gallery was excavated in the extension of the previous one and was interpreted as a central sewer (Zachariadou, 2000). The total excavated length is about seventy meters and this indicates for sure a significant work.

A gallery, almost identical in structure to the previous one, dated most likely in the 3rd century AD, was excavated near the Ilissos River (No 9 in Figure 15.2); it was built of stones and mortar with a vaulted roof and measured 2 m wide and 1.5 m high, including the walls, which were 0.45 m thick; it was founded on solid rock and its roof was 1.3 m below the surface (Alexandri, 1968; Chatzipouliou, 1988). The gallery (G) was situated near the city walls as shown in the Figure 15.5. It is suggested that this type of vaulted gallery (No 9) of heavy and waterproof construction does not correspond to any of the types of sewers previously reviewed in Section 15.2. It was described by the excavators as a conduit and it is suggested by the authors that the possibility of connection with a sally port of the circuit wall cannot be excluded.

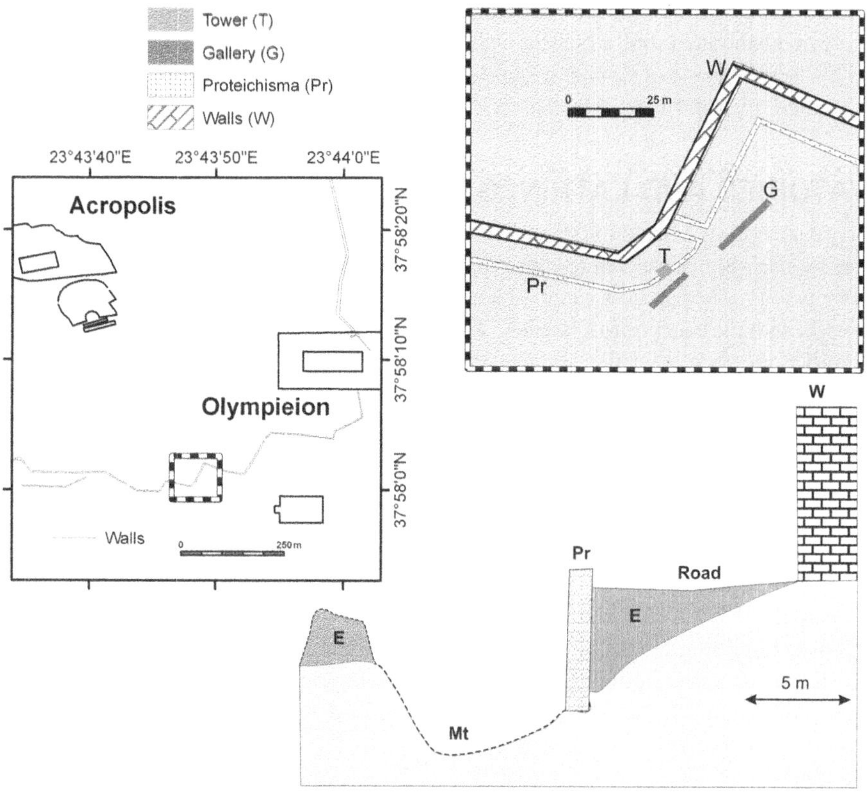

Figure 15.5 The fortification walls near the Olympieion (left), location map of a gallery sections G (top right) and schematic section through the wall (W), the ring road on earth pile (E), the proteichisma (Pr) and the moat trench (Mt) (bottom).

The excavations provided also good evidence on the walls at this site, which is summarized in the following in order to illuminate the function of the moat as a peripheral feature in the drainage of the city. The fortification system consisted of the circuit wall (W), the first defensive line or 'proteichisma' (Pr), the trench of the water-filled moat (Mt), a ring street between the circuit wall and the proteichisma partially constructed on earth pile (E) (Figure 15.5). It is noted that in the 1st century AD, the ancient fortification moat was filled in with earth and the area was used for burials or constructions, such as channels, buildings and cisterns (Alexandri, 1967c; Theocharaki, 2011).

A gallery carved in soft rock, 0.6 m wide, 1.3 m high and about 10 m long, was excavated at the site No 10 (Figure 15.2). The site is projected on the Map of Kaupert (1878) close to a 'Felsrinne', a gallery hewn in the rock. Furthermore, limestone lenses either intercalated or thrust over the schists occur in this area according to the geological map; these data combined suggest that the particular gallery might be a work for water capturing.

Another excavation, No 1 in the Figure 15.2, with significant underground works is described along the Demosion Sema, the street connecting Athens with the Academy (Karagiorga 1978); a pipeline of elliptical tiles 5.60 m long and a manhole have been brought to light 5.60 m below the surface. The tiles were placed in a trench hewn in the schists in the Late Hellenistic times. They were covered with amphorae in two layers, according to the common practice for overburden reduction. In the 2nd century AD, the street seems to be abandoned and a 'sewer' was constructed at the same place, founded 3.40 m below the surface. This 'sewer' made of Roman concrete measured 4.90 m wide, the walls were 0.85 m thick and the roof was vaulted. Evidently, the dimensions of this structure correspond to a cistern.

15.5 LAVATORIES AND LATRINES

Lavatories in private houses existed in Athens since the 5th century BC. They were located inside the house close to the street and were connected to pits along the street below the ground level, where the waste was accumulated by gravity. The pits were emptied periodically and only household sewage was disposed in the drainage network, with the exception of latrines. The characteristics of lavatories in antiquity in the Hellenic territories as well as the ancient written sources on the subject are extensively reviewed by Antoniou (2010).

Thompson (1959) excavated a group of private houses on the gently sloping ground between the northern foot of the Areopagus and the east-to-west roadway that bordered the south side of the Agora. The houses were presumably established in the second quarter of the 5th century BC. As Thompson notes, a feature that would seem to be characteristic of these houses is a rectangular stone-lined pit set down below ground level, along the street; the pits were considered most probably as cesspits which were connected by means of a short length of drain with lavatories immediately inside the house (image/1997.18.0089 ASCSA).

A similar pattern of lavatories and pits along the streets is also suggested by Steinhauer (2009) in Piraeus, in Classical building blocks of the Hippodamean urban design.

The most illuminating description of lavatories in private houses is due to the excavation at the New Acropolis Museum (Figure 15.4). The lavatories measured a few square meters in plan and their pit was about one meter deep and about half a sq. m. in plan. They were placed close to the main streets (Eleutheratou 2008).

Sanitation rules were adapted, the application of which was supervised by pertinent officers. As noted by Thompson (1959), a decree of 320/19 BC appears to prohibit cesspits in the streets of the Piraeus (IG II^{2}380, lines 34–40). The waste of the lavatories was removed by the so called Koprologoi under the command of the Astynomoi, five for Athens and five for Piraeus (Steinhauer, 2009).

Owens (1983) points out that although the collection of garbage was supervised by the state, a large part of the operation may have been in the hands of private entrepreneurs, the Koprologoi, who were able to turn a profit first by collecting waste material and then by recycling and reselling it as fertilizer. In addition, Owens stresses that it was the duty of the Astynomoi to ensure that the Koprologoi deposited the kopros at least 10 stades from the city. Further discussion on the function of Koprologoi is given by Papadopoulos (2006).

As Owens concludes 'at Athens in the fifth and fourth centuries BC there existed a well-organised municipal service for the collection and disposal of sewage and waste. Citizens were forbidden to foul the streets and were obliged to empty waste and sewage into statutory dumps, which were periodically cleaned by the so called Koprologoi. The Koprologoi carried out a private service and collected the waste and then dump it beyond the permitted minimum distance from the city'.

It seems that the same approach was practiced also in other Hellenic cities. Ault (1999) reexamined sizable stone-lined pits, sunk into the surface of the courtyard, in a number of houses of the 4th century BC excavated at Halieis in the southern Argolid. They were originally considered as cellars, but Ault reinterpreted them as cesspits. Furthermore, Ault accepted Owens's views on the organization of urban refuse disposal in Classical Athens and suggested that cesspits at Halieis were used as a source of fertilizers for the cultivation of olive trees.

Only two public latrines of the Roman period are known in Athens, the larger one close to the Roman market (Orlandos, 1940) and the other one in the NW part of the Agora.

The latrine in the Agora was recognized close to the NW corner of the Poikile Stoa (Shear, 1997); the channel under the latrine was flushed out with water at the NE corner and the waste flowed below the Panathenaic Way into the Eridanos River. The site of the latrine was occupied later in the Roman times by part of an enormous bathing establishment. A slab of Pentelic marble, preserving the characteristic keyhole-shaped opening of a communal latrine seat block, found in recent excavations of the ASCSA, should be associated with the above mentioned latrine (Camp, 2007).

15.6 POSSIBLE SANITATION PROBLEMS

Shortcomings of Athens' public sanitation under extreme conditions are examined below to investigate potential drawbacks when normal life is disorganized, as in the case of war. The excavation of the New Acropolis Museum is taken as an urban model, since this quarter of the city is adequately known.

15.6.1 Sanitation and water supply at the excavation for the New Acropolis Museum

The regular water supply from aqueducts could easily be interrupted by enemies, since the water resources were beyond the city walls and the underground works of the aqueducts were accessible through noticeable wells or manholes; perhaps the Peisistratean aqueduct was less obvious. Thus, Athenians under siege, when enemies ravaged the countryside, would have to rely largely on the few springs around Acropolis, on wells and cisterns.

According to Eleutheratou (2006) the fact that the number of wells has increased significantly since the 2nd century AD is interpreted as an indication that even the aqueduct donated to the city of Athens by the Emperor Hadrian could not supply sufficient water. This is a reasonable assumption for the area of the New Acropolis Museum, although the Hadrianic aqueduct or any other aqueduct approaching Acropolis from the east could not supply water by gravity flow at elevations higher than 83.9 m because of topographic limitations. Both the Hadrianic and the Peisistratean aqueducts should pass through a narrow saddle at the elevation of 83.9 m in order to reach the slopes of Acropolis from east to west (Figure 15.6). Therefore, a waterbridge or a siphon would be required for the water to reach at higher elevations; there is, however, no such evidence in the area. Thus, the springs around Acropolis, groundwater from wells and the rainwater stored in cisterns were the only sources of water at the higher garland surrounding Acropolis. An indirect indication for that is the presence of the two spacious Roman cisterns near the Odeion of Herodes Atticus (Figure 15.3, top).

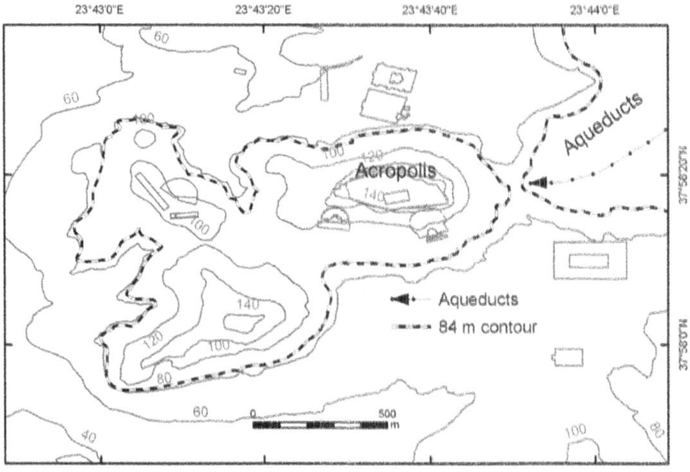

Figure 15.6 Potential route of aqueducts approaching the hill of Acropolis and the contour of 84 m.

A stone built channel (Figure 15.7) constructed for the drainage of the Dionysos Theatre was revealed in excavations (Dörpfeld und Reisch, 1896); wasting this source of water would be extravagant for the water shortage of this particular area, but no associated cistern is known nearby. The pipelines described by Dörpfeld und Reisch at the theatre of Dionysos were uncovered at elevations higher than 84 m and cannot therefore be related to the Peisistratean aqueduct, as suggested by them.

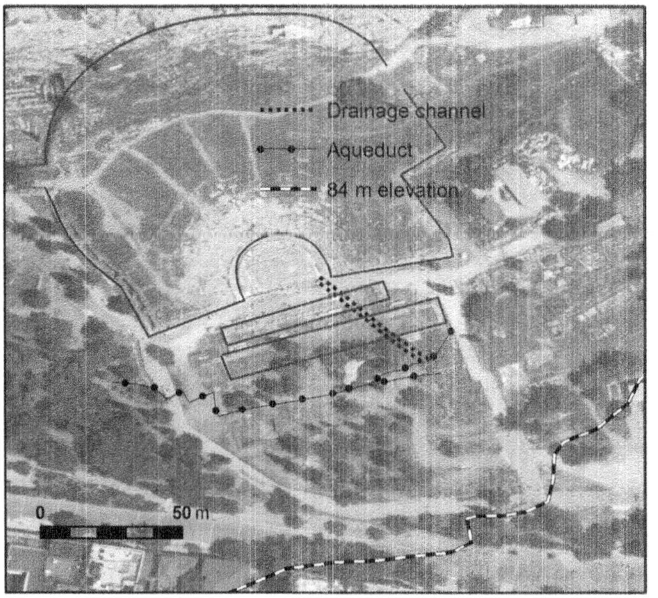

Figure 15.7 The area of the theatre of Dionysos projected on the Ktimatologio air photos, the drainage channel and the aqueduct after Dörpfeld and Reisch (1896).

The steep limestone hill of Acropolis is surrounded by a flyschoid formation, composed of schists with intercalations of sandstones and limestones. In the absence of better water reservoirs, this formation was the only local source of underground water and was therefore intensively exploited by means of wells and galleries. The wells were normally lined with especially designed tiles and this should reduce the pollution near the surface of the ground; however, the water bearing sections at the bottom of the wells would be unlined, exposed to potential pollution.

The proximity of sewers to water supply systems aggravates the danger of pollution in some densely populated areas of the city, like the one excavated for the New Acropolis Museum. Thus, 37 wells ranging in depth from 5.2 to 26.4 m and seven cisterns were uncovered at the excavation of the Acropolis Metro station within an area of 2500 sq. m. (Eleutheratou, 1997). Based on these figures, the average distance between adjacent wells is estimated to 16 m.

No doubt, the sanitation conditions of water supply network through tile pipes along the street close to the network of sewers would be subject to pollution (Figure 15.4). Not only because of the proximity of the sewer at the other side of the street, but also due to the outlets of the domestic waste pipes which pass on top of the water pipeline. In addition, pollution would be possible through the elliptical opening of the water pipe sections (Marinakis, 2000). It is noted that this pipeline (Figure 15.4), found below 84 m, is considered as a branch of the Peisistratean aqueduct and is dated to the second half of the 5th century BC (Eleutheratou, 2006), which means that it might have been in operation during the Plague of the Peloponnesian War. Cisterns were lined with waterproof mortar and would be therefore less vulnerable to pollution: the water of the cisterns would normally be reserved for domestic use.

A rather sophisticated water system in the same area functioned during the Late Roman–Early Christian period (Kalligas, 2000)[2]; it is a combination of old cisterns and wells with the addition of new cisterns (inset map of Figure 15.4). The system was composed of the cisterns C1, C2 and C7 which were connected through galleries and wells lined with waterproof plaster. The cistern C1, dated in Late Roman–Early Christian period, was built of tiles bound with mortar (Eleutheratou, 1997). It is assumed by the excavators that this system collected rain water; it is envisaged though that it should not be excluded that the water could be supplied from a branch of the Late Roman aqueduct (Chiotis & Chioti, 2011).

As mentioned earlier, lavatories were close to wells and constituted, therefore, an additional source of pollution. The waste would be normally disposed outside the city; if however the drainage lines were used instead of the sewers, pollution might have caused serious diseases. Furthermore, the statutory dumps of waste mentioned above by Owens would be a permanent source of pollution, also if the law was upheld. The experience from the beginning of the 20th century is mentioned for comparison: the use of sewage for the irrigation of cultivated fields from the stream of Prophet Daniel in Athens, in the Kifissos plain, has caused many diseases including typhus, dysentery, amoebae and other intestinal diseases (EYDAP at http://www.eydap.gr/index.asp?a_id=164).

A similar hazard might be likely in the ancient Piraeus in case of siege and civil disorder; it seems that city planning by Hippodamus around 460 BC accommodated a net of galleries for rainstorm drainage. If however lavatories were emptied into these galleries due to negligence of the Koprologoi, this could cause pollution of the wells.

15.6.2 The Plague during the Peloponnesian war

An impressive mass burial was excavated recently at the Kerameikos cemetery (Baziotopoulou-Valavani & Tsirigoti-Drakotou, 2000; Baziotopoulou-Valavani, 2002), dated approximately in 430 BC. It was

[2] According to the excavator this period includes the 4th, 5th and 6th century AD (Kalligas, p. 37, 2000).

therefore associated with the spread of the Plague in Athens during the first years of the Peloponnesian War (Baziotopoulou-Valavani, 2002).

The cause of the Plague is a subject of longstanding controversy (Morens & Littman, 1992). Based on molecular evidence, resulting from investigation and analysis of ancient DNA, typhoid fever was considered as a likely cause of the epidemic (Papagrigorakis *et al.*, 2006); this conclusion, however, has been disputed (Shapiro *et al.*, 2006).

The Peloponnesian War between the Athenians and Peloponnesians and their allies started in 431 BC when the Spartans invaded Attica and ended in 404 BC with the victory of the Spartans. The historian Thucydides in his History of the Peloponnesian War described also the Plague of Athens, a terrible epidemic which occurred first in 430 BC. Thucydides himself suffered the illness and survived.

As described by Thucydides in the second Book (2.52), 'An aggravation of the existing calamity was the influx from the country into the city, and this was especially felt by the new arrivals. As there were no houses to receive them, they had to be lodged at the hot season of the year in stifling cabins, where the mortality raged without restraint. The bodies of dying men lay one upon another and half-dead creatures reeled about the streets and gathered round all the fountains in their longing for water. The sacred places also in which they had quartered themselves were full of corpses of persons that had died there, just as they were; for as the disaster passed all bounds, men, not knowing what was to become of them, became utterly careless of everything, whether sacred or profane'. Under such conditions of disintegration 'Fear of gods or law of man there was none to restrain them (2.53)' (translation by Crawley, 1950). Refugees in Piraeus in particular occupied the fortified space between the walls (Conwell, 2008).

Under these conditions, sanitation problems in water supply and sewage disposal at Athens would be inescapable. This would obviously facilitate the dispersion of the epidemic and cause intestinal diseases even to those not infected by the epidemic itself.

15.7 CONCLUSIONS

A well organised infrastructure for water supply, rain water drainage and sewers, as well as a municipal service for the collection and disposal of lavatory waste existed in Athens since the 5th century BC. Drainage and sewerage works were adapted to the topography, the natural network of streams and rivers, the streets and the moat trench of the fortification.

The local drainage basin of ancient Athens is delineated on modern maps and the natural streams are traced and confirmed on the basis of archaeological evidence. It is also documented that these streams were gradually developed into the backbone of an artificial drainage network. Rainwater drainage was of great concern due to the climatic peculiarity of sudden rainstorms which can flood flat areas at once.

The so called Great Drain in the Athenian Agora was the first torrent transformed into a stone built channel in the early 5th century BC, designed mainly to drain rainwater in the flat area of the Agora, the political centre of Athens; the channel was covered with stone plates and served as a street too. More or less, all streams of the flat terrain north of the Acropolis within the city walls were gradually incorporated into an integrated drainage network.

Houses and workshops were drained through tile pipes which transferred the sewage into bigger U-shaped pipelines along the streets or directly into stone-built channels arranged along streams. Alternatively, drainage galleries were excavated at shallow depth in the soft basement rock to drain flat or swampy areas. The development of the drainage network all over the city and the application of rather standardized techniques are noted.

The drainage pattern of the buildings at the rather steep slopes south of the Acropolis hill was based on artificial drainage lines, either below the foundation and the retaining walls or along the streets.

Since the 5th century BC collection and disposal of waste and sewage was well organized in Athens; lavatories built close to the streets were established as a standard feature of the houses. In addition, municipal officers supervised the fulfilment of sanitation rules and restrictions.

However, the proximity of water wells, lavatories and sewers could cause water pollution if the sanitation rules were neglected, as in the case of war; the outbreak of the Plague in the first years of the Peloponnesian war would further aggravate the situation. Therefore under these conditions water pollution and consequent diseases should be considered as a very likely eventuality. Since, 'it is conceivable that the Plague was a combination of diseases' (Salway & Dell, 1955), it is suggested that symptoms associated with water pollution should be differentiated from those of the Plague itself.

City-planning of Piraeus by Hippodamos provided the space both for the accommodation of rainwater drainage galleries along the streets and for underground cisterns at the yards of the houses.

By comparison of the drainage and sewerage techniques at ancient Athens to the elaborated prehistoric drainage systems in Mesopotamia, the Indus Valley and Knossos, a couple of millennia BC, Wilson's conclusion (2000a, p. 178) is justified, that 'the individual elements of drainage systems appear relatively early and remain largely unmodified until the Middle Ages.

The remarkable point with the well organized infrastructure of drainage and sewerage of Athens is, in the authors' opinion, the social structure which supported the regular operation of the system. The 'cheaply built' houses of Athens in the Hellenistic times, in the opinion of the traveller Heraclides Creticus, made use of the same sanitation infrastructure with the 'few houses of higher standard'. In addition, all citizens enjoyed 'the most beautiful sites on earth: a large and impressive theatre, a magnificent temple of Athena, something out of this world and worth seeing' (Austin, 1981, p. 151).

Acknowledgments

Without the publications on the archaeological excavations in Athens over the last decades this compilation would not be possible and the contribution of the excavators is thankfully appreciated. The authors wish also to express their gratitude to the Editor Prof. Andreas Angelakis and to the reviewers of this chapter Prof. Benoit Haut, Université Libre de Bruxelles, Belgium and Prof. Jens Krasilnikoff, Aarhus University, Aarhus, Denmark, for the improvements they suggested. Special thanks are also addressed to Jens Krasilnikoff for the recommendation of additional sources of literature thanks to which the subject was placed into a broader context as well as to the IWA Editor Maggie Smith for the constructive handling on the Contributors Contract. Discussions with James A. Herbst and Bruce Hartzler, both of the American School of Classical Studies at Athens, were very enlightening and are acknowledged with appreciation.

The documentation of this Chapter incorporated data from the website of the American School of Classical Studies at Athens (www.agathe.gr), a rich, open and reliable data base of the excavations at the Agora of Athens supported by the Packard Humanities Institute. The air photos of Hellas, available online by KTIMATOLOGIO S.A., were of great assistance, especially for the compilation of accurate maps by means of GIS applications. The ever increasing possibility of downloading references online from open sources was a great convenience, but our final shelters were the friendly Libraries of the American and the British Archaeological Schools at Athens.

15.8 REFERENCES

Alexandri O. (1967a). Miaouli 8. *Archaiologikon Deltion, Chronika*, **22**, 100–102 (in Hellenic).

Alexandri O. (1967b) Ag. Theklas 10–12. *Archaiologikon Deltion, Chronika*, **22**, 39–43 (in Hellenic).

Alexandri O. (1967c). Voulis and Petraki. *Archaiologikon Deltion, Chronika*, **22**, 66–70 (in Hellenic).

Alexandri O. (1968). Koryzi 6. *Archaiologikon Deltion, Chronika*, **23**, Fig. 16 and p. 67 (in Hellenic).

Alexandri O. (1973). Agatharchou Str. and Lepeniotou. *Archaiologikon Deltion, Chronika*, **28**, 25–26 (in Hellenic).

Alexandri O. (1975). Lepeniotou 5–7. *Archaiologikon Deltion, Chronika*, **30**, 24–25 (in Hellenic).

Angelakis A. N., Koutsoyiannis D. and Tchobanoglous G. (2005). Urban wastewater and stormwater technologies in ancient Greece. *Water Research*, **39**, 210–220.

Antoniou G. P. (2010). Ancient Greek lavatories: operation with reused water. In: Ancient Water Technologies, L. W. Mays (ed.), Springer, Heidelberg, London, New York, 67–86.

ASCSA (American School of Classical Studies at Athens) Excavations in the Athenian Agora. http://www.agathe.gr/ (accessed May 5th, 2014)

Ault B. A. (1999). Koprones and Oil Presses at Halieis: interactions of town and country and the integration of domestic and regional economies. *Hesperia*, **68**, 549–573.

Austin M. M. (1981). The Hellenistic world from Alexander to the Roman Conquest: a Selection of Ancient Sources in Translation. Cambridge University Press, Cambridge, New York .

Axioti K. (2009). Excavation in a building plot in Terpsitheas Square at Piraeus. In: From Messogeia to Argosaronikos, V. Vassilopoulou and S. Katsarou-Tzeveleki (eds), Municipality of Markopoulo of Messogeia, Athens, pp. 489–495.

Baziotopoulou-Valavani E. (2002). A mass burial from the cemetery of Kerameikos. In: Excavating Classical Culture, M. Stamatopolou and M. Yeroulanou (eds), The Beazley Archive and Archaeopress, Oxford, pp. 187–201.

Baziotopoulou-Valavani E. and Tsirigoti-Drakotou I. (2000). Kerameikos station. In: The City under the City, L. Parlama and N. Stampolidis (eds), Museum of Cycladic Art, Athens, pp. 264–275.

Brouskari M. S. (2002). The excavations south of the Acropolis: the sculptures. *Archaolaiologikon Ephemeris*, **141**, 1–204 (in Hellenic with extened summary in English).

C´ Archaeol. District (1964). *Archaiologikon Deltion, Chronika*, **19**, 50–52 (in Hellenic).

Camp II J. McK. (2007). Excavations in the Athenian Agora 2002–2007. *Hesperia*, **76**, 627–663.

Chatzipouliou E. (1988). Koryzi 4, *Archaiologikon Deltion, Chronika*, **43**, 29–30 (in Hellenic).

Chiotis E. D. and Chioti L. E. (2011). Industrial activities in the ancient Agora of Athens in the Late Roman times. In: The Agora in the Mediterranean from Homeric to Roman time, A. Giannikouri (ed.), Ministry of Culture, Athens, Hellas, pp. 181–196 (in Hellenic).

Chiotis E. D. and Chioti L. E. (2012). Water supply of Athens in the antiquity. In: Evolution of Water Supply trough the Millennia, A. N. Angelakis, L. W. Mays, D. Koutsoyiannis and N. Mamassis (eds), IWA Publishing, London, pp. 407–442.

Conwell D. H. (2008). Connecting a City to the Sea: the history of the Athenian Long Walls. In: Mnemosyne, Supplements, History and Archaeology of Classical Antiquity, S. E. Alcock and R. Osborne (eds), BRILL, Leiden-Boston, p. 267.

Costaki L. (2006). The Intra Muros Street System Street System of Ancient Athens. University of Toronto, PhD thesis, vol. 2.

Crawley R. (1950). Translation of the History of the Peloponnesean War by Thucydides. New York, London.

Dontas G. (1962a). Excavation at the Agelopoulou plot. *Archaiologikon Deltion, Meletes*, **17**, 83–95 (in Hellenic).

Dontas G. (1962b). Excavation at the Zacharatou plot. *Archaiologikon Deltion, Meletes*, **17**, 101–103 (in Hellenic).

Dörpfeld W. and Reisch E. (1896). Das Griechische Theater. Athen, p. 396.

Eleutheratou S. (1997). Excavation of the Metropolitan Railway of Athens. *Archaiologikon Deltion, Meletes*, **52**, 34–36 (in Hellenic).

Eleutheratou S. (2006). The Museum and the excavation: findings from the plot of the new Museum of Acropolis, Acropolis Museum, Athens, p. 165 (in Hellenic).

Eleutheratou S. (2008). Urban-planning in the south part of the ancient city of Athens during Roman times. In: Athens During the Roman Period, S. Vlizos (ed.), Mouseio Benaki, Athens, pp. 185–205 (in Hellenic).

Eydap S. A. Athens Water Supply and Sewerage Company. http://www.eydap.gr/index.asp?a_id=164 (accessed May 5th, 2014).

Kalligas P. (2000). Acropolis Station. In: The City under the City, L. Parlama and N. Stampolidis (eds), Museum of Cycladic Art, Athens, pp. 29–39 (in Hellenic).

Karagiorga, Th. (1978). Plataion and Agesilaou, *Archaiologikon Deltion, Meletes*, **33**, 19–21 (in Hellenic).

Kaupert J. A. (1878). Athen mit Umgebung (Map 1:12,500), Berlin.

Ktimatologio S. A. Air photos of Greece. Available http://gis.ktimanet.gr/wms/ktbasemap/default.aspx (last consulted on May 5th, 2014).

Kyparisses N. and Thompson H. A. (1938). A sanctuary of Zeus and Athena Phratrios newly found in Athens. *Hesperia*, **7**, 612–625.

Lang M. (1968). Water works in the Athenian Agora, Excavations of the Athenian Agora Picture Book No 11. AMSCSA, Princeton.

Marinakis S. (2000). Tile pipes of the aqueduct. In: The City under the City, L. Parlama and N. Stambolidis (eds), Museum of Cycladic Art, Athens, Hellas, p. 58 (in Hellenic).

Morens D. M. and Littman R. L. (1992). Epidemiology of the Plague of Athens. *Transactions of the American Philological Assessment*, **122**, 271–304.

Orlandos A. (1940). The purpose of the Roman building north of the clock tower of Andronikos Cyrrestos. *Proceedings of Academy of Athens*, **15**, 251–260 (in Hellenic).

Owens E. J. (1983). The Koprologoi at Athens in the Fifth and Fourth Centuries B.C., The Classical Quarterly (New Series). Vol. 33, Cambridge University Press, Cambridge, UK, pp. 44–50.

Paliatsos A. G., Nastos P. T., Tzavelas G. and Panagiotakos D. B. (2005). Characteristics of precipitation in urban Athens area, from 1891 to 2000. *Fresenius Environmental Bulletin*, **14**, 422–428.

Palinkas J. and Herbst J. (2011). A Roman Road Southeast of the Forum at Corinth-Technology and Urban Development. *Hesperia*, **80**, 287–336.

Papadopoulos J. K. (2006). Sella Cacatoria, a study of the potty in the Archaic and Classical Athens. *Hesperia*, **75**, 1–32.

Papagrigorakis M. J., Yapijakis C., Synodinos P. N. and Baziotopoulou-Valavani E. (2006). DNA examination of ancient dental pulp incriminates typhoid fever as a probable cause of the Plague of Athens. *Interntional Journal of Infectious Diseases*, **10**, 206–214.

Salway P. and Dell W. (1955). Plague at Athens, Greece & Rome. 2nd Series, vol. 2, Cambridge University Press, Cambridge, UK, pp. 62–70.

Shapiro B., Rambault A. and Gilbert T. P. (2006). No proof that typhoid caused the Plague of Athens, a reply to Papagrigorakis *et al.*, *International Journal of Infectious Diseases*, Elsevier, Letters to the Editor **10**, 334–335.

Shear Jr T. L. (1997). The Athenian Agora: Excavations of 1989–1993. *Hesperia*, **66**, 495–548.

Spathari E. and Chatzioti M. (1983). Queen's Sophias Ave. and Herodes Atticus 2. *Archaiologikon. Deltion, Chronika*, **38**, 23–25 (in Hellenic).

Stavropoullos F. D. (1965a). Ag. Anargyron and Katsikoyianni *Archaiologikon Deltion, Chronika*, **20**, 55–56 (in Hellenic).

Stavropoulos F. D. (1965b) Benizelou 5, *Archaiologikon. Deltion, Chronika*, **20**, 62–64 (in Hellenic).

Steinhauer G. (2009). Street network of Piraeus. In: Attica's streets-ancient streets of Attica, M. Korres (ed.) pp. 236–243 (in Hellenic).

Theocharaki A. (2011). The ancient circuit wall of Athens – its changing course and the phases of construction. *Hesperia*, **80**, 71–156.

Thompson H. A. (1959). Activities in the Athenian Agora: 1958. *Hesperia*, **28**, 91–108.

Tigginaga Y. (2008) The invisible architecture of the Hadrian Library. In: Athens during the Roman Period, S. Vlizos (ed.), Mouseio Benaki, Athens, pp. 133–152 (in Hellenic).

Vanderpool E. (1965). The Acharnian Aqueduct. Χαριστήριον εις *Α.Κ. Ορλάνδον* (Honorary Volume in Memory of A. K. Orlandos), **1**, 166–174.

Wilson A. (2000a). Drainage and Sanitation. In: Handbook of Ancient Water Technology, Ö. Wikander (ed.), BRILL, Leiden-Boston-Köln, pp. 151–179.

Wilson A. (2000b). Land Drainage. In: Handbook of Ancient Water Technology, Ö. Wikander (ed.), BRILL, Leiden-Boston-Köln, pp. 303–317.

Zachariadou O. (2000). Herode's Atticus shaft (in Greek). In: The City under the City, L. Parlama and N. Stambolidis (eds), Museum of Cycladic Art, Athens, Hellas, pp. 190–195 (in Hellenic).

Chapter 16

Sewerage system of Diocletian's Palace in Split – Croatia

Jure Margeta, Katjia Marasović, and Snježana Perojević

16.1 INTRODUCTION

Diocletian's Palace in Split (Croatia) is the best preserved Late Roman palace in the world thanks to the continuity of the life among its walls for more than 1700 years. The perimetral walls are very well preserved, together with 3 of its 16 original towers, because, until recently, they served as city walls. The Roman square, the so called 'Peristyle', together with two sacral buildings: Diocletian's tomb, and the small temple are still in use. More than 40 vaulted Roman rooms of Emperor apartment's substructures are explored and put into function. Some other Roman structures are still visible in the streets or in the houses built inside the Palace.

In 305 the Roman emperor Diocletian left his throne in Nicomedia (modern Ismit, Turkey) and went to his homeland in order to spend the rest of his life in his palace built in the bay 5 km from Salona, the capital of the Roman province Dalmatia, Figures 16.1 and 16.2. After the death of the Emperor in 316, the Palace became imperial property and life continued there. In the 7th century refugees from the destroyed town of Salona found shelter inside the Palace protecting themselves from the invading barbarians. There they built new houses and started to live, work and trade. The settling of the Palace by common people is considered the beginning of the City of Split. Today many restaurants, shops and homes still stand within the walls. The city gradually spread around the Palace so that by the 15th century it doubled its surface. Until today it has constantly been inhabited bearing the features of a common Mediterranean town. Split is now a major city of the Split-Dalmatian County and the second largest city in Croatia. Thanks to the well preserved Diocletian's Palace and buildings of all historic periods, it holds an outstanding place in the Mediterranean, European and world heritage. In 1979, in line with the international Convention on Cultural and Natural Heritage, UNESCO adopted the proposal that the historic core of Split built around Palace should be included in the register of the World Cultural Heritage (Figure 16.3).

Figure 16.1 Location of Diocletian's Palace in Split harbour.

Figure 16.2 View of the Palace from sea by J. Marasović.

The Palace lies in a bay on the south side of a short peninsula of the Dalmatian coast, Figure 16.2. At the time of its construction, it was the site of a small settlement, Spalato. The terrain at this location slopes gently seaward and is typical karst, consisting of low limestone ridges running east and west with marl layers between them. The construction of Diocletian's Palace combines elements of a *villa maritima* and a military camp (*castrum*). It occupies an area of 3.3 ha. The shape of the Palace is a rectangle approximately 180 × 220 meters. There are four entrances to the Palace, three from the land and one by sea. The entrances are connected by main perpendicular streets *cardo* and *decumanus*. The *decumanus* connects the eastern and western gates and divides the complex into two halves. The Palace combines the qualities of a luxurious villa in the southern part where the Emperor's family lived, with the public area and religious buildings, and a military camp in the northern part housing soldiers, servants and other facilities. This part of the Palace is divided in two parts by the central street *cardo* from the North Gate (Gold Gate) to the main square Peristyle. The Palace is built of white local limestone, imported marble, bricks and tuff used for vaults. It housed a number of people, depending on whether the Emperor was present or not. The Palace held the necessary infrastructure required for living. Thus, among other things, a water system, or water supply and sewerage system were built, which will be described in more detail in this chapter.

Figure 16.3 Location of the Palace in the city core of Split.

16.2 PALACE WATER SYSTEM

The water supply system was based on the use of water from the Jadro spring, about 7 km from the Palace, Figure 16.4. This spring supplied not only the Palace but also Salona, the largest Roman city on the East Adriatic coast. Water from the Jadro spring flowed by gravity to the Palace. In order to allow gravity flow, it was necessary to build an aqueduct 9,5 km long consisting of channel, bridges and tunnels which have a slope of 0.65–2.66%. The capacity of the system was about 450–700 L/s (Marasović *et al.*, 2007). At the end of the 19th century Roman aqueduct was reconstructed and used for Split city water supply (Kečkemet, 1993). Therefore a great part is preserved and still in function.

Figure 16.4 Aqueduct of Diocletian's Palace.

The system was equipped with all the necessary facilities to ensure a continuous supply of good quality water. In this sense the *Castellum Aquae* (juncture of aqueduct and pressure pipes) was built. Water flowed to the city fountains by lead pressure pipes from where people took it for their needs, and directly to the imperial chambers for the emperor.

The Palace had a drainage system for surface and foul water from the area of the Palace, and for perimeter water around the Palace. This system will be further elaborated in the next chapter. It is important to note that the Palace water system has not yet been fully explored. However, the basic characteristics of the system are known.

16.3 DRAINAGE SYSTEM OF THE PALACE

16.3.1 The concept

The drainage system was of combined type. Surface water and foul water were taken from the Palace by joint channels and discharged into the sea on the west part of the Palace, Figures 16.5 and 16.6. The sewerage consisted of two subsystems: (a) southern subsystem- the Emperor's apartments and public area with religious buildings; (b) northern subsystem soldiers and servants area. The northern subsystem is about 560 m long, of which 250 m have been explored so far (Marasović, 1989). The southern subsystems sewerage is still not sufficiently explored.

Different sectors of the Palace are clearly differentiated by its elevation. The Emperor's rooms were located in the southern part of the Palace over the vaulted substructure at a higher elevation than the northern part of the Palace located directly on the ground.

The main channels of the northern subsystem were located in the middle of *cardo* (north–south) and western part of *decumanus* (east–west). By this layout, collecting water from smaller surrounding streets and buildings was simple by means of relatively short lateral channels. Main channels have a slope towards south or west. The main sewerage channel went out of the Palace below the Western Gate and in a mild curve turned southeast. Some 20 m from the West Gate it ended with a stone portal from where it continued towards the sea by an open channel. Water from the area used by the Emperor was evacuated directly into the sea.

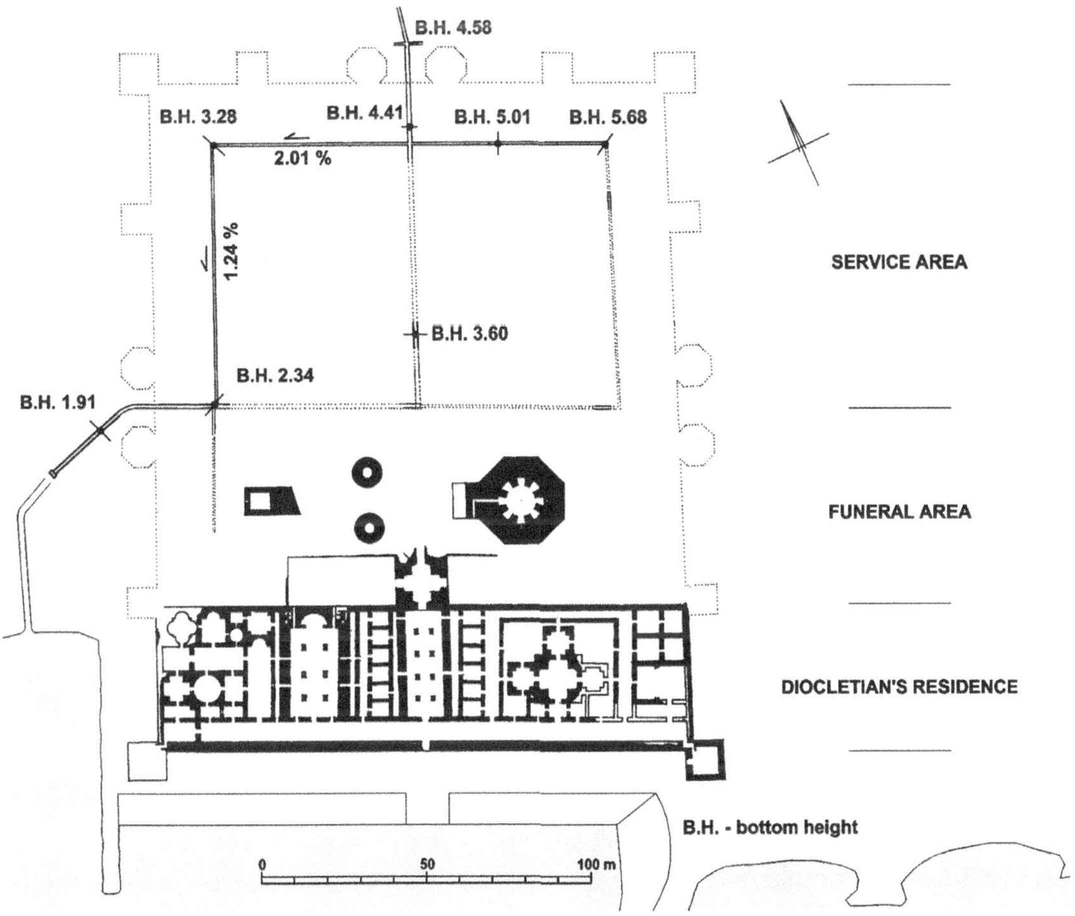

Figure 16.5 Layout of the main sewers of the Palace by J. Marasović.

Surface water from the area of the Palace was collected by stone rain gutter inlets of different shapes, Figure 16.7. Thanks to the significant transversal slope of the streets, the surface water was efficiently directed toward the gutter inlet.

Wastewater from facilities was collected by vertical pipes and by outpouring of night pots and other vessels into the channels through street drains.

In the rainy period wastewater and surface water flowed through the channels, as well as surface water which gravitated from the surrounding terrain towards the walls of the Palace. In dry periods, wastewater and water flowing from the fountains, as well as water supply system overflow flowed through the channels. In this way a significant quantity of clean water flowed through the system in relation to wastewater, by which the system was continuously rinsed and maintained in good hygienic condition with no odor and other negative influences, which would have otherwise caused (especially during periods without rain) deposition of filth at particular points.

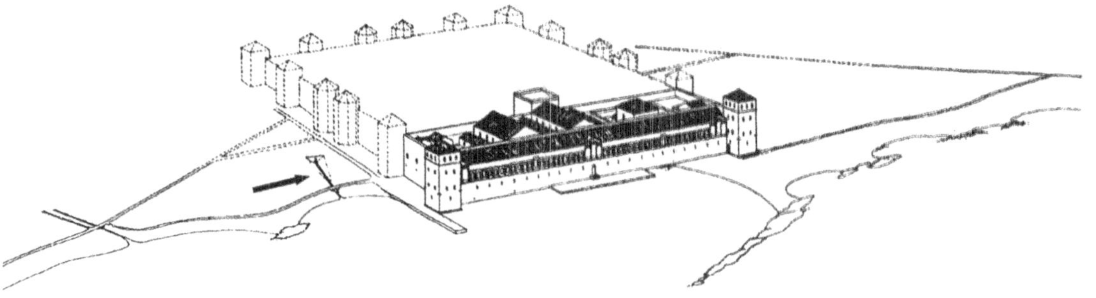

Figure 16.6 Location of the discharge channel into the sea.

Figure 16.7 Different shapes of gutter inlet.

16.3.2 Elements of the sewerage system and its characteristics

The main elements of the system are channels of a big cross-section, transverse channels, vertical channels, inspection shafts and gutter inlets. There are two types of channels of a big cross-section: (a) main channel and (b) lateral channels (Figure 16.8).

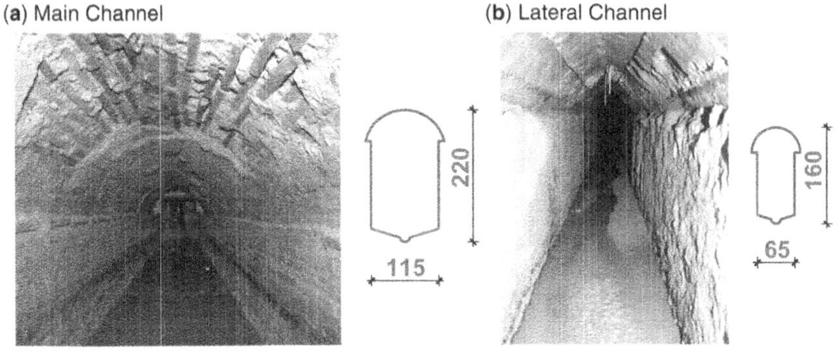

Figure 16.8 Main and lateral channels.

The channel bottom slope is from 1.26 to 2%. The main channel (115/220 cm) as well as its vault were built of brick in a thick layer of mortar, while lateral channels (65/160 cm) of hewn stone and its vault of Roman concrete (*opus caementicium*) where traces of formwork can still be seen. The channels were of mixed type and because of their large size walkable and therefore easily maintained. Analogous to other ancient cities and palaces the overflow from the water supply had to flush the sewer system. It can therefore be assumed that the water supply overflow to the Palace entered from the northeast, at the highest point of the sewerage and from there was to be directed to the west and south in order to flush the whole system.

During research and excavation of the Palace sewer system, shafts for entering the channel, of which three originate from the Antiquity, were found and left accessible, Figure 16.9.

Figure 16.9 Original entrance to the channels.

Transverse channels were instilled into the main channel network, collecting wastewater from buildings and draining the gutters. The streets of the Palace were paved with large stone slabs. Water poured into the street gutters placed along the street edge, interconnected with a channel. From there water flowed into the channels of the main street network via transverse channels. Five stone gutters have been found to the day, in the form of rosettes: four on in the *Cardo*, and one on *Decumanus* near the Peristyle.

From 1995 to 1997 the area to the north of the Palace was excavated in order to determine the position of the ancient waterworks distributor (*castellum aquae*) unfortunately not yet found. On that occasion, a connection of the main *Cardo* channel and north wall drainage channel was found 11.5 m from the Golden Gate (North Door). Drainage channel ran along the north wall and collected water flowing there from the terrain in the north of the Palace, Figure 16.10. This water was additional inflow for system flushing. This drainage channel is 45 cm wide, 50 cm high and was built of small stones of irregular shape with compacted earth bottom. A branch to the west was subsequently interrupted by a Late Antiquity channel from the north.

Figure 16.10 Connection of the Cardo channel with drainage channel north of the Golden Gate.

The northern end of the Cardo channel was built of stone and covered with two 30 cm thick capstones, thus creating a sort of a portal. On the underside of the south capstone are two holes in which a metal grating was probably attached that guarded the entrance to the Palace sewerage system from people and animals (Figure 16.10).

16.3.3 After roman period

Over the time, the system continued to be used and maintained. However due to the gradual devastation of Roman buildings in order to build new medieval houses within the Palace, the system was abandoned and replaced by septic pits and new channels that took water directly into the sea.

Gradual construction of the sewerage system within the historical core of Split that is still largely in function began in the 19th century. During the period of the French rule (1806–1813) a drainage system was built around the Palace. The channel was built of cut stone and covered with stone slabs.

During the second Austrian rule (1813–1918) in the mid 19th century, Vicko Andric, architect and first conservationist worked on street improvements which included paving and building sewers (Kečkemet, 1993). In 1845 he designed the Grota street, which connected the Peristyle (central Palace square) with the coast. There he built a masonry channel 110/135 cm (vaulted by a brick arch) that was demolished in 1961 during the reconstruction of the Peristyle and the central hall of the Roman substructure (Figure 16.11).

Figure 16.11 A 19th century channel on the Peristyle.

Due to the large population growth in the 19th century, many single-family houses and mansions in the historical center of Split were turned into houses with apartments for rent. Pipes of perforated stone elements were incorporated into the facades for drainage of sinks and toilets (Figure 16.12). In places where there was a street channel they connected to the channel and if no channel existed, septic pits were built under the houses. Some 19th century channels are still in function, but under big weight of supply vehicles their vaults sometimes collapse. The responsible authority is making efforts to gradually reconstruct them or replace them with new ones.

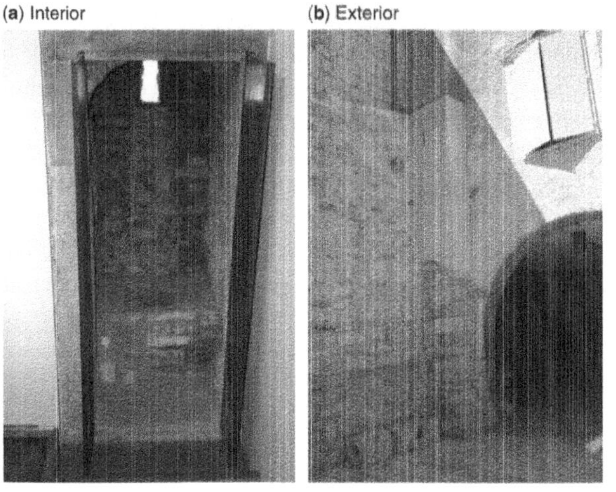

Figure 16.12 Toilet in a Romanesque house in the Diocletian's Palace with direct drainage into the Roman substructure.

Today the explored part of the sewerage system in the Diocletian's Palace is partially under water because underground water penetrates into it and there is no discharge into the sea. In order to solve this problem, an extension of the Diocletian channel to the sea was designed in the 1980s (Margeta), Figure 16.13. During the construction of the collector sewer at the Riva (Waterfront) in 1996, a section from the Square Voćni Trg to the sea was constructed. The section from the end of the Diocletian channel (under the kitchen of the Hotel Central at Narodni trg) to the Square Voćni trg still remains to be done. We hope that this will be built soon allowing full exploration of this well-preserved sewerage system of the Roman palace.

Figure 16.13 Proposed solutions of revitalizing the drainage system of the Palace.

16.4 THE DISCUSSION AND CONCLUSION – LESSON LEARNED

The Diocletian Palace in Split is one of the best preserved Roman palaces. The same applies to the sewer system of the Palace. Although the Palace changed over the time due to the construction of residential buildings, a number of unchanged structures still remain there, depicting the construction from Roman times. This also applies to the sewerage which remained largely preserved and unchanged, because it is under the ground filled with earth material. It is not in function today because drainage to the sea is disabled due to the expansion of the city. Until the present 250 m of the 560 m long main sewer has been explored.

The drainage and sewerage system of the Palace, as well as the protection from external surface water and groundwater, represents an excellent example of good Roman engineering practice and solutions. Based on the available information, it can be concluded that when selecting the location of the Palace, its planners and builders took into consideration good solutions related to water infrastructure, water supply and sewerage system.

This is primarily the adequate location of the Palace in relation to the water intake and sea level. In relation to the water intake for water supply (it is assumed that *castellum aquae* was to the northeast of the Palace (Nikšić, 2012), at the slopes of Gripe at 20 m OD), the location of the Palace is low enough (floor level at 8.20 m in the northeast, 5.20 m OD at the Western Gate, the Emperor's chambers in the southern part of the Palace 8.50 m and its substructure 0.40–0.90 m OD) to allow water supply by gravity. In relation to the wastewater and storm water drainage it is high enough to allow gravity drainage of all water into the sea (it must be borne in mind that the sea level today is at 0.32 m OD and it is assumed that in Diocletian's era it was at 1.40 m below OD). The ratio of higher and lower parts allowed efficient drainage in certain areas due to good arrangement of the streets and channels laid in them. The distribution of channels between these spaces is such that it protected the lower parts from water of the higher parts (underground and surface). (Figure 16.14).

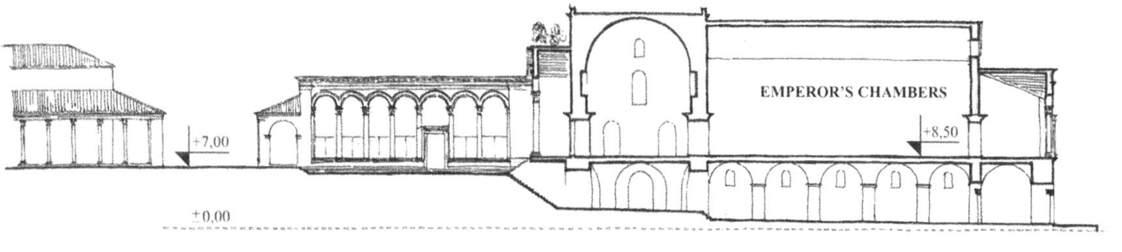

Figure 16.14 North–south section of the middle and southern part of the Palace.

Southern part of the Palace used by the Emperor, his family and personnel was raised about 1.5 m from the northern part of the Palace which enabled protection from surface of surrounding areas. Substructure was drained directly into the sea through openings in the wall of the Palace. In front of the Palace there was a 12 m wide coast where these channels continued. During the archeological excavations a number of Roman stone pipes elements were found in substructures, Figure 16.15.

The position of the Palace within the bay was carefully selected, not only for safe docking of ships by the coast which stretched along its southern front, but also related to water drainage. The position within a valley that slopes gently towards the sea in the north–east to south–west direction allows gravity collection of all surface water and groundwater into the western part of the bay, away from the Palace. In the valley there are no permanent surface flows, but only small amounts of groundwater that drain to the sea. Therefore it was not necessary to build deep trenches around the Palace to prevent water inflow into the Palace. Instead, a small boundary drainage channel was built, through which small amounts of surface water and shallow groundwater were collected and discharged into the main channel of the Palace. Also, the discharge of wastewater and storm water collected within the Palace was situated to the west of the Palace, seemingly to protect the sea in front of the Palace. Specifically, the sea along the Palace flowed in the east-west direction, so that the discharged water flowed away from the Palace, leaving thus the sea around the Palace clean.

Figure 16.15 Roman stone pipes elements found in substructures.

The concept, details and individual solutions can be considered standard practice in sanitary engineering that has not changed significantly since. Today, as in the Roman era, the main task of the sewerage is to take all the waste water as quickly and safely as possible from the man and the place of his residence without adverse impact on residential areas and the environment. The Roman sanitary engineers realized this very effectively in Diocletian's Palace.

The main drainage channels, as well as the streets, were located in the central part of the Palace so that the space around them could be drained effectively. The drainage was done through smaller sewers of the secondary network and by manual work of people who emptied bedpans and used water into street drains. Accordingly, easy accessibility of these places was taken into consideration. The higher part of the Palace was drained mainly by the gravity channels that were placed inside the structure. The lower part of the Palace was drained through shorter channels directly into the sea. The same mainly applied to the part of the Palace where the Emperor and his family resided.

Constant flushing of the sewers, ventilation and maintenance needs were all taken into account. Constant flushing was accomplished by discharging water from the fountains, but probably also by underground water that was collected around the Palace and discharged into the channels as well as small springs located within Palace. However, the continuity of flushing of the main sewer, as well as all the channels, was provided by a steady stream of overflow water from the distribution tank (*castellum*). Specifically, the municipal water system of the Palace was a flow system. Inlets for entrance of storm runoff are placed at the gutters at street crossings but also along the streets and within other open spaces. The curb-opening inlet and gutter inlet has been used. The opening size (capacity) has been different and carefully selected in accordance with expected amount of storm water. The gutter inlet has a depression in the gutter section through which the surface drainage falls, covered by differently shaped grates made of limestone.

Ventilation was provided through gutter inlets as well as special openings for emptying bedpans and other containers with water. Probably there were no special openings for the aeration of the foul water drainage system, which will have to be confirmed by further research.

Channels were built from available materials, designed not only to be structurally sustainable, but to achieve by their shape the good hydraulic runoff conditions in different operating regimes of the system, rainy and dry. The cross-section of the channels and their capacity were foreseen by planned quantities of water, wastewater and storm water. It should be expected that the channel dimensions increased downstream as the area from where water was collected increased but due to construction technology simplification this was not the case in the Palace. It is obvious that the builders knew how to estimate the amount of water that would flow from the Palace, as well as channel capacity according to available inclination and size. All the water that flowed into the Palace and which fell as precipitation on the area of the Palace was collected and discharged from the Palace. As the flow of water for water supply was controlled by a sluice on the distribution tank (*castellum*), the flow of water in the channels could also be controlled. All details were considered. The Palace had private toilets, but also public ones. Their number and exact disposition however is not known. All these structures also had a flushing system to maintain good sanitation and good health conditions within the Palace. Particularly interesting was the imaginative solution for rain grating that fitted with its design into the solution of the central square of the Palace and in main streets. The sewerage system had openings for inspection and maintenance. At the junction of smaller with larger channels cascades were used (Vidović, 2009). The system also had larger openings (on the beginning and on the end) through which materials were brought in for renewal and demolished parts and sediments were taken out. It is expected that further study of this cultural monument will show more interesting details in the field of sanitary engineering. In the long term there are plans for excavating the greater part of the Palace sewerage system, restoring it and making it available to tourists as an example of Roman architectural heritage, and to experts for further study of historical buildings and water infrastructure technologies.

16.5 REFERENCES

Bužanćinć R. (2009). Fontana Dioklecijanove Carske Palače (Diocletian's Palace Fountain), Klesarstvo i Graditeljstvo (0353-7897) 20, Split.

Kečkemet D. (1993). Vicko Andrić, Arhitekt i Konzervator 1793–1866 (Vicko Andrić, Architect and Conservator 1793–1866), Split.

Madirazza V. (2011). Izvještaj o arheološkim istraživanjima u prizemlju hotela Central u Splitu (Report on Archaeological Excavations in the Groundfloor of the Hotel Central in Split), Split.

Magnabal G. (2004). Servitudes Urbaines et Réglements D'urbanisme dans L'habitat de la Rome Impériale (27 Avant Jésu Christ – 192 Après Jésus Christ (Urban Servitude and Urban Regulations in Habitats of the Imperial Rome). Université de Marne la Vallée, Champs sur Marne.

Malissard A. (2002). Les Romains et e'eau: Fontanes, Salles de Bain, Themers, égouts, Aquaducts … (Romans and the Water: Fountaines, Baths, Thermes, Sewers, Aqueducts...) REALIA/Les balles lettres, Paris.

Marasović J. (1982). Le scoperte nel Palazzo di Diocleziano negli ultimi decenni, Antichità Altoadriatiche, XXVI (Discoveries in Diocletian's Palace During the Last Decades), Trieste.

Marasović J. (1989). Research in Diocletian's Palace after 1974, American-Yugoslav Joint Excavations, Vol. Five, University. of Minnesota, Minneapolis.

Marasović J. and Marasović T. (1961–1962). Pregled radova Urbanističkog biroa na istraživanju, zaštiti i uređenju Dioklecijanove palače od 1955. do 1965. (Review of the Research, Protection and Renewal Works in Diocletian's Palace carried out by Urban Planning Institute from 1955 to 1965.) URBS 4, Split.

Marasović J. and Marasović T. (1994). Le ricerche nel palazzo di Diocleziano a Split negli ultimi 30 anni (1964–1994) (Research in Diocletian's Palace in Split During the Last 30 Years (1964–1994)), Antiquité Tardive 2, Paris.

Marasović K. and Perojević S. (2008). Odvodnja otpadnih voda u povijesnoj jezgri Splita (Evacuation of the Wastewater in the Historical Core of Split), Odvodnja otpadnih voda u starim priobalnim gradovima Hrvatske/ Hrvatsko društvo za zaštitu voda, Dubrovnik.

Marasović J., Marasović T., McNally S. and Wilkes J. J. (1972). Dioklecijanova palača, Izvještaj o Jugoslavensko-američkom projektu istraživanja jugoistočnog dijela Palače I (Diocletian's Palace, Report on American – Yougoslav Research Project in the Southeaster Section of the Palace I), URBS, Split.

Marasović J., Marasović K., Perojević S. and Rismondo T. (1999). Kanalizacija i vodovod Dioklecijanove Palače (Sewer and Water Supply of Diocletian's Palace). Dioklecijanov akvedukt, J. Belamarić (ed.), Split.

Marasović J., Marasović K. and Perojević S. (2007). Aqueduct and Sewage of Diocletian's palace. In Second International Conference on Waters in Protected Areas, Zagreb.

Margeta, J. (1989). Projekt o produžetku Dioklecijanovog kolektora (Project for the Extension of Diocletian's Collector), IGH, Split.

Niemann G. (1910). Der Palast Diokletians in Spalato (Diocletian's Palace in Split), Wien.

Nikšić G. (2012). Dioklecijanova palača – reinterpretacija izvorne namjene i arhitekture (Diocletian's Palace – Reinterpretation of the Original Function and Architecture), Niš i Vizantija, Deseti naučni skup, Niš 3–5 June 2011., Z. Radova X, M. Rakocija (eds), Niš.

Perojević S., Marasović K., Marasović J. (2009). Istraživanja Dioklecijanove palače od 1985. do 2005. godine (Research of the Diocletian's Palace from 1985. to 2005.), Zbornik međunarodnog znanstvenog skupa Dioklecijan, tetrarhija i Dioklecijanova palača: o 1700. obljetnici postojanja, N. Cambi, J. Belamarić and T. Marasović (eds), Književni krug Split, Split, pp. 51–94.

Piplović S. (1996). Dioklecijanova palača u Splitu početkom XX. stoljeća (Diocletian's Palace in Split at the Beginning of the XX th. Century), Zbornik radova posvećenih sedamdesetogodišnjici života Danice Božić-Bužančić, Građa i prilozi za povijest Dalmacije 12, Split.

Prina Ricotti E. S. (2001). Villa Adriana – Il sogno di un imperatore (Hadrian's Villa – the Dream of an Emperor), 'L'Erma' di Bretchneider, Roma.

Scobie A. (1986). Slums, sanitation, and mortality in the Roman world. *KLIO*, **68**(2), 399–433.

Sofier S. (1999). The Evolution of the Bathroom and the Implication for Paruresis. http://paruresis.org (accessed 10 June 2012)

Vidović R. and Vidović L. (2009). Antička Kanalizacija Dioklecijanove Palače (Antique Sewer of the Diocletian's Palace), Roterm d.o.o., Split.

William S. (1996). Diocletian and the Roman Recovery (Roman Imperial Biographies) Routledge, 1996.

Wilson A. (1994). Running water and social status in North Africa. In Proc. of North Africa from antique to Islam, Bristol.

Chapter 17

The evolution of sanitation in the rural area of Southwest China: With case of Dai villages of Xishuangbana, Yunnan province

Xiao Yun ZHENG[1] and Yao Yu ZHENG[2]

[1]*Yunnan Academy of Social Sciences, Kunming, China*

[2]*PAYAP University, Mao Kowchiang Mai, 50000, 50000, Thailand, Thailand*

17.1 INTRODUCTION

Sanitation, in general terms, refers to facilities and customs of water related to the daily life of human beings, for example, toilet, bathing, daily water supply including both supply and storage of water, domestic wastewater treatment, rural water environment (water environment surroundings, water quality and water supply conditions and so on) and culture factors which shaped its formation. By use of water, water related facilities, the sanitation, have been getting developed as well as brought producing of related customs and cultures at the same time, enhanced the mankind's quality of life and health as a whole. Therefore, evolution of sanitation is one of the important mark of human civilization. Evolution of sanitation also directly linked with local environment due to sanitation is not only the result of utilization of environmental resource, in particular use of water, but also a necessity to sustain a harmonious relationship with local environment, because that the evolution of sanitation in one hand improved mankind's living quality, on the other hand also imposed great pressure to the environment in terms of water consumption and discharge of wastewater, therefore, sanitation condition is also an important benchmark to evaluate situation of local environment.

China is obvious featured in dual social structure of rural and urban imbalanced development. Compare with urban areas, sanitation change in rural areas has its own particularity, especially for ethnic groups, sanitation development closely linked with their ethnic culture.

The modern sanitation development in the rural areas of China was beginning in 1950s, the residing area of Dai peoples with case in this research. Dai is an ethnic nationality in China with their traditional ethnic culture, especially it is remarkable that their traditional ethnic culture is deeply embedded in water and reflected salient features of water which comparing other ethnic minorities in the region of Southwest China (Xiao Yun ZHENG, 2005). Accordingly, the modern evolution of sanitation was closely linked with the traditional Dai ethnic culture, in particular with water related culture, so the change from traditional sanitation to the present times not only connected with the changes taken place in the living conditions of Dai peoples, but also connected with the change of the traditional Dai ethnic culture.

While sanitation development has been improved Dai people's quality of life, it also brought challenges to the local environment by increased discharge of domestic wastewater. Therefore, observing the change of sanitation and the challenges brought about by sanitation development will help us to better understand water management in Dai villages and find ways to sustaining the water environment. This paper focused on the Dai villages in Xishuangbanna Prefecture of Yunnan Province, China, arms to investigate the sanitation transformation and changes of related culture, in particular the impacts with the local environment by the improvement of sanitation.

17.2 SANITATION IN THE DAI TRADITIONAL SOCIETY

Dai is an ethnic nationality in China, mainly concentrated distributing in the south part of Yunnan Province of China with total population around 1,332,000, (Conditions of Yunnan Province, 2008) 323.596 of them settled in Xishuangbanna Prefecture (The Statistics Yearbook of Xishuangbanna Prefecture, 2011). It is also a trans-boundary ethnic group with close branches of Tai ethnic groups found in Vietnam, Myanmar, Laos, Thailand and India. Dai in China mainly settled in the areas of Xishuanbanna, Lincang, Simao and Dehong, prefectures of Yunann Province.

Most part of the area Dai settled are tropic or subtropical area with plentiful rainfall as well as the Dai concentrated residing near the rivers with nice water environment. Accordingly, traditional Dai village was usually found in where near the water source conveniently which providing conditions for daily life included the traditional sanitation as well as convenient irrigation for paddy fields. In the history, Dai people has cultivated a set of cultural tradition such as, for example, the water festival, collective bathing, well related custom, storage of domestic water, and so on, in line with its traditional sanitation habits and relevant facilities and fit into the local environment and kept to the present times, but in summary, the sanitation facilities in Dai society is comparatively simple due to easy availability and use of water in their daily life. The features of traditional Dai sanitation could be summarized in the following sections.

17.2.1 Personal sanitation

Bathing

bathing is the requisite custom in Dai's daily life as well as also a kind of important culture in their traditional society. Settled in the tropic area, with good water environment is a salient characteristic of the Dai residency, therefore there are always rivers streams and ponds around the villages of Dai as well as many wells built in the Dai house which greatly facilitate use of water for bathing. Dai much like bathing in their daily life, going to bathing everyday become a habit for Dai people from their childhood. For the baby, one or two times bathing will be taken by mother every day, children like bathing in river, pound or use well water inside the house even many times in one day for fun or cool. Aged people also take bathing every day at veranda of house usually. Adult usually take bath in the river or well nearby after work but most typical bathing custom of Dai is they used to meeting together to bathing in the evenings after their daily work in the nearby rivers and streams. Anyway, bathing is not only a need to clean their body but also a collective occasion of daily life, it is combining socializing and entertaining, for example, the villagers taking this occasion to exchange information and chat while bathing and also women washing clothes, vegetables by the way. It also with many customs likes in the

Dai villages of Xishuanbanna, men usually in the upper stream and women in the lower stream while bathing. (Figure 17.1).

Figure 17.1 The Dai traditional bathing in the rivers: (a) a photo in 1980s shows the woman bathing in the riverside and (b) Young girls bathing in the river (with permission of Xiao Yun ZHENG).

Toileting

Before 1960s, evacuation were most done naturally without toilet because most Dai villages are surrounded with the forest and plants, that is, people usually taking evacuation in forest nearby or even back of their house traditionally. Another reason is because livestock growing freely in the nearby forest, human excrement quickly would be eaten by the livestock. For villages located next to river, people also evacuate to the bank, therefore, not much pollution as a whole but in case of high temperature, affection disease also was likely to spread.

17.2.2 Water facilities of daily life

Wells of Village. In spite of the fact that Dai villages were found along the river, in the past, Dai people are not used to drink directly from rivers due to quality of river water is unstable seasonally, or taste not well also because of people take bath in the river, so they only drink from wells because of the quality of the water from well is much better and stable then river water, even today there are many villages still use well for drinking. Accordingly, it is a tradition that every village has one or more wells mainly for drinking that usually built while a village was founded, and people get water from the wells everyday is also a requisite work and an important custom (Figure 17.2). In the past, people used pottery wares, now the buckets are most made of metals. Because the wells of the village are one of most important facility in people's life, it is also considered as a holly place by locals. Villager usually construct a beautiful archıtect above each well to show their respect of the wells, at the same time, relevant customs also formed including forbidden of washing clothes and bathing around the well.

House wells

In Dai villages, most families dig well at their house for convenient water (Figure 17.3). However, according to the traditional customs, water from the house well is only used for bathing, poultry and washing, not for drinking because quality of water from the well is not good enough due to the Dai villages usually located at the foot of a mountain, the aquifer is easy to be polluted by soil also chemic fertilizer from the mountain because of rubble planting in recent years.

Figure 17.2 The well in the Dai villages, it is the important place for the daily life traditionally: (a) a old well structure in the village, (b) a new well structure made in few years ago, (c) a well structure made in the figure of dragon head, and (d) a well structure made in modern style (with permission of Xiao Yun ZHENG).

Figure 17.3 The house well (with permission of Xiao Yun ZHENG).

House water storing

Traditionally, the villagers usually wash clothes, vegetables and other things in the river but get the drinking water from wells every day, they storing drinking water usually in a pottery ware in their house (Figure 17.4). The pottery bucket usually could store water 10~15 L, each house will has two to three pottery buckets to store water for one day use, and as the pottery bucket is good for air ventilation so it is good to keep water quality accordingly.

Figure 17.4 The water storing pottery (with permission of Xiao Yun ZHENG).

17.2.3 Tap water

It is after 1980s that Dai areas in Xishuangbana began to use tap water, we will discuss this change late.

17.3 TRANSFORMATION OF SANITATION IN DAI AREAS

Resultingly convenient water use and small population number compared with land coverage in before 1960s of Dai areas, where simple sanitation is enough to go for years in the past, since 1980s, with the rapid increase of population and economy as well as social progress, sanitation conditions have been greatly changed to getting rid of various spread of infectious diseases and keep a modern living quality in Dai areas.

17.3.1 Water supply

There are two types of the tap water in the Dai area, one is the pipe water from the local dam or well, local people built dam or well where village nearby then use pipe to introduce water into the villages, other one is the tap water from waterworks into each house by the water supply network. The first stage is the dam water usually before 1980s but now some remote villages still use it due to the tap water from waterworks did not reached them yet, or lower cost to use the dam water. The tap water from waterworks was started after 1980s pushed and invested by the governments.

The history of tap water use in Dai areas is short. For example, in Jinghong town, the capital of Xishuangbanna Prefecture, the first waterworks was built in 1976 (Jinghong County Book, 1996). In Menla County, the water supply began in 1970s. At the beginning, the local people dug wells and pumping water into the houses, and later on, in 1981, the first waterworks came into being and started supplying in 1985 (MenLa County Book, 1994). The waterworks in this area all invested by the local governments to improve water use especially inside the towns, but late many villages near the town have had chance to use tap water as well.

Tap water in rural was started in 1980s. At the first, all the tap water systems in the villages were invested by the shares of village and villagers, why villagers lake to do that was due to convenient water use inside the house. The water was from small dams and ponds where village nearby and introducing water into village then houses through water pipes. In year 1988, 65% of the villages of Xishuangbanna Prefecture have been covered by this kind of water supply (Xishuangbanna Prefecture Book, 1998). However, as the water was coming from the natural stream, both unstable in terms of supply and quality, especially, water comes from the pipe was mixed with soil and dirt in the raining season.

Since 2000s, the new water supply has been developing in the Dai area due to some waterworks were built in the towns, so the Dai villages which close to the towns has begun to use tap water, tap water became more stable and quality being improved. However, for the last few years, because of the water price is costly for the villagers, for example, 2 Yuan per cubic meters and 3 Yuan in subarea of Jinhong City (The Statistics Yearbook of Jinghong County, 2011), in addition to unregulated extra charges, many villagers felt unbearable. This explained why many villagers dug their wells to get free water or allocated cost to find water sources around their villages and built pipe system for house water. In subarea of Jinhong City, villagers also drink bucketed water from water companies popularly due to it is easy for drinking with an electric water cooker also better taste then the tap water.

17.3.2 Current situation of water use

Currently, the general situation of water supply in Dai villages is a multiple way including:

Water supply: For recent years, tap water use has been popularized in many Dai villages, and directly linked with water enterprises to supply to the villages. Usually, Dai people only use tap water for bathing or washing, only part of villages use it as drinking water because of the villagers likes drink water from well following their tradition, good taste and stable quality also their idea about village well.

Drinking water: According to the tradition, Dai people used to go to wells for drinking water, this is because of the good water quality in these old wells also its special taste, sweet taste in many old wells. For example, there is an old well in the Manfeilong village, Damonglong township of Jinhong city that offers good quality water, and villagers loved it very much so until to today villagers still go to this well for drinking water. Every two to three days usually, the villagers will drive tractors or motorbikes to carrying drinking water from the well. In contrast, for those villages closed to cities, people usually drink from the bucketed water for sale.

Pipe water: As a result of the costly tap water from the waterworks, many villagers found the way to allocating the cost to construct pipe to introduce water into their house. In this way, they could invest once and enjoy free water for many years. Also at the same time, the water source they found usually with good quality. Today, it is a status that more and more villagers their village near to mountain is choosing to build pipe to introduce free water to their house.

Courtyard well: For freely water use, many villagers built well in their courtyard, but water from the well usually is used only for daily washing or animals drinking.

17.3.3 Sanitation change

After 1980s, with rapidly development of small scale rural manufacture, in particular, the rubber-machining factories as well as plantation of cash crops, rubber and sugarcanes and so on, a lot of fertilizer has been put into use and the drainage of waster water is getting more polluted in rural areas. The water environment in Dai villages are getting worse with most streams and ponds been polluted, this case resulted the traditions of bathing in the nearby stream and river has been abandoned.

But anyway, cash plants has remarkable increased the income of Dai villagers, brought chance to them to improve their daily life, one of the major change is more and more villagers rebuilding their house in recent years.

When the villagers rebuilding a house, they always incorporated with bathing facilities inside their house due to the availability of tap water makes bathing at home more conveniently. In the past, if bathing at home, villager have to go to carrying water from the river, therefore, generally, bathing at home is more difficultly. Since the popularization of water supply facilities, it is convenient to bathe at home to facing the water environment pollution where their village nearby, so family bathing become popular as well. Today, most houses incorporated with bath room that brings bathing conveniently at home after work, totally changed collective bathing previously tradition in the river (Figures 17.5 and 17.6).

Figure 17.5 The new house in a Dai village (with permission of Xiao Yun ZHENG).

Figure 17.6 The solar facility for bathing (with permission of Xiao Yun ZHENG).

17.3.4 Popularization of toilet

Since the beginning of 1960s, toilet has been introduced and put into use in rural areas (Jinghong County Book, 1996). It was required by the governments that one or two toilets have to be built in a village at that time. For convenient of cleaning, some toilet were built near the stream in the village, and introduced the stream water to clean the toilet. This project made the promotion of toilet use popularly in the Dai villages. In fact, the promotion of toilet use in Dai villages do has ups and downs because that in one hand, it changes the toileting tradition and on the other hand it needs new investment. In that period, because of the lack of toilet paper, people use bamboo piece instead of paper. In this way, bamboo pieces piled up in the toilet and took away plenty of space and make the promotion of toilet use difficultly.

Since the beginning of the new century of 2000, economy has been developed rapidly in Dai areas. Also because of plantation of rubber and other cash crops, the villagers's income also increased dramatically so that more and more people began to reconstruct their house mentioned above. In this period, while planning their new house, toilet usually included with the design. Usually, they choose to build toilet on the side or back of the house. But, because of the absent of tap water also not all the house with a well to clean the toilet in many villages, it is difficult to promoting use. Situation in recent years, because of the availability of pottery toilet bowl from the market and convenience of water supply, toilet became popularly in Dai villages not just the villages nearby the town but also in the remote area. Ordinary Dai families used to have one toilet in each floor, decorated with ceramic tile and use pottery toilet bowls, this change totally changed the toileting tradition of Dai (Figure 17.7).

Figure 17.7 The bath room incorporated with the house (with permission of Xiao Yun ZHENG).

17.3.5 The changes of the wastewater management

Traditionally, discharge of domestic sewage was usually done by discharging to the house surrounding. Traditional Dai house are pile dwelling style with two stories in which living in the second floor and the first floor used as storage or animals raising. Kitchen was also on the second floor, therefore, house sewage was discharged to the underground under the kitchen. This kind of sewage discharge could not make much pollution in the conditions of small village, fewer people and not much houses in the past. Another reason for not much pollution was because at that time people used to go to the nearby streams to wash clothes, vegetables and bathing, few washing inside their house exempt of table ware so there was not much discharge of sewage.

After 1990s, with the popularization of rural electricity, in particular, some household electrical appliances, domestic water consumption amount has been surging up. During this period, washing machines also came into rural household in large numbers. Today, washing machine almost has been used in every house of Dai family, and daily washing are completely done at home. Since then, with also the popularization of tap water, the villagers started bathing, washing vegetables and washing clothes and so on at home, which greatly increased the amount of domestic sewage discharge. According to the survey by authors in the Dai village Mangfeilong, Damenglong Township in Jinghong County, the domestic water consumption was 5 L per person per day in 1980s and 20 L in 1990s, but it is increased to 30 L currently. Resultingly, from the early of 2000s, rural sewage discharge has become a serious problem as increased sewage directly discharged to the ground without sewage pipe, resulted in serious surrounding pollution. In the summer time, not only did bad smell coming out, many household wells also been polluted. As a solution, gutter ways were built along each side of the main road in the village, each house discharged their domestic sewage into the gutter way and then further discharged into the nearby rivers. However, this still did not effectively increased improvement of the polluted environment in the villages at all.

Thereby, many villages started to build sewer under the decision of the Villagers' Administrion Group of the village . In the Mangfeilong village, Damenglong Township of Jinhong City, in 2002, the village has allocated fund to construct underground sewer for discharging of domestic sewage, and further discharge it into Liu Sa River one kilometer away. Today, more and more villages have been built sewer system, for domestic sewerage, for example, Mangzhangzai Farmer's Committee in the countryside of Jinhong City is combined with 14 natural villages, 12 of them have completed underground sewer. Since then, whenever new houses being build, the owner has to pay for the sewer system and link house sewerage pipe to the main sewer. In this way, water pollution inside the village has been saved a lot in recent years, wells inside the house yard in these villages has turned to clean again.

A paradigm which brings related data of the currently situation of sanitation from Damenglong Township, Jinghong City is shown in Table 17.1. The date shows that Damenglong Township as a mid level area on the development in Jinghong City, water supply has reached most houses and most of the Dai villages in this area have built waste water facilities, most houses have built incorporated with bath room and toilet so most villagers can take bath in their house. But more than 50% household still get drinking water from the well that follows their tradition, simultaneously the index in the table also did not get 100% by the limitation of their income condition.

Table 17.1 The sanitation situation of the Damenglong Township, Jinghong City.

FC*	Population	Household	With villages	Village with sewerage
Manjinglei FC	4925	1072	9	5
Mannankan FC	2301	469	5	4
Manlong FC	1960	457	4	4

FC*	Drinking well water (household)	Using tap water (household)	With bath\toilet (house)
Manjinglei FC	920	1072	957
Mannankan FC	227	242	450
Manlong FC	320	457	457

*FC = Farmer's Committee, the local administration unit under the township.
(Damenglong Township Government Office, 2012).

17.4 EVALUATING THE IMPACT OF THE DEVELOPMENT OF SANITATION

17.4.1 The social changes

Since 1970s, sanitation in rural Dai areas has significantly improved. Water supply in addition to the availability of electricity enabled convenient use of water pump which greatly increased the amount of water being used at home to make life more and more easily, which also promoted the popularization of home bathing and toilet as well as related sanitation facilities including washing machines and other appliances. At the first, the price was a problem to use the electricity due to imperfect electricity supply network in the villages that cased difficulty to share the cost among villagers equally but a long with the improvement of the electricity supply network and increased income, rising electricity use is a tendency in the villages.

The popularization of toilet and bathing facilities inside house has been changed traditional sanitation style as its natural way in the past. Due to the increased amount of water consumption, also resulted in sewage pollution in surrounding environment. From early 2000s, for solution of domestic sewerage, more and more Dai villages began to construct sewer system and make it with functions of waster water and rain water drainage in the raining season of summer.

Overall, as a result of the development of water supply, wastewater treatment and increasing idea of sanitation, sanitation development in Dai areas has greatly improved the quality of local people's daily life and their health as well as reduced the spread of infectious disease by improved rural environment. For example, Hookworm disease used to rankled Dai areas, and according to epidemiology survey which showed that toileting without toilet was a main cause of the disease. According to the survey of 1960 in Jinhong City of Xishuangbanna Prefecture, 46.3% of the Dai population has been affected by hookworm disease. To tackle this problem, local government took measures to prevent and cure the disease by both applying drug therapy and improvement excrement management, especially, promoting use of toilet. Resulted by this efforts, sample survey done in 1979 showed that the number of patient suffered from hookworm disease has dropped down to 11.43%; while in the survey done in 1993, there was hardly any hookworm patient to be found. Among the most important reasons lay behind were rural toilet popularization and other sanitation conditions improvements (Jinghong County Book, 1996). Improved water supply also should be accountable because it not only improved water quality but also promoted use of toilet and made it possible to improve bathing conditions. All these mentioned above are taken during the sanitation development and progress in Dai areas in past decades which greatly improved living qualities and health of Dai people.

17.4.2 Environmental impacts

In recent years, water supply in rural areas has been developed rapidly, but a result of increasing of population as well as bathing, toilet facilities, washing machine, and washing related items at home, water consumption in rural areas keep on increasing which resulted in increased domestic waster water pollution. At present, there are a small number of villages have installed sewer facilities, therefore, most villages still faced waster water pollution.

In addition to domestic waster water pollution, excessive use of fertilizers and pesticides in scale plantation of cash crops are also factors which caused environment pollution in recent years. The growing of cash crops are usually in the highland slope above residential area, in such a case it easily gets the residential area to be polluted by permeating through the ground surface. At the same time, in Dai rural areas today there are still many small rubber processing factories that charging sewage directly into rivers, which seriously polluted the rural environment. For example, Manjingbao village in outskirt of Jinhong City, due to the sewage charged by local rubber factory, water from the wells is no longer drinkable, resulted an old water well with its traditions so far has been abandoned in this village.

Another problem could not be neglected is that waster water discharged into rivers directly by those installed sewage pipes from the villages also bringing pollution to the rivers. For example, in the watershed of Liu Sa River in Jinhong City, more and more sewage pipes directly discharge waste water into this river, and eventually resulted in ever increased pollution, imposing new challenges to the environment management in Dai areas.

Water environment issue also impacted upon the local people's health and living circumstances. The Dai's life closely linked with the local water environment traditionally, Dai people not only engaged in close contact with waters around, but also garner various kinds of traditional food from natural environment, especially from water. For example, collecting aquatic plants as well as fishing for small fish, shrimp and crab. But resulted by water environment pollution in past decade, it became a threat to health if touch up water or eat polluted food from water. The most serious of all is imposed by drinking water shortage in many villages (Xiao Yun ZHENG, 2005) (Figures 17.8 and 17.9).

Figure 17.8 Good quality water is important for making traditional food (with permission of Xiao Yun ZHENG).

Figure 17.9 Water cultural plants is Dai's favor in food structure (with permission of Xiao Yun ZHENG).

17.4.3 Cultural impacts

The sanitation development in past decades also changed many traditions in Dai society, which include Dai's tradition of water use and water management. It is a fact that water related traditions are not only simple customs, but directly linked with cultural traditions prevailed in the whole society that shaped Dai as an ethnic group with deeply understanding to close to, tend to and protect water. The traditional water related culture of Dai was rooted in their long history, their ancestor used to migrate a lot on the history, but they always chose to settle down in the places where closes to river. Also according to the legend of Dai, everything in the universe came from water, so their universal view, religions, living style, social norms and customs are all closely connected with water, built above the interactive relationship with water.

For example, bathing in nearby river is not only a need to clean one's body, but also an important part of social life traditionally. Again, the public wells in every village also functioned as a facilitating place for communication, reflected in the group of women meet and talking around the well after work. Therefore, women all attach great importance to go to wells everyday traditionally. Before they go, they all dressed up and make themselves look more beautiful. From this example, we could see that going to wells for water has special social intention for Dai women (Figure 17.10). However, with changes took place in terms of water use, the social life of Dai is also rapidly changing, impacted negatively upon water's sustainable use (Xiao Yun ZHENG, 2005) (Figures 17.11 and 17.12).

Figure 17.10 The villagers getting water from the well (with permission of Xiao Yun ZHENG).

Figure 17.11 Binge during the water festival (with permission of Xiao Yun ZHENG).

Figure 17.12 Dropping water during the ritual. (with permission of Xiao Yun ZHENG).

17.5 DISCUSSION AND CONCLUSIONS

China is a typical country with the dual social structures of urban and rural areas. The evolution of sanitation in the cities was almost in common way but it was diversiform in rural areas, especially in the remote rural areas. Southwest China is a largely area with most ethnic diversity and ecological diversity in China, resultingly the evolution of sanitation has been inosculated with the ethnic cultures and local environments, shaped the diversity of sanitation in the ethnic areas before 1950s. This paper with case of the rural area of Xishuangbana Prefecture, Yunnan province to investigate the evolution of situation in Dai's community from their traditional situation to current development, especially the changes after 1950s as well as to discuss the conflict of the sanitation development and environment.

The traditional sanitation of Dai peoples has been inosculated with their culture and local environment. After 1950s, the traditional situation of sanitation has been getting changed by changing water supply situation, and after 1980s, the change of the situation of sanitation was into a rapidly period. More water supply facilities were developed the local governments as well as increased income from local economy brought possibility to people to build or rebuilding their house also incorporated with toilet, bathroom and other water facilities inside it. It is a major change compares their traditional sanitation that improved their living quality in fact but the traditional model of water using and related custom also being changed a lot as well, accordingly brought water pollution to the locals due to increasing water consumption and domestic sewerage as well as imperfect sewerage facilities in this process.

Improvement of modern sanitation in the Dai area was driven by the now model of water supply and the governmental policies, the economic development and the change of lifestyle what people followed outside, but it is also a fact that the water pollution in the village based area has caused impossible for people to keep their traditional way to take bath and use water out of their house, even the traditional culture related water that formed their social relationship and lifestyle are keeping differently due to this change, accordingly current impact of sanitation development on environment is mainly the pollution by wastewater due to incomplete sewer system construction and wastewater control of rural industry, in some areas, it has become the most serious pollutant source, consequently, it is worth to note that worsened water environment will reduce the achievement of sanitation development in past decades if we do not have effective measure to meet it.

To sum up, the development of sanitation in one hand greatly improved the local people's qualities of life and health, on the other hand also exert negative impacts due to changes took place in water environment, that is, while sanitation developed, this development also impact negatively on local people through deterioration of the water environment, including water quality, water pollution and water supply, and so on, With the improvement in sanitation been achieved, the amount of water consumption in urban and rural areas has been surging up, brought dramatic increase of domestic sewage discharge. In rural areas, due to the fragile infrastructures, domestic sewage discharge brought an even great threat to the environment.

Currently, sustainable sanitation development in the Dai areas requires good water environment and water supply, so we have to pay attention to keep good water environment, control domestic and the rural industry wastewater, treat the rural water pollution. It is worth to considering suggestions as bellow:

Promote the construction of wastewater disposal facilities not just in the urban area but also in the rural areas to meet the trend of growing amount of wastewater discharge. The sewer system construction is a real necessary for the rural reconstruction in line with current national policy of New Rural Reconstruction, in this case it is better to institutional invested by governments and villagers. Currently, the sewer system has covered 63% of the rural areas (The Statistics Yearbook of Xishuangbanna Prefecture, 2011), but mainly covered the rural area near the towns, the remote villages still in lower coverage.

Then, the end wastewater disposal facilities are also important to be built, if the sewers only take wastewater into the rivers will pollute the environment at same way. So, more wastewater disposal plants should be built in the towns as well as promote construction of the biological purification facilities in the villages to treat the domestic sewerage.

Acknowledgement

Mainly the data used in this Chapter was from the fieldworks in 2011~2013 by Author.

17.6 REFERENCES

Damounglong Township Government Office (2012). The Rural Statistics of Damounglong Township.

Jinghong County Statistics Bureau (eds) (2011). The Statistics Yearbook of Jinghong County. Jinghong.

The Editing Committee (1994). Men La County Book. Yunnan. People's Press. Kunming, China.

The Editing Committee (1996). Jinghong County Book. Yunnan People's Press. Kunming, China.

The Editing Committee (1998). Xishuangbanna Prefecture Book. Yunnan Peoples Press, Kunming.

The Editing Committee (2008). Condtions of Yunnan Province. Yunnan People's Press. Kunming, China.

Xishuangbana Dai Nationality Autonomous Prefecture Statistics Bureau (eds) (2011). The Statistics Yearbook of Xishuangbanna Prefecture. Jinghong.

ZHENG X. Y. (2005). Water Culture, Water Environment and Healthy Daily Life of Dai Nationality. Globalization and Ethnic Culture. China Book Press, Beijing, China.

ZHENG X. Y. (2005). The Water Culture of Dai Nationality and Its Sustainable Development. The Ideological Front. Vol. 6, 577, China.

Chapter 18

Evolution of sanitation services in the city of Rome between urban development and environmental quality

Renato Drusiani, Alessandro Zanobini, and Gianmarco Margaritora

18.1 INTRODUCTION

The birth and subsequent development of a sewerage and storm water system in the city of Rome is closely bound up with the position and nature of the land on which the first human settlements were built that, according to tradition, in the 8th century BC lay at the origin of the city of Rome. These settlements were situated on the heights close to the Tiber river which formerly marked the boundary between the Latin peoples and the Etruscans. The morphology of the land surrounding the Tiber river bed consisted of a series of short deep transversal ravines that favored the run-off of rainwater. This water was added to that of the springs that, in the lower part of the territory, produced large marshlands fed above all by the periodic flooding of the river, as shown in Figure 18.1.

The requirements of urban development starting from the first settlements thus made it necessary to construct a vast drainage and run-off system to discharge this water into the nearby Tiber. Indeed, during the reign of Ancus Marcius (*ca*. 7th century BC) the city developed on the right bank and had occupied the *Janiculum*, which was linked to the city by the *Pons Sublicius* bridge, the first bridge spanning the Tiber immediately downstream from the Tiberine Island.

The canals dug during this period, technically belonging to the Etruscan school, were actually reclamation canals as they initially collected spring and rain water. Later, with the discharge of domestic effluent they were transformed into sewers. In their workmanship and function these large-scale works rank on a par with the reclamation works of Fucino Plain, located in central Italy and (partly) with those of the the Pontine Plain, ancient marshes located SE of Rome (Imperial age from Augustus to Trajan). In any case these are the works which were to form the basis of the true sewerage infrastructures which were expanded and technologically modernized and are still in place today. These achievements were also accompanied over this long time span by a body of laws and regulations that, although progressively improved, were based on original criteria and intuitions that have remained valid down to the present time.

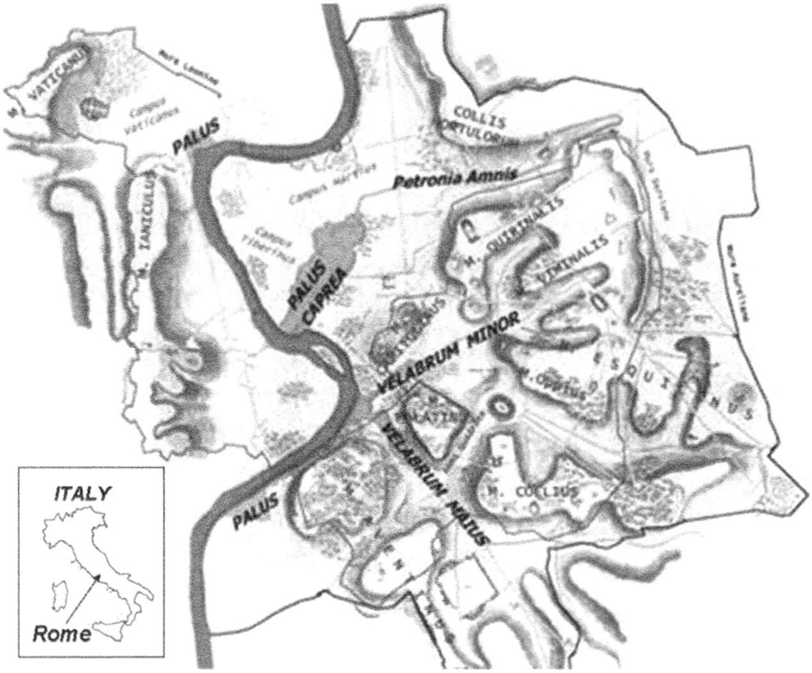

Figure 18.1 Surface waters in ancient Rome (on the basis of maps of G. B. Brocchi, 1820).

18.2 THE *Cloaca Maxima*

It is under the rule of Lucius Tarquinius Priscus that the centuries-old history of the urban drainage system of the city of Rome begins to take shape. This king decided to drain the area around the Forum as well as the depressions lying between the hills by means of sewers with a suitable gradient to discharge into the Tiber as it was not an easy task to drain the water from the plain (Livy *Ab Urbe Condita*, I, 38, 5). The construction of the *Cloaca Maxima*, without doubt the most important infrastructure, took place at the end of the 6th century BC, under the reign of Tarquinius the Proud. The historian Livy describes it as a work that outstripped in grandeur any other construction of the time, that is, the Augustan age (Livy *Ab Urbe Condita*, I, 56, 2-3).

In detail it consisted of a large canal that, probably following the natural contours of the land, was designed to drain off several surface springs, reclaiming the surrounding land. It later had the task of removing both waste water and rain water. It initially consisted of an open canal crossed by wooden foot-bridges to enable people to cross it (Hopkins, 2007) and was later completely covered in as the city gradually became more urbanized. The *Cloaca Maxima*, starting from the *Argiletum*, followed an ancient Roman road linking the district of the Subura (present-day Monti Quarter) to the Roman Forum. It received discharges from the Esquilinus, Viminal and Quirinal. The oldest section started at the Shrine of *Venus Cloacina*, the patron goddess of the sewer. This was followed, along a winding path through the Forum and the Velabrum, by the section of the Vicus Tuscus which then turned towards the Tiber at the entrance to the *Vallis Murcia*, as may be inferred from the layout (Figure 18.2) shown on the *'Forma Urbis'*. This is how the map of the city of Rome was defined at the time of the Emperor Settimius Severus marked on marble slabs and reconstructed in the late 19th century by the archaeologist and topographer Rodolfo Lanciani (Lanciani, 2007).

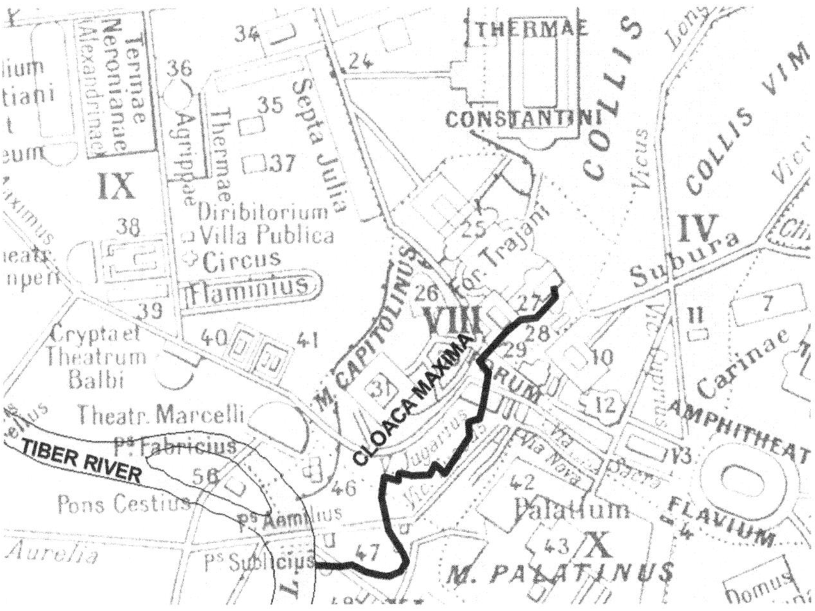

Figure 18.2 The drawing of the *Cloaca Maxima* (drawing of R. Drusiani on the basis of map of Noveau Larousse Ilustré, 1870).

Several secondary canals discharged into the main body. The earliest of these were carved out of the tuff and discharged into the walls while the later ones were made of brick and discharged from the vault. The maximum size of these secondary canals on entering the main body was about 0.60 × 1.20 m. The walls of the *Cloaca Maxima* were mostly made of large blocks of tuff (*lapis gabinus)* 3 ÷ 4 m thick; the bottom was partly paved with lava pentagons. In the oldest sections the vault was made of tuffs (lapis albano and *lapis gabinus*), and in the more recent sections in *lapis tiburtinus* (travertine), *opus latericium* (brick work) and *opus caementicium* (Roman concrete). The canal varied in size from about 2.10 m in width and 2.70 m in height to 3.30 m in width and 4.50 m in height. These great structures, initially designed as a drainage channel, when they began to be used also as a sewer might be considered a kind of water treatment system 'ante litteram' as the domestic waste water was diluted with the large quantities of storm water, especially in the part nearest to the outfall into the Tiber. A series of photos of the main conduit taken on the occasion of an inspection of the *Cloaca Maxima* carried out in April 2010 between the present-day Via Cavour and Forum of Nerva are shown in Figures 18.3 and 18.4.

The following Figure 18.5 shows a secondary canal branching off from the main canal which displays obvious analogies indicative of the same 'school' with the Etruscan sewers of the city of Perugia in central Italy, the urban centre of which dates to the second half of the 6th century BC (Cenciaioli, 1991). The cloaca's outlet into the Tiber river, which is partially covered by vegetation at certain times of the year, is located in the vicinity of the Palatine Bridge (Figure 18.6).

The huge size of this work, although hidden from most people's view, caused the admiration and interest of educated public and scholars throughout the centuries including Giovanni Battista Piranesi, a famous architect and engraver who was defined by his contemporaries as 'the Rembrandt of architecture'. He has left us (Piranesi, 1761) studies and views of numerous architectural works of the Ancient Roman era, including a study of the *Cloaca Maxima* (Figure 18.7).

Figure 18.3 Below the Forum of Nerva (with permission of ACEA).

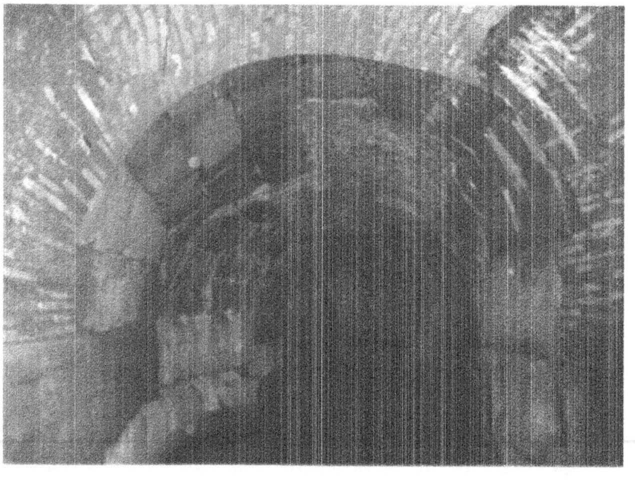

Figure 18.4 Section's change (with permission of ACEA).

It must be noted that even though the prime function of the *Cloaca Maxima* was a stormwater drain, the etymology of the word suggests that it must also have had functions of a hygienic nature. Moreover, the term *cloaca* was possibly derived from the Greek κλυζω which means to wash/clean, and even from the Latin *cluere* meaning to clean but also *colluere* meaning to rinse.

18.3 THE ROLE OF THE ROMAN SEWER SYSTEM

18.3.1 Management aspects

The role of the *Cloaca Maxima* in the urban development of the city of Rome is essentially indispensable from various points of view. This work was constructed to reclaim marshes along the river, to regulate the flow of water from the surrounding hills, rapidly drain the flood water of the Tiber and made it possible to act on the urban fabric by expanding the boundaries of the city of Rome.

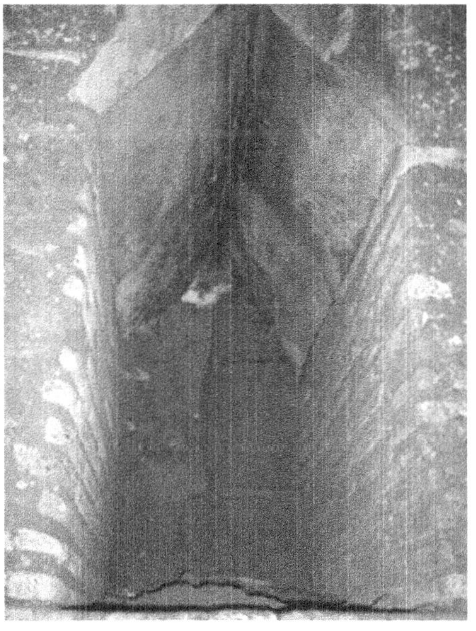

Figure 18.5 Mouth of a secondary branch (with permission of ACEA).

Figure 18.6 Connection of the Tiber river (with permission of ACEA).

It would be a limitation to consider the urban drainage for the only purpose of urban expansion; there is a reason to believe that there were also specific purposes of health and hygiene. While we remember that the Kings of Rome, who built the *Cloaca Maxima* were Etruscans, and there were strong trades and also cultural relations between Etruscans and Greeks, it should not be forgotten the great merit of the Greek school to have set a 'secular medicine' according to which the diseases, that could also result in outbreaks, had natural causes and were not due to the vagaries of the Olympian Gods. In the work of Hippocrates, 'On Airs, Waters and Places' written in the fifth century BC, there is an explicit link between the symptoms of disease and the presence of malarial swamps. Although the Greeks and Romans were ignorant about the nature of microbiological disease (*Plasmodium malariae*) and the species vector (mosquitoes), the correlation between stagnant water and diseases was evident to them. For this reason it was necessary to provide for reclamation of the ponds near human settlements. For further evidence we can cite *Titus Livius* who recalls the deadly epidemics that fatally struck those foreign armies camped in swampy and unhealthy areas. That's what happened in 390 BC to the Gauls led by Brennus during the unsuccessful siege of the Capitol (*Titus Livius Ab Urbe Condita*, V, 48, 2), and later in 212 BC to the Carthaginian troops camped outside Syracuse in the delta of the Himera river during the Second Punic War (*Titus Livius Ab Urbe Condita*, XXV, 26, 8).

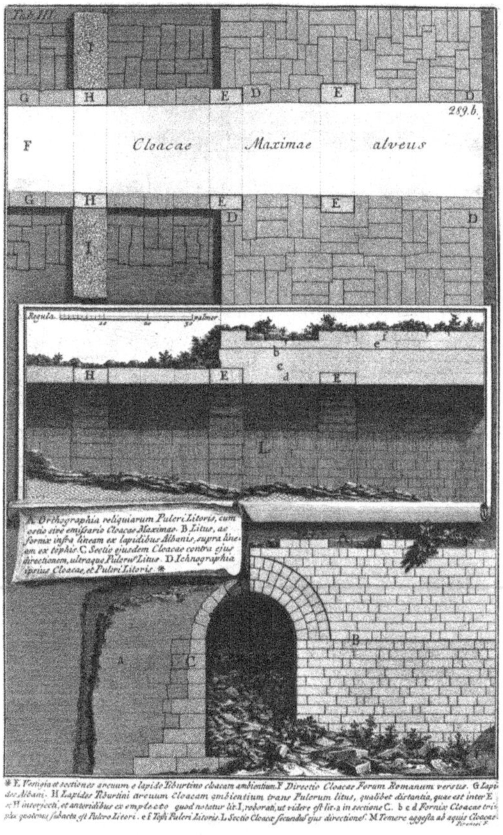

Figure 18.7 Engraving of G.B. Piranesi representing the *Cloaca Maxima* (*De Romanorum Magnificentia Et Architectura, Roma, 1761*).

With the building of the *Cloaca Maxima* it was consequently possible to ensure the security and health of areas that were initially insecure and unhealthy, allowing them to be enhanced with monuments and the paving typical of the more important public areas. The role of this initial sewage system inevitably evolved as the rapid growth of Rome from fortified village to city raised problems that were new for its original population of farmers and shepherds (Jansen, 2000). In particular, the following problems had to be faced:

(a) in a city more densely built up compared with the village structure of the early monarchic period and that was beginning to construct vast paved areas, it was no longer possible just to let the rainwater run off spontaneously or be rapidly absorbed by the soil;
(b) the removal from the streets of dirt, dust, etc. deposited by the wind or produced by the various constantly growing activities needed to be carried out as this dirt gradually accumulated, although this required simple but effective systems to guarantee an acceptable degree of functionality of the streets and thus of civil life itself;
(c) in a village it is easy for someone to dispose of excrement in the open fields, something that cannot be done in a built up area, in particular when it is largely paved. This leads to the need to construct places for this purpose inside and outside the dwellings, together with suitable systems for removing the excreta.
(d) in so far as the population had grown, the city demanded greater quantities of water as the local resources such as springs, cisterns or wells could no longer cope. After the construction of the aqueducts (the first, the Appian aqueduct, was built in 312 BC centuries before the *Cloaca Maxima*) water consumption naturally increased together with the need to discharge greater quantities of water, whether used or not, since the aqueducts of the time were of the continuous flow type.

All this made it necessary to expand the sewer system from the initial main drains, still largely uncovered, into a system of secondary networks that penetrated the urban fabric, scrupulously avoiding counterslopes. The close link between road network and the sewer/drain network immediately below, as shown in Figure 18.8 (Macaulay, 2000) made it possible to use the urban areas also in the case of heavy rain, as well as contributing to a certain extent to the collection and disposal of urban solid waste, which were services spontaneously rendered by the sewer network itself in rainy weather.

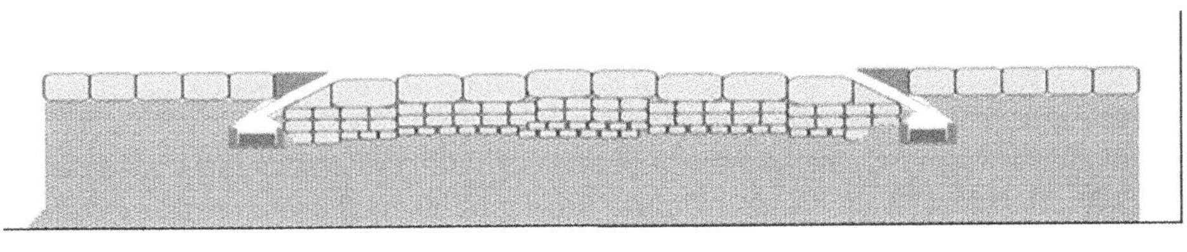

Figure 18.8 Section of a Roman road connected with the sewerage system (with permission of R. Drusiani).

The run-off of rainwater from roads and squares was afforded by special manholes carved with refined artistic taste. One believed to be the most famous is the 'Bocca della Verità' situated inside the porch of the palaeochristian church of Santa Maria in Cosmedin in 1631 on the occasion of the restoration ordered by Pope Urban VIII. This work was originally situated in the church of Santa Maria in Fontana di Roma on the Tiberine Island, as may be seen in a mediaeval guide for pilgrims (*Mirabilia Urbis Romae* 29) dating to the 12th century, that is, in a position quite close to the terminal part of the *Cloaca Maxima*. It consists of a marble sculpture 1.75 m in diameter representing a male human face (see Figure 18.9) and that according to popular tradition bites the hand of any unfaithful woman pushed into the hole corresponding to the mouth.

Figure 18.9 The Mouth of Truth (with permission of ACEA).

The functions of the sewer network included also the drainage of the excess water from the fountains and *castella* scattered over the city. This excess water (*aqua caduca*) was discharged into the sewers to keep them clean and thus reduce the production of bad odours. This use was amply described by Frontinus (*Sextus Julius Frontinus*) who was given the job of *Curator Aquarum* in 97 AD under the Emperor Nerva. In his work *De Aquaeductu Urbis Romae* Frontinus writes that '*Since it is inevitable that some water flows from the fountains it should be reserved not only for the health of our city but also for the purpose of cleansing the sewers*' (Frontinus, *De Aquaeductu Urbis Romae*, 111).

This highlights the close link that existed from these early times between the sewer system and the aqueduct system. However, this does not represent a mere re-interpretation post hoc and in a modern key of a (correct) territorial policy implemented over two thousand years ago, a situation that was clearly perceived as such from the outset.

In this connection it is worth quoting a passage from the geographer Strabo who, on the subject of town planning, extols the approach of the Romans versus that adopted by his Greek fellow citizens. He was in fact born in 58 BC at Amasya, in the North-West of present-day Turkey. Strabo (*Geografia*, Book V. 3. 8.) claims that '..... *but the Romans have added still others, which are the result of their foresight; for if the Greeks had the repute of aiming most happily in the founding of cities, in that they aimed at beauty, strength of position, harbours, and productive soil, the Romans had the best foresight in those matters which the Greeks made but little account of, such as the construction of roads and aqueducts, and of sewers that could wash out the filth of the city into the Tiber*'.

This becomes even clearer when Strabo refers to Greek cities like Smyrna which had paved streets but no sewer system (Strabo, *Geografia*, Book XIV. 1. 37) and notes what he considered to be obvious design flaws: '*But there is one error, not a small one, in the work of the engineers, that when they paved the streets they did not give them underground drainage; instead, filth covers the surface, and particularly during rains, when the cast-off filth is discharged upon the streets.*'

As we have seen a special task performed by the sewer system was to dispose of the discharge from latrines, also in this case revealing a close link with the aqueduct system. Latrines, in particular the public ones characterized by a more frequent usage, required clean running water to continuously flush out the excrements and at the same time to clean the sponges used by the patrons. Latrines were available for a small charge paid to the *conductores foricarum*, and provided an occasion to hold business meetings, to discuss everyday matters, etc. All this took place without any sense of shame or breach of privacy in richly decorated premises which, in winter, were often also heated (Carcopino, 1999). These latrines often housed small temples or rather *aediculae* dedicated to patron saints protecting this particular function, a custom criticized or rather ridiculed by the Father of Church, *Flavius Clemens Alexandrinus* who, albeit at some risk (one more century had yet to pass before the Edict of Constantine that officially put an end to the persecution of Christians), in the early 2nd century wrote '*The Romans, attributing their greatest successes to Tyche and believing this goddess to be very powerful, went and placed her in the latrine*'. (Clemens, *Protreptico ai greci*, Chap. IV)

In so far as they were water-intensive structures (although to a lesser degree than the thermal baths), it is not surprising that their development reached its peak above all starting in the 2nd century AD (Neudecker, 1994), that is, when the aqueduct system was approaching it maximum development in supplying the city. The lists we have of the sites of interest, the monuments and the infrastructures of public utility drawn up in the 4th century AD, the so called Regionary Catalogues (Valentini *et al.* 1940), reveal the presence of some 150 presumably public latrines in the city of Rome. There are nowadays practically no traces of these structures remaining in the city of Rome. Only in 1963, after the accidental collapse of a containing wall in front of the church of St. Peter's in Montorio sited in the Trastevere area (http://www.provincia.roma.it/provinz/news/9562), were the remains of an ancient Roman latrine discovered.

In any case, some idea of what these latrines must have been like may be obtained from the nearby city of Ostia. In the archaeological area of Ostia Antica, just over 20 km from the capital, some well-preserved remains of several public and private latrines may be found (Figures 18.10 and 18.11).

Figure 18.10 Latrines in the Forum of Ostia Antica (with permission of R. Drusiani).

Figure 18.11 Latrine in the house of Fortuna Annonaria in Ostia Antica (with permission of R. Drusiani).

Precisely with reference to the various private latrines situated in Ostia it is possible to admire the rational layout of the latrines and the water canals which allowed discharge into the sewer from both the latrine and the kitchen. It must not however be thought that the connection of the latrines to the public sewers was normally the case. Indeed, the majority of the latter discharged into septic tanks (*lacus*) the contents of which were periodically extracted by the socalled *stercorari* and used for agricultural purposes (Varrus, *De re rustica*, I, 29). Despite the strong impulse given by Cato the Elder during the course of the 2nd century BC when he held the post of Censor, to expand the sewer network in the capital in order to dispense with these evil-smelling tanks, they continued to exist in large numbers still under Emperor Trajan, that is three centuries later (Carcopino, 1999).

Among the by no means orthodox uses of the *Cloaca Maxima* that its huge size allowed mention must be made of that of disposing of 'excellent cadavers'. This was the case in 221 AD when the corpse of the eccentric and hated Emperor Heliogabalus was thrown into the cloaca and obstructed it, only to be pulled out again by the enraged population and cast into the Tiber (D' Onofrio, 1970). One century later, in 304 AD, comes the turn of the martyrdom of St. Sebastian, a victim of the persecution of the Christians unleashed by the Emperor Diocletian. The saint's body was indeed thrown into the *Cloaca Maxima*, a place in which the martyrs' corpses were often abandoned so as to avoid them being found and then buried (Jacopo da Varagine, *Legenda Aurea*, Chap. XIII); this episode provided inspiration for an oil painting by the famous Bologna baroque school painter Ludovico Carracci (Figure 18.12).

18.3.2 Organizational and regulatory aspects

The management of these infrastructures of such importance for the life of the city was entrusted to specific offices. In particular, it was at the beginning of the imperial age, under the Emperor Augustus (Suetonius, *De vita Caesarum*, Augusto 37), that the organization of public services such as water supply, road system, waste disposal, food procurement and fire fighting, at least according to the evidence available to us, that a comprehensive regulatory system was set up or at least a sufficiently credible functional framework created.

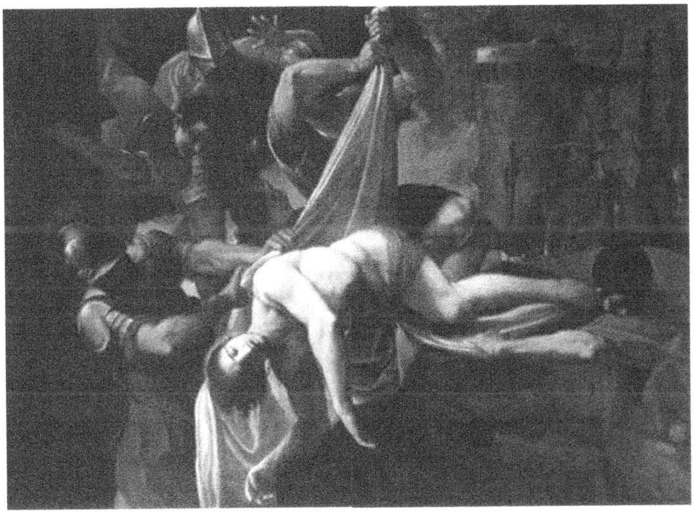

Figure 18.12 St. Sebastian thrown into the *Cloaca Maxima*. Oil painting by Ludovico Carracci (1612), 65 × 91 cm, Los Angeles, J. Paul Getty Museum.

Moreover, the intense activity of urban rehabilitation carried out during Emperor Augustus' long reign (from 27 BC to 14 AD), which represented a period of peace and stability that was almost unique in the history of the Roman empire, meant that the city could be provided with a more substantial water supply. This period saw the construction of the infrastructures that would bring *Aqua Virgo* and *Aqua Alsietina* to the city, the first for drinking purposes and the second for other uses, including in particular the supply of the great *Naumachia* (in Roman period, basin in which entertainment naval battles took place) in the Trastevere area.

The organization of these services in the capital called for a strong administrative body that was highly motivated and that could rely on top quality professional skills able to guarantee the functioning of the works and their development over time. The areas of competence ranged from water supplies to the maintenance of the Tiber river banks and bed and the sewer system. These areas of competence were interlocking and over time gave rise to different forms of organization. One significant organizational phase must certainly be attributed to Trajan who was particularly concerned with protecting Rome from flooding as well as with the various kinds of discharge, including that of the sewers, which might alter the quality of the Tiber river waters (Le Gall, 2005). The comprehensive organization of the overall services of public interest in the city and the surrounding territory (Cantù, 1839; Gatto, 1999) shown below for water-related services points to organizational criteria and a breakdown of responsibilities that are not dissimilar from those of the present day:

(a) *Comes portus*. Monitors the port movement from the mouth of the Tiber (including the port of Ostia) to the city centre.
(b) *Comes riparum et alvei Tiberis et cloacarum*. Regulates river activities in the city section of the river and controls the normal flow of the river water and the proper functioning of the sewers discharging into the river.
(c) *Comes formarum*. Responsible for the maintenance of the aqueducts and keeping the surrounding land clean.

Consularis aquarum. Allocates water to the various uses and collects the relative fees.

It should be borne in mind that the important water-related posts described were headed by a comes which literally means companion but was originally understood as the companion or counselor of the Emperor and thus someone who had undertaken an administrative career. Conversely, water distribution was directed by a person who had been a consul. In many cases, these posts were generically referred to with the term curator, for example, responsible for the activity.

Working under the functionaries responsible for the water services were engineers, architects, administrative functionaries, almost all with a military background, assessed by specialist and unskilled workmen (often slaves) for the performance of the more material tasks. Frontinus himself, before being appointed *Curator aquarum*, was a general commanding the *II Legio Adiutrix* in Britain.

With reference to the criteria regulating the management/safeguarding of the water infrastructures, as well as public health and hygiene, consideration must be given to a set of rules that were shaped and perfected over nearly a millennium. These range from the laws of the XII tables (*duodecima tabularum leges*), a corpus of laws established during the mid 4th century BC, down to the *Digesta*, as the collection of laws performed under Justinian I of the civil and criminal codes in the form of ordered fragments of Roman jurists from different periods was called. This 50 volume work was promulgated in 533 BC, but even before this, during the imperial period, the laws governing water and public health and hygiene had already been written.

In this connection mention must be made of the *Lex Quinctia de aquaeductibus* approved in 9 BC at the time of Augustus by the consul Titus Quinctius Crispinus Sulpicianus. This law contains provisions to safeguard the aqueducts, at the same time imposing heavy fines for those not respecting them. In particular, it was prohibited to damage aqueducts in any way but also to withdraw water from them for farming or other purposes without permission. This fundamental law regulating the water services of ancient Rome came down to us thanks to Frontinus who, in his treatise (Frontinus *De Aquaeductu Urbis Romae*, 129), makes explicit reference to it. In any case the steps taken by the authorities to cope with the problem of urban hygiene in Rome (as well as of any other city in the empire) were codified into various different rules ranging from the penalties applied, to maintenance and to taxation policies.

With regard to the penalties imposed the example may be taken of the regulations governing responsibility for the discharge of excrements and relative liquid containers from premises facing on to public streets, a problem that would afflict Rome well beyond the Middle Ages. In the Digest (*Digesto* 9, 3, 5) of Eneus Domitius Ulpianus, a Roman jurist of the 2nd century AD, the various profiles of the respective responsibilities of owners, tenants and sub-tenants vis-à-vis the innocent citizens covered with excrement or worse still struck by the heavy pots containing it. Again thanks to Ulpianus we are in possession of a whole set of regulations governing the proper functioning of the sewers which, as seen above, if necessary could also allow the discharge of waste material. The *praetor* supervised the activities of repair and cleaning of the sewers as demanded by public health and the proper conservation of the buildings. He took action in the case of unlawful discharges into the public sewers such as to alter their functionality. In such cases the person responsible for the unlawful behaviour was obliged to repair the damaged structure (*Digesto* 43, 23.1.15).

Furthermore, the construction of sewers was considered a compulsory chore for Roman citizens, at least during the Republican period (Milazzo, 1993). The extent to which this was deemed to be exhausting and above all demeaning is shown by the fact that some citizens even committed suicide to avoid this chore. Moreover, the cleaning of the sewers, the so called *purgationem cloacarum* is explicitly included (Pliny, *Epistole*, X, 42) as a possible alternative forced labour punishment of convicts.

As far as the specific financial arrangements aimed at defraying the cost of the service are concerned, the Digest, again in the part drawn up by Ulpianus (*Digesto* 30, 39, 5), refers to the *cloacarium*, namely an ad hoc tax imposed in order to pay for sewer maintenance. In any case, when dealing with the issue of the taxes imposed for public hygiene services it is necessary to mention at this point the innovations introduced by the Emperor Titus Flavius Vespasianus, who reigned from 69 to 79 AD.

One of Vespasian's main concerns and the reason for which he entered history is his constant pursuit of financial solidity and the proper functioning of the state machine. The reason for this policy of the Emperor, apart from his natural inclination (he was the son of Titus Flavius Sabinus, a tax collector), was also the result of the disastrous situation regarding the imperial finances that he inherited. Indeed one year earlier, with Nero's (assisted) suicide, saw the end of a period of extravagant expenditure due to the latter's megalomania, followed by a period of severe instability with as many as four Emperors in a single year which was very detrimental for the collection of taxes, as we learn from Tacitus himself (Tacitus, *Historiae*, IV, 9).

Vespasian thus marked the beginning of a period characterized principally by other values such as thrift, accumulation and saving – principles that would become central, not only for the individual, but also as a guide for the logic of the empire. In the constant attention he paid to the financial equilibrium underlying the huge imperial machine, Vespasian, as well as various other savings contributing to the finances, introduced a tax that had been previously unthinkable, namely that imposed on public latrines. Noteworthy in this connection (Suetonius, *De vita Caesarum*, Vespasianus, 23) is the famous anecdote about Vespasian who, reproached by his son Titus who considered it improper to fill the state coffers in such a way, put under Titus' nose the first cash obtained from the new tax saying *Pecunia non olet* that is 'money (whatever its origin) doesn't stink'.

18.4 DEVELOPMENT OF THE SEWERAGE-WATER TREATMENT SYSTEM AFTER THE FALL OF THE ROMAN EMPIRE

18.4.1 Sewer network development from the Middle Ages until the end of the 19th century

A 'dark age', also for the effects on the Capital's water system starts in 537 AD when Vitiges, King of the Ostrogoths, during the Greco-Gothic War, cut the eleven aqueducts supplying the city with water to facilitate his siege (which in any case failed). After this serious damage to the city of Rome, in the centuries that followed, only the Aqua Virgo aqueduct (most of which ran underground) continued to function, obliging the inhabitants of Rome to use wells and Tiber river water. More than one thousand years were to pass until, in 1587, a new source of drinking water was made available by the construction of the Felix aqueduct using the springs and part of the Aqua Alexandrina aqueduct.

In any case, during this long period, because of the chronic lack of water, in 1122 Pope Callixtus II decided to build an artificial canal to obtain water not especially for drinking purposes (flour mills, irrigation) using the water that used to supply the ancient Roman aqueducts of Tepula and Julia. This canal, denoted as Marrana, followed the course of the ancient aqueducts and descended towards Rome (Angeli *et al.* 2007). From this time on this type of canal, which had existed also in previous periods, took the generic name of *marrana*. These artificial watercourses ran through the urban area (an example in Figure 18.13) and one way or another backed up the sewerage system, extending and completing it, although for a different purpose.

Certain stretches of these canals were also covered over. It should be noted that, as far as their technology is concerned, as we glean from the cross sections contained in Figure 18.14 (Narducci, 1889), the Republican era canals and those of the 17th century were practically the same: 'cappuccina' roof, flat bottom, side drains extensively described and quite similar to those of the *Cloaca Maxima*. They were also constructed using wicker basketwork filled with crushed bricks: known as 'viminate' they became carbonized over time and, after more than two thousand years, the system incredibly still functions today.

Figure 18.13 Marrane near Porta Capena (engraving by S. Du Perac, Pianta di Roma editor Antonio Lafréry, 1577).

Figure 18.14 Comparison between a Roman sewer and a sewer of of the seventeenth century (Narducci, 1889).

The restoration of the ancient sewers and the construction of new ones, especially in areas where the natural ground level had increased considerably, began in about 1230 under Pope Gregory IX, and continued under Gregory XV, accelerated by the aftermath of serious epidemics in 1621. Urban VIII, around 1630, was the last pope to build important works in this sector. After this, and until 1870, the only further efforts made were to satisfy the most pressing needs by restoring or building small stretches of sewers linked to the large existing collectors.

The end of the 19th century was the period in which a start was made on the most extensive work to be done on the city's sewerage system after the construction of the *Cloaca Maxima*. This work involved the construction of two new sewer mains on the right and left banks of the Tiber (Figure 18.15) and is an integral part of the embankment project designed to protect the city from flooding (Box 18.1).

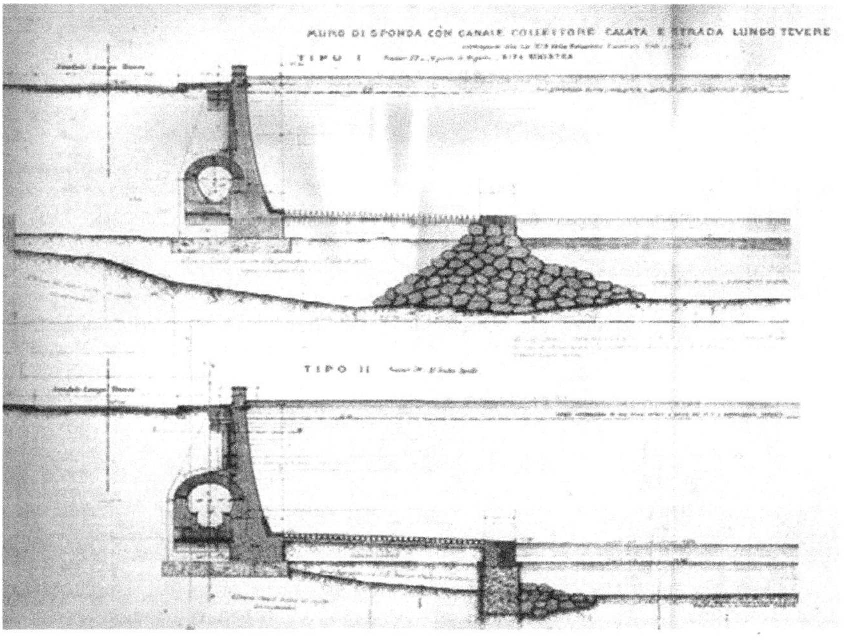

Figure 18.15 Section of the walls with sewer (Canevari, 1875).

These mains were connected to all the sewers and thus conveyed the domestic effluent and rainfall downstream from the city. The undeniable hygienic advantages of this work, although highly appreciated later, were not particularly popular at the beginning as they were conceived of as protection from flooding, including that caused by the Tiber water flowing back up through the sewer drains that often plagued the city. The city drain network continued to be a mixed network with overflow canals able to collect large quantities of storm water however it was drained off and provided with suitable overflow devices. In view of the energy levels available at the time (the 'hills' on the left of the Tiber and the high ground of the Janiculum, Monte Mario, etc. on the right) was one the best and most up to date systems that could be conceived of at the time. Still today, the network still has potential for updating so as to continue to satisfy new health, hygiene and environmental needs.

One important update was introduced in view of the anthropic development of the territory and led to the reduction of the sewer flow rate in dry weather (above all in view of the drop in the water table levels) and thus to the lesser dilution and less oxygen available for the domestic waste water in the sewers (the ratio of black to white water in the case of increased rainfall in Rome is 1/80 ÷ 1/100). The lesser dilution has led to the need to convey this water to the final receiving body as quickly as possible.

BOX 18.1 TIBER FLOODING IN ROME

The fruitful coexistence between the City of Rome and the River Tiber that flows through it is occasionally interrupted abruptly and the river, no longer a peaceful dispenser of benefits, can turn into a force of destruction. Suetonius narrates the frequent flooding that occurred at that time (Suetonius, *De vita Caesarum*, Augustus 30). Pliny the Elder shows that the Tiber is subject to frequent flooding, but that is just in Rome that this occurs with greater intensity (Pliny, *Naturalis Historia*, III, 55). Tacitus narrates the reaction to the disastrous flooding at the time of Tiberius (Tacitus, *Annals, book 1, 79*) in the year 15 AD. As far as the episode described by Tacitus is concerned, it is significant that the measures proposed to reduce the risk of flooding, such as the deviation of watercourses or the controlled flooding of certain areas, met with the compact opposition of persons determined to safeguard their own economic interests who used arguments not unlike those put forward in our times. They claimed, for example, that '*Nature had made the best possible choice for the good of mankind by assigning rivers their sources, their beds and, like springs, so should the mouths like the religious feelings of the allies who had consecrated worship, woods and altars to the motherland's rivers; also the Tiber itself could not accept to flow in the absence of the tributaries surrounding it with lesser glory*'. From 414 BC (the first flooding mentioned by Titus Livy) until 1870 some 132 flood events occurred (Lanciani 1975). The worst, believed to be that of 1598, also caused the destruction of a bridge (never rebuilt and thereafter known as the 'Broken Bridge') and caused 150 deaths. The Popes always had this problem in mind, for example, recognizing the fundamental importance for the hydrography of the Tiber, Benedict XIV ordered it to be made after the 1742 flood. Between 1180 and 1870 a catastrophic flood took place every 31 years, such an event being defined as a level exceeding 16 m at Ripetta while flooding involving extensive overflowing of the banks had a frequency of one every 18 years. However, inside the city it was a problem whenever the river level reached 12 m and sewers began to overflow: this happened once every 2 years. This age-old problem was solved after 1870 through the implementation of three projects – one based on a proposal already made by Julius Caesar (cut-off between Prati and the Sublicius Bridge); another made by Garibaldi (a diversification downstream from the Aniene junction as far as the downstream part of the city): the third was then that of the present-day 'muraglioni' embankments. The latter were built of bricks to a height such that since their completion in the early 20th century there has been no further overflowing. These embankments also contain two outfall drains serving the unitary city sewer system on the left and right banks of the Tiber which prevent backflow during flooding and have had the great environmental merit of conveying the sewer discharge downstream from the city to where, until the middle of the last century, the self-cleaning power of the river was sufficient. Today the discharge takes place after the Roma Sud treatment plant which serves the above two outfall drains.

Now that the final receiving body, when it is not raining, is reached through the treatment plant, this need is even greater. It is a matter of having sewer mains in which, for low flow quantities, the velocity of the fluid is such as to prevent the onset of putrefaction phenomena. Flat bottom mains were therefore replaced by mains with a 'V'-shaped bottom, as shown in Figure 18.16 (Corso Vittorio sewer built in 1884) which illustrates how the classic mixed water canal was now in current use. Owing to the growing street network on the surface the utility vault began to be used to lay and inspect water and gas pipes. This vault afforded a more rational management of the utility services but unfortunately is rarely present in the work performed in Rome after 1920. Also the sewer mains incorporated in the embankments had the same shape to allow fluid flow at low flow rates. Nowadays the hydraulic sections of modern mains are still elliptical in shape for the same reasons. The larger ones are provided with an embankment to further speed up fluid flow at low filling levels (Figure 18.17).

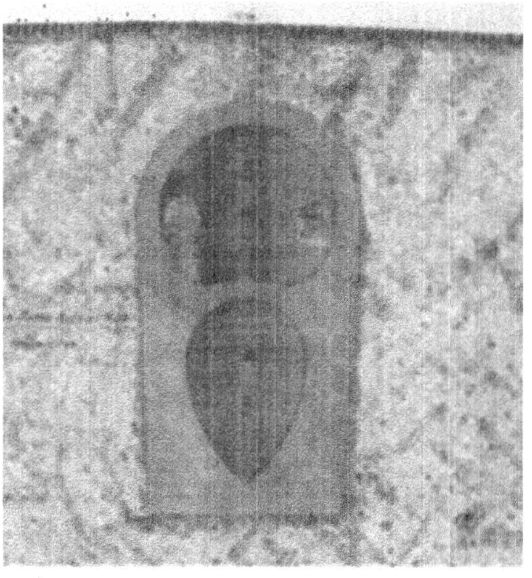

Figure 18.16 Sewer of Corso Vittorio Emanuele (Narducci, 1889).

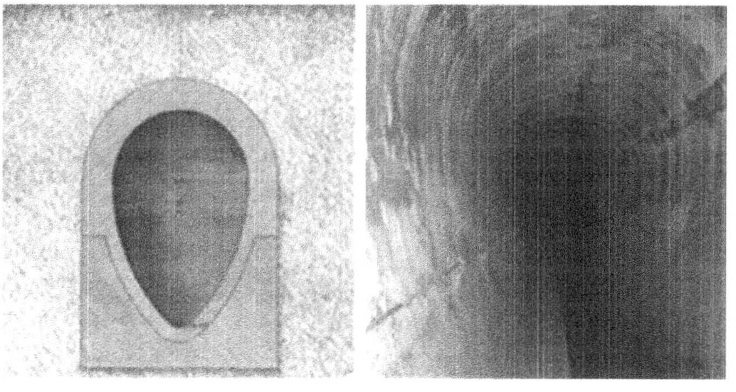

Figure 18.17 Examples of ovoid section of sewers dating (Narducci, 1889 and with the permisisone of ACEA).

18.4.2 Present-day configuration of the sewer system of the Capital

Until the beginning of the 20th century the most wide-spread sewage systems to carry off the city water consist in its dilution in waterways or the spreading on farmlands while biological oxidation process with the use of trickling filters (Cappa, 1907) were being introduced. The biological filtering systems would spread in Italy especially after the 70's it is indeed with the national law 319/1976, the so-called Merli law for the protection of the water quality, that specific chemical/biological limits are introduced to wastewater.

The Rome sewer system was conceived of as being served by four treatment plants with a subdivision into three networks inside the ring road. Figure 18.18 shows the location of the large treatment plants and the two 1870 mains on the right and left banks of the Tiber. The choice of the system with combined sewer in these three areas depended on the fact that it was not possible to completely alter the existing system which, by now, included even the recently urbanized areas. The sewage network used for the filtering Roma-Ostia plan, was instead foreseen as a separate sewage system which in free flowing area would have been at a level lower than of that of the final bodies of water and also of the same surrounding water table.

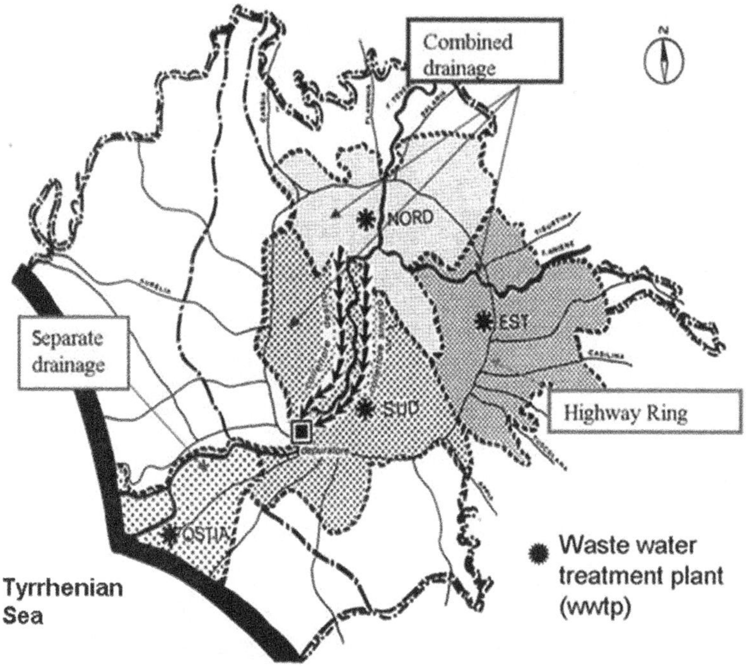

Figure 18.18 The nature of the sewer system in the Rome's area (with the permission of ACEA).

Construction work began on the treatment plants in the 1970s when tenders were called for a system of modular development, that is, an articulation consisting of successive sections to be built as funds became available as the predicted expansion of the city took place.

These plants, the overall characteristics of which are set out in Table 18.1, were (and still are) subject to constant expansion and upgrading both to increase and enhance their treatment capacity and above all to allow for an ever-evolving regulatory system.

Table 18.1 Rome wastewater treatment plants.

Plant name	Drainage system	Start up (yr)	Served inhabitants	Treated flow (cm/sec)	Type of treatment
Rome Ostia	Separate	1974	350,000	0.80	Activated sludge
Rome East	Combined	1976	900,000	3.00	Activated sludge
Rome North	Combined	1981	780,000	3.30	Activated sludge
Rome South	Combined	1985	1,100,000	10.00	Activated sludge & biofiltration

After the various interventions carried out in the course of the years, these plants have now attained a treatment potential that allows them to guarantee the treatment service for the entire population of the Municipality of Rome. Figure 18.19 shows a satellite view of the largest of the four plants, that of Rome

South. The management of the wastewater treatment service was entrusted by the Municipality of Rome to ACEA SpA in 1985. That of the sewer system was also entrusted to ACEA later, in the second half of 2002, in accordance with the provisions of national law 36/1994 (the socalled 'Galli Law') which made the integrated management of the water cycle mandatory.

Figure 18.19 Satellite view of the waste water plant of 'Rome South' (from Google Earth).

18.4.3 Future development of a sustainable urban drain system for the city of Rome

During the 1990s a consultancy group set up by the Municipality of Rome with the participation of Ministries, Universities, and Research Centers, in addition to the management agency ACEA, laid down the new guidelines for urban drainage in the 3rd millennium. The specific points on which it was based are as follows:

(a) Upgrading of treatment plants.
(b) Treatment of the larger part of the white water in the catchment areas served by the Rome East, North and South plants, and
(c) Reuse of white water for irrigation and recycling in the aquifer, especially in the coastal zones.

As far as point (a) is concerned the technical upgrading has largely been completed; the most of the future developments will be related to points (b) and (c). As far as point (b) is concerned, it must be noted that the storm water drainage system of the city of Rome is influenced by a number of negative and positive factors. The negative factors include heavy rainfall, a phenomenon that is likely to increase (Kundzewicz *et al.* 2006) due to the climate change, as well as the difficulty/impossibility of selecting suitable areas for the construction of shuttlecock basins. The positive factors include the city's orography which is characterized by distinct neighboring catchment areas that allow rapid displacements of the rainfall during wet weather. Added to this is the 'reservoir effect' produced by (historically) over-sized mains that can be used as a shuttlecock basins in rainy weather.

The solution then consists in the installation of switches and shut-off devices enslaved to a system of remote sensing rainfall-hydrometric, in the nodal points of the drainage network. In the sewer basin where the energy level required for runoff is attained by pumping systems, the dynamic control of the water volumes in the case of rain can readily be performed by pumping systems. The gravity operated sewers require the remotely controlled mechanical structures regulated according to the receiving capacity of the network downstream. These systems are triggered using methods based on the latest developments in the micro-meteorology field.

We must point out that, although the dynamic regulation of the water volume gathered in the drain lines, is a step ahead compared to the traditional management of the combined sewer network, it also involves considerable risks, compared to the fail-safe system currently in use in Rome. In order to make such regulations feasible, the current water service management is working to develop an efficient telemetric system, equipped with self-diagnosis, and with adequate redundance, so to increase security factors in the execution of maneuvers automatic regulation. Managerial strategies of control in real time, which represent an optimal compromise between the maximization of the water from purified raid and the minimization of the hydraulic risk, are then being implemented.

With regard to the point (c) it should be noted that during the past century the intense urbanization resulted in a series of alterations in the Rome system that caused reduced infiltration, greater velocity of surface runoff, quality differences in the collected water (domestic waste, industrial effluent, rain water). The effects of climate change should be considered as well. The water tables are almost everywhere at levels lower than in the past (AA.VV. 2011) and this facilitates the entrance of salty water in the same tale. To contrast this phenomenon we may introduce in the table the treated sewage water, but this operation requires that the final treatment is able to produce water of adequate quality.

The future Rome sewer/drain system is shown in Figure 18.20. Potentially 7 m^3/sec will be captured to irrigate farmland between Fiumicino and Civitavecchia west and northwest of Rome respectively. This approach will eliminate (or at least reduce) withdrawals from the aquifers and thus offset the saline seepage into the water table. This is the new meaning of the concept of 'reclamation': in the past, it meant draining land while nowadays it can also mean the addition of freshwater as a mean to offset salinization.

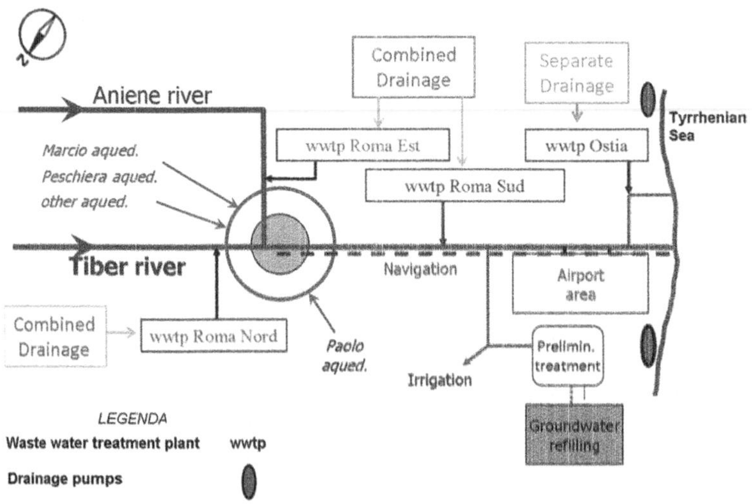

Figure 18.20 The current Roman sewer system (with the permission of ACEA).

18.5 DISCUSSION AND CONCLUSIONS

The history of Roman sewerage systems from Ancient Roman times up to now is longer than the history of Roman Aqueducts. The first water infrastructures had to deal with the urban growth of the Rome area. Important structures like the *Cloaca Maxima* (VI century BC) were necessary to drain the area at the foot of the Palatine Hill and discharge it into the Tiber river.

For the above mentioned reasons there is reason to believe that the implementation of this choice put into operation three centuries before the great aqueduct (acquedotto Appio) there were also sanitary reasons; these sanitary concerns are also present in the hydraulic and urbanistic Etruscan culture, probably inspired by the Greeks. Similar considerations may be made regarding the construction techniques used for several hydraulic works and in particular the tunneling techniques that we find in the underground networks in the Etruscan cities.

On the other hand it was typical of the Romans to assimilate, incorporate and often perfect the knowledge and technology possessed by the peoples they conquered and Romanized. In any case the results were excellent and the non-Roman chroniclers of the time remarked that the water management solutions adopted in an urban environment made Rome a unique example among other cities, particularly Greek as Strabo in *Geografia,* mentioned earlier. The credit also goes to the management and maintenance based on trained technical corps set up in accordance with a precise organization scheme and dedicated to guarantee the functionality and decorum of the city.

It is also interesting to note that a set of rules governing wastewater management (and that of water in general) in Ancient Roman times were significantly similar to approaches adopted in later and even recent eras. This is true for instance of the 'polluter pays' principle in its various forms ranging from the application of taxes to that of fines.

In this case we are facing a happy synthesis between a 'pragmatic vision' typical of the romans of that period, that wanted to discipline, in the best way, the management of the public infrastructures, with the law school, from which so-called roman law, that has greatly influenced, especially in the western world, the legislations of many countries has derived.

The fall of the Roman Empire led to a substantial decline in the functionality of many of the water works as a result of the breakdown in the civil organization having the task of ensuring the maintenance and development of the infrastructures. The robust nature of the infrastructures affected and the intrinsic protection afforded by the fact that they were underground limited widespread destruction caused by the barbarian invasions, allowing these works to continue to make a positive contribution to the life of the city even during the Middle Ages and down to our times. Indeed they were adopted as a paragon of difficult imitation by artists and architects between the Renaissance and the Baroque periods.

Indeed it may be claimed that only towards the late 19th century there was a strong refocusing of attention and a renewal of investments in urban drainage above all to protect the city centre from Tiber River flooding. In the second half of the 20th century under the combined effect of strong urban growth and the availability of modern purification techniques the idea of improving the environment was strongly reappraised. It was during this period that wastewater treatment plants serving the city of Rome and surrounding territory were constructed.

The actual technical characteristics of the sewage system were adapted to suit the different needs of the city. From a simple initial lowland drainage, to a sewage and rainwater container and finally, maintaining the unitary system in the historic part of the town equipped with modern real-time control and with wastewater treatment plants, which conveys purified water to the Tiber downstream from the city. The planned developments, already being implemented, are designed to consolidate and enhance this model.

18.6 REFERENCES

AA.VV. (2011). Roma Capitale: Relazione sullo Stato dell'Ambiente (Roma Capitale: Report on the State of the Environment), Roma, Italy.

Angeli F. A. and Berti E. (2007). Nascita di un Fiume: la Marana (Birth of a River: The Marana). Associazione Culturale SestoAcuto, Roma, Italy, http://www.medioevo.roma.it

Cantù C. (1839). Delle Legislazioni. Appendice alla Storia Universale (On the legislation. Appendix to the Universal History). Editori Pomba, Torino, Italy.

Cappa S. (1907). Idraulica Pratica (Practical Hydraulics), Editrice Carlo Pasta, Torino, Italy.

Carcopino J. (1999). La Vita Quotidiana a Roma. Editori Laterza, Roma (The daily life in Rome), Italy, pp. 52–53.

Cenciaioli L. (1991). Cunicoli di Drenaggio a Perugia da 'Gli Etruschi Maestri di Idraulica' (Drainage tunnels in Perugia, from 'The Etruscans Masters of Hydraulics'). Electa Editori Umbri, Perugia, Italy.

Gatto L. (1999). Storia di Roma nel Medioevo (History of Rome in the Middle Ages). Newton & Compton Editori, Roma, Italy.

Hopkins J. N. N. (2007). The Cloaca Maxima and the monumental manipulation of water in archaic Rome. The Water of Rome num. 4, March, Rome, Italy.

Jansen G. C. M. (2000). Systems for the disposal of waste and excreta in Roman cities. The situation in Pompeii, Herculaneum and Ostia in Sordes Urbis da La eliminaciòn de residuos en la ciudad romana. Actas de la reuniòn de Roma, Rome, Italy, pp. 37–49.

Kundzewicz Z. W. Radziejewski, M. and Pínskwar, I. (2006). Precipitation extremes in the changing climate of Europe. *Climate Research*, **31**, 51–58.

Lanciani R. (1975). Le acque e gli acquedotti di Roma antica (The waters and the aqueducts in ancient Rome). Edizioni Quasar, Roma, Italy.

Lanciani R. (2007). Forma Urbis Romae. Edizioni Quasar, Roma, Italy.

Le Gall J. (2005). Il Tevere, Fiume di Roma Nell'antichità (The Tiber River in ancient Rome). Ed. Quasar, Roma, Italy.

Macaulay D. (1978). City: A Story of Roman Planning and Construction. Houghton Mifflin Company, Boston, Massachusetts, USA.

Milazzo F. (1993). La Realizzazione Delle Opere Pubbliche in Roma Arcaica e Repubblicana (The Realization Of Public Works in Archaic and Republican Rome). Esi Editore, Napoli, Italy.

Narducci P. (1889). Sulla Fognatura Della Città di Roma. Forzani e c. Tipografi del Senato (On the Sewerage of the City of Rome), Roma, Italy.

Neudecker R. (1994). Die Pracht der Latrine. Zum Wandel öffentlicher Bedürfnisanstalten in der kaiserzeitlichen, Stadt, Pfeil, München, Germany.

Piranesi G. B. (1761). Romanorum Magnificentia et Architectura. Roma, Italy.

Valentini R. and Zucchetti G. (1940). Codice Topografico Della Città di Roma I (Topography Code of the City of Rome I). Tipografia del Senato, Roma, Italy.

Chapter 19

History of the sewerage system in Barcelona, Spain: From its Origins to Garcia Faria Plan

Montse Martinez, Pere Malgrat, Cristina Vila, Roman Llagostera, and Miquel Salgot

19.1 INTRODUCTION

The water is basic for human life and as a consequence it has been usual to establish human settlements near water sources. When those settlements became permanent, water sources needed to have the capability of maintain regular flows all around the year to supply the needs of dwellers and in this way ensure the continuity of the villages and towns. It can be said then, that water is basic for urban life (Conillera *et al.*, 1986).

Water availability and use in a continuous way caused the permanent generation of polluted water issued from towns: wastewater. From very old times, and due to the characteristics of wastewater, the dwellers needed to find ways to convey wastewater out of the towns and sewerage systems appeared: this was the case in Minoan times and later on in Greek and Roman times (Angelakis *et al.*, 2012). In the Middle age the hygienic measures disappeared in several places and it was common practice the use of cesspools, and even streets were used for discharging excreta with noxious consequences for the health of the inhabitants.

Walled cities were suffering specially these circumstances due to the overpopulation in reduced areas and the difficulties to evacuate waste as well as the scarce importance conferred to sanitation practices. Along the history of Mediterranean towns, epidemics of water related illnesses appeared in relation with polluted drinking water and wastewater disposal on streets, and history of the cities was pointed by those episodes which marked their evolution and even lead to the disappearance of several of them. Lately, the development or recovery of sewerage systems helped to reduce the mentioned negative health-related circumstances and for this reason the lack and the evolution of wastewater related infrastructures can help to explain historic health-related changes as well as decreases or increases in urban population.

Barcelona, a Spanish town located in the Mediterranean seashore, in the northwest of the Iberian Peninsula, did not avoid the mentioned circumstances. Nevertheless, as an important part of the documentation, relative to Barcelona before the 11th century, was destroyed by the Saracen troops of Al

Mansur during their razia against the town in 985 AD, there is a lack of information on the town for the first millennium of the Christian era. For this reason the available information on the city infrastructures comes in part from the archaeological research.

The origins of the town are confused and from before the Roman domain, the area is preserving remnants from the end of the Neolithic and beginnings of the Calcolithic. Afterwards, the Laietani (Iberian people) culture was developed in the area and certain commercial activity existed. During the second Punic war (218–202 BC) Cartago occupied the site and many papers establish that this is the date when the town was founded. Summarising, the place seems to have been ruled along the centuries by Iberians or Laietans, Carthaginians, Romans, Jews, Romanised Visigoths and Muslims, which named the town Barkeno, Barcino, Barcinona, Madîna Bar^Giluna and finally Barcelona.

Apart from the discussions on the date of the first establishment of people in the present area of Barcelona, there is a consensus on the existence of a Roman settlement named *Colonia Iulia Augusta Paterna Faventia Barcino* in the time of the Roman Emperor Augustus (Voltes, 1967), and is generally accepted that the Roman town was founded between 15 and 5 BC (Hernández-Gasch, 2006) over older settlements.

At the Roman times that colony was less important than other neighbouring towns, such as Tarraco or Betulo (the present Tarragona and Badalona, respectively). Since this founding or transformation by the Romans, on the tiny summit of Mons Taber, Barcino has become a city where water plays a crucial role. As the town gained importance all along the centuries, becoming the capital town of Catalonia, the need for water increased accordingly.

After Romans, Visigoths and Muslims remained around one century each, and few traces of those times are found. In the beginning of 9th century, Franks created the *Marca Hispanica* as a barrier against Muslims. Finally the House of Barcelona was the senior of several counties (whose leaders were all related by family ties) and a dynasty governing Catalonia appeared, being afterwards in close relation with the kingdom of Aragon.

The political changes, as well as the growth of economy and commerce and sanitary problems, caused an uneven development of the town along the studied period (Salgot & Angelakis, 2012).

Barcelona has been and still is a maritime city, bounded by two great rivers, the Llobregat in the South and the Besòs in the North, and the nucleus of the colony was framed by two streams originating in the Collserola small range: Cagalell and Merdançar (Catalan names that directly refer to the functionality of these two water courses as places for wastewater disposal).

Back in history, and in relation with water, it is to note that the development of the town, yet in Roman times, soon required new water supplies from distant places when it was clear that the wells were not enough to satisfy the needs of good quality water. Consequently, Roman engineers (Miró and Orengo, 2010) developed a water supply infrastructure – the classical aqueducts – whose remnants are still present in town (Figure 19.1). The supply of important amounts of water inside the walls, mainly dating from 4th century, required ways to eliminate it safely and as usually happened in Roman Towns a sewerage system was constructed.

The physical features of the city have ben marking the development of the supply and sewerage systems along the centuries. Besides from the institutional buildings located in the top of a small hill (Mons Taber) the old town and its extensions are spread in an alluvial flat area formed along the centuries by the sediments of several temporary streams and over two deltas, from the present Llobregat and Besòs rivers. Quite all the materials settled on the piedmont of the Collserola range, laying a few kilometres inland, during the Pliocene which favoured further sedimentation from the materials carried by the marine streams. Apart from the wetlands which appeared in the interface land-sea, the Mons Taber remained in the centre of the plain which was where the town was built later on.

Figure 19.1 Badly preserved arches of the Roman aqueduct, Barcelona (M. Salgot, with permission).

The present city location among the sea and Collserola range, is physically characterized by a 4% slope from the highest point, the Tibidabo Hill (565 m high, part of Collserola), up to a line defined by one of the main streets, parallel to the sea, at a distance of it of around 1 km. From this avenue to the sea, the slope is around 1%. Another hill (Montjuïc) is located southeast; beside the sea and the Llobregat river, near the harbour, with a castle on the top, remnant of the ancient wars.

Then, and as a complement of the indicated before, it is to note that the alluvial plain, runoff from the neighbouring Collserola was forming a number of small temporary streams transformed sometimes in drainage and sewerage infrastructures. The two permanent rivers – Besòs and Llobregat – and the marshy characteristics of the coastline indicate again a close relationship of the town with water. Consequently, Barcelona is characterized by a large number of water-related structures documented archaeologically, absolutely inseparable from the urban planning, from its founding to the present days.

The historical analysis of the Barcelona sewer system included in this paper covers from its Roman origins to the foundation of the modern sewerage planned and established by the engineer Ildefons Cerdà in 1860 and continued by Pedro Garcia Faria.

19.2 ROMAN SEWERAGE

Since its Roman times, when an organized network was built, Barcelona had sewer systems that evolved over the pace of the city. The sewerage consisted of pipelines that went through the middle of the streets and apart from the excreta and part of rainwater recovered the excess water from the two aqueducts entering the town and merging in the current Bisbe (Bishop) Street. Finally the sewerage system was conveying wastewater into the Mediterranean Sea.

Without forgetting the first facilities from the Minoan and Mycenaean civilisations (Angelakis *et al.* see the corresponding chapter of this book), the Romans were the first civilization that expressed greater interest in sewers. This is evidenced by the tax created for the conservation of sewers (*tributum cloacarium*) and the figure of *curatore cloacarum*, officials specialized in this task (Barcelona City Council, 1986).

Remains of Roman sewers have been found in all the provinces of the Empire, Barcelona is also an example, where remains of a Roman street can be found in the Palma de Sant Just Street (Figure 19.2). This drain was constructed from stone-factory material, the tile floor is flat and the vault was made with stone blocks, with all connections and drains spaced along its length. Although no remains have been found, this sewer was probably extended through the *Decumanus Maximus* to the walls of the city, to finally flow into the beach. A branch of this sewer is preserved in Bellafila Street (Barcelona City Council, 1991). Nevertheless, later papers discuss the role of the mentioned water conduits as a *cloaca* attributing it to a supply for a thermal establishment (Miró and Orengo, 2010).

Figure 19.2 Roman remains of sewers in Palma de Sant Just street (a) and Tapineria street (b), Barcelona City Council (1986 and 1991).

The City History Museum of Barcelona is showing remains of Roman Barcino, specifically from a district of workshops and factories with their corresponding sewerage system, from the 2nd century AD, with remains from a *fullonica* and a *tinctoria*, where clothing was laundered and cloth and garment dyed, and a *cetaria*, a 3rd century AD factory for salting fish and producing fish-based sauces, such as *garum*.

Other remains are located next to the Roman walls, in the basement of the street that now bears the name of Tapineria (Figure 19.2). They consist of two channels of rectangular section covered with slabs laying in both side walls and flowing into the extramural Merdançar Stream, today Via Laietana street (Barcelona City Council, 1991).

Hernández Gasch (2006) in a study on the Roman Castellum area explains that archaeological surveys detected there structural remains, including sewers. Nevertheless, the successive strata of the town make it difficult to define the real sewerage system of the time, which clearly drained used water to the neighbouring sea. It also remains unclear which part of the Roman town relied to sewerage and how the villas located near the town managed its water supplies although water recycling and reuse was typical of the Roman settlements of the time.

19.3 MEDIEVAL BARCELONA

After the fall of the Western Roman Empire, cities in Europe, especially the ones in the western side of the continent, fall usually into a deep decline, and there were an almost total neglect of the Roman building techniques and routines, while the population rusticated.

Afterwards, the sewers, aqueducts and terms of Barcelona continued in operation at least until the 6th century, although in its later years the pipes no longer drained out into a network and finally to the sea but into cesspools. Nevertheless, other authors (Voltes, 1967) indicate that the aqueducts were operating until the 10th century and that the building of the Romanic Cathedral changed all the urban structure of the area, where a number of water infrastructures were located. The author establishes that the hydraulic infrastructures built by the romans were operating until an unclear moment of the 9th or 10th centuries when the water distribution system changed.

In the Middle Ages, characterized by frequent wars which forced to confine towns inside walled areas, inhabitants of cities like Barcelona, which were beginning their development and heavily and quickly increased the number of people living inside or near their walls, were prone to infections due to extremely high population density and the consequent hygienic and water supply problems, resulting into periodic epidemics. Disasters, floods, diseases, fatalities... were accepted as having divine origin, and so, no practical questions were formulated regarding their origins, especially for the diseases. That religious idea of fatality is why the maintenance of the sewer system was neglected and sanitation was reduced to cesspools or septic tanks which in turn were polluting existing wells inside the town.

With the feudal-related disturbances and riots of the 11th century, the County of Barcelona was no longer the political centre of the territory and Catalonia began to fragment. Later on, the spectacular growth that the city experienced up to the mid-14th century was mainly due to its role as a commercial and handicraft centre. This created the need to increase water supplies for houses, industries and agriculture; and consequently the generation of more wastewater, not only from houses but also from industries.

Former streams, like the mentioned Merdançar, were real open sewers that flowed up to the sea. In some sections, the polluted streams went parallel or crossing at different levels the water supply systems like the Rec Comtal (Count's Channel) which supplied fresh water to the city, and to mills and industries based in its vicinity. This conduit was gradually covered, and in the 12th century it was indeed used as and called 'Major Sewer'. This was caused by the development of the textile and other water demanding industries which characterized and favoured the development of the city but needed ever increasing water supplies (Salgot & Angelakis, 2012). The lack of adequate infrastructures to carry used waters and the difficulty to control discharges favoured the use of clear water conduits as waste and wastewater disposal facilities and subsequently the appearance of water-related illnesses. It was from then on (13th and 14th Centurics) that part of the old Roman water supply system was recovered.

Other references from this period include the sewer of the street presently named Mestres Casals i Martorell, known then as 'The Sewer,' consisting on clay pipes, located open-cast, and covered by paving stones, or the channelling and covering of the Riera (stream) d'en Malla, that was transformed into a large collector, that nowadays still exists and that flows from the former Porta d'Isabel II (Isabel II Gate) up to the still existing Drassanes (The Dockyard).

In 1406 the level of the Rec Comtal (Counts' Channel) was lowered to facilitate wastewater disposal into that 'Major (Greater) Sewer'. The convenience of having such a neighbouring water outlet together with the water supply from the Rec, caused that in the banks of this channel-sewer lots of tanning industries were established, as evidenced by the streets that currently receive related names (Blanquers and Assaonadors – Bleachers and Tanners) (Carreras-Candi, 1918).

Cagalell Stream, currently under La Rambla (Rambla meaning a temporary stream: torrent or gully; later on became synonym of avenue, usually built by covering a gully or torrent in the middle of the town), was an urban boundary in Barcelona besides the walls. With the building of a second wall surrounding the town, which begun to be built under the auspices of the king Jaume I, this non-permanent stream was diverted and channelled among most of its course, from the present Pla de la Boqueria (Boqueria Plain) up to the Drassanes, where it flowed opencast up to its mouth. New sewers were continuously led to this stream. Actually, the work, completed in 1366, was so splendid that it was taken as Roman for a long time. Thanks to this work, the lower part of Raval neighbourhood could be protected from occasional flows, while the city was growing westwards, thus demanding a new extension of the walled area. At that time, walls were defending the town and consequently the citizens from the Arab Razias (11th century), later on from North African pirates and among other from armies in the civil wars (up to the 18th century), but were also used (Figure 19.3) to contain and divert the water courses during torrential storms, thus keeping water out of town (Barcelona City Council, 1991). As Conillera *et al.* (1986) also establish, other non-permanent streams have been covered along the time to manage overflows and later became roads and streets which sometimes keep the old water-related name (e.g. Riera Blanca: White Stream). At present, stormwater is drained by the sewerage system with some occasional uncontrolled flooding well until mid-20th century.

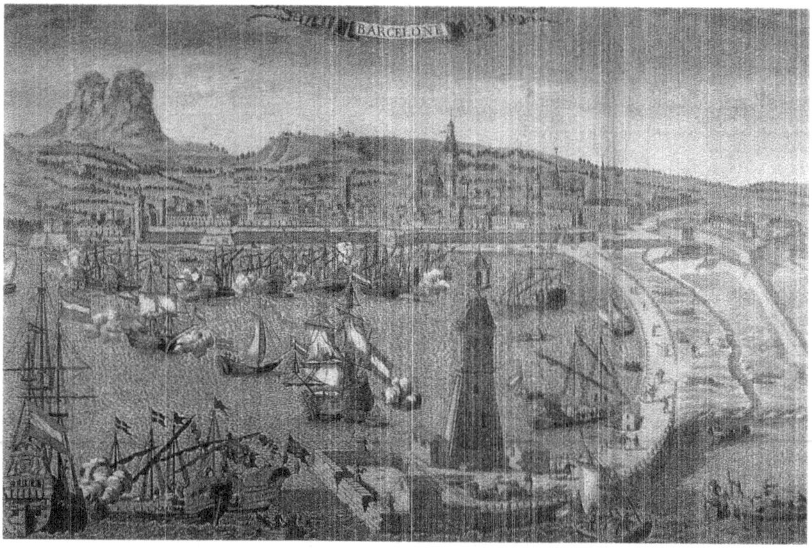

Figure 19.3 Barcelona harbour in 18th century, with the mouth of Rec Comtal/Major Sewer on the right (Barcelona City Council, 1991).

During the 13th and 14th centuries, the sewer system of the city gained importance. It was designed to properly receive rainwater and to drain wastewater, and secondarily was carrying out the additional horticultural and industrial functions planned. Nevertheless, the high water demand caused scarcity and quality problems and even civil disturbances when water was stolen from public conduits and did not reach downtown because was used before (Salgot & Angelakis, 2012).

According to the references from some travellers who passed through the city of Barcelona between the 15th and 16th centuries, it seems that the cleanliness of the streets was the pride of the town; cleanliness achieved with the help of the sewers (Barcelona City Council, 1991). Nevertheless, this situation did not last for long time and other descriptions indicated later on really different conditions.

19.4 18th AND 19th CENTURY

In the 18th century the hygienic conditions of the Barcelona streets were really bad and the City Council decided to implement measures to improve the situation. Applying the Roman jurisprudence, the Council decided that before starting any new construction the works on new sewers and paving roads must be built and landowners must pay for the implementation, on the grounds that properties will become more valuable.

This was the reason why sewers were renewed or rebuilt and drainage pipes implemented for the management of wastewater, thus being removed from houses. For example, when the vegetable patches of El Raval (The Suburb) were urbanized and new roads were built, the construction of sewers was required before the building of new houses (Barcelona City Council, 1986).

The type of sewer normally used at that time was a channel with straight walls covered with slabs. Then, due to the problems caused by the fracture of these slabs, and the continuous input of stones, it was required to use vaults instead of the traditional lintel.

Simultaneously, must be also taken into account the high number of streams and rivers that in the second half of the 18th century still crossed the plain of Barcelona draining water from the neighbouring Collseroal range and were even creating wetlands. In 1789, after a severe flooding, the General Captain, the Count of Lacy, ordered to prepare a project on the diversion of the Riera d'en Malla (Malla's stream), which didn't really materialize until the 19th century, forming what are now the sewers of Ronda de Sant Pere, Sant Joan-Almogàvers and Bogatell (Saint Peter, Saint John-Almogavars and Bogatell) to the sea (Barcelona City Council, 1986), partly following the old city walls which at that time were demolished as happened with a lot of European towns walls. In fact, the name of 'Ronda' stands for the roads built were the walls were lying, in a similar way of the French 'boulevards'.

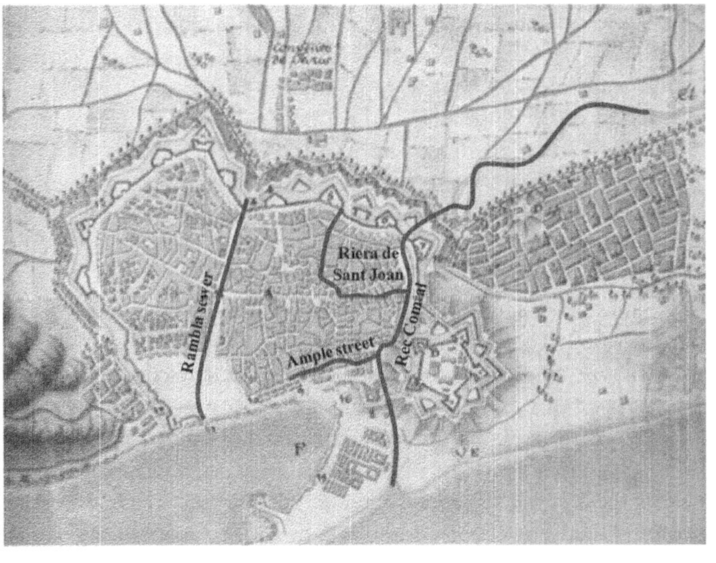

Figure 19.4 Scheme of the main sewers existing in Barcelona at the end of the 18th century (Barcelona City council, 1986).

In the late 18th century, the main sewers existing in Barcelona were: Rec Comtal (transformation of the old freshwater channel into sewer); Riera de Sant Joan (Saint John`s stream); La Rambla, Ample street (Large Street which flowed into the Rec Comtal); Rondes (as indicated, from Catalonia Square to Bogatell Street and Vila Vilà Street and the port), Passeig de Sant Joan (St. John's Avenue), Casp Street and Gran Via (Great Road) Street (Figure 19.4).

Evidently, hygiene and sanitary conditions were deplorable in the city, but people did not relate the presence of diseases to the wastes. It was not until the 19th century that the science and technology advances were able to link both issues.

It is to note that the existing slopes in the Barcelona Plain favoured the existence of drainage systems based on gravity flow. This is the case not only for the old town (located on a hill) but for its enlargements in a way that quite all the sewerage system is working without the need to pump wastewater.

19.5 THE PROBLEM OF SOIL'S AND WATER'S PUTREFACTION: THE MIASMAS

Between the 17th and 18th centuries, in Barcelona, as well as in most European cities, the medical and scientific studies gained prominence as a relevant part of Science. In this sense, the miasma theory dominated the medical studies and finished consolidating the hygiene as a particular discipline within medicine (Francisco de Assis da Costa, 1999). It was believed that the miasmas, emanations from unclean soils and waters, were the cause of infections, and the stench, the putrid odour, was the immediate symptom before real infection developed. To avoid the putrid threat, it was necessary to avoid superficial and stagnant waters, and their mephitic emanations (Barcelona City Council, 2011). In fact, a relevant characteristic of textile and leather industries was the extremely bad odours generated by the facilities. It was then immediate and easy to establish a relationship between bad odour and infection since waters were heavily polluted.

All the rulers of the towns with industrial facilities creating odours tried then to define the hygienic problem and to find a solution. As a result, several premises were established: against overcrowding, low density; against the slums and streets of short width, immense promenades conveniently located far away from the sea in drier areas and with a lot of sun to counteract the miasmas. To avoid risk in the public areas, for example, it was prescribed to facilitate air and water circulation, increase the number of fountains, renew sewers and paving and frequently watering streets. In the same way, streams either permanent or temporary were generally covered. An example is the high number of requests and reports that proposed the desirability of covering the Rec Comtal 'for the benefit of public health' in Barcelona (Historical City Archive of Barcelona, 1818).

Moreover, until 1824 rainwater from roofs and terraces poured directly on the streets. From this date, it was conveyed to the sewers through downspouts and private drains. In addition to the rainwater, wastewater from households and various establishments (often unhealthy) was also conveyed to sewers, but sometimes after being previously disposed of into the street. When the cesspools of houses were full, its overflow (black water) was also connected to the public sewers. Nevertheless, due to the low slope of the sewers, the heavier material was accumulated inside the conduits and sewers were obstructed thus becoming black spots. Cerdà advocated for the renewal of sewers, but in no case he was in favour to convey raw sewage and transport it into agricultural fields, as occurred in London. At the time, several European inland located towns started to use neighbouring fields as places to dispose of untreated wastewater. 'Sewage fields' were active for decades in France (Paris) and Germany (Berlin and other towns), being the practice known as *épandage* in France, and *rieselfelds* in Germany.

The basis of the practice was the consideration that this 'end-product' was a valuable agricultural fertilizer, available at short distance to the vegetable gardens which supplied the town. Obviously, it was a source of sanitary problems, favouring epidemics and water-related illnesses because of the recycling of pathogens to population through vegetables.

Long time before, during the Middle Ages this was accomplished with hand-carried buckets or small carts (Figure 19.5) but inland the sewers performed later on the same function.

Figure 19.5 Cart used to empty cesspools. Los perfumes de Barcelona (Barcelona scents), 1875.

Subsequently, the expansion of the cholera, that reached a lot of European cities in the early 1830s, coinciding with the resumption of densification of cities, extremely sharpened health crises and hygienic concerns. With the new densification of cities, the practice of placing disposal wells in the courtyards of houses was also widespread. Houses that had grown in height and in the number of residents did not renew nor enlarge old cesspools. The cesspools or septic tanks were not stagnant, and as indicated they were usually located in the courtyards, where usually wells of fresh water (for supply) were also placed. At the same time, and because water distribution pipes were not watertight, the pollution of fresh water due to filtrations from wastewater conduits was almost inevitable. The situation was even worst because the habit of throwing garbage and faecal wastes directly to the street, which started centuries before, was maintained for a long time.

The inconveniences arising from the latrines and cesspools affected both domestic and public space. In the private spaces it was a stinking focus that could invade the courtyard and the stairs inside the building, facilitating at the same time the spread of infections. The action of emptying the cesspools also affected the public space. Hence, for emptying them was necessary to ask for permission to the foreman of the city, and the schedules for this task were strictly established. Fetid emanations of flushing and transport operations were very unpleasant and disturbing from a health perspective (Barcelona City Council, 2011).

These facts, in addition to the problems of dirtiness and flooding caused by the streams, generated a series of cholera outbreaks (the years 1833–1843, 1854 and 1865) which caused many victims (El País, 2012) in town and created a problem difficult to manage by the health authorities. Drought problems, especially in summertime, and the consequent lack of water resources aggravated by the population increase and the subsequent enlargement of density inside the walls, made the problem worsen.

19.6 CHARACTERISTICS OF THE SEWER SYSTEM IN BARCELONA BEFORE THE CERDÀ PLAN

Barcelona's inhabitants, although the town experienced two enlargements of its walled area (the second finished in the 16th century), were asphyxiated due to the lack of space inside the walled precinct. The population density reached at the beginning of the 19th century was inhumane, with peaks of about 1000 inhabitants per hectare. The industrial revolution was requiring a city with improved services: large roads and transportation systems as well as water supplies and sewerage (Barcelona City Council, 1991).

In 1854 the walls of the city were demolished, except from the ones of the side facing the harbour. At the time, between the walls and the closest building the army forced to have a wide area. The free distance was the same that common cannons can reach; this allowed executing the new enlargement without having to eliminate existing buildings or infrastructures.

In 1859, a year before the publication of an enlargement plan – named after his author: Cerdà Plan – a complete review of the real condition of all sewers and paved streets of Barcelona was performed (Daniel & Garriga, 1859). This study was showing that 90% of streets had already sewerage systems, but only 57% of them were considered to be in good condition. In the Barceloneta (a neighbourhood besides the sea, eastward of downtown) however, only nine streets (19%) had sewerage system, and the remaining 81% had none.

There were basically two types of sewers: the traditional ones 0.39 cm wide and 0.58 high, covered just by cobblestones or slabs about 1 meter long, 20 cm wide and 15 cm thick, slightly separated to drain rainwater, and the sewers of more recent construction, 78 cm wide and 78 high, vaulted, with inlets every 3 meters and covered with cobblestones, which made them very difficult to clean. Precisely in 1796 an agreement was drafted that forced municipal sewers to be built compulsorily in vault instead of with the old straight and covered walls.

The City Council, in 1860, convened an urbanism contest to choose the most suitable project for the Eixample implementation (widening of the city; the present area with perpendicular squares) extending the town towards all the neighbouring fields, dedicated to agriculture with the exception of few buildings for services yet existing. Two of the points referred on the basis of the mentioned contest were related to urban health and sewerage. One was dedicated to the derivation of streams and the other one to the need of creating new sewerage works with the aim to eliminate the unsafe system of latrines and cesspools.

The winning proposal was the one by Ildefons Cerdà, who raised a totally innovative project for the Eixample, also dealing with sanitation problems, channelling and diverting streams and laying the foundations of the modern city and of a new science, the Urbanism.

19.7 FROM CERDÀ TO GARCÍA FÀRIA

The Cerdà Plan rose in an innovative dual approach: the need for water supply and for evacuation of the correspondingly generated wastewater for each cell (housing) of the urban city. Nevertheless, the plan presented to the City Council did not yet include the sewer system. Later on, Cerdà prepared large detailed drawings with longitudinal profiles of the sewerage where, in addition to the current ground level and to the proposed level for every street in the Eixample, he also stated the level for the corresponding sewer.

One of the most recognised merits of the new infrastructure was the construction of two sewers to save the Torrent de l'Olla (a rain fed stream). Nevertheless, the rest of the planned works were disconnected and instead of solving problems, generated and sometimes caused them.

Regarding the necessary diversion of the 'natural' streams, the engineer Cerdà envisaged two big channels which flowed towards and into the Besòs River and Riera Blanca respectively, intercepting all waters flowing down from Collserola; the range parallel to the sea which is limiting the present town by the northwest.

In 1862, a great storm devastated Barcelona. La Rambla became again a stream of tumultuous waters, as it was reflected in a drawing of the time (Figure 19.6). Cerdà, then, began the task of building a sewer collector to catch water from the Riera d'en Malla (currently Las Ramblas) and conveying it from the Rondes (boulevards) to the Bogatell. At the same time, he proposed to build the Ramblas' collector (diversion channel already included in the Cerdà Plan) for which it would be necessary to buy lands belonging to the channels that crossed the Eixample. Nevertheless, the Administration ignored the proposal as happened with other projects.

Figure 19.6 The disaster of La Rambla in 1862 (Barcelona City Council, 1991).

The lack of a drainage system within the Cerdà Plan was due to fact that at the time it was written, there were no regulations concerning the elaboration of Eixample's projects. A subsequent law (1876), with its regulations (1877), inspired by the experiences and reflections of Cerdà, specified all the required elements (longitudinal and transverse profiles, the depth to build the sewers and install water services and gas networks, etc.).

In the old city, some sewer projects were also initiated within the partial reform plans. This is the case of the project proposed by Miquel Garriga i Roca for the Rondes boulevard, dated November 10th, 1862, which contained all the elements at that time considered essential for sanitation, such as the section of the roads that should be convex in order to drain rainwater into the sewer, first because of the slope of the pavement through the curb and then directing water to the gutters (inlet to the sewerage system). The man-hole connections to houses and to sewers were also described, and this infrastructure was supposed to be built as closed as possible.

Another important project was the urbanization of the streets in the old town, which then suffered from high traffic which is not the case at present, like Carme (Carmen) Street, Riera Alta (High Stream) Street, Sant Antoni (Saint Anthony) Street, and so on. The author of the project, Francesc Daniel i Molina, suggested the urbanization of that area defining two categories of streets and still maintaining for the second type of streets, the traditional concave profile, stream-like. Another interesting aspect of this project was the segregation of the service gallery from the sewers, both with independent manholes.

Nevertheless, all the mentioned projects were devoid of systematization, because of the lack of consistent integration among them. Meanwhile, several cholera epidemics hit the city and the World Expo was approaching. Barcelona badly needed a sewer system, or at least its project, in order to be capable of being compared to the other metropolis of that time, especially Paris, in view of the upcoming event.

19.8 'SALUS POPULI SUPREMA LEX EST': THE UTOPIA OF GARCIA FARIA

In the last quarter of the 19th century, Barcelona went through a period of economic euphoria, known as 'The Gold Rush'. Because of the rapid enrichment of the bourgeoisie due to industrialization, agricultural exploitation and speculation; the Eixample was growing around Passeig de Gràcia (Gracia Avenue), between Barcelona downtown and Gràcia district, an independent municipality at that time.

Nevertheless, health conditions in town were still appalling. The cholera caused 6,419 deaths in 1865 and 3765 in 1885. The proven success of hygiene in reducing mortality in the most progressive cities (first of all London, then Paris) made unbearable the health situation of Barcelona, which, with the Universal Exposition of 1888, had the aspiration to become the showcase of the civilized world.

In 1885, the City Council created a committee to establish the basis for the reform of the sewerage, the conclusions of which were approved in 1886. The Secretary of this committee was the engineer (28 years) Pedro Garcia Faria.

Garcia Faria was early concerned about the sanitary problems related to hygiene. In 1885, he analysed the streets where high mortality rates appeared, all of them in the oldest districts of the town. Then, he issued some significant ground plans of houses where deaths had occurred. The lack of municipal ordinances necessary to fix the minimum criteria for habitability was then denounced, causing the greed of landlords.

The Eixample was not free from infectious diseases. Some houses were still using untreated groundwater for supply, although septic tanks were located near the existing wells; then sanitary problems soon appeared. In case there was a sewer, it is to note that at the time only admitted domestic wastewater. Rainwater, due to a lack of a proper drainage system, flooded unconstructed areas, which brought up a new disease: the malaria, hitherto unknown in the old quarters.

In 1888, just before the World Expo, the project of Garcia Faria for the Vila de Gràcia was approved. It channelled and urbanized the Riera d'en Malla (currently la Rambla de Catalunya), up to Catalunya Square, where the new underground stream was connected with the collector of les Rondes. Garcia Faria took then the position of Chief Engineer of the Barcelona City Council, and he prepared and issued the sewer project entitled 'Proyecto de Saneamiento del Subsuelo de Barcelona' (Project of the Barcelona Underground Sewer System) which appeared the same year, 1888.

With this project, Pedro Garcia Faria, not only provided a solution to a particular problem but was also the forerunner and the first expert of a new science, the Sanitary Engineering. With this project, this engineer highlighted the need for an interdisciplinary understanding of urban water related problems. As part of his hygienist concern, he has been working using a new scale, beyond the scope of the municipality, through the establishment of the concept of greater cities area of influence (conurbation) and an approach considering costs and benefits of new infrastructures at a greater scale, in this case the sewerage system.

Three data sets were used in the development of this project. The first one was an exhaustive topographic plan of the present greater Barcelona (the Barcelona plain and the Llobregat delta: Metropolitan Barcelona) including the original Barcelona municipality. The second one consisted on rainfall data, and on the hydro-geological data obtained from a careful study of a series of wells located in different parts of the studied area. And last but not least, the medical topographies, which were indicating the areas of urgent action.

This information was the basis of the strength to the project, exactly delimiting the problem and providing the considered best available solution.

The general design of the sewerage network was oriented towards the treatment-use of wastewater as a resource to irrigate the Llobregat Plain as it is explained later on. The network was therefore organised into three big areas involving 23 sub-basins. An interesting aspect of this project is the economy of resources, accomplished through the longitudinal layout of the collectors above the Gran Via Avenue. Finally, it should be also noted that this structure allows the integration of the growth of the surrounding 'added' towns, such as Gracia or Sants, among others. The planned network, considering also the renewal of the existing sewerage system (about 32 km), increased up to 212 km.

Within the catalogue of constructive sections were included the new so-called 'monolithic sections', which used concrete. The large collectors designed by Pedro Garcia Faria still continued to be built using stone works and bricks for the vaults; but the plaster, ditches and central channels were made with this new material.

Two important factors were considered when dimensioning, that is determining the size of, each section. The first one was the imperviousness of this new material and the second one the construction of manholes. In fact, the minimum height of a sewer was fixed by this requirement: each gallery had to be capable of receiving workers, both to make its regular maintenance and to transport waste. This led also to look for the safety of the staff, fixing a maximum slope of 5%.

During that time, the Commission for the Treatment of Wastewater from Paris set standards that can be summarized in the sentence *Tout à l'égout!* In the project of García Faria, this can be seen clearly reflected in one of its sheets (see Figure 19.7), where there is a drawing of a typical building including the downspouts from roofs and patios and the conveyance of domestic waters and, particularly, wastewater from toilets towards the sewers in the streets.

This philosophy required a municipalized supply system, with a minimum water supply for each house, each floor and each home. Meanwhile, the public administration was doomed to undertake major sewer works. There were new elements in the domestic landscape, such as ceramic toilets, urinals, the bucket with chain that discharges a volume of water on the matters to be dragged, bathtubs for the wealthy and showers for the less wealthy or with less available space, sinks and individual washers in kitchens, and so on. As for the basement, when occupied for living or other purposes, it was required to have a minimal amount of external air entries and daylight.

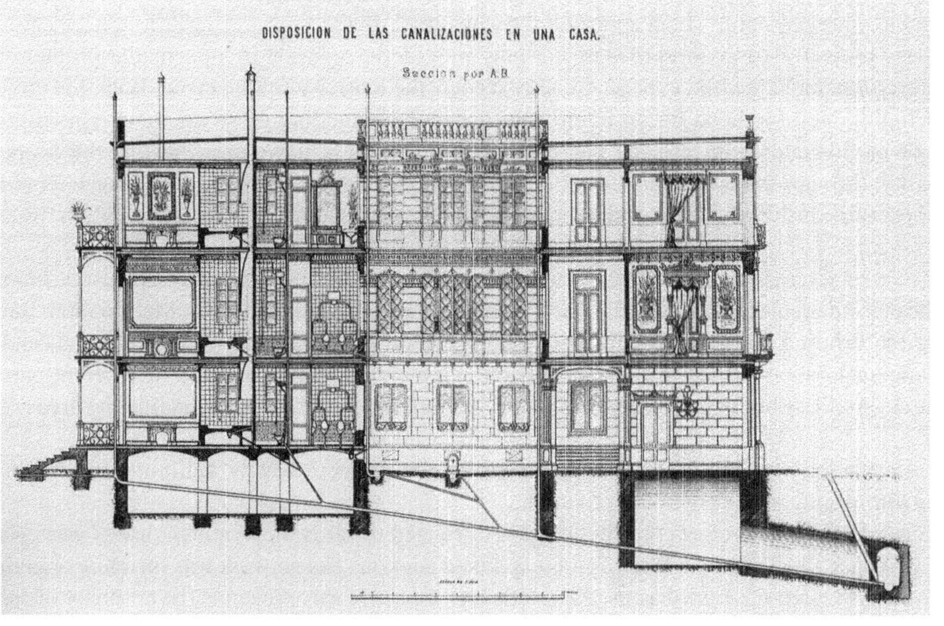

Figure 19.7 Sheet n° 10 in 'Project of the Barcelona Underground Sewer System' (Barcelona City Council, 1991).

All these elements, nowadays considered as common, defined at that time a new concept of quality of life, a new type of relationship with the chaos and the society as an environment.

In the Figure 19.7 is also easily noticeable that the section of pavement is slightly convex as introduced in the previous project of Garriga; in order to easily convey the water to the drains and afterwards to the sewer system. This allowed to quickly washing the streets and draining rainwater.

Again, the manholes for sewers did not only involve the connections under the building, but also a large cavity below them, to enable the monitoring of downspouts.

Additionally, the Garcia Faria's Plan included aspects which were unusual at that time, trying to solve not only the hygienic problems, but also the rationalization of service's supply and proposing a complete project for the removal of waste. Another important aspect of the project is the use and recycling of waste and wastewater produced by the city. For wastewater he projected the use of discharge pumps, which allowed the elevation of the water produced under the Gran Via Street towards the outlet of the Llobregat River, and its use, along with those obtained in the Eixample and its bordering villages, for irrigation. These approaches intended to make part of these works profitable, not only from a social point of view of improving public health, resulting in a mortality decrease, but also from a practical point of view, like irrigating a large area of the Llobregat delta and treating the organic elements in suspension and dissolution carried by wastewater.

This is why the vision of the engineer Garcia Faria was prophetic. The health problems derived from wastewater went beyond the jurisdiction of a single municipality; it was enlarged to the entire basin (in this case the Baix Llobregat). Only nowadays something similar has been achieved, when unifying the wastewater treatment services of different municipalities.

Nevertheless, and with the pretext of the urgency and necessity of executing the sewer network's project, Garcia Faria was 'elegantly' put aside of his position. That prevented him from addressing and leading the ambitious project concerning the water distribution system to Barcelona with the objective of rationalising water consumption.

After the approval of the sewerage system project in June 1891, problems began to arise. Barcelona, which had already demonstrated their ingratitude towards its modern planner, the civil engineer Cerdà, started to do the same with Garcia Faria. After refusing a proposition of bribery (for the equivalent of one hundred thousand euros), Garcia Faria fell into disgrace.

Garcia Faria was subjected to an insistent smear campaign, and at the same time his contract with the City Council was cancelled. Afterwards, he began a long legal battle to get a compensation for his work (he had not yet been paid for the development of the project), a process that he finally won. His progressive and honest character crashed against the corrupted government prevailing at that time.

Finally, from the entire project, just the part of the sewer system comprised within the already urbanized area was executed between 1902 and 1907 and was finished towards 1914 (Figure 19.8). This includes the area between the Compte d'Urgell (Urgell Count) Street and Passeig Sant Joan (Saint John Avenue), and between Diagonal Avenue and the Rondes. After that, the real construction of the sewers disturbed even more the original philosophy of the project. Except from the Magoria Stream, which flows through Tarragona Street, the rest of the sewers reverse the direction of the basins, flowing into Bogatell instead of into the Llobregat River, as planned by Garcia Faria. The issue of wastewater treatment and further reuse was completely forgotten by the municipality.

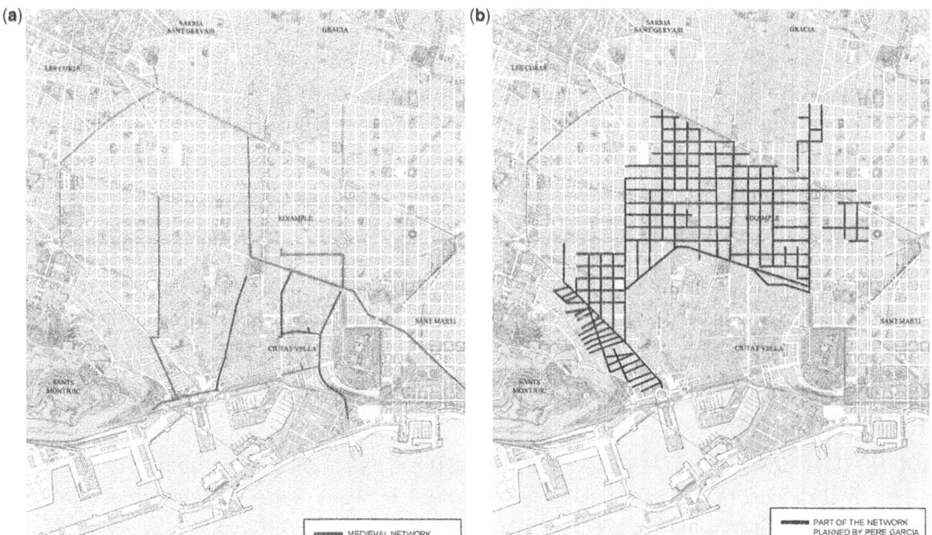

Figure 19.8 Sewer network before the project of Garcia Faria (a) and partly executed between 1902 and 1914 (b) (Barcelona City Council, 1991).

Even as late as 1914, another epidemic took 2036 lives, and in 1920 the mortality in Barcelona remained as one of the highest in Europe, with a declared 27.60%.

After the Garcia Faria's project, and regarding sewer plans, there is a later reference, the project from Jaussely Leon, presented in 1907, which complemented the Cerdà Plan. In general, the philosophy of the new additions planned in the project continued the ones from Garcia Faria, but extending the spill basins both towards the Llobregat and Besòs Rivers, involving all the municipalities covered by this plan for the Greater Barcelona.

From here on, it will not be until 1954 that Barcelona developed a new sewerage plan, 'El Pla General de Sanejament i Clavegueram de Barcelona' (The Barcelona's General Plan for Sanitation and Sewerage), which involved Barcelona and the municipalities of its area of influence: the Greater Barcelona.

19.9 CONCLUSIONS

Barcelona's sewer system evolved following the pattern common to Mediterranean towns of Roman origin: from comparatively advanced facilities dating from roman times to unsafe structures in the middle age, which remained more or less in use until the destruction of the fortifications typical from medieval towns. The space gained and the enlargement of the town occupying agricultural areas surrounding the old cities allowed the building of new infrastructures which dramatically improved the health conditions of the citizens.

As already stated, the Romans were the first civilization that expressed greater interest in the sewers and initially constructed this kind of infrastructure in Barcelona. With the fall of the Roman Empire, during the middle ages, Barcelona started a stagnation period regarding the evolution of the sewerage system. In that period, wastewater management was mainly based on cesspools and on the use of natural water courses as sewers.

The improvement and the implementation of the modern sewerage system was associated with the works undertaken for the enlargement of the city, based on the planning performed by the engineer Ildefons Cerdà, after 1860; few years after the demolition of the city walls.

Ildefons Cerdà raised a totally innovative project for the Eixample of Barcelona, addressing sanitation problems, channelling and diverting streams and laying the foundations of the modern city and of a new science, the Urbanism. Nevertheless, the plan presented to the City Council did not include a specific project for the sewer system, and it was not until 1888 that another engineer, Pedro Garcia Faria, completely concerned about the problems related to hygiene and that were causing thousands of deaths due to cholera, developed a complete project for the sewerage of Barcelona.

With this project, Garcia Faria highlighted the need for an interdisciplinary understanding of urban problems, trying to solve not only the hygienic problem, but also the rationalization of infrastructures for services and the treatment and recycling of wastes and wastewater produced by the city.

Nevertheless, despite the innovative proposal, and due to economic interests of the corrupted government prevailing at that time, the project of Garcia Faria was only partially executed.

19.10 REFERENCES

Angelakis A. N., Mays L. W., De Feo G., Salgot M., Laureano P. and Paranychianakis N. (2012). Water and wastewater in ancient civilizations. In: Global Trends and Challenges in Water Science, Research and Management: A Compendium of Hot Topics and Features from IWA-SG, H. Li (ed.), IWA, London, UK, pp. 90–94.

Barcelona City Council (1986). Barcelona Sub: El Clavegueram de Barcelona (Barcelona Sub: The sewerage of Barcelona). Ajuntament de Barcelona, Barcelona.

Barcelona City Council (1991). Sota la ciutat (Under the city), Ajuntament de Barcelona and Col•legi d'Enginyers de Camins. Canals y Ports de Catalunya, Barcelona.

Barcelona City Council (2011). La revolució de l'aigua a Barcelona (The water revolution of Barcelona). In: De la ciutat preindustrial a la metròpoli moderna (From the preindustrial city to the modern metropolis), 1867–1967, M. Guàrdia (ed.), Ajuntament de Barcelona, Museu d'Història de Barcelona, Barcelona.

Carreras Candi F. (1918). Geografia general de Catalunya (General geography of Catalonia). A. Martin (ed.), Barcelona.

Conillera P., Llabrés A. and Parés M. (1986). Descobrir el Medi Urbà. 4. L'aigua a Barcelona (Discover the urban environment. 4. The water in Barcelona). Barcelona City Council, Barcelona.

Daniel F. and Garriga F. (1859). Cuaderno Demostrativo de la Medición y del Estado de los Empedrados y Cloacas de la Ciudad de Barcelona y de su Barrio de la Barceloneta (Demonstration notebook about the measurement and state of paving and sewers in the city of Barcelona and its Barceloneta neighborhood). signed by F. Daniel i Molina and F. Garriga i Roca (AMCB-2443-3/1).

El País (2012). Quadern. La ciutat de sota (Notebook. The city beneath), El País, Thursday, January 5th of 2012, number 1.426, pp. 1–3.

Francisco de Assís da Costa (1999). La Compulsión por lo Limpio en la Idealización y Construcción de la Ciudad Contemporánea (The compulsion for cleanliness in the idealization and building of the contemporary city). Unpublished PhD thesis, Universitat Politècnica de Catalunya.

Hernández-Gasch J. (2006). The castellum of Barcino: from its early Roman empire origins as a monumental public place to the late antiquity fortress. *QUARHIS*, **época II**(2): 74–91.

Historical City Archive of Barcelona (1818). Acords (Agreements), Ajuntament borbònic, 1DI-98.

Miró C. and Orengo H. A. (2010). El cicle de l'aigua a Barcino. Una reflexió entorn de les noves dades arqueològiques (The water cycle in Barcino. A reflection about the new archeological data). *QUARHIS*, **época II**(6), 108–133.

Salgot M. and Angelakis A.N. (2012). The Historical Development of Water Supply Technologies in Barcelona, Spain. In: A. N. Angelakis and J. B. Rose (eds), Evolution of Water Supply Throughout Millennia, IWA Publishing, London, UK, Chapter 19, pp. 497–520.

Voltes P. (1967). Historia del abastecimiento de agua a Barcelona (History of water supply in Barcelona). Sociedad General de Aguas de Barcelona, Barcelona.

Chapter 20

A brief history of the sewerage system in Prague and the role of William Heerlein Lindley

Jan Palas, Jaroslav Jásek, and Ondrej Benes

20.1 A SHORT HISTORY OF SANITATION IN THE CITY OF PRAGUE

20.1.1 Prague – history and first sewers

The first real agglomeration in the place of current city of Prague is documented around 200 BC and was built by the Celts (Boii), who created an oppidum Závist. In the 6th century the nation of Slavs took place of parting German tribes and according to legends, the city Prague was founded in the 8th century by the Czech duchess and prophetess Libuše and her husband, Přemysl and by the year 800 there was a simple wooden fort, later modernised to a stone castle in late 9th century.

One of the earliest found written notes, confirming an existence of sewer systems in Prague dates back to 1310, when first 'drainage systems' were constructed in the Provost's House in Ostruhová Street (now Nerudova Street) in Malá Strana even though this was rather an exception of the unsatisfactory sanitation situation in that period of time. The matter of centralise sewage removal played no role in the first few centuries of existence of Prague and, as in the rest of Europe, domestic waste, both solid and liquid, was thrown out into the street and waste was 'cleared up' mainly by the wind and the rain. A ban on releasing pigs into the street in the Old Town dates back to 1380. The situation was similar in all population centres of the time. We are given an impression of the situation during the reign of the king Charles IV (1346–1378) from a complaint written by his Chancellor, Johann von Neumarkt, about the conditions in Imperial Nuremburg, where ... *due to the rains, such a quantity of refuse grows so that riders on a horse cannot pass through the city without fear that either their horses will fall into the muck, get smeared with excrement and stink like pigs, or they become smeared by other horses*... (Ruth, 1903).

Care of public sanitation however slightly improved during the reign of Charles IV, who had been brought up in France. During the foundations of Prague's New Town, he let his engineers to employ some basic sanitation principles which he came across at the court of Philippe VI in Paris. These included *quinettes*, which were stone drainage gullies running down the middle of paved streets with essential conduits through the walls of the castle with outflow into the moat or into the River Vltava.

The first real modern sewer was built in the year 1660 at the Jesuit seminary, the Clementinum. Water from the fountain flushed away faeces from the latrines and waste from the kitchens and washrooms straight into the river Vltava (Havrda a Kovář, 2011).

20.1.2 Progress over next centuries

A change in the concept of sewerage did not occur until the 18th century. In 1782, the moat between the Old and the New Town was arched in where Ferdinandova (today Národní) Avenue is located. The real impulse came with a decree issued in 1787 by Emperor Josef II for the construction of drainage sewers. Advances in military engineering passed into the civil sphere thanks mainly to František Antonín Hergeth (1741–1800), professor at the engineers' guild school, the forerunner of the Prague Technical College. Hergeth also worked as the supreme construction director for the Czech Lands and ran state construction for twenty years. Around 1790, his pupils, engineers Oppelt and Lutz drew up the technical plan of Prague which became the basis for construction of the first *systematic sewerage system*. Up until the slump in the sector of public construction following the Napoleonic Wars, around 19 km of sewers were built, channelling sewage into the river Vltava.

Another wave of construction occurred over the years 1818 to 1820 at the impulse of the Supreme Burgrave of the Kingdom of Bohemia, Count Rudolf Chotek. This man's achievements are far-reaching; he is responsible for the development of streets in Prague and saving the Royal Deer Park in Bubeneč and its transformation into an urban park (now known as Stromovka). Under his leadership, 44 km of sewers were created, draining sewage into the Vltava without treatment in 35 different places. In the mid 19th century, this system was already dilapidated, blocked and also there were complaints about pollution of the Vltava which was the source of drinking water up until 1914, when combined water from artesian wells and natural infiltration wells from Káraný was channelled into Prague. If we quote from an edict of the Prague City Corporation of 8.4.1914 (Jásek, 2006):

> '…*It is a matter for Prague inhabitants to – regardless of misconceptions – avail themselves in their own interest of all the advantages of the uncontaminated water of Káraný, thereby avoiding possible infection from private wells, however much it is revered, where freedom from harm of a medical nature cannot be guaranteed…*'

Problems reached dangerous proportions with the economic recovery of the 1860s and 70s. However, neither speed nor plenitude of funds for public construction were characteristic of the Austrian, or rather Austro-Hungarian situation. Conviction of the need of building a modern sewerage system ripened in professional and thereafter in administrative circles only in 1883.

20.2 TOWARDS A MODERN CONCEPT OF SANITATION SYSTEM

20.2.1 The first tender for the Prague master plan

The initial search for the right approach took place on the territory of the Association of Architects and Engineers, a respected guild, where the advantages and disadvantages of various sewer systems were discussed as well as basic methods of treating urban waste water. Several interesting memoranda have survived from that time. In the file '*On the matter of flushing sewerage*', the physiocratic society lobbied for a barrel system, claiming that it would be a waste to flush the great fertilising wealth and to pollute the river to boot. A report by professor of medicine, Popper, came in reply: '*On cleansing and drainage of the City of Prague*' which introduced to the contrary of the necessity of immediate sewerage of waste water for medical reasons. The respected Captain Liernur also hastened to Prague to propagate his vacuum sewers. He succeeded in having a local experiment performed, probably due to his connections, in one of the Prague barracks. However, the greatest domestic expert in the field, the engineer Jan Kaftan, probably the first Czech author of a book on sewers, dated 1880, came out adamantly against Liernur. Based on a detailed study, he rejected vacuum sewers for their shortcomings, especially the smell from toilets, blocked pipes and the operational complexity of the system using for example following arguments.

> *'If the pipe between the toilet and the street conduit becomes completely or partially empty, it can be very dangerous for the house as far as health considerations are concerned, faeces from the piping from neighbouring house, either due to being over-full or by being sucked in, without proceeding to the street conduit, their fumes wafting up the empty domestic piping.'*

Kaftan claimed that at the Ferdinand Barracks in Karlín fitted with this sewerage system, illness in the soldiers had increased. After this crushing criticism, further mention of Liernur in Prague is not to be found. Significant in this was also a speech by the High Inspector of the Buštěhrad Railway, M.Eng. Polívka who firstly emphasised the necessity of addressing drainage together with villages (now city quarters) then still standing outside Prague, namely Žižkov, Královské Vinohrady and Smíchov and secondly he rejected treating waste water by using it for irrigation, supporting his rejection by simple comparison of the sheer quantities of waste on the one hand and available areas to spread them on the other. Based on the expert recommendations by the Association and on the urging of the Governor of the Kingdom of Bohemia, Count Thun-Hohenstein, the City Corporation issued the first tender in 1884. A total of five set of plans were submitted, a description of which survives in 'The Rationale for the Decisions of the Panel regarding the Competence and Merits of the Plans for Cleansing and Draining the Royal Capital of Prague' from 1885.

The Kaumann project, submitted by the civil engineer Mr. Kaumann from Wroclav, the creator of the sewers there, included, in addition to Prague and Vyšehrad the suburbs of Bubny, Holešovice, Vinohrady, Žižkov, Smíchov and Karlín, that is, an area totalling 2845 ha. He proposed to channel the sewers towards the then existing Holešovický Island (area on the right bank between Pelc-Tyrolka and Troja). The whole island was to be built up with a steam-powered sewage farm occupying an area of 43 ha, sufficient for the practices of the day for connecting up approximately 350 000 people. Because a wastewater treatment plant of almost twice the size was needed for the whole of the area served by sewers, the project considered later expansion onto Císařský Island too.

The Frisch gewagt ('A fresh start') project was to equip only part of Prague with sewers, on the right bank and some suburbs: Vinohrady, Žižkov, Karlín and Smíchov. The area served by sewers was to be 765 ha and the sewer pipes ran along the right bank of the Vltava to Libeň, where it was to discharge without treating into one of the river's side channels there. Significant parts of the city, such as Holešovice-Bubny, were to have no sewers.

The Sine munditia nulla sanitas ('No good health without cleanliness') project planned to provide sewers for the territory of Prague, Vinohrady, Žižkov, Karlín and Smíchov, with a total area of 1177 ha. It left three independent sewerage catchments in Prague, each ending in a collection tank. The method of handling the sewage in the tanks is not completely clear from the project. The project considered either chemical coagulation in the collection tanks themselves, or pumping it to a mechanical treatment plant 10 km away.

The Divissione ('Division') project submitted by the Rella´s Neffen firm from Vienna introduced two revolutionary elements. Firstly, it proposed *separate sewers* with rainwater drains taking the shortest route towards the river to discharge, and secondly a main sewer tunnel running from the Old Town was to tunnel under the riverbed of the Vltava through a huge inverted siphon, thereafter to continue through a bore driven through to Bubeneč and end discharging into the watercourse below Císařský Island, with a view to construction of an irrigation treatment station on the island. One unusual aspect of the project was use of small diameters to reduce construction costs. For the same reason, used sewers were to be run through the Rudolf gallery. The project was to serve the territory of Prague,

Vinohrady, Žižkov and Smíchov with the total area served being 2311 ha. Karlín and Holešovice were not considered in this project.

The project designed by M.Eng. Kaftan, Praga caput regni ('Prague, Head of the Kingdom') proposed draining Prague and Vyšehrad, with an area of 627 ha and part of Žižkov, Karlín and Smíchov, the area unspecified. It was to channel sewage along combined flushing sewers from the right bank to Holešovice, where it was to unite with the left bank branch. *'So that the matter contained in the sewage so valuable for farmers should not go to waste and also to abate the fears of pollution of the Vltava, we have decided for these ends to mix the sewage with a solution of lime and argillaceous sulphur salts and to let it settle in special tanks before releasing it into the river. The considerably treated drain water shall be discharged into the river, while the sediment shall be drawn off by special sludge pumps and used for making compost. The most suitable place to locate the tanks would be on Holešovice Plain whence the separated manure could be taken away on boats cheaply and easily.'*

Not one of the projects succeeded in passing through the filter of demanding criteria. The panel recommended that none of the competitors receive the first prize of 6 000 Austro-Hungarian gulden. It identified the Kaumann project as the best, awarding it three thousand gulden, and the Praga caput regni project received 2 000 gulden. In addition, the City Corporation was advised to buy the Divissione project to have it at their disposal, since it contained valuable elements for the solution. The tender was, therefore, unsuccessful.

20.2.2 The second master plan tender

A substantial reason for the first round being unsuccessful was seen to be insufficient background materials and preparative works. So, in 1888, the *Sewers Office* was established, with engineers Josef Václavek and Vincenc Ryvola being placed in charge, the latter having collaborated on the Kaftan project. At the recommendation of Karl von Kořistka, M.Eng. Máslo, who subsequently became famous between the wars for his modern treatment plant project in Řež, was made the first surveyor of the Office. The Office was required to collect all necessary materials, in particular to perform detailed measurement of geodetic elevations in Prague, compile data on the frequency of rainfall, flooding and groundwater and also to formulate conditions which the author of the new project must satisfy. When all had been performed, for reasons unknown, the City Corporation did not commission it to produce a project, meaning that the Office ceased de facto to exist. M.Eng. Kaftan, Dr. James Hobrecht, building councillor in Berlin and M.Eng. Kaumann all expressed their interest in proposing their projects. During further negotiations, Dr. Hobrecht succeeded with his proposal that instead of three sets of plans, only one should be produced, proposed by himself and M. Eng. Kaftan taking in account his perfect knowledge of the local conditions. The Prague City Council unanimously approved this remarkable proposal in October 1889 and drew up a contract. Work on the plans took place between February 1890 and March 1891, the finished project being displayed at the Regional Jubilee Exhibition. The Kaftan – Hobrecht project directed the sewers from the right-bank parts of the city to meet at the lower end of Karlín. There they continued by inverted siphon to the Holešovice side of the river, joining up with left-bank main interceptor sewer running from Malá Strana beneath the slopes of Letná. The combined main sewer received the main interceptor sewer from Holešovice-Bubny and was to flow into the Vltava below the bottom edge of Holešovice. The plans did not include treatment of waste water. It stated only that on request of the authorities, 'cleansing tanks' with chemically assisted sedimentation could be placed on the northern edge of the Royal Deer Park. Also the option of irrigation on Císařský Island was suggested. The total budget of the project was 3.72 million gulden with a sewerage area of 1651 ha.

Václavek and Ryvola were clearly not pleased with the steps taken by the city council and so, in protest, compiled a completely private and basically secret project plans which they subsequently ostentatiously donated to the city. Submitted on 5th March 1891, only three days after the first project, it provoked a sensation and undoubtedly too the animosity of some municipal functionaries.

> '*...the aforementioned experts at their own risk and keeping exceptional office hours, they worked on a rival project without the knowledge of civil engineer Kaftan. Almost simultaneously to submission of Mr Kaftan's project, the City Corporation received a special communiqué which the experts mentioned above submitted for assessment, requesting the City Corporation to exhibit that plan before the public without a word of any remuneration...*' (Jásek, 2006).

Since the commissioned project thereby faced serious competition, it was necessary to discuss both proposals. Václavka and Ryvola's project worked on the principle of two elevation zones of sewers serving a total area of 2202 ha. The backbone of the upper zone which served the historical heart of the city on both banks was an inverted siphon under Letná leading to Bubeneč. The lower zone, made up of Petrská Quarter, Karlín and Holešovice, included the Karlín inverted siphon and a main interceptor sewer running along the Holešovice bank with the option of emptying into the river or continuing to a treatment plant in Bubeneč. This was to be made up of basic pre-sedimentation in *grit traps*, and chemically assisted sedimentation in round tanks. Water wheels were to be installed at the intake and outflow to drive items of machinery, thereby saving on the expensive work performed by steam engines. The budget of costs for this project was estimated at 3.05 million gulden.

The qualities of this competing project could not be ignored and so both were compared and assessed. This thoroughly objective matter soon became politicised (Václavek and Ryvola were supported by the Young Bohemians Party, Kaftan was a member of this party) and complaints from Doctor Hobrecht regarding failure to satisfy the city's conditions threatened to provoke an international incident. It can be assumed that all of the capacities of that time in that field had a conflict of interest in one way or another, and so the City Corporation decided upon the only possible solution, despite the shocked looks on the faces of the Czech professional community: it called in an independent expert from abroad who was meant to adjudicate between the two camps.

> '*...the Prague community, faced unexpectedly with two plans for sewerage cannot conclusively decide on one or the other, and therefore it is necessary to call in an authority to make a definite statement on the issue.*' (Jásek, 2006).

20.3 THE ARRIVAL OF W. H. LINDLEY

20.3.1 When two are fighting ...

The authority that used his expertise to evaluate the projects was the building councillor in Frankfurt am Main, of English origin, M.Eng. William Heerlein Lindley, builder of modern sewerage and water works structures in Europe. After familiarising himself with the Prague terrain and reconnoitring the Rudolf gallery, Councillor Lindley evaluated both projects in detail. He faulted the conservative Kaftan project for being unable to drain low-lying quarters when the river level was high. He praised the other project for the idea of a tunnel under Letná, but pointed out that Václavek and Ryvola failed to solve the same problem as Kaftan, despite the fact that a deep-positioned tunnel would provide the solution. Lindley also criticised both projects for the limited scope of the districts served by sewers, and insufficient depths and diameters of the sewers. He stated that a city the size of Prague cannot rely on self-cleansing in the river and therefore it needed a decent treatment plant. He supported its location in Bubeneč, but did not trust the untested concept of round tanks. Lindley did not recommend either project to be implemented.

After such an assessment, it was expected that the city would again reopen the sewer office and commission it with adapting the Václavek – Ryvola project according to Lindley's remarks. Commissioning

Lindley himself with the project was, therefore, understood as an expression of lack of trust in domestic experts. Lindley submitted his project in July 1893. His proposal openly exploited the ideas of both rejected projects, although he perfected many aspects and made numerous additions. The combined system was divided into two elevation zones, reducing the risk of sewer blockages. Full exploitation of the tunnel under Letná enabled drainage of the lowest-level blocks even during runoff from torrential rainfall. The advantages of the tunnel gradient were exploited, enabling sewers to be laid deep down in order to drain cellars. Widening the diameter of sewers facilitated later enlargement of the area served by sewers and was therefore an important element for long-lifetime of Lindley's project. Considerable priority was given to laying sewers on municipally owned land, limiting complicated negotiations with landowners and compulsory purchase. The cost budget for Lindley's project reached the astronomical figure of 6.5 million gulden, with a sewage catchment area of 2588 ha.

At that time, construction was already underway of isolated sections of sewers, especially in Holešovice, where the newly-built abattoir site had to be fitted with drainage. On 21st April 1894, Lindley's project was accepted by the City council, on 2nd May the board of community elders approved it and on 6th June, it was passed up to the governor for approval. Planning permission was granted in January 1895, and afterwards the hitherto unsurpassed construction was launched, its scale and complexity being comparable to today's construction of the metro.

In the Spring of 1896, a heated debate began again to rage among the Czech technical community. This time it concerned who was going to lead the construction work, in other words, who would be head of the reopened sewer office. Native experts were very offended when the City council entrusted this office to Lindley. The reaction of the Association of Architects and Engineers expressed bitterness regarding the fact that the skills and merits of Czech engineers had been overlooked and the shortcomings involved in Lindley's controlling construction performed at a distance from Frankfurt. Lindley's salary of 10 000 gulden also seemed high, being three-times higher than the salary of the previous master of the planning office. In the Association Memorandum dated 14th May 1895 (Jásek, 20006), we can read:

> *'Celebrated City council of the Royal capital of Prague, we beg you to reconsider your resolution for creation of the plans and control of sewer laying works to a foreign expert, but to entrust both to the Sewers Office which would be reactivated and re-staffed with a new workforce.'*
>
> *'... with reference to the reasons mentioned in the sewer issue, the Association reiterates and declares to the general public:*
>
> *... that domestic technicians, having completed the initial project (whence Lindley drew inspiration for his project), could also manage to produce an implementation plan, ... neither construction of sewers nor the inverted siphon nor the Letná tunnel nor the wastewater treatment plant are such difficult tasks that Czech experts could not figure them out, ... as far as responsibility is concerned, control would be best granted to a domestic technician ... that if control of the sewerage project is entrusted to a foreigner, Czech technicians must and will consider this to be an expression of lack of confidence.'*

Various tirades and defamations were to be heard at official soil, to which the respected functionaries had to deal with. As we can read in a speech made by Dr. Kühn, second secretary to the mayor: *'Aspersions have been cast regarding Mr Lindley that he is a German and a Jew, but he is in fact a Dane, his wife is English and he is of Anglican faith.'* Most importantly, Lindley could not be criticised for his professionalism. Prof. Kristian Petrlík, a community elder, said in a speech at on Association soil on 11th March 1896: *'I have known building councillor Lindley since 1885, I know his works in Frankfurt, which I again and again inspected with my students, I know some of his newer works, the sewers in Warsaw, I know his work at international conferences on non-sea waterways, I know him for his works and literary activities and on this basis I respect Councillor Lindley as an illustrious engineer who is a credit to his rank.'* And so the decision was made.

20.3.2 Building the new sewers of prague

In May 1896, the sewer office was quietly opened again with one main task – to produce implementation plans for construction of the *combined sewerage* and later of the wastewater treatment plant in Bubeneč. Preparations continued at a fast pace thanks to a well-coordinated team of experts. Work was performed by M.Eng. Kořínek, Ing. Heinemann, Ing. Ryvola, Ing. Máslo and Ing. Mašín. The basic plan of the sewer network of an approximate length of 90 km, the territory of Prague was divided into four categories:

(a) Inner City, outflow of sewage 1 L/ha.s, which was about 150 L/inh.d.
(b) Outer City (Vinohrady, Vyšehrad, Podolí, part of Holešovic etc.), outflow 0.75 L/ha.s.
(c) Steeper districts of the Outer City (Bubny, Hradčany, Žižkov etc.), outflow 0.67 L/ha.s. and
(d) Suburbs and countryside districts such as Nusle and Michle, outflow 0.5 L/ha.s.

The territory, served by proposed sewers, comprised of two areas. The right-bank historical town and adjacent quarters to the south were drained by interceptor sewer A. Its most important junctions were two *linkage chambers* – below Old Town Square and on the embankment before moving into the *inverted siphon*. Interceptor sewer A crossed the river though the Old Town inverted siphon, joined by waste water from the left from Smíchov, Malá Strana and Hradčany. Afterwards it continued through the tunnel to Bubeneč towards the planned wastewater treatment plant. The second main interceptor sewer, B, began in Karlín, where it was also joined by sewers from Žižkov. To cross the Vltava, it continued through an inverted siphon under Rohanský Island to Holešovice, where sewers from Bubny and Letná joined it and reached Bubeneč via the Royal Deer Park. The smaller interceptor sewer C, into which sewer D emptied near to the wastewater treatment plant, was built subsequently draining Dejvice, Vokovice, Střešovice, Bubeneč and their surrounding districts (see Figure 20.1).

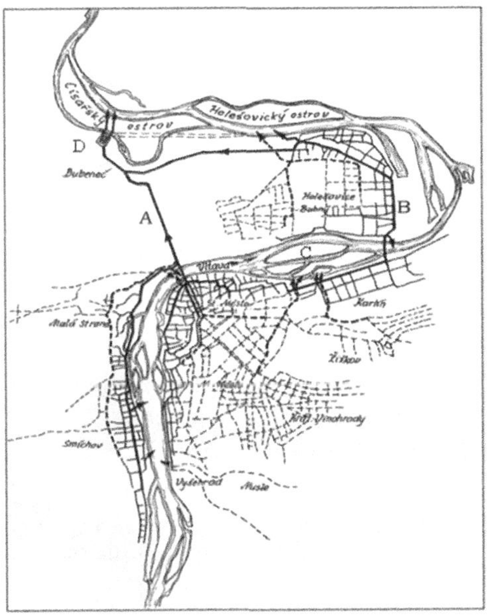

Figure 20.1 The proposed two sectional system including A – inverted siphon Svatopluk Čech Bridge, Letná tunnel, B – trunk sewer conducted through Karlín, inverted siphon of Karlín, C – CSO Vltavská and D – WWTP Bubeneč.

Both areas were then divided into two elevation zones whose main interceptor sewers were channelled separately into *rainwater overflows* into the Vltava (Figure 20.2). This arrangement protected the lowest quarters against flooding. Because in times of heavy rain and thaw and so on, excessive water were to run off the sewer catchment area and no zones existed, the sewers on higher ground would be at a higher pressure than the lower sewers and would occupy the capacity of the inverted siphons and main interceptor sewers. The water in the lower-lying areas would not drain away and flooding would occur despite the protection of raised river embankments. In Lindley's ingenious solution, at high-risk flow rates when there was excessive water in both the river and in the sewers, it was possible by simple operation of sluices in *combined sewer overflows* to release all water from the upper zones into the river, thus reserving the capacity of the main interceptor sewer for the lower-lying parts of Prague under threat.

Figure 20.2 Combined sewer overflow chamber Vltavska.

In February 1897, Lindley visited several brick kilns in the surroundings of Prague which were to become the suppliers of the approximately eight million sewer bricks needed. Afterwards, professors from the Prague Polytechnic, Goller and Slavík performed detailed tests on the raw materials from selected areas. March, the first order of bricks was made: from Zákolany 300 000 bricks, from Uhříněves 3 250 000 bricks and from Bližejov, 1 500 000 bricks for a price of approximately 30 gulden per thousand bricks. In August 1897, the City council decided that for construction of the Old Town inverted siphon and asked a company based in (Buda) Pest – Gregersen and sons, which earlier had performed well during a complicated repair of Charles Bridge, damaged by floods. The Old Town inverted siphon of a length of 174 m, laid in a line running parallel to Čechův Bridge, comprises two cast iron pipes, 1 metre in diameter. The inverted siphon was, due to the brittle nature of the material, laid painstakingly in stages into a trench dug on the bed of the Vltava. The connecting tunnel under Letná is 1200 m long in its straight section, its oval cross-section measuring 180 × 260 cm. At a gradient of 1:1200, it has a capacity of up to 5 m^3/s. Digging and walling was performed by the company Kress & Bernard for 415 000 gulden. During construction they came upon rich stocks of water in fissures, which, after the drainage tunnel had been built, several wells in Bubeneč lost their water and those affected had to be compensated.

In June 1898, inverted siphon structures were concreted and walled on the embankments and on the river bed, a lunette for laying the pipes was dug out using the Priestmann excavator. In August, further bricks were ordered: from the Hála Brick Kiln in Satalice, Urbánek Brick Kiln at Jenerálka, Schwarzenberg Brick Kiln in Zliv and again in Uhříněves. Digging of the Letná tunnel was completed in September and then walling work started. Both major structures, the inverted siphon and the tunnel, were ready by the end of 1899. At the same time, hectic construction of sewers went on across the sewer catchment area. It is a little-known fact that in connection with drainage of Hradčany, Lindley designed the adaptation of Chotkova Road to allow for the tramway to be built and make Jelení Moat accessible.

On 13th June 1901, Emperor Franz Josef I came to visit the sewer construction works. After the inevitable speeches by Prague dignitaries, he saluted his bust and climbed down into the tunnel 3 m below the level of the Vltava and was given a demonstration of flushing by river water. *'The water gushed with a load roar so fiercely that the potentate expressed wonder over its vehemence.'* Once construction of the combined circular and outflow sewers around the future wastewater treatment plant was completed on 30th August 1901, the inlet sluice valves were opened and wastewater from the centre of Prague began to drain into the Vltava downriver of Císařský Island.

20.4 THE FIRST WASTEWATER TREATMENT PLANT

20.4.1 Construction and operation start-up

Wastewater treatment plant location in Bubeneč was ideal both as far as distance and elevation were concerned. The height difference between wastewater treatment plant and river where the treated wastewater was discharged was sufficient even at high river water levels. Conditions were improved by construction of a navigation channel and regulation of the riverbed which also underway at that time. In 1896 and 1897 land purchases for the wastewater treatment station were made. In some cases agreement could not be reached between the owner and the city due to excessive asking prices, so in April 1897 the board of community elders proposed applying compulsory purchase of the disputed plots of land.

Under the contract between the city and Councillor Lindley, completion of the implementation plan for the wastewater treatment plant was set for 15th May 1899. In January 1900, a counsel of official experts from the fields of medicine and chemistry appraised the project plans and in February water rights commission proceedings were launched. The Agricultural Commission for the Kingdom of Bohemia attached their standpoint which demanded that ferrous salts be not used in operations *'... since that would reduce the value of sludge as manure'*.

In March 1901, Councillor Lindley made a speech concerning the wastewater treatment plant project at the Association of Architects and Engineers in the Kingdom of Bohemia, following on preceding lectures by Ing. Heinemann on the sewer network. Lindley gave a comprehensive explanation of the diagrams and also mentioned budget costs which were to reach 1 850 000 crowns. In subsequent discussion, objective remarks and queries were raised – how are the sludge tanks on the island protected against floods and whether they would offend visitors to the Royal Deer Park with their smell. Engineer Kalousek expressed his foresighted wish that the project would bear in mind *'bacteriological treatment which can certainly be expected in the future.'* (Jásek, 2006). Failure to respect this remark was one of the few shortcomings of Lindley's design which manifested itself in the late twenties – early thirties with the need to find a location for a new mechanical and biological treatment plant.

The model for Prague's wastewater treatment plant was Lindley's sedimentation wastewater treatment plant in Frankfurt am Main (Jásek and Palas, 2006). It was modern not only for the technology used, but it was also progressive as to locating all odour emitting operations underground, so it did not offend the surroundings and the land saved could be put to other use. What an advance in comparison to other methods such as extensive irrigation treatment!

The mechanical pre-treatment apparatus (Figure 20.2), for example, coarse screens, *grit trap* and fine screens, were hidden in the inner parts of the operation building, while the sedimentation tanks were located below the plot between the building and the navigation channel. A total of 400 litres of waste water per second flowed to the plant through the three main sewers. The largest of them, labelled with letter A, had an outfall about 3 metres above the level of water in the treatment plant and so it first flowed through a chamber where the descent of the sewage was used to power a water wheel which drove ventilation of the underground spaces. All three sewers then entered the '*grit trap* cathedral' together. Here gravel, sand and mud settled at a flow rate of 90 mm/s in a 34 m long and 6 m wide tank. The suspension was drawn off by a centrifugal pump and forced into the sand separator trough on the wastewater treatment plant grounds. The trapped sand, called 'wash out' in treatment vernacular, was bought by farmers for soil improvement and the coarser elements were also sold as a construction material.

Floating debris was captured on the fine screens with teeth spaced 7 mm apart, lining the full length of the rim of the trap. The screens were unclogged using special rakes and the caught material – screen rakings – were transferred by a paddle to a barrel lift which transported it to ground level and then taken by narrow-gauge wastewater treatment plant service railway to the storage area on Císařský Island. About 4 tonnes of screen rakings were caught daily.

An apparatus was located at the outlet from the *grit trap* for dosing and mixing milk of lime and aluminium sulphate solution to support the effect of subsequent sedimentation in sedimentation tanks (Figure 20.3). This extremely advanced apparatus, however, ran only in trial operation and soon after was decommissioned. Used chemicals were not only expensive, but also reduced the suitability of use of the sludge in agriculture, so customers (mostly farmers) refused it.

Figure 20.3 Mechanical pre-treatment step.

The final stage of the treatment process was sedimentation of sludge in ten underground *decanters* almost 90 m long with a volume of 1200 m^3, where the waste water from the *grit trap* distributed through a feed gallery using sluices. The bottom of the tanks was sloped against the current with a sludge sump in the deepest place. The sewage moved through the tanks very slowly at speeds of around 1 cm/s at which the suspended grey-black sludge sank to the bottom. The treated water trickled from the tanks to the outflow gallery and then into the Vltava through two discharge sewers 200 × 250 cm in diameter with inverted siphons below the navigation channel and under Císařský Island. The wastewater treatment plant could be disconnected in the event of breakdown and the untreated waste water diverted directly into the river through two by-pass sewers 200 cm in diameter.

Sedimentation produced about 100 m^3 of sludge daily, which was sucked up from the bottom of the tanks by two three-plunger pumps located in a separate engine room below the operations building. It could be pumped onto the sludge fields on Císařský Island (in the winter) or into wooden sludge barges which, towed by a rented steamship, took it for final drying outside Prague. Two 'sludge wells' were located next to the operations building, serving for temporary storage and thickening of sludge during manipulation. In the sludge pump machine room there stood two smaller pumps for Vltava river water used for washing and preparation of coagulants. The technological arrangement of Lindley's wastewater treatment plant meant removal of around 70% of insoluble pollution from waste waters when using chemically assisted sedimentation.

Construction of the wastewater treatment plant was entrusted to the builder, Quido Bělský, who started the construction on 9th September 1901 and by first frosts the outlet he completed the inverted siphons under the navigation channel and ground works for sedimentation tanks. By the end of 1902, the tanks were already standing. The foundations for the walls of the *grit trap* cathedral were ready and the field railway bridge had been completed and approved by the building authorities. A year later, the whole cathedral and three intake sewers, access staircases to the basement were ready, foundations of the outer walls of the building and backfill of the vaults of the settling tanks. In 1903 a tender was issued for complete delivery of the machinery. In this tender, three significant Czech companies participated: The First Czech-Moravian Machine Works (*První Českomoravská továrna na stroje*) in Libeň, Machine Works Limited Company (*Akciová společnost strojírny*), formerly Breitfeld, Daněk and Co. (*Breitfeld, Daněk a spol*) from Karlín, and finally the Märky, Bromovský, Schulz machine works from Hradec Králové. It was officially Českomoravská that won the tender, but because all competitors were all bound by a cartel agreement, they divided up the large order as it suited as follows: Českomoravská supplied all pumps, piping and bridging and crane constructions, Breitfeld & Daněk manufactured the steam engines and boilers, including the necessary pipe fittings and Märky, Bromovský, Schulz produced all the transmissions with consoles, belt wheels, crank shafts and other parts. The machinery for the wastewater treatment plant cost 250,000 crowns.

At the end of 1905, the construction stage of the wastewater treatment plant (Figure 20.4) was completed and on 27th June 1906, trial operation was launched. After fine tuning of operational defects, the official approval process took place in two parts, on 21 May and 11th June 1097.

Figure 20.4 Old Lindley´s WWTP in Prague.

The actual costs for building the wastewater treatment plant reached 2 million crowns, the land plots were bought up for another 550,000 crowns. The cost of the whole new sewerage system for the city, built between 1893 and 1907 was almost 15 million crowns.

Although the origins of the wastewater treatment plant are comprehensively charted, one thing remains unclear. The original plans for the exterior appearance of the building, produced by the offices of the builder, Bělský, were changed fairly extensively literally at the last minute. If we compare plans from 1903 with the actual appearance of the building, we find that somebody pushed through significant expansion of rooms for workers and administration, completely different division of windows with added decorative plaster window surrounds, a mansard roof, rustication on the corners, gables with volutes, alternating convex and concave cornices, an onion-shaped roof for the staircase and so on. And this all occurred when all underground parts of the wastewater treatment plant including outer and supporting walls of the building itself were already standing. Lindley's budget had already been exceeded and economising was taking place on all aspects. Perhaps a separate tender had been issued for rendering the façade and the layout of the rooms in the building, but what is certain is that thanks to this fact, the architectural value of the old wastewater treatment plant rose into a higher class.

The whole of the wastewater treatment plant site is a perfect example of textbook brickwork. Vaulted ceilings were built with masterly expertise, openings in complicated surfaces, corridors, niches, sky-lights and tunnels. The operation building was designed to be almost symmetrical, with side wings joining the higher, central section. At the corners of each wing, a 30 m tall chimney stands with a decorated crown. The right-hand one was used for ventilation of the basement described above, while the left-hand one was for smoke, carrying away fumes from the steam boilers. The whole wastewater treatment plant building is remarkable for its architecturally balanced harmony of the white plasterwork with the bare brickwork. The structure gives a majestic impression which is enhanced by the sandstone blocks used for the window surrounds, skirting and corners. The most interesting part of the operations building is the basement. Beneath a massive barrel vault of the *grit trap* cathedral which finishes in an apse leading to a portal made of specially shaped bricks, the visitor really does feel like they are in a temple of worship. The three main sewers entering here from the south are vaulted, made of bricks which unusually model the space. The unusual spell is enhanced by the lighting when light and shadow from various angles alters the space even more. Above the *grit trap*, on ground level, there is a 16 m high central hall with a floor area of 300 m^2, looking outsized when compared with the other spaces in the building. It has, however, an important static purpose, because it weighs down on the vertical walls in the basement which the dome of the cathedral leans against. The administrative part in the front façade, where offices, laboratories and technicians' rooms are located, links onto the hall, the cross-ways built section of the building. Operation of the wastewater treatment plant was at first handled by a nine-man shift made up of a foreman, machine engineer, boiler man and sixteen manual workers.

The steel heart of the wastewater treatment plant was a steam engine room with the boiler room in the west wing. Steam for the machines was produced by two fire-flue boilers built to operate at a pressure of 11 atm. The steam was heated to a temperature of around 300°C. The coal store next to the boiler room could hold up to one hundred tonnes of coal, while the daily consumption of this fuel was around 2 tonnes. The engine room is designed as a two-story atrium, where two centrifugal pumps in case of basement flooding stand on its lower floor, while on the upper floor, carried by five arches, two Schmidt condenser steam engines, each with an output of 90 horse power. Transfer of the engines' torque to the driving machinery was ensured by leather belts running from the engine room through the whole building. In addition to the flood pumps, sludge and water pumps, a grit pump in the basement, a small generator supplying electricity for operations, the rakings lift and chemical operations equipment, that is, coagulant mills and mixers which took up the whole eastern wing of the operations building, were also run by steam engines.

The new technology was an important source of knowledge for the Czech professional community. Numerous excursions visited the wastewater treatment plant and its operation was monitored permanently by the water management authorities. They investigated the properties of the sludge and the most suitable methods of processing it, compared the content of fertiliser substances in the water before and after treatment, the laboratories monitored the quality of groundwater in the surroundings of the treatment plant and so on.

The smooth-running system was interrupted after seven years by the First World War. Two thirds of the staff were mobilised and operation of the treatment plant suffered for lack of personnel. Money for essential repairs was not received, suspension of operations occurred more and more frequently and in 1918, the treatment plant stood inoperative for almost one month due to a lack of coal for the steam engines.

20.4.2 Between the wars

In the 20s, the so-called Greater Prague was created by making surrounding boroughs to the centre, and this meant new times for the wastewater treatment plant. Although the quantity of sewage flowing into it grew only slowly, the quantity of rakings captured increased to 16.3 m^3/d, sand from the original 20 m^3 to double that amount and in particular amounts of sludge increased to 270 m^3 /d. Modernisation measures for the wastewater treatment plant therefore had to be implemented over the years.

In 1921, electricity was introduced, meaning the end of continuous operation of the steam engines, their being used only in the spring flood season. Electric motors gradually took over the job of driving the sand pumps in the *grit traps* (1921), the sludge pumps (1924) and rakings lift (1927).

After twenty years of operation, the facilities no longer suited the new situation, a fact which was reflected also in their condition. We can read in the report (Jásek & Palas, 2006), drawn up by the City council fact-finding committee which performed a surprise inspection of the wastewater treatment plant at the beginning of 1926:

> *'Overall, the current state of the wastewater treatment station is very poor ... inside it was observed that the rooms have not been painted since construction was finished, the walls are black, windows smashed ... the engine room is poorly maintained and contaminated, tools lying around on the ground, the boiler room pipe conduit insulation has completely disintegrated, hanging down from the pipes in many places ... in the coal store we found 15 dovecotes belonging to the boiler man ... the lime milling room is used to store odds and ends ... in the hall the lift is visibly eaten away by rust, needs repair to avoid collapse ... the primitive method of removing rakings from the grille, propelling it along troughs to a lift which is completely disintegrating is striking ... the surroundings have been tainted by the Feca cooperative which stores its end product manure by the building, the front garden is overgrown and neglected ...'*

> *'We also discovered that numerous strangers use the station's bridge structure as a thoroughfare without being challenged. A stop must be put to this, so that the ownership rights of the Prague municipality are imposed anew...'*

The report ends with a list of measures and budget for repairs, which were performed over the subsequent two years. As regards water management tradition in the Czech lands, it should be noted that in the same year, this country's first *'Laboratory for testing sewage waters and experiments on the processing of sludge and slurry'* was founded due to the efforts of Dr. Ferdinand Schulz, professor at ČVUT and master of the Water, Fuel and Illuminant Research Institute. Setting up and management of the laboratory was entrusted to Ing. Dr. Julie Hamáčková, soon replaced by Ing. Dr. Vladimír Maděra. The laboratory performed bacteriological analyses of Vltava river water below the wastewater treatment plant, monitored the parameters of sewage and sludge and slurry and led extensive semi-operational experiments with *anaerobic sludge stabilisation*, where the resulting biogas was used for heating utility water for wastewater treatment plant personnel.

In 1927, extension of the wastewater treatment plant was started. A new screen facility, a three-section *grit trap* and four sedimentation tanks grew on the site. It was assumed that this would be a temporary measure which would be replaced by a new treatment plant within only a few years. Neither negotiations nor a call for tendering plans for the project in 1933 which received 15 entries resulted in any concrete steps.

'Very few are aware of the difficulties the present wastewater treatment plant must tackle. Awareness is limited only to occasional complaints concerning specifically the inadequate method of sludge despatch. Proper urban sewage cleansing has so far been ascribed only a subsidiary role in Prague, although this should have priority while solving the majority of issues involved with large cities. This can be the only explanation for a large proportion of the wastewater treatment plant being so old that it could be exhibited in a technical museum', we can read in a report by the master of the construction department of the municipality (Jásek, 2006), Ing. Vondráček from the year 1936. If he had only known that the wastewater treatment plant was to be in operation for another 30 years!

Sludge processing was becoming an ever more urgent matter. The main customers, farmers and gardeners from around Prague, soon realised that the wastewater treatment plant had problems with sales of sludge and began dictating drastic conditions for purchasing it, including even pricing terms. The whole matter were also complicated by ever more frequent repairs of the sludge barges since 1909 and so the wastewater treatment station had to give in to the pressure. Although the purchase of a Škoda steam sludge tanker in 1926 reduced the problems, but did not bring any fundamental improvement in operational problems. The city council was determined to solve the situation by construction of new sludge fields near Veltrusy and Hostín. Under the expert leadership of Ing. Krouza, professor at Czech Technical University in Prague, complex sludge pumping experiments were conducted to gain as much background material for a sludge pipe project. Three pumping stations were to be built along the route, the first at the wastewater treatment plant in Bubeneč, in Řež (on the site of the planned new treatment plant) and in Veltrusy. Sludge wells in Bubeneč were to facilitate sludge homogenisation while in Řež, 4 reinforced concrete digestion tanks, 18 m in diameter and 21 m high, with heating and mixing gear were to be built. Use of resulting biogas was planned for producing electricity by using a gas motor with an output of approximately 500 hp. Plans were completed in 1938, but due to the political situation in Czechoslovakia, the project was never implemented.

The areal expansion of the modernised treatment plant laid increased demands on transport access and so, in 1928, the field railway was extended. This small railway with a gauge of 700 mm, built simultaneously with the wastewater treatment plant in 1906, was in fact a comprehensive internal transport system which provided cheap and reliable transport between all work stations: the operations building, the new screen facility, the new *grit trap*, grit separator, sludge barge dock and storage areas on Císařský Island. The railway, originally 815 m long, in places had a fairly extreme 1:25 gradient and a minimum track radius of 14 m, which later precluded use of locomotives with more than two axels. Because the railway crossed the navigation channel, a 34-metre long steel bridge with a load capacity of 340 kg/m^2 was ordered from Pražské Mostárny (Prague Bridge Construction Company). The bridge's load bearing test was performed on 19. 9. 1902 together with approval inspection of the railway.

To begin with, the trucks were pushed manually, but, with increasing demands on transportation, this became beyond the physical abilities of the workers. In 1909, the first locomotive was purchased, powered by a petrol engine, supplied by the Wohanka company, Příbram. After decommissioning in 1922, a tractor was used for a time to tow the trucks and between 1924 and 1925, another petrol-driven, German-made locomotive was tried out, but did not fulfil expectations. The tried and tested Wohanka company in 1927 delivered another petrol-driven locomotive, joined ten years later by a modern engine built by ČKD with a three-cylinder diesel engine with an output of 27 horsepower. This was one of the first diesel locomotives constructed by this firm. At first glimpse, it seems surprising that the Planning Office did not purchase any of the small steam engines which were commonly available second hand from building companies and for which the fleet of engines of the wastewater treatment plant was perfect. The reason was the intermittent use of the railway. A steam engine cannot be simply switched off during breaks at the turn of a key in the ignition and that meant a permanent supply of fuel, water and supervision by staff, which would have made it uneconomical. It can be seen in historical photographs that separate journeys were made with rakings which were then covered over with earth and ashes and separately with sand which was left in piles and purchased by customers. Although each side-tipping truck with a load volume of 0.75 m^3 had a screw-down handbrake, the brakeman was only on the last truck, each train containing a total of only 3–4 trucks, so the locomotive would be able to brake by itself.

Only in 1955, the treatment plant purchased two Stavoloko BNE 25 locomotives with Gebus system of diesel-electric power transfer to replace the decommissioned Wohanka and ČKD engines. Each engine weighed 4.5 tonnes and could tow a train of up to 89 t. In 1970, both locomotives were sold to a fire-clay brick kiln in Zliv, where they survived over 30 years. Today they can be found in private collections.

Some fragments of the treatment plant railway have survived. The truck lift and part of the track system can be found inside the building. On the upper yard, there is a passing loop, a turntable and a swivel track leading to the lift. The rest of the track has been asphalted over, so it can be identified at first glance only by an embankment leading to the bridge. Railway buildings surviving also include a double-track locomotive depot with an inspection put and a tower loader from the new screen facility. Six of the original twenty tipping trucks still exist today.

20.4.3 Post-war years

Almost no records have survived regarding operation of the wastewater treatment plant during the Second World War. It can be assumed that apart from a reduction in staff numbers, no significant changes occurred.

The post-war fate of the wastewater treatment plant somewhat copied pre-war development – it was expected that it would soon be decommissioned and so only provisional changes to technology were made. Permanent lack of finances, however, delayed construction of a modern treatment plant to such an extent that Lindley's project became one of the oldest functioning mechanical-chemical wastewater treatment plants in the Europe as in most agglomerations the old plants were upgraded or modified mainly using the emerging technology of activated sludge (Benidickson, 2011).

In the spring of 1947, the Prague Planning Office presented a new project for reconstruction of the screen and old *grit trap* facilities. This adaptation, part of a two-year plan, shortened the trap to half its length, which meant that the sand pumps had to be moved. New protective grilles were installed at the intake, after which the flow was split along two 3-metre wide concrete troughs, finished with closable sluice gates. The troughs were fitted with electrically wiped Dorr screens with 12 mm gaps. The screens had a time switch which could be set to wipe every 5 to 60 minutes, depending on the quantity of water flowing in. The rakings were ground in a small electric grinder made by Sulzer, in the shaft next to the screens and diverted by to the intake. A new conveyor belt making it simpler to load the field railway trucks which could be positioned right next to the screens. Provisional adaptation of the traps relieved the hard and dangerous work linked with the unhygienic despatch of rakings to the storage area. The *grit trap* cathedral has survived in the state as it was in 1947 to this day.

In 1957, work on the first stage of construction of the new Central Waste Water Treatment Plant on Císařský Island commenced. In connection with the nearing decommissioning, great changes were made in the course of the sewers in Papírenská Street. Three quarters of the decorative fencing, mosaic paving and a row of trees fell victim to a huge trench of up to 6 m in depth. The old gatehouse was replaced by a standardised block with a flat roof. In 1964, the state-owned design company Hydroprojekt compiled an improvement study proposing demolition of the whole site, luckily exceeding the state financial resources of the time.

In the spring of 1965, the steam engines started to run during the floods and the water was pumped back into the river. In June, the grand opening of the new treatment plant took place, although due to non-functional pre-cleansing, it was closed down after the end of the official event and the old wastewater treatment plant had to take over again. Elimination of defects took two years. Only then was the aged wastewater treatment plant decommissioned (Palas, 2010).

During the first stage of improvements, the wooden sand separation building was demolished using a Breitfeld-Daněk dredger, the new sedimentation tanks and new *grit trap* disappearing under the debris. An oxy-acetylene welder was used to cut the Geiger screens, dreg lift and electric engine room out of the screen facility, still standing to this day.

Due to difficulties with the using the new treatment plant, the historical building remained as a back-up in case of emergency. At the beginning of the 80s, the original sedimentation tanks were still being used as handling reservoirs for sludge from the new treatment plant, reliably pumped by the eighty-year-old (!) První Českomoravská sludge pumps. And it is this endless long service which is the main reason that the old wastewater treatment plant survived until better times.

The deserted site slowly neared a state which demanded a fundamental decision: either modest investments had to be made or complete devastation would occur. The forgotten building was luckily discovered by enthusiasts in the mid-1980s who began to return the tarnished shine to it. The mechanical equipment was renovated, performed in field conditions, made more difficult due to many missing components and also non-existence original documentation. In those days it was almost impossible to find a corporation willing to make a new piston rod out of quality material for one of the steam engines which had been severely damaged by film-makers. Repair was successful in the end, by even so, some of the scars from filming are visible today.

On the initiative of ČSOP (Czech Union for Nature Conservation) for Prague 7, the old wastewater treatment plant was declared a cultural monument on 26.4.1991, thereby establishing a legal framework for further activities. The idea of using this unique technical structure as a museum came to fruition.

20.5 EPILOGUE

It is well noted that wastewater collection&treatment professionals that ever visit the old Prague wastewater treatment plant usually recognise that the city of Prague has another significant monument next to all historical buildings – the old Prague wastewater plant. Similar structures constructed elsewhere were generally demolished between the wars and those that survived cannot boast such completely preserved equipment. This is, therefore, a site of exceptional value both from the point of view of water management and civil and mechanical engineering, which rightfully holds a leading position amongst our industrial monuments. Numerous visits by experts in these fields just go to confirm this fact. Another monument is the 'old' brick-layed sewerage system that even in present creates a backbone of the Prague sewerage system. In 1996 the old wastewater treatment plant was opened to public a museum and so far it attracts the interest of the wide public, which is the best guarantee for this technical jewel to survive as a monument for coming generations.

20.6 REFERENCES

Benidickson J. (2011). The Culture of Flushing: A Social and Legal History of Sewage. UBC Press, Vancouver, Canada.

City of Prague (1914). Edict on Use of Sewerage Systems. Documenta Pragensia Monographia, Czech Republic.

Havrda J. and Kovář M. (2011). Contribution to the Discussion About the Building Originsa of the Dominican Monastery of St. Climent in Prague. Staletá Praha, Praha, Czech Republic, XXVII/2011/2, pp. 74–86.

Jásek J. (2006). William Heerlein Lindley Prague Sewer System History. Scriptorium, Prague, Czech Republic.
Jásek J. and Palas J. (2006). 100 Years of New Prague Sewer System. Pražské vodovody a kanalizace, Praha, Czech Republic.
Palas J. (2010). About History of Prague Sewer System. Ekotechnické museum, o.p.s, Praha, Czech Republic.
Ruth F. (1903). The Chronicles of the Royal City Prague and Neighbouring Municipalities. Körbr publishing, Praha, Czech Republic.

APPENDIX I

Terminology

- *Decanter*: A sedimentation tank with start-stop operation. Waste water fills it slowly for several hours or days, after which the intake is closed, the layer of water is pumped out while the sludge settled out at the bottom is pumped into a sludge tank and used for example, as a fertiliser.
- *Combined sewerage*: Sewers draining rainwater from the streets, pavements, roofs and so on, together with domestic or other types of waste water; opposite to separate sewerage which consists of double piping, one to drain rainwater into the nearest watercourse, while the sewage is delivered to the treatment plant by the other.
- *Systematic sewerage system*: Sewer network for a whole area facilitating connection of further structures; the opposite of non-systematic sewerage which serves to drain only one structure or site.
- *Combined sewer overflow*: Structure in the sewer facilitating automatic discharge of some of the sewer water into the river to avoid sewer overloading during high flow rates (e.g., in times of heavy rain).
- *Grit trap*: Sedimentation tank for capturing heavy mineral particles carried along the sewer, in particular sand, grit and mud.
- *Inverted siphon*: Complex structure on sewer network for avoiding obstacles in the terrain, most often rivers and streams or infrastructure networks.

APPENDIX II

Biography of Sir William Heerlein Lindley

W. H. Lindley was born on 30th January 1853 in Hamburg and died 30th December 1917 in London as a one of three sons of the famous British engineer William Lindley. W. H. Lindley became civil engineer and worked together with his father on a number of projects including for example, design of the Pest sanitation system. In 1879 he took over father's responsibilities and established himself as a respected engineer. In 1980 he designed the sanitation system in Düsseldorf. Between 1881 and 1889 he oversaw the construction of Warsaw waterworks, designed by his father in 1876-8. He also evaluated alternative variants of sanitation in the city of Prague and later won with his own compiled project the design of a modern sanitation system including advanced wastewater treatment plant in the Prague quarter Bubeneč (in use from 1907 until 1967). He oversaw the construction of the sewage system in Prague between 1895 and 1906. The main structures of the sewer system are in use until today with no major problems experienced over more than 100 years due to the quality of the design and materials used. The wastewater treatment plant in Bubeneč was decommissioned in 1967 and now serves as a museum of Prague's sewage system and Lindley´s work. Next to these major achievements he for example, designed the water/sanitation systems in Tbilisi, Giessen, Würzburg, Bucharest, Łódź, Sankt Peterburg, Kaunas or Beograd.

Chapter 21

Waterborne diseases in the Hippocratic treatise *Airs, Waters, Places*

Maria E. Galanaki

21.1 INTRODUCTION

The Hippocratic Corpus, that is, the collection of ancient Greek medicine works attributed to Hippocrates (Figure 21.1) or students of his, covers a variety of medical issues, such as prognosis, therapeutics, dietetics and surgery. There are, though, some theoretical treatises which elaborate on basic principles which should govern the art of medicine and determine a physician's work; the treatise *Airs, Waters, Places* is one of these. Specifically, this very interesting and insightful work stresses the importance of the climate, the seasonal changes, the airs, the waters and the locations of the various cities to the human health. The writer of the treatise, who is not certain to be Hippocrates himself or some student of his (Jouanna, 1999), examines all the above factors and their influence on the diseases which affect the people who inhabit areas of very different characteristics, stressing especially the role of the waters which flow in the various places. The writer emphasizes on the need for the physician to be aware of this kind of knowledge, in order to be able to better understand the causes of a disease and, subsequently, to find the right remedy for the cure of the patient.

In this paper, it will be examined how, according to *Airs, Waters, Places*, the different kinds of waters, which flow in various kinds of areas, affect people's health and can provoke various diseases. A detailed account of these diseases and their causes will be given, since Hippocrates considers them as a necessary knowledge for the physician who arrives at an unknown city and has to practice his art on the people who live there. Moreover, the text implies the importance of the prognosis, especially for a younger physician, who needs to be very careful and accurate, not only when diagnosing a disease, but also when watching the evidence which could lead to a disease. In this treatise, the evidence is provided by nature, which helps the physician to make a prognosis about forthcoming diseases, combining the climate with human health (Edelstein, 1931).

It should be noted here that bibliography concerning the environmental and climatological observations in *Airs, Waters, Places* is very limited. It should also be mentioned here that Hippocrates will be referred to as the writer of this treatise, because it is beyond doubt that the views exposed are totally consistent with the spirit of Hippocratic medicine, that is, the cooperation of the empirical and the dogmatical way of thinking, that is, theory together with practice (Smith, 2002; Heidel, 1981), as well as the respect to the power of nature in maintaining and restoring human health (Jouanna, 1999).

Figure 21.1 *Hippocrates*, engraving by Peter Paul Rubens, 1638. (National Library of Medicine 2006: http://en.wikipedia.org/wiki/Hippocrates#CITEREFNational_Library_of_Medicine2006).

21.2 CITIES EXPOSED TO WARM WINDS – PROFUSIVE AND SALTISH WATERS

In passage 3 of the treatise (Littré, 1961), Hippocrates begins with the description of waters in the cities that lie between the wintery rising and the wintery setting of the sun, that is, in the orientation of South East towards South West. It is worth mentioning here that sources from ancient literature about the way that Ancient Greeks oriented themselves are very scarce. They were most probable to use the East or West as orientation points, due to the apparent sun movement. This is obvious as early as in the Homeric poems, such as in Book 8 of the *Odyssey*, where King Alcinous speaks to the Assembly about Ulysses, the stranger who had found his way to Alcinous' palace, either from East or West (Von der Mühll, 1962). Another enlightening passage comes from Strabo's *Geographica*, Book 10, Chapter 2.12, where the writer speaks about Ithaka and its position according to the four cardinal points (Meineke, 1877). It should also be made clear here that, as far as the wintery rising and the wintery setting of the sun is concerned, the Ancient Greek philosophers Plato, Eudoxus and Aristotle had studied the sun, the moon and the planets and had come to the conclusion that they had a natural circular motion. More specifically, Eudoxus assumed that the sun made two circular movements, one around the earth and one around itself, thus he justified the twenty-four hour long day and night and also the difference in the sun positions during summer and winter (Fowler, 2008). When Hippocrates speaks in *Airs, Waters, Places* about the wintery rising and the wintery setting of the sun, he most probably has this in mind.

The waters in this type of cities are profusive and saltish, hot in summer and cold in winter. This causes the heads of the inhabitants to be full of phlegm, which disturbs their bellies and makes them flabby. Phlegm, according to the Hippocratic Humoral Theory, lies in the head. In these cities, where the winds are warm, phlegm is excessive and runs from the brain down to the rest of the internal organs, that's why it causes belly disorders.

In this point it should be noted that, according to the Hippocratic physicians, phlegm is one of the four basic humors of the human body, together with blood, yellow bile and black bile (Nutton, 1993). These humors run in balance within a healthy body and derive from the four basic human organs: brain, heart, spleen and liver (Figure 21.2).

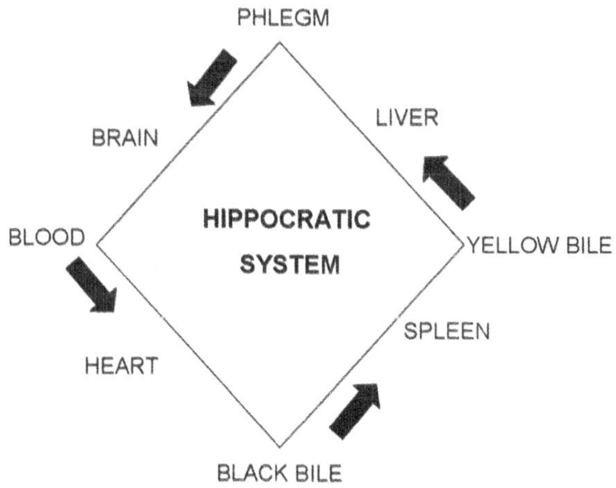

Figure 21.2 Hippocratic Humoral System (based on Kontopoulou – Marketos 2002).

If one humor becomes overmuch or meager, the balance is disturbed and diseases will come about (Carrick 2001). The humoral theory associates the four humors with the four basic elements of the universe (Diels – Kranz: *Empedocles*, fr. 1,181), earth, fire, air and water (Figure 21.3) and each humor has the qualities of the element with which it is connected, that is, phlegm is associated to water, thus it is wet and cold. The humors are also connected with the four seasons (Figure 21.4), because the Hippocratic physicians noticed that the diseases which affect people during the seasons are mainly caused by the excess or lack of one humor; in every season, the humor which causes the disease is different. This leads to the conclusion that Hippocratic medicine paid a high amount of respect to nature and its forces and kept the strong belief that the human body is a reflection of the environment and the universe (Nikolova, 1999), so that the physician must initially study nature in order to cure a disease.

The diseases in the cities lying between the wintery rising and the wintery setting of the sun are as follows: women are sickly, suffer from excessive menstruation, become unfruitful when they get sick, not by birth, and have frequent miscarriages. Infants often suffer from spasms and asthma, which, in Hippocrates' time, were thought to be symptoms of epilepsy, 'the sacred disease'. Men suffer from *dysentery*, diarrhea, fevers with shivering and chronic fevers, *epinyctis* and hemorrhoids. Also *ophthalmies* of humid character occur, but are not very serious. Men over their fifties have 'liquids', due to the excessive phlegm, running down from the brain, which paralyze them when suddenly exposed to sun or cold.

These are the endemic diseases in these cities, caused by their locality. It is worth noticing that most of the diseases that Hippocrates mentions, concerning the cities in the orientation of South East towards South West, are caused by the excess of phlegm in the human body. Since phlegm is the humor which predominates in winter (Nutton, 1993), the point made here is that these cities, having warm waters in summer and cold in winter, disturb the balance of phlegm in the human body and cause all the above diseases. Apart from those, the inhabitants also suffer from the epidemic diseases, which appear due to the rest of seasonal changes.

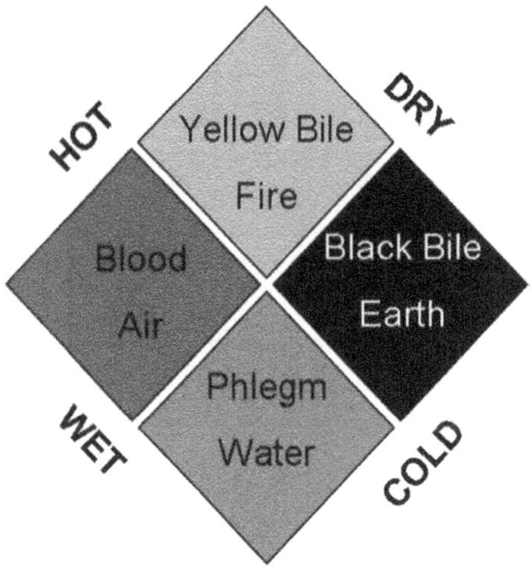

Figure 21.3 The Humoral Theory of Hippocrates: the four basic humors of the human body, their association with the four elements of the universe and their qualities (http://en.wikipedia.org/wiki/File:4_body_fluids.PNG).

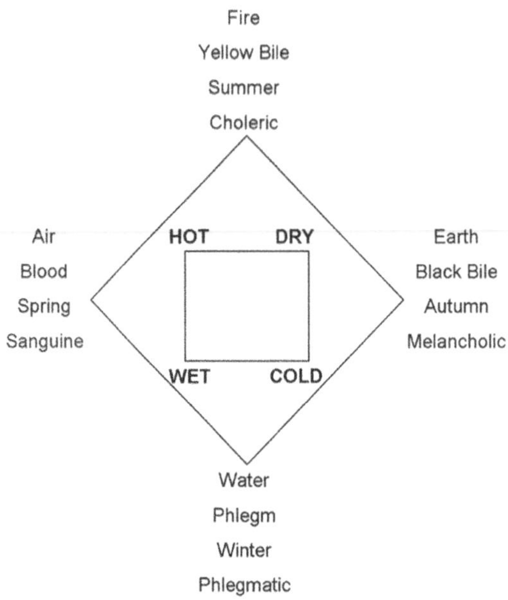

Figure 21.4 Schematic representation of the Hippocratic Humoral Theory (http://matrix.msu.edu/hst/hst425/hst425.php?page=session1).

21.3 CITIES EXPOSED TO COLD WINDS – HARD AND COLD WATERS

In the next passage of *Airs, Waters, Places*, Hippocrates writes about cities which are exposed to cold winds, that is between the summer settings and the summer risings of the sun – from North West to North East. These have hard and cold waters, which cause men to be stout and slim. The main humor in their bodies is bile, not phlegm. This means that cold waters have caused an imbalance in the bodies, by making bile excessive, instead of phlegm, which predominates in wintery climates (Nutton, 1993). The diseases that affect them are mostly pleurisies and acute diseases; in the treatise *On Acute Diseases*, Hippocrates names as acute all the diseases which share the common characteristic of constant fever. Due to the dryness and the coldness of the waters, these people suffer from internal bleedings, stiffness of the body and hardness of the intestines. Subsequently, they have to eat much and drink a little. *Ophthalmies* and epilepsy are rare, but rather violent.

As far as women are concerned, they become uptight due to the cold, hard and indigestible waters. They have small amounts of painful menstrual discharges and difficult childbirths. When they give birth to their children, they are unable to breast feed them, due to the waters' constitution. *Phthisis* is frequent to women after they give birth, due to the violence of the throes, and little boys suffer from *dropsy* in the testicle, which goes away as they grow older. Puberty arrives late for these young men.

In general, according to Hippocrates, these people enjoy long living, in spite of the diseases they suffer from, compared to those who inhabit other areas.

21.4 CITIES WHICH LIE BETWEEN THE SUMMER AND THE WINTER RISING OF THE SUN

21.4.1 Cities lying towards sunrise

Hippocrates goes on with describing the characteristics of the cities that are exposed to winds between the summer and the winter risings of the sun, that is from North East to South East, and of the opposite direction too. Firstly, the cities that lie to the rising of the sun (passage 5) are generally healthier than those turned to the North or those exposed to hot airs, because their climate is mild and the heat or the cold are not excessive. This makes the waters clear, fragrant, light and delightful to drink, due to the sun which makes them fresh. The inhabitants have healthy appearances, loud and clear voices, better temperaments and they are more intelligent than people who live in locations turned to the North. All in all, these cities seem to enjoy the constant spring, their inhabitants do not suffer from harsh diseases and the women, especially, are very fertile and give birth to their children with little suffering. According to the Humoral Theory, blood is the strongest humor in spring, but it is also the humor which makes people vigorous and stout (Nutton, 1993); this justifies why Hippocrates considered this type of cities as the healthiest one.

21.4.2 Cities lying towards sunset

Hippocrates describes this kind of cities as the unhealthiest one (passage 6), due to the fact that hot and cold north winds blow sidelong to them and the eastern winds do not approach them. Firstly, the waters there are not clear, because the sun is not shining in the morning above them, thus creating a mist, which gets mixed with the waters. These special conditions make the inhabitants very frail and susceptible to all the diseases that were previously mentioned. Their voices are rough due to the unhealthy winds. Concluding, these cities seem to have autumn as the main season, because the temperature difference between day and night is rather big.

21.5 OTHER KINDS OF HEALTHY AND UNHEALTHY WATERS

Hippocrates so far has described the characteristics of the various kinds of waters which flow in different areas. His aim was to give the categories of the cities, according to their location, and, subsequently, to describe their climatological conditions, including the waters, so as to come to conclusions about the kinds of diseases that affect the inhabitants. The next part of the treatise *Airs, Waters, Places* (passages 7–9) is concerned with the elaboration on other kinds of waters and their effects on human health, which Hippocrates considers as crucial.

21.5.1 Stagnant waters

The first kind of waters studied next is that of the lakes, when they are stagnant and swampy. These are harmful to the human health and cause excessive bile in the body, since they are hot, thick and smelly in summer and, during rainfall, they mix up with rain water; this makes them unclear, due to the fact that they cannot be renewed. In winter, they get frozen and blurred and cause excessive phlegm and hoarseness.

People who drink that kind of water have always large and swollen spleens, hard, thin and hot bellies and they are very thin in their shoulders, collar bones and face. Their slenderness is caused by the fact that their fleshes 'melt', in order to nurish their spleens; that's why they are constantly hungry and thirsty. Their bellies are very dry, both in the above and below area, all year round, and this is very frequent to them, thus they need the strongest pergative medicines. Fatal *dropsies* are often and, in summer, they are affected by many *dysenteries*, diarrheas and long-standing *quartan fevers*, which, if they stay for a long time in people with that kind of body constitution, may cause *dropsies* and, consequently, death. These are their summer diseases.

In winter, young men suffer from *peripneumonia* and delirium and, the oldest ones, from ardent fevers, due to their bellies' hardness. Women are subject to *oedemata* and *leucophlegmasiae*. They have difficult conceptions and deliveries. Their babies are large and swollen and, when they breast eat, get very thin and sickly. Womens' lochial discharges after birth are not regular. Children often suffer from *hernia* and men from varices and ulcers on their legs, which make them grow old very fast. Lastly, women often have pseudocyeses, due to *dropsy* of the uterus.

Due to all the above diseases, Hippocrates concludes that stagnant waters from lakes and swamps are very unhealthy for every use. This is a very interesting conclusion, because it testifies that the Greek physician had already realized that swamps produce undrinkable, harmful waters.

21.5.2 Waters flowing from rocks and from soils which produce thermal waters

Hippocrates considers the waters that come from soils like these as unhealthy, due to their heat, hardness and the large amount of metals that they contain. These are the reasons which make them provoke constipation to people who drink them. The best of them are the waters that flow from elevated areas and earthy hills, because they are clear, sweet and can be mixed with only a little wine. In winter they are hot and in summer cool. Hippocrates especially recommends those waters which flow to the summer sunrise, as being clearer, fragrant and light.

21.5.3 Salty waters

Salty waters are generally, according to Hippocrates, unsuitable for drinking, because they are hard to digest. The best of them are those who derive from the east, then come those who have their fountains between the summer sunrise and the summer sunset, at best those who are closer to the sunrise. After them, they come those between the summer and winter sunset. The worst of them are those who flow to the South, between the winter sunrise and sunset, and are more harmful to people who inhabit south areas, than to those who live in the North.

Hippocrates argues that healthy and strong people should drink every kind of water indiscriminately. But people who suffer from a disease should always drink the suitable for them kind of water; next, Hippocrates gives his piece of advice on this matter: people with hard and hot bellies should drink the sweetest, lightest and clearest waters, because these are solvent and make the bellies loose and moist. People with soft, humid and pituitous bellies are benefited from hard, indigestive and not very salty waters, because they absorb the excessive humidity from the body and make the organs as dry as needed.

Hippocrates finally states that salty waters are generally, but falsely, thought to be purgative, whereas they are the exact opposite, due to their hardness and unsuitability for cooking. That's why they make the bellies tight and cause constipation.

21.5.4 Rain and snow waters

According to Hippocrates, the rain waters are the lightest, the sweetest and the clearest of all. He proves this by mentioning that the sun attracts the thinnest and lightest parts of all kinds of lake, river and sea waters, as well as of the rain waters, after rainfall. Then he explains how rainfalls are created, through the condensation and accumulation of the clouds, which are formed by all the sweet ingredients of the earth waters that they have absorbed. That's why these rain waters are the best, according to Hippocrates, but they have to be boiled, in order to reacquaint a good smell, otherwise they may cause a hoarse voice to those who drink them. At this point, Hippocrates once again causes a surprise to the reader of today, because he had already perceived the need of boiling, so as to make water drinkable.

Snow waters are considered by him to be the most damaging for the human health, because, if once frosted, then they lose their initial constitution and what stays in them is only their blurred and heavy part. Thus Hippocrates comes to the conclusion that all waters coming from snow or ice are unsuitable for any use.

21.5.5 River and lake waters

These waters cause a large amount of diseases to people who drink them, thus Hippocrates does not recommend them as drinkable, although he stresses the fact that they are not harmful to every person who drinks them. At first, he mentions the diseases that these waters may cause, which are *lithiasis*, kidney inflammation, *strangury*, sciatica and *hernia*. He explains that the river and lake waters are unhealthy because they are a mixture of many different kinds of other waters, which end up flowing into them, such as brooks run into larger rivers or streamlets run into lakes. Thus, the rivers and lakes contain a lot of water ingredients of doubtful origin, sometimes from distant areas, which are totally different in relation to one another; consequently, when they all mix up, their ingredients may not unify, so their strongest part will dominate upon the others, and this is different every time, because the winds also play an important role on the constitution of these waters. What is reflected here is the Hippocratic notion that a disease is caused by an imbalance in the human body, due to an excessive or lacking quality (Jouanna, 1999). People who are mostly prone to the above mentioned diseases, by drinking these waters, are those who have very hot bellies, thus very hot bladders, which cause their urine to be divided into two parts, a thin and clear one which is discharged, and a thick, condensed part, which stays in the bladder and forms a stone. These people will suffer if they drink river and lake waters, whereas people with normal bellies and bladders are not in the same danger.

21.6 CONCLUSION

The treatise *Airs, Waters, Places* is a Hippocratic work which examines the effects of the climate and the environment, in general, to the human health. In the Hippocratic medicine, every disease has its natural logical causes (Moon, 1979; Wright, 1920), which are either internal, deriving from a man's bad body constitution or unsuitable dietetics, or external, thus coming from the environment. The role of nature is crucial in the Hippocratic medicine, since the human body is part of the environment and the physician helps the patient to return to the initial, natural healthy condition (Figure 21.5). Hippocrates elaborates here on the crucial role of the locality, especially of the kinds of waters which lie in the different kinds of areas, because their qualities may preserve or destroy the human health. Specifically, what is shown here is that diseases are less frequent in cities with an eastern aspect, since the waters which flow there are considered to be healthier and more suitable for drinking. On the contrary, cities lying to the West have the worst kinds of waters and they are thought to be the most unhealthy areas. Cities exposed to warm or cold winds do not have waters with balanced qualities, that's why they destroy the balance of the body qualities as well, causing excessive phlegm or bile, and, consequently, provoking a large amount of diseases. Hippocrates is also skeptic about the sanitary character of other kinds of waters, such as stagnant, thermal, salty, rain, snow, lake and river waters and he mentions boiling as a way of making them drinkable. He states, though, that a healthy man should drink all the kinds of waters that he has access to.

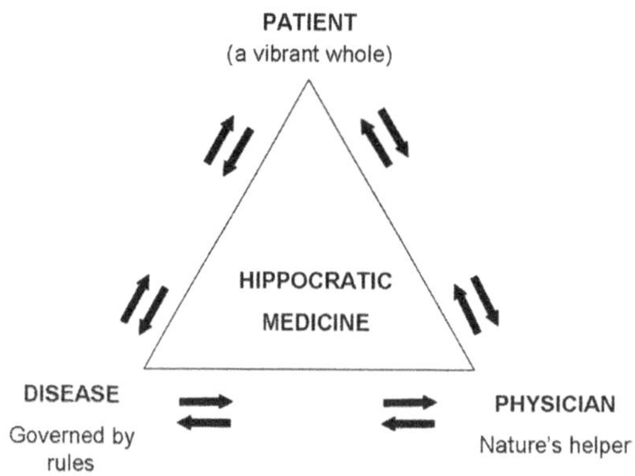

Figure 21.5 Three interacting elements of the Hippocratic rational system.

What should be emphasized on here is that *Airs, Waters, Places*, as well as all the other Hippocratic treatises, made a strong contribution to the scientific evolution of the later centuries, mainly because they suggested basing the explanation of natural phenomena on material, deterministic causes, rather than on supernatural forces (Calame, 2005). On the other hand, *Airs, Waters, Places* is the only work of the Greek Antiquity which elaborates on the effects of climate not only in human health but on ethics too. All in all, the treatise *Airs, Waters, Places* is not merely a medical work, but also a work on political geography, and it made Galen call Hippocrates the 'Prince of Philosophers' (Moon, 1979).

21.7 APPENDIX OF HIPPOCRATIC MEDICAL TERMS

dropsy: an accumulation of liquid in body parts

dysentery: an inflammatory disorder of the intestine that results in severe diarrhea with mucus, blood and pain

epinyctis: a rash which appears mostly during the night
hernia: the protrusion of an organ
leucophlegmasia: a white painful inflammation
lithiasis: the production of stones in organs, such as the kidneys or the bladder
oedema: a swelling of bodily parts
ophthalmy: a disease of the eyes, usually accompanied with liquid discharges
peripneumonia: a lung inflammation
phthisis: the consumption of the body powers
quartan fever: a fever which appears after four days or comes back every four days
strangury: painful, frequent urination in small drops

(Mitropoulos 1982, translation by Maria Galanaki)

21.8 REFERENCES

Calame C. (2005). Masks of Authority. Fiction and Pragmatics in Ancient Greek Poetics. Translated from the French by P. M. Burk (ed.), Cornell University Press, Ithaca.

Carrick P. (2001). Medical Ethics in the Ancient World. Georgetown University Press, Washington, D.C.

Diels H. and Kranz W. (1972). Die Fragmente der Vorsokratiker: Griechisch und Deutsch (The fragments of the Pre-Socratics). Weidmann, Berlin.

Edelstein L. (1931). *Περί ἀέρων* und die Sammlung der Hippokratischen Schriften (*Περί ἀέρων* (Peri Aeron) and the Collection of Hippocratic Writings). Weidmann, Berlin.

Heidel W. A. (1981). Hippocratic Medicine. New York.

Jouanna J. (1999). Hippocrates. In: Hippocrates, M. B. DeBevoise (ed.), John Hopkins University Press, Baltimore, Md.

Kontopoulou, Th. and Marketos Sp. (2002). Homeostasis. The Ancient Greek Origin of a Modern Scientific Principle. *Hormones*, **1**(2), 124–125.

Littré É. (1840). Oeuvres Complètes D'hippocrate. Vol. 2, Baillière, Paris, (repr. Hakkert, Amsterdam, 1961).

Meineke A. (1877). Strabonis Geographica. 3 vols. Teubner, Leipzig.

Mitropoulos K. (1982). Glossarion Hippokrateiou Syllogis (Glossary of the Hippocratic Corpus), Athens.

Moon R. O. (1979). Hippocrates and his Successors in Relation to the Philosophy of their Time. AMS Press, New York.

Nikolova V. (1999). Homeopathy vs. allopathy in Hippocratic writings. In: Aspetti della terapia nel Corpus Hippocraticum: atti del IXe Colloque international Hippocratique, Pisa, a cura di Ivan Garofalo, Lami, A., Manetti, D. and Roselli, A. (eds), 25–29 Settembre 1996, L.S. Olschki, Firenze, pp. 89–105.

Nutton V. (1993). Humoralism. In: Companion Encyclopedia of the History of Medicine, W. F. Bynum and R. Porter (eds), vol. I, Routledge, London, p. 281.

Smith W. (2002). The Hippocratic Tradition. Electronic edn, revised (First published 1979 by Cornell University Press)

Von der Mühll P. (1962). Homeri Odyssea. Helbing & Lichtenhahn, Basel, pp. 1–456.

Wright J. (1920). The origin of Hippocratic theory in some of the science of the nature philosophers. *The Scientific Monthly*, **11**(2), 127–140.

Websites

Michigan State University – Online Courses: History 425: Introduction to Late Medieval Medicine. http://matrix.msu.edu/hst/hst425/hst425.php?page=session1 (accessed 12 February 2013).

Fowler M. (2008). Lectures, Greek Astronomy, How the Greeks Used Geometry to Understand the Stars, 9/16/2008. University of Virginia, http://galileoandeinstein.physics.virginia.edu/lectures/greek_astro.htm (accessed 1 July 2013).

Wikipedia, the Free Encyclopedia. http://en.wikipedia.org/wiki/Hippocrates#CITEREFNational_Library_of_Medicine2006 (accessed 4 February 2013) http://en.wikipedia.org/wiki/File:4_body_fluids.PNG (accessed 4 February 2013).

Chapter 22

Ancient Greek and Roman authors on health and sanitation

Heikki S. Vuorinen

22.1 INTRODUCTION

In this article the relationship between sanitation and health is scrutinized as presented by ancient authors. The concept of sanitation is used to mean '*The establishment of environmental conditions favorable to health*' (Dorland's Illustrated Medical Dictionary, 1965, p. 1341). In particular, this article is focusing on the importance of water with regards to population's health, and on the measures needed to guarantee delivery of drinkable water.

The time period covered in this article is approximately one thousand years: from the fifth century BC to the 7th century AD. The referred treatises together with their respective dates and some comments are presented in the Appendix. Almost all of the medical authors (the Hippocratic writings, Diocles of Carystus, Galen and Paul of Aegina), originally wrote in Greek, with the exception of Caelius Aurelianus who wrote in Latin. On the contrary, non-medical authors (Vitruvius, Celsus, Pliny, Frontinus), whose texts were reviewed, wrote in Latin. The quotations in the article are taken from the modern English translations of the original ancient texts.

Because of the strong influence of Galen in late Antiquity and medieval times, the sources are biased (Nutton, 2004, p. 5). There is almost a gap in the surviving medical literature from 4th century BC up to the times of Galen in the second century AD. Only fragments of texts have survived even from the most famous 3rd century BC Alexandrian physicians, Erasistratos and Herofilos. Fortunately, the non-medical authors Vitruvius, Celsus, Pliny and Frontinus provide some useful information that helps us to fill this long gap. The surviving medical literature is heavily influenced by the Galenic humoural theories of disease causation. This is reflected on literature describing health and the role of sanitation.

22.2 THERAPEUTICS ABOVE ALL

Throughout antiquity medical authors give evidence that diseases, which are closely associated to unhygienic conditions, like dysentery and different kinds of diarrhoea must have been catastrophic to the population (Vuorinen, 2007, 2010). In the Hippocratic writings (written mainly in the late 5th and early 4th century BC, see Jouanna, 1999, pp. 373–416) a variety of intestinal diseases are described.

However in retrospect, diagnoses are not always straightforward and the culprit agents are difficult to identify. Summer and early autumn, when water resources were low in the Mediterranean world, the risk of drinking water contamination was high, therefore intestinal diseases were widespread, as presented in several passages in the Hippocratic writings (e.g., *Airs, Waters, Places.* 7; *Aphorisms.* III, 11, 21, 22; *Internal Affections.* 26, 45). The mortality of children, especially recently weaned, must have been high, which is probably described in the following passage of a Hippocratic author: '*It is mostly children of five years that die from this disease* (dysenteries), *and also older ones up to ten years; other ages less*' (*Prorrhetic II.* 22).

The poor level of waste management, including wastewater, was a cause of major risk to public health during antiquity. For instance, toilet hygiene was poor (Vuorinen, 2010). Public toilets were common to all; they were cramped, without privacy, and there were no wash hand facilities. The Romans did not use toilet paper. They most probably used sponges or moss or something similar, which was moistened in the conduit in front of the seat and then used to clean themselves. Private toilets were without running water and they were commonly located near kitchens. All this created an excellent environment for the spreading of intestinal pathogens. As a consequence water-borne infections were among the main causes of death. A modern reader would expect that ancient physicians could have paid attention to the link between sanitation and health. However, none of the medical authors mentioned anything about toilet hygiene.

Throughout antiquity the writings of medical authors – the Hippocratic writings, fragments from Diocles of Carystus, Galen, Caelius Aurelianus, Paul of Aegina – together with the *De Medicina* of Celsus and the *Naturalis Historiae* of Pliny suggest that medicine during antiquity was principally dealing with therapeutics: the description of diseases or symptoms and their treatment, including medicines, surgery and regime (food, drink, exercise, massage, baths, sleep, etc). In most of the texts there are none or just a few scattered remarks which make a shallow approach to the possible role of sanitation to the health of people. Ancient physicians were primarily interested in the correct identification, course, and treatment of the disease of the suffering individual. The cause of the disease appeared to be less important.

Caelius Aurelianus (floruit around year 400 AD), who belonged to the Methodist sect of physicians, and who referred heavily to Soranus of Ephesus (floruit early 2nd century AD) is illustrative. The way he presents the cause of a disease, which during antiquity was called cholera, is a representative example: '*And so, to quote Soranus, the disease of cholera is a state of looseness in the esophagus, stomach, and intestines, involving acute danger. We hold that the following may be antecedent causes of the disease: excessive drinking of wine, the taking of a harmful drug, the drinking of waters of hot springs, and the unsettlement caused by the tossing of the sea when one unaccustomed to sailing makes a first voyage. But a more violent attack of disease results from prolonged indigestion due to overeating or the eating of strange or highly seasoned food. To be sure, an understanding of these circumstances is a proper basis for a reasoned account of the causes of the disease, but is neither useful nor necessary for treating it or for describing its nature*' (Caelius Aurelianus, 1950, p. 419). The brief but quite systematic way in which Caelius Aurelianus presented the cause of diseases in *On Acute Diseases* (Celerum vel acutorum passionum) and *On Chronic Diseases* (Tardarum passionum) can be considered to be among the more extensive reports of medical authors of that time.

Although from Caelius Aurelianus one can find hints that he identified the role of hygienic conditions as causative factors of cholera – '*drinking from hot springs*' and '*eating of strange or highly seasoned food*' – that did not trigger him to propose any sanitation measures. The above quotation makes it clear that for Caelius Aurelianus, and most probably for all ancient physicians, the most important thing was the correct identification and treatment of a disease. The knowledge of the causative agents was of minor importance.

22.3 *AIRS, WATERS, PLACES*

Only one treatise among the more than sixty Hippocratic writings devotes a substantial section on the relation between water, health and disease (Jouanna, 2012a). This treatise, called *Airs, Waters, Places* was written by an unknown Greek physician in the second half of the fifth century BC (Jouanna, 1999, p. 375). *Airs, Water, Places* was a fundamental text about the effects of water on the health of people and remained unrivalled during antiquity (Jouanna, 2012a). However, even in this very influential treatise the focus was on the influence of the environment on human health and not the connection between people's health and sanitation.

The ancient Greek author of *Airs, Waters, Places* deals with the different sources, qualities and health effects of water at length: '*I wish now to treat of waters, those that bring disease or very good health, and of the ill or good that is likely to arise from water. For the influence of water upon health is very great. Such as are marshy, standing and stagnant must in summer be hot, thick and stinking, because there is no outflow; and as fresh rain-water is always flowing in and the sun heats them, they must be of bad colour, unhealthy, bilious. … Such waters I hold to be absolutely bad. The next worst will be those whose springs are from rocks – for they must be hard – or from earth where there are hot waters, or iron to be found, or copper, or silver, or gold, or sulphur, or alum, or bitumen, or soda. … So from such earth good waters cannot come, … The best are those that flow from high places and earthy hills.… Rain waters are the lightest, sweetest, finest and clearest. … Such waters are naturally the best. But they need to be boiled and purified from foulness.… … Waters from snow and ice are all bad. … I am of opinion that such waters derived from snow or ice, and waters similar to these, are the worst for all purposes… Stone, kidney disease, strangury and sciatica are very apt to attack people, and ruptures occur, when they drink water of very many different kinds, or from large rivers, into which other rivers flow, or from a lake fed by many streams of various sorts, and whenever they use foreign waters coming from a great, not a short, distance. … Such waters then must leave a sediment of mud and sand in the vessels, and drinking them causes the diseases mentioned before*' (*Airs, Waters, Places,* 2004, pp. 83–95).

The starting point in this discussion is that especially in summer, marshy, standing and stagnant waters are dangerous to people's health. Following that several points are presented that show that the ancient physician understood that the quality of the water may be influenced by the activities of people. Firstly, water coming from earth rich in metals is of a poor quality, secondly the importance for rain water to be boiled and purified from foulness, and thirdly the observation that people who drink water of many different kinds, or from large rivers, or from a lake are prone to diseases. These observations show that already by the fifth century BC in ancient Greece a lot of experience has accumulated concerning sensible measures to guarantee the quality of water in a human community.

22.4 GALEN

For later generations the best known and most appreciated ancient medical author was Galen, who was born in 129 and died during the first decades of the third century AD. He was a very active writer and had a keen interest in the causation of diseases as revealed for example, in *On the Causes of Diseases*, *On Antecedent Causes*, and *The Best Constitution of Our Bodies*. His ideas of cause of diseases are generally incompatible with modern beliefs. Health for Galen was usually the right mixture of the qualities hot, cold, dry and wet, but he also referred to the four humours: yellow bile, black bile, blood and phlegm (e.g., Galen, 1997, pp. 291–292; 2006, pp. 159–170). The following statement in *The Best Constitution of Our Bodies* summarizes his typical approach to the external cause of diseases: '*There are two causes of harm to our bodies: external influences and excretions from food. External influences are, for example, when someone is heated, cooled, moistened, or dried beyond the appropriate level. Exhaustion, grief, insomnia, worry, and all such matters should also be put in this category.*' (Galen, 1997, p. 292)

Galen focussed on hygiene in *De Sanitate Tuenda*. However, he has only few references to suggest that sanitation might be required for people's health. The detrimental influence of unsanitary conditions for people is most clearly expressed in a short statement in *De Sanitate Tuenda*: '*I consider the best air that is absolutely pure. And pure air is that which is not defiled by the exhalation from pools or marshes or from a pit giving off deleterious vapour, as around Sardis and Hierapolis and many other places in the world. And likewise whatever air is defiled from any sewer of those draining any large city or populous camp is harmful; and harmful too is that which is contaminated from any putrefaction of animals or vegetables or oils or manure; and it is not good which is cloudy from a neighbouring river or swamp; and likewise that which, in a hollow place, surrounded on all sides by lofty mountains, receives no breeze, for such air is stifling and foul, like that in certain closed houses in which it collects from putrefaction and lack of ventilation. Such air is injurious at all ages, just as absolutely pure air is beneficial at all ages*' (Galen 1951, pp. 35–36).

In the quotation Galen expresses clearly the idea that sewers, stagnant water and putrefying animal and vegetable matter can render the air harmful to human health. Galen's influence was without exaggeration massive in the following centuries and his ideas developed the miasmatic theory of disease causation. However, Galen's focus in the *De Sanitate Tuenda* was on the health of an individual, not the general population. He gives no advice how an individual could promote his health by sanitation. Instead of suggesting sanitation he advises people to adopt a careful regime (food, drink, exercise, baths etc.) in order to promote their health according to their individual circumstances.

Besides *De Sanitate Tuenda* Galen also made short references in other treatises to harmful factors, which prove that he was sensitive to the hygienic conditions of the environment. In the *On the Causes of Diseases*, for example, he writes: '*In the eyes and throat roughness is not from these things alone, but also occurs from acrid vapours, dust or smoke, just as, in my opinion, in the gullet, belly and intestines, from superfluities arising in the body itself and from the quality of what is eaten and drunk in which there are also deleterious things*' (Galen, 2006, p. 176).

22.5 SANITATION AFTER ALL

The ideas introduced in *Airs, Waters, Places* that running, tasty or tasteless, cool, odourless and colourless water was considered good drinking water, and stagnant, marshy water was to be avoided, were adopted until the end of antiquity (Vuorinen, 2007). These recommendations were the most systematic expression of the value of clean water for the benefit of people's health expressed by the medical authors. As a result of this, the source of water for an aqueduct in a Roman town was chosen with care. If the water did not satisfy their quality requirements the Greeks and Romans used different methods to improve it. From written sources and archaeological excavations, we know that the use of settling tanks, sieves, filters and the boiling of water were among the methods used during antiquity.

Boiling, which was recommended by the medical authors during antiquity, was an efficient sanitation measure and diminished the biological risks of poor quality water. The author of *Airs, Waters, Places* recommended boiling to purify rainwater. There is also a short and confusing passage on boiled water in another Hippocratic treatise (*Epidemics* VI.4.8). Besides boiling, the fourth century BC Greek doctor Diocles of Carystus recommended even more elaborate methods for improving water: '*You will make water weakest by boiling off a third of it; you should boil off whitish water in the same way, and add to it lumps of dry potter's earth until they are soaked, about a twelfth of a measure per amphora; when you have boiled it, drink it. You can expel the hot smell of the water in the following way, by beating it with the hand against the wind and exposing it to sun and air in a wide-mouthed vessel and by pouring it into many vessels, in small portions*' (Diocles of Carystus, 2000, p. 383). The famous medical encyclopedist Paul of Aegina in the 7th century AD also reported that marshy, stinking water could be made drinkable by boiling it (Paul of Aegina, 1914, p. 39).

As far as the non-medical authors are concerned, Celsus (early 1st century AD) in his *De Medicina* made only a short reference to the drinking of boiled water when dealing with the treatment of epilepsy (*De Medicina* III.23.7) and again when dealing with the treatment of an intestinal problem (*De Medicina* IV.19.4). Pliny in the middle of first century AD had a more detailed recommendation about boiling: '*At any rate it is agreed that all water is more serviceable when boiled, … It purifies bad water to boil it down to one half*' (Pliny, 1975, p. 403).

The available sources do not permit the estimation of how often boiling or other methods were used or how effectively they improved the quality of the water (Vuorinen, 2007). As a consequence, it is not possible to evaluate how efficient of these methods were on public health. Although the boiling of water might have been feasible from a hygienic point of view, it was not ecologically and economically feasible for extensive use since firewood and other combustibles would sooner or later become a scarce resource around the more heavily populated coasts of Mediterranean.

The scattered evidence advising the avoidance of unfamiliar water is a proof of the value of good quality of water for the health of people, together with the recognition of the fact that we get used to the type of water we commonly drink. An early fourth century BC Hippocratic treatise called *Internal Affections* contains a statement that a change of water could cause disease: '… *it* (erysipelas) *may also attack as the result of eating meat, or from a change of water*' (*Internal Affections*, 1988, p. 91). The dangers of a change in one's habits are clearly expressed by Diocles of Carystus: '*One should not drink water one is not used to as it comes to hand, for this is bad and risky* …' (Diocles of Carystus 2000, p. 309). Galen also suggested relying on water that people were familiar with: '*It is safest, then, to judge such water* [that people drink] *by experience*' (Galen, 1951, p. 35). However, the ancient authors were not unanimous in this recommendation: The author of *Airs, Waters, Places* was of the opinion that '*A man in health and strength can drink any water that is at hand without distinction,* …' (*Airs, Waters, Places*, 2004, p. 89).

During antiquity it was believed that epidemic diseases (pestilences) were caused by a change of air which then led to disease. In the Hippocratic writings *Breaths* and *Nature of Man* the principal factor causing pestilence was corrupted air (miasmas). Both treatises have been dated to the late 5th century BC (Jouanna, 1999, p. 378, 400). Later on, the belief of corrupted air (or emanation) as a cause of diseases was used in medical texts throughout antiquity, although the word miasma(s) was quite rare (Jouanna 2012b). Galen had a slightly different standpoint with regards to pestilences and he also admitted the role of food and water in causing of epidemics.

22.6 NON-MEDICAL AUTHORS ON SANITATION

Non-medical authors are important sources for many aspects of hygiene during antiquity. For instance, Frontinus (circa 30–103/4 AD) informs us that the abundant supplies of water coming from aqueducts were used to flush the streets and sewers of Rome to make the capital more salubrious (*De aquaeductu urbis Romae*. 88, 111). It is assumed that the Romans did not use taps like modern people, but kept the water running all the time. The constant flow of the water meant that in most towns supplied by aqueduct(s) there was plenty of 'waste' water to flush out the toilets, sewers and streets. Although we lack any evidence we may assume that this measure might have had some positive effect on the health of people in urban areas (Vuorinen, 2007).

The largest body of evidence that ancient people realized the beneficial effect of sanitation measures comes from non-medical authors (Frontinus and Vitruvius) and from archaeological evidence. Vitruvius' manual of architecture *De Architectura* in the late first century BC is the best written source about sanitation during antiquity. In *De Architectura* Vitruvius has a whole book about different aspects of water, and he suggested several factors that needed to be considered when choosing a water source as well as methods for examining the healthiness of water or food. For his part, Frontinus classified the aqueducts of Rome according to the suitability of their water for drinking (*De aquaeductu urbis Romae*, 89–93).

A lot of the water in a Roman town or city was consumed in baths connected to the aqueducts. Ideally shining marble walls and limpid water were considered a feature of a bath in Rome, the cleanliness of which was supervised by aediles. Baths were probably beneficial for public health in urban areas where there was an abundance and rapid turnover of water (Vuorinen, 2007). However, in towns where water was in short supply, cisterns had to be used, and the turnover of water was slow, baths might well have had a negative effect on public health.

22.7 EPILOGUE

The ideas of sanitation and health expressed by the ancient authors, especially by Galen, were adopted and developed further by medical authors in the Middle Ages. Galenic influence was particularly strong in the Islamic world as revealed, for instance, by ibn Ridwān (floruit 11th century AD) when he described conditions in Egypt: '*Epidemic diseases have many causes that may be grouped into four kinds: a change in the quality of the air, a change in the quality of the water, a change in the quality of the food, and a change in the quality of physic events. ... The water may create epidemic illness if the water is excessive in its increase or decrease, or if a corrupt substance mixes with it. The people are forced to drink it, and the air surrounding their bodies is corrupted by the water as well. This corrupt substance may mix with the water, either in a nearby or distant place, when the water's course passes by a battlefield where many dead bodies are found. Or the river passes by polluted swamps, and it carries and mixes with this stagnant water*' (Ibn Ridwān. 1984, pp. 112–113).

Ibn Ridwān referred especially to Galen, but he expanded the idea of sanitation in a more practical and detailed direction and had in his treatise a chapter entitled 'On the Means of Improving the Badness of the Air, Water, and Food in Egypt'. Concerning water and sanitation he has, for instance, the following advice: '*It is desirable to skim the purified water and, then, to drink it. Clarification is accomplished by putting the liquid in ceramic vessels, earthenware, or skins, and removing what is filtered from it by secretion. If you wish, you may heat the liquid by fire, place it in the night air until it is pure, and skim what is clarified. If it appears to you that it has a noticeably bad quality, cook it on a fire, cool it outdoors in the cold of the night, and purify it with one of the potions that I have mentioned. This water is made better by clarifying it several times. For example, heat or cook it and cool it in the night air, then, cook what is purified again and clarify it with some potions, Take what is pure and put it in vessels that filter it in the cold of night. Then, you can take the filtered water and drink it*' (Ibn Ridwān, 1984, pp. 135–136).

In another chapter, titled 'On the Means of Preventing Injury from Epidemic Diseases in Egypt', Ibn Ridwān had the following advice: '*Likewise, concerning water, if it differs from ordinary, you should not risk drinking much of it. Water is improved by boiling; it should be boiled if it is spoiled or if much corruption is mixed with it. Then it is purified by what opposes this corruption, and it should be protected from the putrid air. Its containers should be fumigated with mastic and washed with cyperus and sandalwood*' (Ibn Ridwān, 1984, p. 138).

As a conclusion, although ancient texts, which have survived until modern times, contain excellent remarks on the role of the environment with regards to people's health, there are only scattered remarks concerning the role of sanitation on the health of people. It was not until the late 19th century when the role of personal and public hygiene on people's health was clearly understood.

22.8 REFERENCES

Airs, Waters, Places (2004). In: Hippocrates Volume I, with an English translation by W. H. S. Jones (ed.), The Loeb Classical Library. First published 1923. Harvard University Press, Cambridge, Massachusetts, pp. 70–137.

Aphorisms (2005). In: Hippocrates Volume IV, with an English translation by W. H. S. Jones. The Loeb Classical Library. First published 1931. Harvard University Press, Cambridge, Massachusetts, USA, pp. 98–221.

Breaths (2006). In: Hippocrates Volume II, with an English translation by W. H. S. Jones. The Loeb Classical Library. First published 1923. Harvard University Press, Cambridge, Massachusetts, USA, pp. 226–253.

Caelius Aurelianus (1950) *On Acute Diseases and On Chronic Diseases*. Edited and Translated by I. E. Drabkin. The University of Chicago Press, Chicago, USA.

Celsus (1948). *De Medicina*, with an English translation by W. G. Spencer, in three volumes. The Loeb Classical Library. First published 1935. Harvard University Press, Cambridge, Massachusetts, USA.

Diocles of Carystus (2000). A collection of the fragments with translation and commentary. Volume one, text and translation by Philip J. van der Eijk. Brill, Leiden, The Netherland.

Dorland's Illustrated Medical Dictionary (1965) 24th edn, W. B. Saunders Company, Philadelphia and London, UK.

Epidemics VI (1994). In: Hippocrates Volume VII, edited and translated by Wesley D. Smith. The Loeb Classical Library. Harvard University Press, Cambridge, Massachusetts, USA, pp. 218–291.

Frontinus (2004). *De aquaeductu urbis Romae*. [The Aqueducts of Rome] Edited with introduction and commentary by R. H: Rogers. Cambridge University Press, Cambridge.

Galen (1951). A translation of Galen's Hygiene (*De Sanitate Tuenda*) by Robert Montraville Green. With an introduction by Henry E. Sigerist. Charles C. Thomas, Springfield, Illinois, USA.

Galen (1997). *The Best Constitution of Our Bodies*. In: Galen Selected Works. Translated with an Introduction and notes by P. N. Singer. Oxford University Press, Oxford, pp. 290–295.

Galen (1998). *On Antecedent Causes*. Edited with an introduction, translation and commentary by R. J. Hankinson. Cambridge University Press, Cambridge, UK.

Galen (2006). *On the Causes of Diseases*. In: Diseases and Symptoms, J. I. Galen (ed.), Cambridge University Press, Cambridge, UK, pp. 157–179.

Ibn Ridwān (1984). On the prevention of bodily ills in Egypt. In: Medieval Islamic Medicine. Ibn Ridwān's Treatise 'On the Prevention of Bodily Ills in Egypt'. Translated, with an Introduction, by M. W. Dols Arabic Text edited by Adil S. Gamal. University of California Press, Berkeley, USA, pp. 75–149.

Internal Affections (1988). In: Hippocrates Volume VI, with an English translation by Paul Potter. The Loeb Classical Library, Harvard University Press, Cambridge, Massachusetts, USA, pp. 70–255.

Jouanna J. (1999). In: Hippocrates, Translated by M. B. DeBevoise (ed.), The Johns Hopkins University Press, Baltimore.

Jouanna J. (2012a). Water, health and disease in the Hippocratic treatise airs, waters, places. In: Greek Medicine from Hippocrates to Galen by Jacques Jouanna, Translated by N. Allies Edited with a Preface by P. van der Eijk (eds), Brill, Leiden – Boston, USA, pp. 155–172.

Jouanna J. (2012b). Air, miasma and contagion in the time of Hippocrates and the survival of miasma in Post-Hippocratic medicine (Rufus of Ephesus, Galen and Palladius). In: Greek Medicine from Hippocrates to Galen by Jacques Jouanna. Translated by N. Allies (ed.), Edited with a Preface by Philip van der Eijk. Brill, Leiden – Boston, USA, pp. 121–136.

Nature of Man (2005). In: Hippocrates Volume IV, with an English translation by W. H. S. Jones. The Loeb Classical Library. First published 1931. Harvard University Press, Cambridge, Massachusetts, USA, pp. 2–59.

Nutton V. (2005). Ancient Medicine. Routledge, London and New York, USA.

Paul of Aegina (1914). Paulos von Aegina des besten Arztes Sieben Bücher. Uebersetzt und mit Erläuterungen versehen von I. Berendes. E.J. Brill, Leiden [An English translation by Francis Adams *The Medical Works of Paulus Aegineta*].

Pliny (1975). *Natural History*, in ten volumes. Volume VIII, by W. H. S. Jones. The Loeb Classical Library. First published 1963. Harvard University Press, Cambridge, Massachusetts, USA.

Prorrhetic II (1995). In: Hippocrates Volume VIII, edited and translated by Paul Potter. The Loeb Classical Library. Harvard University Press, Cambridge, Massachusetts, USA, pp. 218–293.

Vitruvius (1983). *On Architecture*; in two volumes; translated into English by Frank Granger, The Loeb Classical Library. First printed 1931. Harvard University Press, Cambridge, Massachusetts, USA.

Vuorinen, H. S. (2007). Water and health in antiquity: Europe's legacy. In: Environmental History of Water-Global Views on Community Water Supply and Sanitation, P. Juuti, T. Katko and H. S. Vuorinen (eds), IWA Publishing, London, UK, pp. 45–67.

Vuorinen H. S. (2010). Water, toilets and public health in the Roman era. *Water Science & Technology: Water Supply*, **10.3**, 411–415.

APPENDIX I

Treatises Referred to in the Text

Author/treatise	Date	Comment
Airs, Waters, Places	Late 5th century BC	Hippocratic writing
Aphorisms	4th century BC	Hippocratic writing
Breaths	Late 5th century BC	Hippocratic writing
Epidemics VI	Around year 400 BC	Hippocratic writing
Internal Affections	Early 4th century BC	Hippocratic writing
Nature of Man	Late 5th century BC	Hippocratic writing
Prorrhetic II	Late 5th century BC	Hippocratic writing
Diocles of Carystus	4th century BC	Collection of fragments
Vitruvius: *De Architectura*	Late 1st century BC	Non-medical author
Celsus: *De Medicina*	Early 1st century AD	Encyclopedist
Pliny: *Naturalis Historiae*	Middle 1st cent. AD	Encyclopedist
Frontinus: *De aquaeductu urbis Romae*	Around year 100 AD	Non-medical author
Galen: *De Sanitate Tuenda*	Late 2nd century AD	Author later very influential
Galen: *On the Causes of Diseases*	Late 2nd century AD	Author later very influential
Galen: *On Antecedent Causes*	Late 2nd century AD	Author later very influential
Galen: *The Best Constitution of Our Bodies*	Late 2nd century AD	Author later very influential
Caelius Aurelianus: *On Acute Diseases*	Around year 400 AD	Methodist physician
Caelius Aurelianus: *On Chronic Diseases*	Around year 400 AD	Methodist physician
Paul of Aegina	7th century AD	Medical encyclopedist
Ibn Ridwan: *On the Prevention of Bodily Ills in Egypt*	11th century AD	Galenic influence

Chapter 23

Historical development of sanitation from the 19th century to nowadays: Centralized vs decentralized wastewater management systems

Paolo Roccaro, Antonella E. Santamaria, and Federico G.A. Vagliasindi

23.1 INTRODUCTION

From 19th century to nowadays sanitation has had a profound development thanks to the scientific progress, the improvement of economical condition, the industrial revolution, the implementation of novel technology, and the growing attention for the environmental issues. In fact, during the 19th century sanitation systems did not include any treatment process and were realized by using simple static and dynamic systems to collect and dispose the wastewater. Later, the urban development, the increasing population density in specific areas and the new concept of environment protection have resulted in complex sanitation systems, which include a sewer followed by a centralized wastewater treatment plant. A different approach based on the segregation principle is under consideration. Indeed, several decentralized wastewater treatment systems are used or under investigation. These on-site systems often distinguish brown, yellow and grey water in order to optimize the recovery of water, nutrient and energy following the eco-sanitation principle.

The gained experience has demonstrated that the scientific progress is not always linear, but has developed through forward and backwards steps. This is in agreement with the Kuhn's theory of the scientific progress (Kuhn, 1962). Because scientific progress is not a linear accumulation of new knowledge, it is necessary to undertake a 'paradigm shift', in order to break off the 'normal science' and allow the transformation. This 'paradigm shift' may be the case also for sanitation systems; however this is a challenging transformation which should involve a change in the uses of the society and not only in the technical system.

Therefore, the objective of this chapter is to analyze the historical development of sanitation systems from the 19th century to nowadays, in order to define the evolution of treatment systems, the technologies available today and to discuss the needs and perspectives considering also a 'paradigm shift' as a starting point for the scientific revolution which may bring to a completely different sanitation systems in urban and rural areas.

23.2 WASTEWATER COLLECTION AND TREATMENT IN THE 19th CENTURY

Since the end of 3000 BC there was a clear need to remove excreta from residential areas to avoid sanitary hazards. Mesopotamian Empires of Assyria and Babylonia, and their antecedent Sumerian and Akkadian populations made great advances in civilization with the realization of complex drainage systems (Gray, 1940). Sewage drainage systems, public wells, and private and public baths, dated 2500 BC have been found, through the excavation of Mohenjo-Daro (Gray, 1940). Also in the Bronze age in Crete, advanced water management and sanitary techniques were practiced in several settlements. These advanced technologies began in the Minoan Crete and subsequently were expanded to Mycenaean and then Archaic and Classical Greece (Angelakis, 2010).

A different water management attitude was showed by the Romans (800–300 BC) who used water as a matter of luxury and prestige, building mega water projects using aqueducts to transfer water to their public fountain and baths (Mays, 2010). In this period the Cloaca Maxima, the so-called sewers of Rome was realized (Gray, 1940). When the Roman Empire fell (Dark age), the decline led to a lack of hygiene and spread of disease until early 19th century.

In the 19th century better sanitary conditions were recovered and sanitation systems consisted mainly in the collection and disposal of urine and faeces by using different techniques, such as simple latrines, privy vaults, cesspools, and pail systems. However, in low-density areas, only one closet was used by different families. Indeed, the presence of a closet for every family would have led to a pipe system that would put in contact environments lived by different families with the possibility of the spread of diseases. Most schools and institutes of education lacked of sewer systems for economical reasons. However, by the end of 1800 more attention was given to the realization of such systems, in particular to ensure hygienic conditions. In general, two types of wastewater collection systems were used: static systems and sewer systems. Static systems were the first generation of wastewater collection technologies, while centralized sewer systems were canalization and piping systems and are the second generation of sewerage, still used nowadays. Sewer systems replaced the decentralized (e.g., latrines, privy vaults, cesspools, and pail systems) during the second half of the 19th century (Gray, 1884; Spataro 1887; Burian *et al.*, 2000).

Where there were not latrines, people resorted to defecation in the open. This may be indiscriminate or take place in special places such as defecation fields, rubbish and manure heaps, or under trees (Franceys *et al.*, 1992). In particular, manure heap was used in most rural houses and in many small towns. It consisted of piles of wastes realized near the houses. The stabilized waste would then be used as fertilizer. The practice of leaving the waste in the open air in the years was abandoned, due to the spread of many diseases and it was replaced by closed and sealed containment systems. The simplest realization consisted in a space, raised from the ground, made with bitumen or other waterproof material covered by a simple roof.

The evolution of these rudimentary methods was the 'latrine' that introduced the possibility to confine excrete. It consisted of a superstructure, where the user interface, and a pit were located. In some latrine there were two user interface, one of which for children Figure 23.1. The main types of latrines included 'single pit latrines' and 'ventilated pit latrines' with or without the use of ash.

Pits could be absorbent (soak pit) or watertight (vault latrine). The 'absorbent pits' were not lined. This system did not require spurges or a separate channel for water because waste materials were dispersed in the ground, however it resulted in the contamination of soil and groundwater thus resulting in the spread of disease. In 'watertight pits' gathered sewage needed periodic emptying. Simple applications consisted in pits where wastewater and ashes were discharged. These pits, waterproofed through the use of clay, were completed by walls and the roof was made of wooden and tar or iron, elevated so as to permit aeration (ventilated pit latrines) or complete closure. At the bottom, a bed of 15–20 cm of fresh litter or sawdust was placed in order to absorb the liquid in the pit. The addition of ash could be done either directly from the latrine or externally.

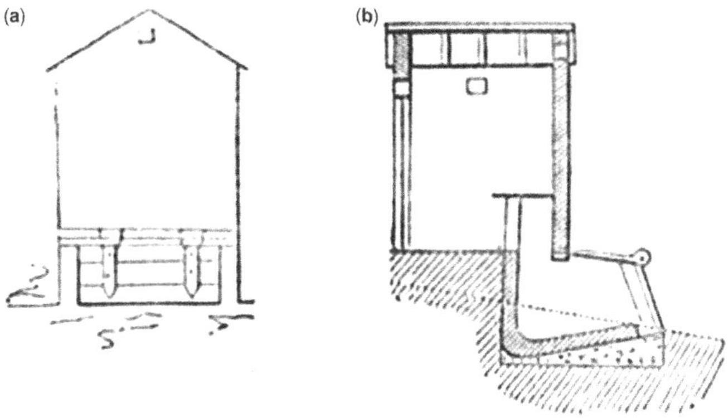

Figure 23.1 Single pit latrine: (a) front and (b) section (Spataro, 1887).

More complex watertight pits were built out of tuff masonry, cemented with hydraulic lime, not less than 50 cm thick, with a basement of 40 cm thick and a vault 14 cm high. The opening was at the center of the vault, with a diameter not less than 60 cm and was equipped with closing in stone. The communication between the watertight pits and users interface occurred through vertical pipes made of iron, baked clay or stone with an inner diameter of 20 cm. The main problem of this type of system was attributable to losses due to not entirely effective waterproofing system and the required frequent cleaning. Over the years, it was realized that tuff, bricks and cement were not material suitable for waterproofing. Then, it was proposed to utilize a jet of concrete mixed with hydraulic mortar. Therefore, the pits were realized by a double wall with clay interposed and equipped with ventilation pipes. These types of pits included the 'Ghirardi System', in which the pit has egg-shape with the major axis vertical and the 'Damiani System' in the shape of a frustum of cone with the smaller base at the bottom (Gray, 1884; Spataro, 1887). The spurge of the pits was performed manually with buckets or by pneumatic pumps that could produce vacuum inside a metal cask. The cask was connected to the pit by means of pipes made by zinc and rubber and was equipped with a lower opening, connected with the pipe, and an upper opening as shown in Figure 23.2.

Figure 23.2 Collection system from watertight pit (Spataro, 1887).

Another way to confine the excreta was the use of buckets, in which sewage was covered with absorbent material or disinfectant and equipped with a tight closure (Figure 23.3). This dry system, named 'bucket latrine' did not employ the use of water and piping. Excreta removed in this way were often termed 'night soil' because they were used as fertilizer and required to be removed periodically (Franceys *et al.*, 1992). Figure 23.4 shows some type of van used for night soil transport in the 19th century. The containers (buckets) were made of wood with tar or galvanized iron, round in shape, with a capacity less than 45 L.

Figure 23.3 Buckets (Spataro, 1887).

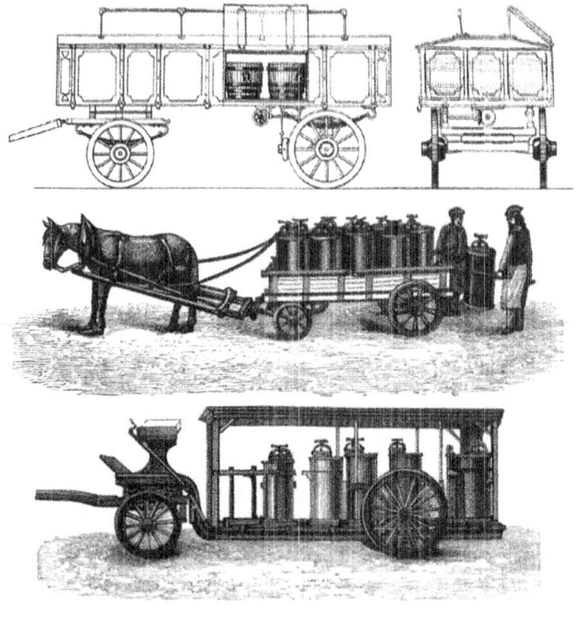

Figure 23.4 Night soil vans for night soil transport (Gray, 1884; Spataro, 1887).

The 'earth closet' followed the bucket latrine (Figure 23.5). These could be made inside or outside buildings and were based on the mixing of wastewater and dry earth. The earth, properly dried (in oven or under the sun) was contained in the back of the seat and was dosed through a system named Mouleu's apparatus. The mixture (earth and wastewaters) was discharged in a pit or in a closed container made of metal (bucket). The pit was made of bricks covered with cement with dimensions (1.20) × (1.20) × (0.60 – 1.20 m). The earth could be again dried and reused thereby acquiring the best features for fertilization. In some cities there were services organized to collect used buckets and deliver the dried earths.

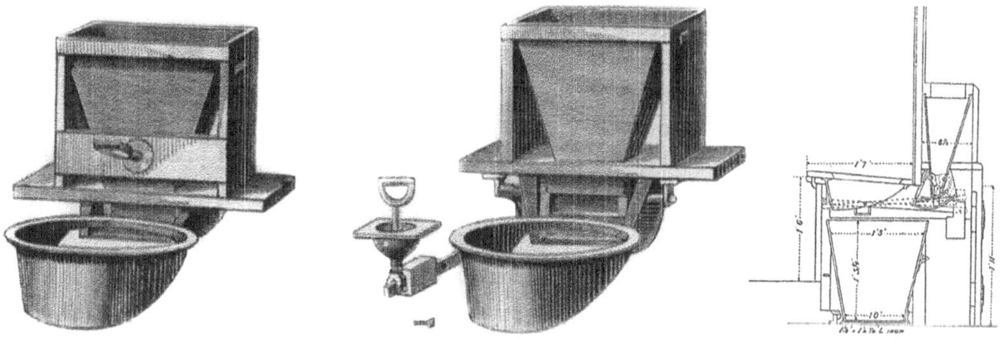

Figure 23.5 Earth closet: Moule's apparatus (Spataro, 1887) and section (Moore, 1909).

The ever increasing need to remove the excrement from the domestic area, involves the evolution of systems like 'bucket latrine' and 'earth closet' through a piping system (drains) and the use of water to wash pipes. This system, a precursor to the sewer system, was considered the most hygienic and was also used in public buildings in the 19th century (Figure 23.6). The absence of a system of sewers involved the use of receptacles, barrel or oak staves, circled in iron and covered in tar or sheets of galvanized iron with a hole at the top of 0.40 m that permitted the connection through pipes with the user interface. During transportation, the hole of the barrel was sealed with a stopper made in wood plastered with clay and closed by a band of iron nailed to the bottom of the barrel with a hinged opening. The transportation was carried out using equipped carts. The barrels were on the average 1 m high, had a diameter of 1.5 m and were contained in special rooms in the building. However, this system had different disadvantages:

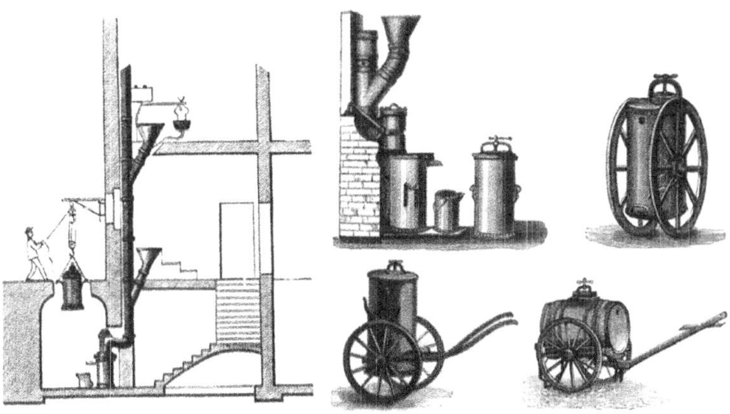

Figure 23.6 Sanitation system precursor of the sewer system (Spataro, 1887).

(a) formation of odours;
(b) incomplete absorption; and
(c) presence of insects in the vicinity of the receptacle.

The need to ensure a closing trough the user interface and the pipe system involved the realization of hydraulic (like siphon water system) or mechanical systems accessories to the closet. As a result of the Industrial Revolution, a rapid industrialization and urbanization during the second half of the 19th century occurred in many European and North American cities. The increase of the population and the use of plumbing fixtures (e.g., water closet), producing more wastewater, lead to crowding and increasing concerns about public health. Indeed, both lined and unlined decentralized systems were unable to manage urban wastewater effectively because the lined ones required too frequent cleaning to be cost effective over a long term, and the unlined ones contaminated groundwater and the surrounding soil (Burian *et al.*, 2000). In order to control outbreaks of disease, a network of piping, named sewer system, was realized with the purpose of removing excreta immediately from inhabited areas before the beginning of fermentation (Gray, 1884; Seeger, 1999). It is noteworthy that in some city, drainage systems existed and were often used to discharge wastewater (Burian *et al.*, 2000).

Pneumatic and mechanical operating vacuum sewage collection systems were firstly introduced in the second half of the 19th century. Indeed, in 1866 the engineer Liernur (past Captain in the US Army) patented in England and The Netherlands his vacuum operating sewage. Liernur systems were successfully installed in several European cities such as Amsterdam, Dordrecht, Hanau, Leiden, Prague, Stansted, and Trouville. In the latter French city the Liernur system was operated until the 1980. Other pneumatic systems were also developed by the designers Shone and Berlier (Gray, 1884). Excreta collected trough the vacuum sewer system were often spread out over land as a fertilizer (IWA, 2008).

The first comprehensive planning, designing, and building of a sewer system was carried out in Hamburg (Germany) in the 1840s by Lindley, a distinguished English engineer. This experience, which became soon as the model for other cities (in Europe and the United States), resulted from the opportunity of re-building a large part of the city after that the older half of Hamburg burned. For instance, the first such systems in the United States were built in the late 1850s in Chicago and Brooklyn. Later, the centralized sewer systems were unanimously considered the most long-term cost effective wastewater management system and also the most effective measure to control the disease outbreaks. The piping system was made of different materials like clay, cement, cast iron, lead, iron. Initially these systems discharged sewage directly to surface waters or soils without treatment (Burian *et al.*, 2000).

Overall, in the 19th century, the treatment of wastewater was limited to natural fermentation in pits (e.g., vault latrine) or to soil treatment capacity (e.g., open defecation, soak pit, bucket latrine, earth closet and sewer system). The advantages and disadvantages of the sanitation and wastewater treatment systems used in the 19th century are summarized in Table 23.1. Since these systems had different disadvantages, such as the incomplete oxidation of organic matter, odours and possible groundwater contamination (Table 23.1), several studies were conducted in order to identify suitable wastewater treatment techniques. The results showed bad performance of filtration and chemical processes, while the natural treatment on irrigation fields was found to be an efficient treatment option with advantages for agriculture. Initially, the method used, consisted of releasing wastewater directly on the soil, with a continuous flow, through an irrigation system. However, due to the rapid saturation of the soil capacity, this method was not considered valid for ensuring appropriate hygienic conditions (Spataro, 1887; Wiesmann *et al.*, 2007).

Table 23.1 Advantages and disadvantages of sanitation practice and wastewater treatment systems in the 19th century.

Sanitation practice, wastewater treatment system	Advantages	Disadvantages
Open defecation	No cost	Health hazard, degradation of environment
Absorbent pits	Low cost and simple realization and manage	Incomplete oxidation of organic matter, smell and possible groundwater contamination
Vault latrine	Low environmental hazard	Efficient infrastructures required, health hazard during collection
Bucket latrine	Low cost	Health hazard, smell, require collection
Earth closet	Low initial cost	Health hazard, smell, require efficient managing and collection
Sewer system	Excreta removal before the beginning of fermentation, good hygienic and environmental conditions	High cost and complex realization, efficient infrastructures required
Irrigation	Benefits for vegetation	Needs of large areas, hygienic conditions not always guaranteed, possible management issues
Discontinuous irrigation	Small areas of application, good hygienic conditions	Possible management issues
Sub-irrigation	Hidden system, good hygienic conditions	Particular attention in the choice of the ground and the drainage system
Intermittent slow filtration	Good treatment efficiency	Smell and flooding

Another method used was therefore the discontinuous irrigation. This system consisted of the subdivision of the soil in several batches, the preparation of the surface, and then the wastewater was released on each batch in a discontinuous manner in order to ensure oxygenation. Since the results obtained were positive in terms of hygienic conditions, this method was used and was combined with soil irrigation. The management of these types of system was particularly difficult, when the surface of the soil was not leveled or when the wastewater was rich in solids resulting in stagnation and putrefaction with formation of odours. To overcome these difficulties the sub-irrigation system was developed, namely the releasing of wastewaters through an underground system of pipes. This system was widely used at the end of the nineteenth century in England and North America. It was first used in 1868 by Reverend H. Moule for an isolated house, and was subsequently improved by R. Field. It was based on the natural degradation through oxidation and nitrification of organic matter in the soil. A necessary condition for a good performance of the process was the presence of oxygen, useful to the microorganisms for the degradation. Therefore, the water was lost in the subsoil, while organic pollutants were removed by means of the surface of the soil pores and, coming in contact with oxygen, were degraded by microorganisms. To operate the sub-irrigation it was necessary that the soil was well ventilated and not saturated with wastewaters, so that the system had to be operated discontinuously and the soil had to be well drained and porous. The method of sub-irrigation still requires particular attention in the choice of the soil, it had to be located far from households and natural sources and in the opposite direction to that of the main winds and the drainage system. A necessary condition for a good performance of a sub-irrigation system was an appropriate flow rate of the wastewater. Therefore, between house discharges and the sub-irrigation system a 'outflow tank' useful to collect the effluent for the time necessary for the soil aeration was placed. The outflow tank mostly used was that from R. Field (Figure 23.7), which included two different units; the tank and a siphon. The first unit was a sedimentation step useful to stop solids, and therefore to avoid a bad performance of the siphon; the second was an outflow unit. In the sedimentation unit, the hydraulic closure occurred by means of the pipe inlet coming from the house, while the activation of the siphon in the outflow unit required atmospheric pressure, ensured by holes or channels realized in the tank. Later, this system was modified in order to ensure hydraulic closure and a continuous flow, useful to avoid odors by the elimination of the first tank. Moreover, the siphon was placed outside the tank to allow inspections. In the same century, the first experiments on intermittent slow filtration were conducted in England by Edward Frankland (1868–1871). The first plant was built by Baily Denton, a student of Frankland, and at the end of 1800, this treatment system became very popular, but the operation highlighted some issues related to the design, odours formation and flooding (Metcalf & Eddy, 2002).

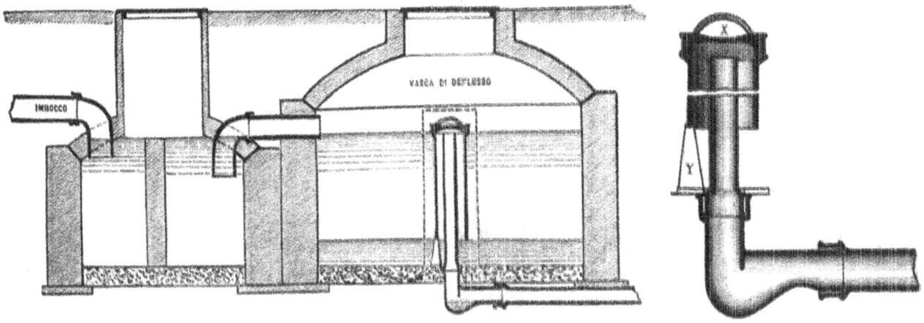

Figure 23.7 Parts of outflow tank: (a) the tank and (b) the siphon (Spataro 1887).

23.3 CENTRALIZED WASTEWATER TREATMENT PLANTS: THE GOLD TECHNOLOGY

Issues like safe water drinking, sewerage, wastewater treatment and stream pollution led to the beginning of biological treatment which started in 1890 with experiments carried out at the Lawrence Experiment Station in Massachussetts (Hendricks, 2011). The development of biological treatment represents a fundamental step in the history of wastewater treatment, because of the mineralization of organic matter due to degradation by microorganisms. In the same historical period (1890–1891), in order to reduce the area required by intermittent slow filtration, the trickling filter technology was developed by Col. George E. Waring. Mainly, trickling filters were of two types: standard and high-rate filters, with the first large-scale plant built in Stahnsdorf near Berlin (Wiesmann *et al.*, 2007; Woodbury 1906). The first full- scale installation of high-rate filters was made about in 1936 (Woodbury, 1906). In 1906 Dr. K. Imhoff developed the Imhoff technology, the first combined treatment system (Seeger, 1999; Metcalf & Eddy, 2002). The main advantages of this technology, that allows the degradation of organic matter, are: the presence of two units in a single tank, and the separation of sludge/effluent which allows for more complete settling and digestion. The application of this technology has removal rates of 30%, 90% and 60%, respectively, for organic substrate (BOD and COD), sedimentable suspended solids, and total suspended solids (Masotti,

2002). Subsequently, in 1914 a publication of Arden and Lockett reported the results of experiments on the oxidation of sewage, launching activated-sludge practice, a new artificial biological method, which was employed first in Germany in the sewage plant of Essen-Recklinghausen in 1925 (Seeger, 1999). Initially, it was based on the 'assimilative capacity' of receiving waters, but after the Water Quality Act in 1965, this concept was dislodged, introducing the concept of water quality standard. Metcalf and Eddy (1972) provided an important insight to the activated-sludge process and design with the introduction of two variations of conventional treatment using diffused air (Hendricks, 2011). In 1942, Monod defined an empirical equation, developed on the basis of an experimental activity, still applied in modern bioreactor theory. Subsequently, in 1954 Fair and Geyer revealed the biological character of activated sludge and introduced several sizing parameters for the aeration system. Starting from 1950, the 'modern' history of water pollution control has taken place. In these years, activated sludge reactor modelling begun and several studies were conducted by researchers to increase our knowledge about the process and its design. Some of the biggest names are:

(a) Eckenfelder, who in 1954 proposed a mathematical model for activated sludge (Eckenfelder & O'Connor, 1961).
(b) McKinney, who in 1962 proposed a completed mixing model (McKinney, 1962).
(c) Pearson, who in 1968 described the materials-balance model like approach to reactor analysis (Pearson 1968).
(d) Lawrence and Carty, who in 1970 proposed the concept of sludge age (Lawrence & McCarty, 1970).
(e) Metcalf and Eddy, who in 1972 incorporated the reactor theory approach (Metcalf & Eddy, 1972). And
(f) The IWA Task group, who developed in 1986 a comprehensive computer model (Henze *et al.*, 1987).

Standards of treatment, identified by wastewater regulations and management issues highlighted from the operation of biological units highlighted the need to consider not a single treatment, but different treatments together, introducing the concepts of pre-treatment, primary treatment, secondary treatment and tertiary treatment. Not indifferent is the role of disinfection and sludge treatment. Therefore, depending on the characteristics of the area in which the effluent is to be disposed and the required standards of treatment, it is possible to define the flow chart, starting from simple up to complex treatment trains.

Imhoff tank is often used before sub-irrigation for isolated houses or as primary treatment before of biological trickling filter in small wastewater treatment plants (WWTPs) (Metcalf & Eddy, 2002). In the latter case the treatment trains is shown in Figure 23.8 and sludge anaerobic digestion occurs in the Imhoff reactor which receives the primary and secondary sludge. The stabilized sludge can be periodically extracted from the Imhoff reactor to be dewatered and disposed. The treatment performance of the Imhoff-Trickling filter-Settling as BOD removal is lower than that for an activated sludge system and depends on the filter rate (Metcalf & Eddy, 2002). In the case of small communities, the secondary sedimentation may consist of a collecting slump. Other similar treatment trains are employed for small-scale plants (Masotti & Verlicchi, 2005).

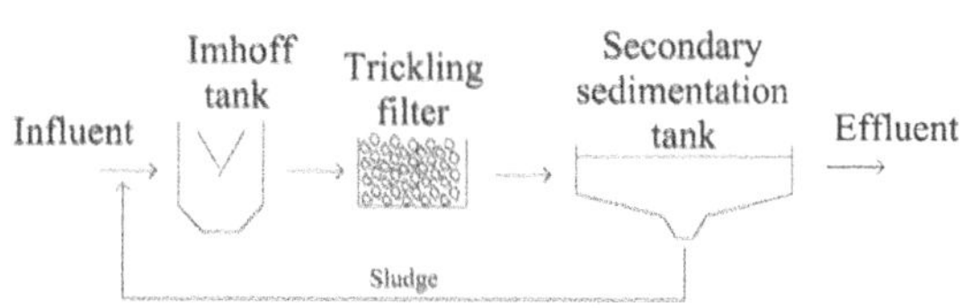

Figure 23.8 Treatment train with Imhoff–Trickling filter – settling units.

Conventional activated sludge (CAS) plants are widely used for large communities and represent the 'standard technology' in the wastewater treatment field. CAS systems consist of a flow chart that includes preliminary treatments, primary sedimentation, activated-sludge reactor, secondary sedimentation and final disinfection (Figure 23.9). A separate treatment train is included to treat the excess sludge. CAS is designed to remove mainly the organic matter and to some extent nutrients (nitrogen and phosphorus) and pathogens. The treatment performance of CAS varies with the organic load (food/microorganisms) ranging from 85 to 97% as BOD_5 removal (Masotti, 2002; Metcalf & Eddy, 2002). The CAS system has also been modified in order to adapt to different treatment needs and system requirements. Specifically:

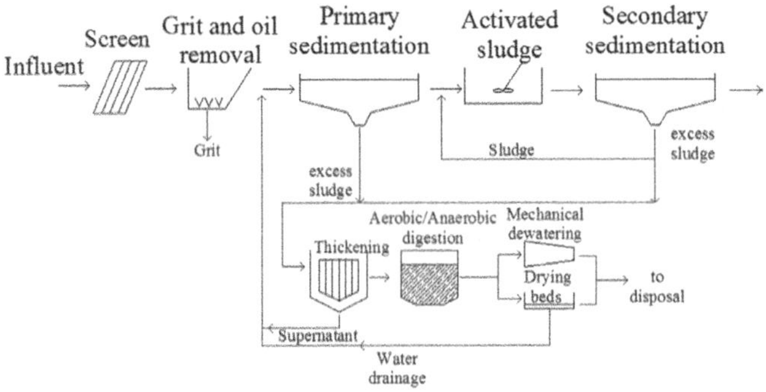

Figure 23.9 Flow chart of conventional activated sludge (CAS) system.

(a) Conventional CAS with Imhoff tank like primary sedimentation in order to reduce construction costs and simplify the management.
(b) Simplified flow chart without primary sedimentation and aerobic digestion of the excess sludge, in order to limit the construction costs, limit the areas of encumbrance, plant safety, simplified management.
(c) Activated sludge with primary sedimentation and aerobic digestion of sludge with aerobic digestion of excess sludge in order to remove any substances toxic to microorganisms and/or allow the construction of a conventional plant in two phases of which this is the first. And
(d) Extended aeration with elimination of primary sedimentation and attribution of high retention times in the aeration compartment and recirculation of sludge finalized to an aerobic stabilization.

Specific treatments are needed for the removal of nutrients, usually required by the regulation when the receiving water is sensitive to eutrophication. The nitrification and subsequent denitrification are often carried out by using biological processes. Nitrification (aerobic conditions) can be achieved in the same activated sludge with higher sludge retention time (combined nitrification) with benefits in construction costs or in a separate reactor (separate nitrification) with higher costs of construction and operation, but better treatment performance. Denitrification (under anoxic conditions) can be realized as post-denitrification, often adding an external source of carbon, such as methanol, or as pre-denitrification by the recirculation of the mixed liquor rich in nitrates. The removal of phosphorus can be achieved by biological or chemical processes. Biological processes require an anaerobic reactor as the first biological reactor in the treatment train, which leads to the growth of phosphorus accumulating bacteria (POA). The Bardenpho process is an example of a complex biological process for the removal of nitrogen and phosphorus and it includes an anaerobic reactor followed by a pre-denitrification, nitrification, post-denitrification (Metcalf & Eddy,

2002). Chemical phosphorus removal is based on the coagulation process by using aluminum or iron based coagulants and can achieve higher removal (95%), but results in higher sludge production and related disposal costs (Metcalf & Eddy, 2002).

Water related problems and the needs to reuse wastewater have prompted researchers to identify useful treatments to remove contaminants still present downstream primary and secondary treatments in order to enable the reuse of water (GEC, 2005; Metcalf & Eddy, 2002). On the basis of numerous technologies available for the tertiary treatment, different treatment processes are available nowadays, the combination of which guarantees the fulfilling of high water quality standards even for direct wastewater reuse. Some possible tertiary treatment trains are reported in Figure 23.10 and can be applied based on the required quality of produced waters. For instance, the filtration process, possibly following a coagulation step, could be applied to treat municipal wastewater to be reused in agriculture, while other advanced treatment processes can be employed to produce water of high quality and remove some trace toxic contaminants. These advanced processes include adsorption on granular activated carbon (GAC), advanced oxidation processes (e.g., ozonation, H_2O_2/UV processes), biological activated carbon process (BAC), membrane filtration (e.g., micro-, ultra-, nano-filtration, reverse osmosis) and are also reported in Figure 23.10. These treatment trains represent the 'gold technology' for wastewater treatment and reuse. Other innovative biological treatments have also been developed such as membrane biological reactors (MBRs), moving bed biological reactors (MBBRs), biological aerated filters (BAFs), which can be employed alternately to the CAS.

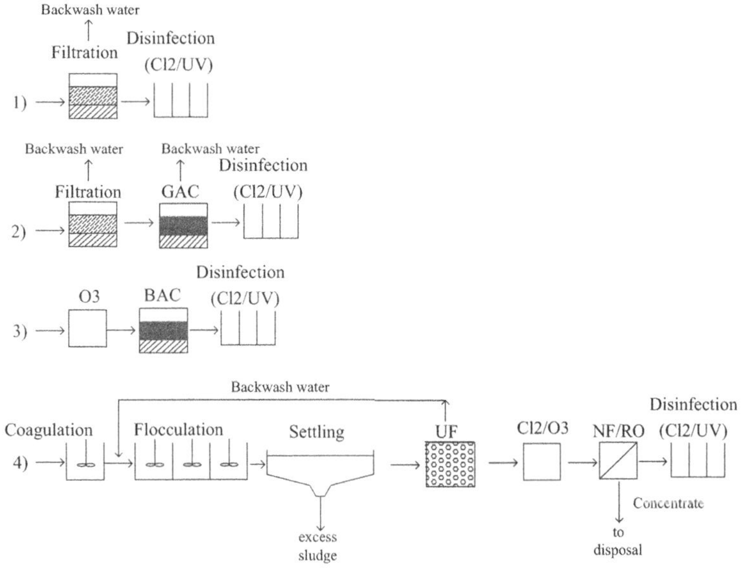

Figure 23.10 Tertiary treatment to fulfil high water quality standards.

Based on the experience gained on centralized systems, the following could be highlighted:

(a) Centralized systems allow the control of parameters and therefore the achievement of high performance of treatment.
(b) Centralized treatment facility, while having a high capital cost, allows to contain the costs of management that can be partially or completely automatic.

(c) The structural and operational safety can be controlled and thereby ensured in large plant. And
(d) The treatment processes that take place in centralized plants allow an excellent performances for the mineralization of the organic substance.

The twentieth century showed a significant progress together with the rise of different unresolved issues. Population growth, contamination of natural resources, the water crisis and the short durability of environmental engineering projects represent some issues that led the scientific world to review the direction of technological development in water and wastewater sector (Angelakis *et al.* 2008; Angelakis *et al.*, 2010).

Centralized sanitation systems, which are considered the gold technology, have showed different disadvantages such as:

(a) A high total cost (Hophmayer-Tokich, 2006).
(b) A localized environmental impact.
(c) Possible public opposition and difficulties of stakeholders in locating (IWA, 2012a; IWA, 2012b).
(d) Specific treatment and disposal measures for output materials of advanced treatments such as reverse osmosis.
(e) Difficulties in realization in areas with low population density, scattered settlement or in certain topographic conditions (IWA, 2012a; IWA, 2012b).
(f) High cost for the production of high quality reclaimed water for reuse purposes.
(g) High cost for the recovery of materials and energy from wastewater. And
(h) Questionable sustainability of current wastewater treatment plants.

What happened in the past has been the object of an analysis conducted in order to identify the reasons for current issues. Some studies have shown that the technological progress belongs to ancient populations, since Minoans, Greeks to Romans, who used current principles with particular regard to nature, to safety and security and therefore to sustainability and durability (Angelakis *et al.*, 2010). Today the sustainability and durability of environmental engineering projects has been missing (Angelakis *et al.*, 2008; Mays, 2010).

The progress has also involved the scale of environmental engineering projects. During the Romans' time it has gone from small water projects, such as wells and cisterns, to large water and wastewater projects. Subsequently, the civilization resulted in a dominance of large and mega-scale projects. However, in developing countries the construction of centralized systems has been very difficult due to rural life of population, the rapidity of urban planning growth, culture of people together with the high cost and the unfeasibility (National Centre of Competence in Research 2011; Massoud, 2009).

Especially for sanitation, the search for solutions to identified issues, both in developed countries and even more for the developing ones, has led the scientific community to consider again the application of decentralized systems. In fact, aspects such as durability, sustainability, environmental and economical, and flexibility have been the turning point for progress.

23.4 DECENTRALIZED SYSTEMS

The concept of a decentralized system refers to the need for dividing the territory to small portions, each served by a sanitation system. The dimensions of such decentralized systems can therefore be variable and under the most restrictive conditions coincide with the individual household. In this context, therefore, decentralized sanitation systems consist of either a collection and storage/disposal system or a collection and treatment system. They can be classified in terms of localization, water demand, utilized technology and costs. In addition, each technology is characterized by input materials and outputs like wastes and/or products (IWA, 2012a; IWA, 2012b). The choice of appropriate technologies must be based on the local environment, culture and resources and specific systems, like combination of technologies, could be used depending on the engineering perspectives (EAWAG, 2008).

Decentralized sanitation systems are characterized by a greater development of technologies for developing countries, which require solutions capable of ensuring good hygienic and sanitary conditions. In this context different systems have been developed that include the entire cycle of wastewaters from the production to the use or disposal, through the collection, transport and treatment. The developing countries need to design sanitary systems with particular attention to issues such as the non-existence of a network collecting the wastewater. This does not always correspond to the possibility of applying such systems also in industrialized countries, due to the difficulty in changing well established habits and due to the need for large spaces not always available in urban areas.

In this chapter, sanitation systems are classified according to the application level: household, neighbourhood and urban areas. The smallest possible decentralized system serves an individual household (IWA, 2012a). Furthermore, each system has specific characteristics that make it suitable for an application in a specific area. Table 23.2 shows the advantages and disadvantages of the main decentralized sanitation systems.

Table 23.2 Advantages and disadvantages of decentralized sanitation systems.

Sanitation Systems	Advantages	Disadvantages
Single Pit Systems	Useful for rural and peri-urban area	Need large space Not applicable in areas with heavy rains or flooding Sludge require a treatment High maintenance costs
Waterless System with Alternating Pits	Useful for water scarce area Not need large space Permanent It's not necessary a treatment of outputs	Require an extended storage period Need to control parameters of degradation process
Pour flush system with twin pits	Allows water and greywater use	Needs of low groundwater tablet Specific issues in function of storage/treatment technology chosen Needs a separate treatment for dry cleansing materials
Waterless system with urine diversion	Useful for rocky areas, scarce water areas or areas where there is a high groundwater	The performance depend on efficient separation of urine and faeces Needs the acceptation of urine use Need the separation of solid cleansing materials and anal cleansing water
Blackwater treatment system with infiltration	Useful for areas where desludging services are available	Need sufficient space and suitable capacity of soil absorption Require constant source of water High investment cost
Blackwater treatment system with sewerage	Useful for dense, urban areas Allows anal cleansing water	Moderate investment costs Necessary maintenance of sewer network and treatment facility
Sewerage system with urine diversion	Useful for dense urban and peri-urban areas	High capital cost Require water source

A sanitation system to be applied at household/neighborhood level is the *Waterless system with alternating pits*. It is based on alternating use of pits without addition of flushwater. The recommended user interface is only 'dry toilet', as it does not require water and should be excluded anal cleansing water. Excreta from the 'dry toilet' are collected, stored and treated with a 'double ventilated improved pit', 'fossa alterna' or a 'composting chamber'. The compost/ecohumus generated from this system could be utilized in agriculture as soil conditioner. 'Double ventilated improved Pit' consists of systems of two pits to be used alternately, while the contents rest, drain, reduce in volume and degrade (EAWAG, 2008). The ventilation system is realized by means of a pipe, useful to reduce flies and odours. It can be realized in peri-urban areas and does not require a large space, but has low/moderate reduction in pathogens. This technology is similar to 'fossa alterna' with the only difference being that 'fossa alterna' produce humus and require the addition of soil, ash and/or leaves. A 'composting chamber' is a biological unit, in which microorganisms degrade the organic matter. It consists of a reactor, a ventilation unit to ensure sufficient oxygenation, a leachate collection system and an access door to remove the mature product. The performance of a 'composting chamber' depends on appropriate parameters such as oxygen, humidity, temperature and carbon to nitrogen ratio.

The water usage is a basic characteristic of the *Pour flush system with twin pits*, that utilizes the Pour Flush Toilet as a user interface. Blackwater is then collected and sent to a storage/treatment unit. It's a Twin Pits for Pour Flush or an anaerobic biogas reactor. The 'twin pits for pour flush' are two alternating pits with a structure of simple realization and unlimited life, but low performance for the reduction of pathogens. They are based on the absorption capacity of the soil and the degradation of organic matter. This system is appropriate for very densely built houses, but not for area with a high groundwater table or areas frequently flooded.

A system that does not require the use of water is the *Waterless system with urine diversion*. In this case, urine and faeces are separated in the user interface, with the urine diverting in a dry toilet or urinal, and are collected, stored and treated differently. Faeces are collected in 'double dehydration vaults' that allow to obtain dried faeces, while urine is stored in a Storage tank, before the use and/or disposal through land application, irrigation or soil infiltration through a soak pit. Dried faeces, which retain a human risk health, could be utilized on soil with surface disposal or a simple application. 'Double dehydration vaults' is a technology based on dehydration of faeces. It is inspired by technologies used in the past by improving the materials and construction methods. Also, in this case, there are two pits to be used alternately which allow a long use. Such a system requires the addition of ash, sand or lime, but it does not require addition of water. It is suitable for all types of users but requires a step of education for better usage.These systems are applicable to small communities, in particular for household/neighborhood level, allowing the decentralization.

In addition to the described systems, the literature also reports other systems that require a subsequent treatment of the effluent and/or of the sludge and therefore are not applicable for household, but can be considered valid for neighborhood. Among these, a sanitation system suitable for rural and peri-urban area is the *Single Pit System*. It is based on a dry toilet or a pour flush toilet, like user interface, and a single pit or a single ventilated improved pit, like storage system. Then the Pit can be filled and covered with soil or the outputs (faecal sludge) can be removed, transported and treated in a (Semi-) Centralized Treatment System.

Similar to the Single Pit System, the *Blackwater treatment system with infiltration* requires a (Semi-) Centralized Treatment of faecal sludge before use and/or disposal. This system consists in a flush toilet, like 'pour flush toilet' or 'cistern flush toilet', and in a storage system useful to store large quantities. A first treatment is realized through an 'anaerobic baffled reactor' or an 'anaerobic filter'. The effluent, suitable for direct use, is diverted to the soil through a Soak pit or a Leach field.

The faecal sludge, highly pathogenic, is then treated in a dedicated faecal sludge treatment facility or co-treated with blackwater.

The *Blackwater treatment system with sewerage* differs from the Blackwater treatment system with infiltration only in the management and processing of the effluent of the Collect and Storage/treatment. The first system provides a treatment, with or without faecal sludge, in a treatment facility, while in the second the effluent can be used and disposed directly.

Finally, the *Sewerage system with urine diversion* represents a system applicable not only for urban areas in developing countries, but also for those industrialized. It is based on a 'Urine Diverting Flush Toilet', a special toilet that allows the separation and collection of urine and brownwater (faeces with flushwater), a sewer system and a treatment facility. The urine, separated at the user interface, is stored and then applied on agricultural lands. Brownwaters are transported by the sewer system to a treatment facility.

Systems like the single pit system, blackwater treatment system with infiltration/sewerage and sewerage system with urine diversion require a subsequent centralized or semi-centralized treatment system for sludge and/or effluent. Table 23.3 summarizes the applicability of these centralized or semi-centralized systems for wastewater or sludge treatment to different scales. Among these systems, 'constructed wetlands' (free-water surface, horizontal surface flow, vertical flow) and biogas reactors can be utilized at household level, while others can be utilized at neighborhood level or at larger scale.

Table 23.3 Application level (legend: – not suitable, x less suitable, xx suitable) of centralized and semi-centralized treatment technologies (adapted from EAWAG 2008).

		Application level		
		Household	**Neighborhood**	**Urban areas**
Faecal sludge treatment facility	Sedimentation/thickening ponds	–	x	xx
	Unplanted drying beds	–	x	xx
	Planted drying beds	–	x	xx
	Co-Composting	–	x	xx
	Anaerobic biogas reactor	xx	xx	xx
Wastewater treatment facility	Waste stabilization ponds	–	x	xx
	Aerated Pond	–	x	xx
	Constructed wetlands	x	xx	xx
	Trickling filter	–	x	xx
	UASB	–	x	xx
	Activated sludge	–	x	xx

The main faecal sludge treatment facilities are listed in the following:

- *Sedimentation/thickening ponds,* which are settling ponds that allow the sludge to thicken and dewater (EAWAG, 2008). This type of treatment must be appropriate for the characteristics of the sludge to be treated: high or low strength. Therefore high-strength sludge requires a prior stabilization to allow anaerobic degradation, while low-strength is easily degraded.
- *Unplanted/planted drying beds* which are permeable beds. The system produces leachate which is collected through a system of pipes from the bottom, and the dried sludge. In such a case, the use of vegetation allows to increase the transpiration rate.

- *Co-composting* which allows simultaneous treatment of faecal sludge and organic matter from solid waste by means of an aerobic degradation. The co-composting can be accomplished outdoors or indoors by placing the material in piles periodically turned over to ensure proper aeration and therefore degradation.
- *Anaerobic biogas reactor,* which treats faecal sludge, wastewater and biodegradable wastes by an anaerobic degradation. The reactor degrades the organic material, releasing biogas and a digested slurry. These systems can be directly connected to indoor toilets.

Waste stabilization ponds are large water bodies, in which wastewaters are treated by a biological process. They could be classified in three types: anaerobic, facultative and aerobic. They are utilized singularly or in a series to obtain higher BOD removal (up to 60% for anaerobic unit and up to 75% for the facultative unit). The aerobic waste stabilization ponds instead have the purpose of pathogen and nutrients removal if used with algae and/or fish. These technologies require large space and therefore are suitable for rural areas. If wastewater is mixed in an aerobic reactor, the treatment is an *Aerated pond*. In this case, the space required is less than for waste stabilization ponds, making it suitable for rural and peri-urban areas. Also, mechanical aeration allows to obtain a high rate of organic matter and nutrients degradation.

When wastewaters move through or in contact with filter media and with the contribution of aquatic vegetation, it is referred to as *Constructed wetland*. It can be done in three different ways:

(a) Free-water surface Constructed Wetland when wastewater flows above ground.
(b) Horizontal subsurface flow Constructed Wetland when wastewater flows horizontally. And
(c) Vertical flow Constructed Wetland when wastewater flows vertically down thought the filter bed.

The role of the vegetation is to transfer the oxygen from the atmosphere to the root zone. This allows to promote the degradation of organic matter, to take up nutrients like nitrogen and phosphorus, to maintain the permeability and to provide a habitat for microorganisms. Horizontal subsurface flow systems have found an increased application over the years, thanks to the better control of odours and insects. Such systems are particularly suitable for the removal of suspended solids by sedimentation and filtration, of BOD through the bacterial metabolism, of nitrogen through denitrification (Masotti & Verlicchi, 2005). Contrary to the horizontal subsurface flow system, the vertical flow system enables a good nitrogen removal (about 80% if the load specific maximum superficial of nitrogen is less than 6–12 g $N\text{-}NH_3/m^2d$) as characterized by an intermittent flow and therefore saturation and unsaturation conditions (Masotti & Verlicchi, 2005). In free-water surface constructed wetlands the treatment processes are similar to those that occur in aerated ponds to which the contribution of the vegetation is added. In such cases, the presence of oxygen is maximum and therefore the biological processes that are carried out are of aerobic type. Also, if the vegetation is very dense removal of pollutants by filtration is also observed. Such a treatment can manifest the development of insects and odours (Masotti & Verlicchi, 2005).

Another treatment of wastewater and greywater is the *Upflow Anaerobic Sludge Blanket Reactor (UASB)*. It allows, compared to normal digesters, a higher removal of the organic load, with minor hydraulic residence times and therefore smaller volumes. The water to be treated, released from the reactor bottom, comes into contact with the anaerobic sludge degrading and releasing biogas. Subsequently in the upper section of the reactor three phases are separated: liquid, gas and sludge. This system is suitable for large communities and can ensure a constant water supply and electricity.

Various biological and chemical-physical treatment systems are reported in the literature in order to treat greywater. In particular, the complexity and degree of treatment employed for greywater can vary with the type of greywater and reuse options (Li *et al.*, 2009). While more simple treatment processes can be employed in rainwater harvesting systems.

Finally, conventional centralized treatment technologies, such as *activated sludge* and *trickling filter* can be used also for decentralized systems. Indeed, industrialized countries have conventional centralized supply systems that can treat wastewater of millions of urban residents. However, the population increasing within the long implementation times of new centralized treatment plants and their poor flexibility, makes centralized system unsuitable in growth-periods. Also decentralized solutions, treating wastewater on site at the household level, are solutions not appropriate within areas characterized by high population densities because of the limited space available and sanitation habits of population. Therefore, for countries characterized by a rapid urban growth are proposed semi-centralized supply and treatment systems (Cornel & Bieker, 2006). These systems work in smaller district units and are located close to urban housing areas. Semi-centralized supply and treatment systems can perform as an integrated treatment of water, wastewater and wastes within new techniques and a flexible implementation. In this case, the input of the system can be greywater, wastewater and solid waste, while the output can be treated wastewater, service water, fertilizer, stabilized waste. These systems are often characterized by no collection and storage units; in fact the user interface is directly connected with the semi-centralized treatment system by a sewer network (EAWAG, 2008).

In the future, it is anticipated that hybrid wastewater management systems will occur in order to achieve sustainable solutions with smaller footprint (Tchnobanoglous & Leverenz, 2013). For instance, Figure 23.11 shows different approaches, processes and technologies which can be employed in sustainable sanitation systems proposed by the Deutsche Gesellschaft für Internationale Zusammenarbeit (GIZ). These sanitation options can result in a better recovery of water, materials and energy from waste. However, what is unknown at the present, is what is the most sustainable configuration and physical scale of the facilities needed (Tchnobanoglous & Leverenz, 2013).

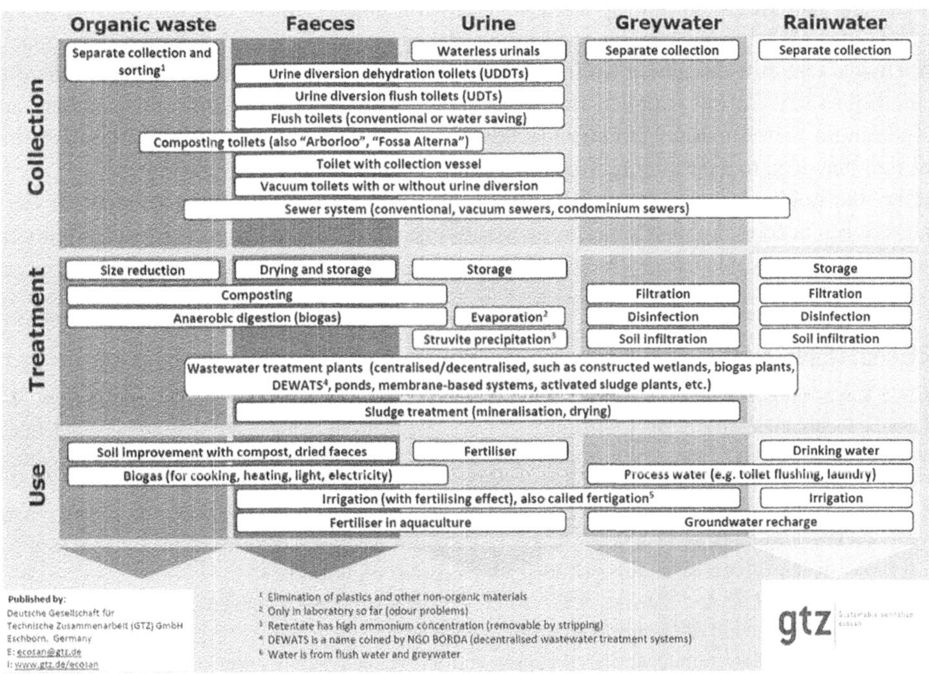

Figure 23.11 Possible approaches, processes and technologies for sustainable sanitation proposed by GTZ.

23.5 KUHN AND THE SCIENTIFIC REVOLUTION: IS BACK PROGRESS THE FUTURE ECO-SANITATION?

As highlighted by the historical events and the evolution of sanitation systems, wastewater treatment from the early nineteenth century to the present exhibited a cyclical tendency. Indeed, from the beginning of the nineteenth century decentralized systems were adopted due to the distribution of population and the absence of collection systems, while in the middle of the nineteenth century and in the twentieth century centralized systems were dominant with huge benefits for both the environment and health. This was supported by the development of new treatment techniques and the definition of treatment cycles useful to achieve high standards required for the environmental protection. These treatment systems are reaching even higher removal efficiency thanks to the more advanced treatment processes, in order to recover water, materials and energy from wastewater. However, the concept of separating what was mixed at the user interface seems not sustainable.

The evolution of wastewater infrastructure was often based on easily implementable, pragmatic solutions, with little regards to the unintended consequences of the designs. Today much of the existing infrastructure is well beyond its useful life and will need to be adapted to future conditions, therefore new design approaches are needed in order to avoid the mistakes of the past and to reduce the impact of the new infrastructure and optimize the recovery of water, materials and energy. In this context, the decentralized approach to wastewater management can play an important role (Tchnobanoglous & Leverenz, 2013). Indeed, decentralized treatment systems, which are often adopted in the developing countries, are in principle more sustainable than centralized systems. For this reason, during the past few decades, renewed interest in decentralized management alternatives has risen. For instance, the scientific world has tried to find new solutions for developing countries that could guarantee adequate sanitary conditions, while containing capital costs of treatment facilities and allowing local management. Some of these decentralized solutions can be also applied in the industrialized countries, even though they may require a revolution in the modus vivendi.

Inspired by the scientific thought of Kuhn, progress is not always linear but requires periodic revolutions called 'paradigm shift'. In fact, defined the characteristics of a given paradigm, researchers apply the scientific paradigm to the natural world and check its validity. During the application can then arise anomalies that may lead to a paradigm crisis and therefore to a 'scientific revolution'.

Therefore, the question is: what could be the development of wastewater treatment in the future? Probably, there is a need to develop a 'scientific revolution' that starting from the eco-sanitation principle can lead to a more sustainable management of wastewater.

23.6 REFERENCES

Angelakis A., Koutsoyiannis D. and Mays L. W. (2008). The 1st IWA International Symposium on Water and Wastewater Technologies in Ancient Civilizations: Overview and Conclusions. Newsletter May 2008, International Water Association (IWA). http://www.iwahq.org/contentsuite/upload/iwa/Document/SG%20newsletter_ancient%20civilization_april2010.pdf (accessed 06 June 2013).

Angelakis A., Salgot M., Paranychianakis N. V. and De Feo G. (2010). Prolegomena. Newsletter April 2010. International Water Association (IWA), pp. 3–4. http://www.iwahq.org/contentsuite/upload/iwa/Document/SG%20newsletter_ancient%20civilization_april2010.pdf (accessed 06 June 2013).

Burian S. J., Nix S. J., Pitt R. E. and Durrans S. R. (2000). Urban wastewater management in the United States: past, present, and future. *Journal of Urban Technology*, **7**(3), 33–62.

Cornel P. and Bieker S. (2009). Semicentralized supply and treatment systems – integrated infrastructure solutions for fast growing urban areas. KFW Water Symposium 2009: Financing Sanitation, October 2009, Frankfurt, Germany, pp. 38–45.

Eawag (2008). Compendium of Sanitation Systems and Technologies. Swiss Federal Institute of Aquatic Science and Technology, Dübendorf, Switzerland, ISBN 978-3-906484-44-0.

Eckenfelder Jr., W. W. and O'Connor D. J. (1961). Biological Waste Treatment. Pergamon Press (Macmillan), New York.

Franceys R., Pickford J. and Reed R. (1992). A Guide to the Development of on-Site Sanitation. Water, Engineering and Development Centre. Loughborough University of Technology, Loughborough, England, World Health Organization, Geneva.

GEC (2005). Water and wastewater reuse. An Environmentally Sound Approach for Sustainable Urban Water Management. United Nations Environment Programme.

Gray H. F. (1940). Sewerage in ancient and medieval times. *Sewage Works Journal*, **12**(5), 939–946. Reprinted with permission from *Sewage Works Journal*. Copyright 1940 Water Environment Federation, Alexandria, VA.

Gray S. M. (1884). Proposed Plan for a Sewerage System, and for the Disposal of the Sewage of the City of Providence. Providence Press Company, Providence, Printers to the City, 1884.

Hendricks D. (2011). Fundamentals of Water Treatment Unit Processes, Physical, Chemical and Biological. IWA Publishing, London & CRC Press, Boca Raton, FL.

Henze M., Grady C. P. Jr., Gujer W., Marais G. V. R., Matsuo T. (1987). A general model for single-sludge wastewater treatment systems. *Water Research*, **21**(5), 505–515.

Henze M., van Loosdrecht M. C. M., Ekama G. A. and Brdjanovic D. (eds.) (2008). Biological Wastewater treatment: Principles, Modelling and Design, IWA Publishing, London, UK.

Höphmayer-Tokich S. (2006). 'Wastewater Management Strategy: Centralized vs. decentralized technologies for small communities'. CSTM Studies and Reports nr. 271 ISSN 1381-6357, pp. 1–27.

IWA Specialist Group (2012a) Decentralized wastewater management: An overview. Global Trends & Challenges in Water Science and Management, International Water Association (IWA), pp. 68–70. http://www.iwahq.org/ContentSuite/upload/iwa/all/Specialist%20groups/members%20only%20resources/IWA_Web.pdf (accessed 06 June 2013).

IWA Specialist Group (2012b) Resources-oriented sanitation. Global Trends & Challenges in Water Science and Management. International Water Association (IWA), pp. 64–67. http://www.iwahq.org/ContentSuite/upload/iwa/all/Specialist%20groups/members%20only%20resources/IWA_Web.pdf (accessed 06 June 2013).

Kuhn T. S. (1962). The Structure of Scientific Revolutions. 1st edn, University of Chicago Press, Chicago, p. 168.

Lawrence A. W. and McCarty P. L. (1970). A unified basis for biological treatment design and operation. *Journal of the Sanitary Engineering Division, American Society of Chemical Engineers*, **96**, 757.

Li F., Wichmann K. and Otterpohl R. (2009). Evaluation of appropriate technologies for grey water treatments and reuses. *Water Science Technology*, **59**(2), 249–260.

Masotti L. (2002). *Depurazione delle acque. Tecniche ed impianti per il trattamento delle acque di rifiuto* (Wastewater treatment. Technologies and plants for wastewater treatment). Calderini, Bologna, Italy. ISBN 88-506-2786-6.

Masotti L. and Verlicchi P. (2005). *Depurazione delle acque di piccole comunità. Tecniche naturali e tecniche impiantistiche* (Small wastewater treatment plants. Natural and intensive technologies). Biblioteca tecnica Hoepli, Hoepli, ISBN 88-203-2963-8.

Massoud M. A., Tarhini A. and Nasr J. A. (2009). Decentralized approaches to wastewater treatment and management: applicability in developing countries. *Journal of Environmental Management*, **90**, 652–659.

Mays L. W. (2010). Ancient Water Technologies, Traditional Knowledge, and Water Resources Sustainability. Newsletter April 2010. International Water Association (IWA), pp. 15–18. http://www.iwahq.org/contentsuite/upload/iwa/Document/SG%20newsletter_ancient%20civilization_april2010.pdf (accessed 06 June 2013).

McKinney R. E. (1962). Microbiology for Sanitary Engineers. McGraw-Hill, New York.

Metcalf & Eddy Inc. (1972). Wastewater Engineering – Collection, Treatment, Disposal. McGraw-Hill, New York.

Metcalf & Eddy Inc. (2002). Wastewater Engineering: Treatment, Disposal, and Reuse. 4th edn, G. Tchobanoglous, F. L. Burton and D. H. Stensel (eds), Mcgraw-Hill, New York, USA.

Moore E. C. S. (1909). Chapter I: conservancy systems. Sanitary Engineering, E. J. Silcock (ed.), Vol. I, 3rd edn., J. B. Lippincott Company, Philadelphia, pp. 5–15.

National Centre of Competence in Research (2011). Local Solution for Sanitation. Research evidence for policy, n. 5, Paul Mundy.

Pearson E. A. (1968). Kinetics of biological treatment. In: Advanced in water Quality Improvement, E. F. Gloyna and W. W. Jr. Eckenfelder (eds), University of Texas Press, Austin, TX.

Seeger H. (1999). The history of German waste water treatment. *European Water Management*, **2**(5), 51–56.

Spataro D. (1887). *Igiene delle Abitazioni.* (Sanitation in households). Fognatura Domestica, vol. 1, Ulrico Hoepli, Milano, Italy.

Tchnobanoglous G. and Leverenz H. (2013). The rationale for decentralization of wastewater infrastructure. In: Source Separation and Decentralization for Wastewater Management, T. A. Larsen, K. M. Udert and J. Lienert (eds), IWA Publishing, London, UK.

Wiesmann U., Choi I. S., Dombrowski E. M. (2007). Fundamentals of Biological Wastewater Treatment. WILEY-VCH Verlag GmbH & Co. KGaA, Weinheim, pp. 1–23.

Woodbury J. F. (1946). The trickling filter and its operation. *Sewage Works Journal*, **18**(1), 89–91.

Chapter 24

The history of land application and hydroponic systems for wastewater treatment and reuse

V. A. Tzanakakis, S. Koo-Oshima, M. Haddad, N. Apostolidis, and A. N. Angelakis

24.1 INTRODUCTION

Land application has a long history with the first evidence dating to Minoan Crete (Hellas) and Mohenjo-Daro (Indus valley) approximately 4000 years ago (Asano & Levine, 1996; Angelakis *et al.*, 2005; Tzanakakis *et al.*, 2007). The Hellenes and Romans continued this tradition; however with the fall of the Roman Empire, this practice declined and only re-emerged in force in the mid-nineteenth century following the great epidemics. During this period authorities recognised the need for sanitation and this led to the development of effluent application practices, known also as 'sewage farms' as a means to protect public health and to control water pollution (Stanbridge, 1976; US EPA, 1979). For the first time, sanitation legislation criteria with reference to 'sewage farms' were developed.

The advent of the twentieth century brought significant technological innovations and growth in wastewater treatment plants that were able to handle large volumes of wastewater (Metcalf & Eddy, Inc., 1991) for direct discharge to waterways and the ocean. These plants were widely adopted by many major urban centres around the globe as they were compact and did not require large areas for treatment compared to sewage farms. Population growth and urbanisation combined with limited water resources led to a wider range of reuse applications being considered by urban communities. Land application schemes re-emerged as an environmental friendly and low-cost technology suitable mainly for small communities, clusters of homes, institutions, and isolated industrial units (Angelakis, 2001).

Recent advances in wastewater treatment technologies, in particular membranes, have allowed the cost effective production of high quality effluent for a wide range of urban applications such as toilet flushing, garden watering, industrial use, cooling towers, fire-fighting and direct and indirect potable use. In water scarce regions the value of recycled water is such that it is no longer viable to give it away to agriculture. In those regions it makes more sense to use the recycled water in urban areas as a substitute for water; otherwise they would have to transfer water from a long distance away or from other sources such as the sea and brackish groundwater (US EPA, 2012). In water scarce countries such as Australia we have seen competition for recycled water from agriculture and the urban communities (Apostolidis *et al.*, 2011). Governments have even changed over the issue. As the world's population grows and access to fresh water resources become even more constrained the value of recycled water will be better appreciated and the demand and competition for this valuable resource from urban and agricultural users will increase.

Hydroponic systems for wastewater treatment and reuse are not a new idea; it is deep rooted in history. Egyptians, Chinese and Indians used it almost 4000 years ago. This may even include the Hanging Gardens of Babylon about 660 BC (Pattenson, 2010). During the modern times, the first insights into hydroponics date from the mid-seventeenth century but basic principles were established after the first half of the nineteenth century. The term 'hydroponics' was first introduced in the late 1920's and early 1930's being drawn from the Hellenic term 'hydroponos' or working-water or working with water (Thiyagarajan *et al.*, 2007). From that period and until the 1960s numerous projects were undertaken particularly in the US, most of them, however, were unsuccessful due to poor construction techniques and operating practices (Studymode, 1999). Since the seventies advances from research programs along with the increased needs for alternative water resources resulted in the improvement of hydroponic system performance. This led to significant worldwide growth in hydroponic systems involving a range of applications, including water quality control, pollution removal as well as energy and food production (Asano, 2002; Bertoldi *et al.*, 2009).

This Chapter provides a brief overview of the evolution of land application and hydroponic systems for wastewater treatment and reuse over the last 4000 years. By understanding the practices and solutions of the past we are better placed to meet the challenges of the future. For the purposes of this paper land application includes wastewater collection and treatment and relates to all forms of applying effluent and biosolids to land for beneficial uses.

24.2 EVOLUTION OF WASTEWATER COLLECTION AND LAND APPLICATION AND HYDROPONIC TREATMENT SYSTEMS

24.2.1 Bronze ages (*ca.* 3200–1100 BC)

The first historical evidence of effluent reuse for irrigation goes back to Minoan Crete (Hellas) and Mohenjo-Daro (Indus valley) approximately 4000 years ago (Angelakis *et al.*, 2005). The outputs of drainage and sewerage systems in Minoan palaces and towns, such as Knossos, Phaistos, Malia and Zakros, seemed to be somewhat similar. Thus, the sewage and rainwater in the palace at Knossos were disposed in the torrent of Kairatos passing in the eastern side of the palace. In the palaces of Zakros and Malia the sewage and rainwater were disposed partially to the sea and partially to the surrounding agricultural land. Finally, in the palace of Phaistos, rainwater and sewerage were mainly disposed in the lower southwestern agricultural land, instead of the Geropotamos river passing from the north side of the palace. Part of the central sewerage and drainage system of the Phaistos palace was directed towards the agricultural land downhill in the Mesara valley instead of the Geropotamos river as shown in Figure 24.1a. In addition, in the Phaistos palace rainwater was collected from the major yards and stored in cisterns for household use. Similar techniques for the collection of rainwater in other towns and palaces of the Bronze Age have been reported (Angelakis & Koutsoyiannis, 2006). A typical example is a rectangular rainwater storage cistern (dimensions of $1{,}6 \times 2{,}0 \times 6{,}0$ m^3) in the Agia Triada villa, located west of the Phaistos palace (Figure 24.1b). The water from this cistern was probably used for washing or other household uses. Also, there are indications that in the Agia Triada villa, apart from using rainwater, wastewater effluents for irrigation were probably available to fertilize agricultural land. As is known, Crete has relatively high water availability. However, palaces and towns in the central and eastern parts of the island are located in areas under very low water availability. Also, Crete is characterized by very high climatic variability; thus, several periods of the Minoan civilization have been under serious water shortage. Therefore, the reuse of wastewater was a necessity. It appears that bath water could be reused to irrigate gardens (Crouch, 1996).

Figure 24.1 (a) Part of the central sewerage and drainage system of the Phaistos palace and (b) A typical rectangular rainwater storage cistern in the Agia Triada villa, located west of the Phaistos palace (with permission of A. N. Angelakis).

24.2.2 Indus valley civilizations (*ca.* 2600–1900 BC)

An elaborate sanitary and drainage system, a hallmark of ancient Indus cities, is in evidence everywhere. Almost every house unit at Harappa, Mohenjo-daro, and Lothal was equipped with a private bathing area with drains to take the dirty water out into a larger drain that emptied into the sewage and drainage system (Figure 24.2). The most unique aspect of planning during the Indus Valley civilization was the system of underground drainage, with U-shaped bottoms lined with loose brick easily taken up for cleaning. The main sewer, 1.5 m deep and 0.9 m wide, connected to many north-south and east-west sewers (Kenoyer, 1998), was made from bricks smoothened and seamlessly joined together. The expert masonry kept the sewer watertight. Drops at regular intervals acted like an automatic cleaning device.

Tunnels carried the waste liquids to the main channel connecting the dockyard with the river estuary. Common houses had baths and drains that emptied into underground soakage jars. The Indus architects designed sewage disposal systems on a large scale, building networks of brick effluent drains following the lines of the streets. In a few cases, drains were located close to wells, which could cause a possible contamination, but most of the time drains were located far from wells. It has been mentioned that the street drains flowed into the nearest river or stream (Kenoyer, 1998); however, because of the presence of cesspits and 'settling vessels', the sewage systems can be considered as a sort of wastewater treatment (De Feo *et al.*, 2013).

Agriculture was well developed in the Indus valley. However, due to the nature of the topography and the intensity of floods along the Indus and its tributaries, it was not possible to maintain major canals for irrigation. Most cultivation in the alluvium was based on adequate rainfall and opportunistic agriculture where crops could be planted along the banks of oxbow lakes and slower streams. Along the piedmont zone of Baluchistan, some Indus settlements constructed diversion canals for directing floodwaters to fields, and there is some evidence for the construction of small irrigation canals near the site of Shortughai in northern Afghanistan (Kenoyer, 2006). Thus only indirect reuse of wastewater effluent was possible in Indus Valley civilizations.

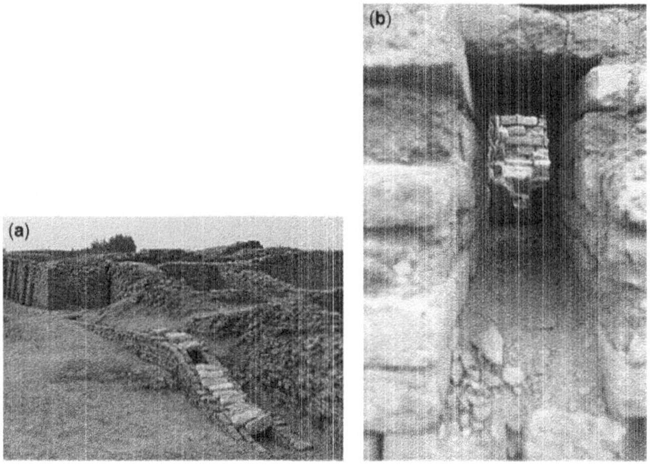

Figure 24.2 Drains in Mohenjo-Daro: (a) Covered drain emptying down slope on major street and (b) close-up view of ground floor drain outlet from the street side (http://www.mohenjodaro.net/index.html).

24.2.3 Historical times (*ca.* 800 BC–330 AD)

Etruscan civilization (ca. 800–100 BC)

Rome had extensive street washing programs (water supplied by aqueducts, the first being built in 312 BC). Only a few homes had water piped directly from the aqueducts. The vast majority of the people came to fountains to gather their water. Even though not many homes were directly plumbed into the sewers, when the wastes were thrown into the street the street washing resulted in most of the human wastes ending up in the sewers anyway. Direct connection of homes to the sewers was not mandatory until nearly 100 BC, due to the cost of the connection. Also imposing such a connection was considered an invasion of private life.

Sewage resulting from the public baths and lavatories was discharged into sewers. It is worth noting that the Etruscans recognized the value of their water (which was transported to the city via aqueducts, often over a distance of 50 km); as such, any wastewater from the public bath facilities was often reused, frequently as the flushing water that flowed continuously through the public latrine facilities (De Feo *et al.*, 2014). From the latrines, it flowed to a point of discharge into the sewer system.

The Classical and Hellenistic periods (ca. 480–67 BC)

The Hellenes, following the tradition of the Minoan and the Mycenaeans, developed very advanced technology regarding sanitation, including baths with flowing water and underground sewers and drains. The drains were large pipe or channels constructed from roof tiles and/or large U-shaped channels constructed of stone slabs. Due to the flooding events, drains are essential in densely settled areas. Drains are commonly associated with catchment basins, surge chambers, manholes, urinals and toilets, cisterns, and laundry slabs and basins (Crouch, 1993).

Sewers in ancient Athens delivered storm water and human wastes to a collection basin outside of the town. From the basin, the storm water and wastes were conveyed through brick-lined conduits to fields in order to irrigate and fertilize fruit orchards and field crops (Schladweiler, 2002). In addition ancient Hellenes in Athens, Dion, Katerini, Sparta and Cyprus, and other places practiced wastewater irrigation of agricultural crops to ensure sanitation and benefits of the nutrients contained in wastewater for increasing yield (Tzanakakis *et al.*, 2007).

Traditional knowledge about water and wastewater management was widespread in the culture, as evidenced from the behavior of the Hellenes when they left their mother cities to establish colonies in the beginning of the eighth century and continuing into the fourth century, though at a reduced rate. They planted scores of new cities northeastward toward the Black Sea and southwest ward into Sicily. Traditional criteria developed for site selection included defense, abundance of foods, climatic and environmental conditions, and water availability. A characteristic example is the archaeological site of Morgantina in east central Sicily, southern Italy. It is 60 km from the coast of the Ionian Sea, in the province of Enna. The agora at Morgantina, an important feature from the *ca.* mid-fifth century BC, had a road along the north side, under which stone drains carried the runoff from the hill to the north. There were springs at the northwest and northeast corners from which deep drains in a Y-pattern converged on the South Gate which led to the road down to the valley (Crouch, 1993). A drain approximately 0.75 m wide, along the west side of the House of the Official, Morgantina, bonded to the house foundations made of the same limestone, and hence built at the same time, *ca.* 4th–3rd century BC, is presented by Crouch (1993).

Also in Hellenistic Athens, Hellas the drains and sewers delivered storm water and wastewater to a collection basin outside the city, implementing for that also the Eridanos river (De Feo *et al.*, 2014). From that basin, the effluents were conveyed through brick-lined conduits to irrigate agricultural fields located downhill the basin (Figure 24.3a). Possibly, the epidemic of Athens *ca.* 430–426 BC, was due to the enhancement of the drainage and sewerage system of the city (Schladweiler, 2002). In addition, in Makrygianni site, located southwestern of Acropolis in Athens, Hellas, the undersurface, central sewer is following the whole length of the central road (dating from ca. the middle of 4th century BC through the 7th AD). The sewer was used to collect both the stormwater from the theatre of Dionysos (located in the south slope of Acropolis) and its surroundings and the wastewater of the houses and workshops in this area (Figure 24.3b). The stormwater wastewater effluents were discharged to southern area of Makrygianni site, where it was probably used for irrigation of agricultural fields.

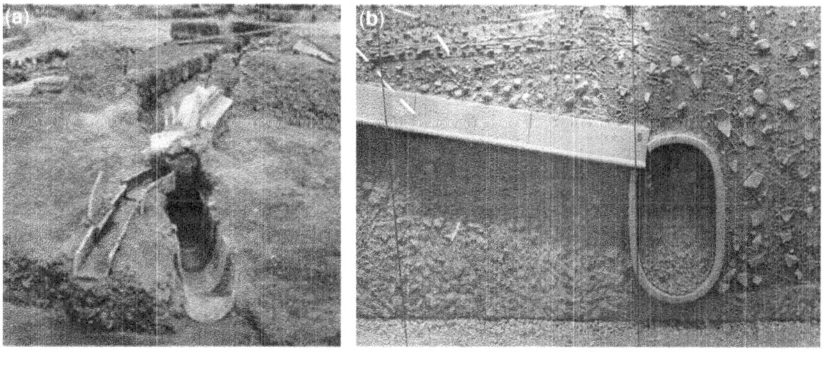

Figure 24.3 Remains of sewers and drains in Hellenistic Athens: (a) Drains south of the Middle Stoa in Hellenistic Agora and (b) Central sewer southeast of Acropolis and clay made drain which conveyed the waste of surrounding houses and worshops to the central sewer.

Crouch (1993) hypothesized that in Classical Hellas stormwaters were used to flush not only latrines but also public sewers, and the combined effluent was used to irrigate crops, especially trees. The Hellenes probably utilized wastewaters in Aidone (near Morgantina) and other places where they spread sewage out on what are called 'sewage farms'. Given the prevalent sunshine, the anaerobic bacteria are quickly removed as well as the offensive smell. This is an example of a simple, low-technology solution which was available to the ancient Hellenes. Crouch (1993) would like to see an excavation of parts of the south slope at Morgantina, to search for the remains of irrigation channels connected to the evident drainage channels near the city gates.

Another evidence of the advanced know-how in terms of collection and reuse of water in ancient Hellas is coming from the archeological findings in the area of Lavrion dating from the ca. 5th century BC (Hellenic Ministry of Culture and Archeological Receipts Fund, 2007). In this area a large number of cisterns, with a storage capacity varying from 300 to 1000 cubic meters of water, supported by an extensive network of channels (Figure 24.4a) are preserved. Usually, nearby these cisterns there is a smaller one called a 'pre-cistern' (Figure 24.4b), that was used as a settlement or sedimentation tank and it was channeled by pipes into the large cistern. In addition, various engineering projects are preserved, such as the building of dams on streams, waterproofing of natural depressions or the facing of winter torrents, indicative of the efforts made by engineers to make possible the collection of water and reuse it efficiently to adjacent mines and, probably, to crops.

Figure 24.4 Water cisterns: (a) Cistern of collecting and storage rain water and (b) a pre-cistern' located nearby the main cistern, used as settlement or sedimentation tank.

In conclusion, Hellenistic societies seem to have achieved a sound ecological balance between man and environment. Ancient Hellenes developed the ability to live lightly on the land, contributing to its renewable resources, observing and utilizing natural cycles, in harmony with their environment.

The historical times have also shown insights into hydroponics; Hieroglyphic records from Egypt, dating back several hundred years BC, describe the growing of plants in water and the Hanging Gardens of Babylon built around 600 BC by King Nebuchadnezzar for his wife, Queen Amyitis who was homesick for the mountains of her homeland in Media, Persia (Pattenson, 2010). There is also evidence of hydroponic farming in ancient China. However, there is no compelling evidence found for the use of hydroponic systems for waste water treatment and reuse prior to modern times (before the 15th century).

The Roman period (ca. 67BC–330 AD)

The Romans developed very advanced technology for sanitation, including baths, with flowing water, and underground sewers and drains, mainly by enlargement of their scale. Remarkable designs in remnants of Roman lavatories can be found throughout the Roman Empire. However, only in rich and well developed urban region of Roman civilization a typical street included a burier lead water pipe and an underground sewer. Sewage from the nearby buildings and homes was brought into the sewerage system through pipes, most of which were drained into the rivers or the sea in the cases of coastal urban areas (Schladweiler, 2002). The 3.35×3.65 m^2 Cloaca Maxima, the 'Main Drain' for example, that drained into the Tiber river was completed in 510 BC and is still in use. Limited information is available on disposal of sewage and rainwater in the land during the Roman period.

Most Romans lavatories did not have advanced plumbing for flushing. They usually used pots that emptied outside. For this purpose, extensive street washing programs were applied in most Roman cities. Thus, most of the water used for washing the streets eventually ended up in the sewers (Schladweiler, 2002). It could be hypothesized that in the Roman period indirect water reuse from the rivers and sewage disposed to land in inland areas, far away from rivers, were somewhat practiced.

Although not by design, the Romans produced a linkage of urban water supply and urban drainage by way of the aqueduct overflow into the sewers. The Roman linkage of the urban water supply and drainage systems marks one of the earliest examples of establishing an urban water cycle. Previously, other civilizations (e.g., the Minoans) had constructed water distribution and urban drainage systems in the same city applying the same principles, but not on the scale of the Romans. The urban water cycle became common during the late nineteenth century in Europe and the United States with the widespread construction of piped-in water supplies and water-carriage sewer systems (Burian & Edwards, 2002).

24.2.4 Medieval times (*ca.* 300–1500 AD)

During the Medieval times very little progress was made from AD 300 through the early nineteenth century. The Emphasis was rather on wars than on civilization and on sanitation. In general sanitation reverted back to the basics (in the best of cases). It was very primitive in most towns. As a result, diseases were commonplace; epidemics decimated towns and villages. Twenty-five percent (or more) of the ancient European population died of disease (cholera, plague, etc.) (Schladweiler, 2002).

During the Dark Ages few technological advances were made, let alone implemented, in Europe. Consequently, urban infrastructure elements including urban drainage systems were not improved. The prevailing public perspective of urban drainage during this period was unnecessary. In Medieval Europe, urban drainage practices were limited because most people lived close to streams, rivers, or other bodies of water. Residents lived close to waterways because water was not commonly brought into the urban area via aqueducts or pipes as had been done in antiquity. Urban stormwater runoff and industrial (e.g., tanners and dyers) wastewater were the primary waste discharges into local streams and rivers (Burian & Edwards, 2002). Human faeces were collected and used in backyard gardens. Other garbage and household trash were typically stockpiled near the city or fed to pigs.

The disposal of human faeces gradually became an issue in large cities during the Middle Ages as populations expanded. Waste disposal for example, in Paris was unregulated for the most part until a decree in 1530 required property owners to construct cesspools in each new dwelling. In general, each community and neighborhood had a self-centered attitude towards urban drainage and municipal services. Citizens were willing to pay for sewers to drain only their neighborhood into the next (Burian & Edwards, 2002). Tzanakakis *et al.* (2007) report references in the Bible for wastewater application to the land and a new concern regarding health and water issues during the Renaissance, evidenced by cesspools which were used as sedimentation tanks and liquid soil infiltration practices.

24.2.5 Aztecs (*ca.* 1100–1521 AD)

The major center of the Aztec culture was called Tenochtitlan. Aztecs built the city of Tenochtitlan on an island in the lake of Texcoco. The Spaniards built Mexico City over Tenochtitlan. Efforts to control flooding led to the drainage of most of the lake, leaving a much smaller lake Texcoco east of the city, surrounded by salt marsh. Tenochtitlan and Mexico City have suffered from periodic floods since the Aztecs times.

The drainage system consisted of several drains that crossed the city from west to east, emptying in the lake of the Texcoco. The system operated by floodgates that allowed stormwater to evacuate in the mornings and prevented the access of waters from the lagoon to the city during the afternoons. It also included pipes in which the black waters of the city discharged, but without solid wastes; they had special depots for solid remainders which were spilled later to the culture systems (Becerril & Jimenez, 2007).

Although the Aztecs had no citywide drainage system, with much of the wastewater ending up in the lake surrounding the city, they had a system to handle human waste by means of privies in all public places and many private dwellings from which excrement was collected in canoes. The excrement was applied as fertilizer on chinampas (or parcels of arable land) or sold in the market to be used for tanning animal hides. Urine was collected in pottery vessels to be used later as a mordant for dyeing cloth. The Tenochtitlán environment was obviously healthy for its time, especially in comparison to European cities. Public and personal hygiene contributed to minimizing the incidence and severity of illnesses. Since that time, the Aztecs recognized the importance of recycled water, nutrients, and other compounds contained in wastewater (Becerril & Jimenez, 2007); in addition the Aztecs had sewer systems that would recycle human waste to fertilize crops.

Chinampa as earlier mentioned is a method of ancient Mesoamerican agriculture which used small, rectangle-shaped areas of fertile arable land to grow crops on the shallow lake beds in the Valley of Mexico. They were created by staking out the shallow lake bed and then fencing in the rectangle with wattle. The fenced-off area was then layered with mud, lake sediment, and decaying vegetation, eventually bringing it above the level of the lake.

In conclusion, Hilary Hansen, Field Museum exhibition project (the Aztec world exhibition at a glance 2008–2009) manager, observes, *'The Aztecs were ingenious and resourceful, building on other peoples' ideas, such as aqueducts, terracing, and creating artificial islands (chinampas) for crop production. They learned new ways of growing food, had clean water, and built ecologically correct sewer systems that recycled human waste as fertilizer. These were problem-solving people'* (http://fieldmuseum.org/about/aztec-world-exhibition-glance).

24.2.6 Modern times (*ca.* 15th to the present)

Land application and reuse of wastewater

In more recent history, the earliest documented 'sewage farms' (e.g., the wastewater application to the land for disposal and agricultural use) were operated in Bunzlau (Germany), in 1531, and in Edinburgh (Scotland) in 1650, where wastewater was used for beneficial crop production (Reed & Crites, 1984; Shuval *et al.*, 1986; Tzanakakis *et al.*, 2007). The epidemics of cholera which occurred in England in the mid-nineteenth century brought 'sewage farms' to the fore again to control water pollution and protect public health. One of the first attempts to organize a sewage farm was by Jame Smith, a Stirlingshire, cotton mill owner. He thought that by taking the excrement from his factory that he would improve his farm yields (Stanbridge, 1976). He moved to London in 1842 and introduced the ideas of distributing sewage on the agricultural lands by hoses and jets. A whole range of ideas modifications and process designs were used over the next fifty years (Cooper, 2001). However, in most cases 'sewage farms' operated as disposal sites whose aim was to maximize the volume of wastewater applied per unit surface area rather than to recycle it efficiently for crop irrigation (Stanbridge, 1976; US EPA, 1979). In the following years in many rapidly growing cities of Europe and the United States, 'sewage farms' were increasingly seen as a solution for the disposal of large volumes of the produced wastewater, some of them are still in use (Table 24.1). Paris was a typical example with the first sewage farms to be established at Gennevilliers in 1872 handling, after an extension, the wastewater of the whole town. At the beginning of the last century, the sewage farms reached their maximum extent, established in four different areas; in Gennevilliers (900 ha) and the Achères (Achères plain, 1400 ha, Pierrelaye, 2010 ha and Triel, 950 ha) supplied with raw wastewater by the Colombes pumping station (Brissaud, 2013). Also, a large 'sewage farm' was established in Melbourne, Australia in 1897 (Reed *et al.*, 1995).

Table 24.1 Selected early land treatments systems (adapted from Metcalf and Eddy, Inc., 1991 and Jiménez, 1995; Reed and Crites, 1984).

Location	Date started	Type of system	Area (1000ha)	Flow (m^3/d)
In Europe				
Berlin, Germany	1874	Sewage farms	2.7	N/A
Braunschweing, Germany	1896	Sewage farms	4.4	60.0
Croydon-Beddington, UK	1860	Sewage farms	0.25	17.4
Leamington, UK	1870	Sewage farms	0.16	3.4
Paris, France	1869	Irrigation	0.64	30.3
Wroclaw, Poland	1882	Sewage farms	0.80	10.6
In US				
Calumet City, MI	1888	Irrigation	0.005	
Ely, NV	1908	Irrigation	0.16	6.1
Fresno, CA	1891	Irrigation	1.60	10.6
San Antonio, TX	1895	Irrigation	1.60	75.7
Vineland, NU	1901		0.006	
Woodland, CA	1889	Irrigation	0.07	15.5
Others				
Melbourne, Australia	1897	Irrigation	4.16	189.3
Mexico city, Mexico	1896	Irrigation	40.00	650.0

The use of the land treatment systems continued into the twentieth century and the last system in the UK continued to be used until the 1980s. However, by the end of the first half of the century these systems were spreading at a slower pace due to certain drawbacks, such as large area requirements, field operation problems, and the inability to achieve the higher hygiene criteria requirements (Cooper, 2001; Asano *et al.*, 2007) as well as to the development of mechanical treatment plants (e.g., trickling filter and activated sludge plants) that were capable of treating large volumes of wastewater requiring much less land. According to a survey conducted in 1939, 125 municipalities used land application in the US (US EPA, 1979). The number increased to 2,200 by 1964 including treatment of domestic wastewater, food industry effluents and petroleum byproducts (Hill *et al.*, 1964). In another survey, in 1972, 571 municipalities serving 6.6 million people plus 1300 industries used land application systems (Thomas, 1973). Of these 571 systems, 316 were identified as crop irrigation systems whereas the 255 as infiltration systems were not defined.

The increasing water needs in the late 1950s combined with the lack of alternative water supplies brought renewed interest in land application in the US. Thus, this practice started to expand again at greater rates than previously. Indicative of this renewed interest was the great number of field, column, and review studies, conducted during the 60s and 70s. These studies emphasized the fate of nutrients and pathogens (e.g., bacteria and virus), the role of soil and vegetation, and the impacts on the quality of groundwater (Matlock *et al.*, 1972; Schmidt, 1972; Bouwer & Chaney, 1974; Thomas, 1973; Tofflemire & Van Alstyne, 1974). Previously, land application was primarily aimed at reducing health risk and enhancement of crop production, however, these studies showed that environmental impacts were also an important issue; as until then little information was available in terms of the consequences on surroundings, such as impacts on soil properties or groundwater pollution from nutrients. Hutchins, in the 30's, was among the first who, along with considerable details on pre-treatment systems, soil, and crops, demonstrated the problem of groundwater pollution (Thomas, 1973). The need to remove nitrogen from wastewater started to come to light in the early 40s when it was considered responsible for eutrophication phenomena in aquatic ecosystems.

The Clean Water Act of 1972 also played a crucial role in the expansion of land application of wastewater. The number of land treatment systems at this period reached about 3400 representing 10 to 20% of wastewater treatment facilities under operation in the US (US. EPA, 1979). Since then the land application systems continued to increase, providing treatment and reuse services for a wider range of wastewaters such as landfill leachates, food processing industry effluents, meat processing wastewater, olive oil mill wastewater, agricultural drainage, liquid dairy waste, and contaminated groundwater (Tzanakakis *et al.*, 2007). Resent research projects contributed significantly to the recognition and expansion of land application. They provide valuable information on the role of soil and vegetation on treatment and reuse potential of land application as well as offering selection criteria for candidate plant species and suitable management practices (Sparling *et al.*, 2001; Walker & Lin 2008; Tzanakakis *et al.*, 2011, 2012; Tsiknia *et al.*, 2013)

In the last few decades the use of treated wastewater for irrigation purposes is increasingly practiced particularly in arid and semi-arid regions as alternative source of water and nutrient values for cultivations. Although that recycled water is a relatively small component of water supply overall, in some countries it has a prominent role, especially for agriculture. For instance in Kuwait reused water accounts for up to 35 percent of total water extraction. In California, US, 31 percent of reclaimed water is used for crop or landscape irrigation (FAO, 2010). In Mexico, most of the wastewater from Mexico City is used in irrigation districts surrounding the city (Scott *et al.*, 2000). The UN FAO has estimated that at least 20 million ha in 50 countries are irrigated with raw or partially diluted wastewater, around 10 percent of total irrigated land. About 525,000 ha are irrigated with reclaimed water. The use of reclaimed water for agricultural irrigation has been reported in at least 44 countries with a total use of over 15 Mmεκθέτης/d (Jiménez & Asano, 2008). The wide array of crops grown with untreated and treated wastewater is shown in Table 24.2.

The use of hydroponic systems for waste water treatment and reuse

The earliest published work on growing terrestrial plants without soil was the 1627 book, Sylva Sylvarum by Sir Francis Bacon, printed a year after his death (Pattenson, 2010). In 1699 John Woodward published his water culture experiments with spearmint. He found that plants in less-pure water sources grew better than plants in distilled water. The work of the German botanists Julius von Sachs and Wilhelm Knop, in the years 1859–65, resulted in a development of the technique of soilless cultivation which was called solution culture (Douglas, 1975). They independently showed that terrestrial plants could grow in nutrient solution without soil. They demonstrated that plants could supply all their needs from inorganic elements and sunlight only (Pattenson, 2010).

Table 24.2 Crops irrigated with municipal wastewater (Asano *et al.*, 2007; Lazarova and Bahri, 2005; Pescod, 1992; Pettygrove & Asano, 1985).

Types	Examples of crops
Field crops	Barley, corn (maize, Zea mays), oats, wheat
Fibre and seed crops	Cotton, flower and vegetable seeds
Vegetable crops that can be consumed raw	Broccoli, cabbage, cauliflower, celery, chllli pepper, green tomato (tomatillo), lettuce, pepper, tomato
Vegetable crops that will be processed before consumption	Artichoke, asparagus, beans, onion, peanut, potato, spinach, squash, sugar beet, sunflower
Fodder and forage crops	Alfalfa, barley, clover, cowpea, hay, maize, pasture
Orchards and vineyards	Fruit trees, apple, avocado, citrus, lemon, peach, pistachio, plum, olive, date palms, grapevines
Nurseries	Flowers
Commercial woodlands	Conifers, eucalyptus, poplar, other trees

The term hydroponic was named thus by William Gericke in the late 1920's and early 1930's and was drawn from the Hellenic term 'hydroponos' or working-water or working with water. Gericke planted tomatoes in his back yard using nutrient solution only (Thiyagarajan *et al.*, 2007). He published in 1940 the first edition of a book entitled: The Complete Guide to Soilless Gardening (Gericke, 1940). Hydroponics was used during the1930s on Wake Island in the Pacific Ocean, which was a refuelling stop for Pan American Airlines. There they grew vegetables for the passengers because there was no soil, and it was expensive to airlift in fresh vegetables (Wikipedia, 2009). During the late 1940's, another hydroponic method was developed at Purdue University by Robert B. and Alice P. Withrow and called Nutriculture. Their system alternately flooded and drained a container with gravel which was used as a rooting medium (Studymode, 1999).

Numerous experiments by Hoagland and his associates demonstrated at UC Berkley, in 1938 that many plant species, normally grown in soil, thrive well in properly aerated solution cultures; in fact, their whole life cycle can be completed, apparently normally, when grown in nutrient solution. After World War II a number of commercial installations were built in the United States, mainly in Florida. Poor construction techniques and operating practices caused many of them to be unsuccessful and their production inconsistent (Studymode, 1999).

In the seventies hydroponic systems were used with wastewater in various applications including water quality control and pollution removal as well as energy and food production (Wolverton *et al.*, 1976; Lewis *et al.*, 1978). During the eighties and nineties (1980's–2000) research and work emphasis was placed on chemical and microbiological processes within the hydroponic system treating wastewater (May *et al.*, 1990), suitable plant types and their yield (Budenheim, 1991), integrating and combining treatment processes (Ayaz, 1996; Chaves *et al.*, 1999; Diver, 2000), water quality aspects and impacts related to the hydroponic system treating wastewater (Burau *et al.*, 1987; Asher & Loneragan, 1995; Boydena & Rababah, 1996), and specific use of the hydroponic system (Schlick & Bubenheim, 1996).

Since 2000 the scope and application of hydroponic systems in various types of wastewater treatment has widened and been extended (Asano, 2002; Berndtsson, 2001; Austin *et al.*, 2003; Ash & Truong, 2003; Anderson *et al.*, 2004; Diver, 2006; Andersson & Norström, 2007; Baldwin & Butcher, 2007; Feng *et al.*, 2011). Hydroponic system production and efficiency has been optimized and monitored (Guo *et al.*, 2002; Haddad *et al.*, 2009; Haddad *et al.*, 2012), and the crop types planted have been extended and now include commercial crops (Bertoldi *et al.*, 2009).

24.3 ENVIRONMENTAL CONCERNS AND PUBLIC HEALTH

The Hellenes and Romans were among the first who were aware of the importance of sanitation, developing innovative hydraulic technologies and reuse practices evidenced by findings in several archeological sites (e.g., Knossos, Cyclades, Athens, and the mainland of Hellas) (Koutsoyiannis & Angelakis, 2006; Angelakis *et al.*, 2005; Cahill, 2003). Indicative of the awareness of ancient Hellenes about sanitation was also the development of theories around the origin and the impacts of microorganisms on the health of civilians, such as the theory of spontaneous generation (384–322 BC). Hippocrates, Hellenic philosopher and physician, in *ca.* 400 BC, declared that diseases do not come from gods but rather they are a result of the existing environmental conditions, diet, and living habits. For a long period the sanitation concept was gradually lost but great epidemics of cholera spread to Asia, Africa, and Europe during the fourteenth and eighteenth centuries again motivated the communities and stressed the need for measures, such as the development of 'sewage farms', to protect public health (US EPA, 1979). However, the first legislation or development of criteria was recorded in the mid nineteenth century in England.

The discovery of the role of microbes by the scientific community occurred in the second half of the nineteenth century following the great epidemics of cholera and typhoid fever in England during 1830–1850. These events were found to be associated with the pollution of water sources with raw wastewater. This discovery made clear the need for sanitation and the protection of water resources motivating health agencies to implement sanitation rules and environmental policies to protect public health (Paranychianakis *et al.*, 2011; US EPA, 1979). These laws allowed the 'sewage farms' to come back to the fore as an attempt to protect public health and to control water pollution (Paranychianakis *et al.*, 2011; US EPA, 1979).

Since the beginning of the twentieth century different regions and governmental agencies, both in the United States and globally, developed and adopted a variety of water reuse criteria for use of reclaimed water for crop irrigation (Table 24.3). These rules and regulations have been developed primarily to protect public health and water resources. In England, in 1912, the 'Royal Commission on Sewage Disposal' adopted for the first time standards for effluent discharge that included limits of 20 mg/L for BOD and suspended solids (US EPA, 1979). A few years later, in 1918, California State Board of Public Health set up the first of the regulations for use of sewage for irrigation purposes. They were modified in 1978 and 2000 and now stipulate a total coliform (TC) quality standard of 2.2/100 ml for effluent used to irrigate vegetable crops eaten uncooked (State of California, 1978; Title 22 of the state Code of Regulations). The California Water Recycling and Reuse criteria requires as a minimum, a filtration and disinfection treatment process for production of recycled water for unrestricted food crop irrigation. The California regulations were replicated or used as a basis for similar regulations in many parts of the world, including developing countries.

The first Guidelines for Water Reuse by US EPA was published in 1980, which was updated in 1992, to support planners, utilities, and regulatory agencies in the development of appropriate water quality and regulatory requirements in the US (US EPA, 1980 and 1992). In these publications guidelines covered the domain of agricultural reuse along with the other types of reuse, including non potable urban, groundwater recharge, environmental, and industrial and large-scale commercial reuse. In 2004, the US EPA updated the 1992 Guidelines presenting and summarizing recommended water reuse guidelines, along with supporting information, as guidance for the benefit of the water and wastewater utilities and regulatory agencies in order to develop water reuse standards and to revise or to expand existing regulations (US EPA, 2004). In this version reclaimed water quality, along with treatment requirements, for different types of reuse are presented and most common parameters for which water quality limits are imposed are biochemical oxygen demand (BOD), total suspended solids (TSS), and total or faecal coliform counts. Recently, it has released another update of the guidelines to incorporate new applications and advances in technologies relevant to the different types of reuse as well as to update state regulatory information (US EPA, 2012). This version provides information for best practices in reuse projects, international experience, and factors promoting safe and sustainable water reuse throughout the world.

Table 24.3 Historical evolution of the treated wastewater reuse quality criteria (adapted from Paranychianakis *et al.*, 2011).

Year	Data and quality criteria
1918	Before 1918 wastewater reuse for irrigation was practiced in various regions since the Minoan Era, but there are not any criteria. California State Board of Public Health set up the first regulations for use of sewage for irrigation purposes.
1932	In Reus (Spain), an association of farmers for irrigation with treated wastewater published its statutes
1973	The first guidelines for effluent reuse by WHO (100 FC/100 ml, 80% of samples)
1978	State of California regulations: 2.2 TC/100 ml (State of California, 1978)
1978	Israel regulations: 12 FC/100 ml in 80% of samples: 2.2 FC/100 ml in 50% of samples
1983	World Bank Report (Shuval *et al.*, 1986)
1983	State of Florida: No detectable *E. coli*/100 ml for crops consumed raw
1984	State of Arizona: Standards for virus (1 virus/40 L) and *Giardia* (1 cyst / 40 L)
1985	Feachem Report (Feachem *et al.*, 1983)
1985	Engelberg Report (IRCWD, 1985)
1989	WHO Recommendations for wastewater reuse: 1000 FC/100 ml; <1 nematode egg/L (WHO, 1989)
1990	State of Texas: 75 FC/100 ml
1991	Sanitary French recommendations based on WHO guidelines
1992	U.S. EPA Guidelines for water reuse: No detectable FC/100 ml (7 d median. No more of 14 FC/100 ml in any sample)
1999	Revised Israel regulations: Unrestricted irrigation <1 FC/100 ml and a multi-barrier approach
2000	Australian guidelines
2000	State of California regulations are revised (Title 22)
2003	WHO State of the Art Report on Artificial Recharge of Groundwater with Recycled Water (Aertgeerts and Angelakis, 2003)
2003	Italian criteria for water recycling
2004	The revised guidelines of U.S. EPA for Water Reuse (indirect potable use)
2006	The second revision of WHO Guidelines 'Treated Wastewater in Agriculture: Risk analysis and management'
2006	'Australian guidelines for water recycling: Managing health and environmental risks': Risk analysis and management (NRMMC-EPHC. 2006)
2006	Portuguese Standard NP 4434
2007	Spanish reuse of recycled water: Quality criteria
2010	French quality criteria
2011	Hellenic regulations (Common Ministerial Decision 2011)

WHO published wastewater reuse guidelines for the first time in 1973 (WHO, 1973) setting a guideline value of 100 coliforms per 100 ml for unrestricted irrigation. The guidelines also made recommendations on treatment processes of wastewater dependent on reuse of the effluents, including irrigation, industrial, and municipal reuse purposes. In 1989, WHO published the guidelines for the 'Safe use of reclaimed wastewater in agriculture and aquaculture' dealing almost exclusively with microbial pathogens (for unrestricted irrigation it has set limits of 1 human intestinal nematode egg/L and 1000 fecal coliforms per 100 ml) (WHO, 1989). The guidelines were later improved, jointly with the FAO, in the 2006 edition where risk management approaches under the Stockholm Framework were applied. This provided a harmonized framework for the development of health-based guidelines and standards in terms of water and sanitation related microbial hazards (WHO, 2006). An analysis of the health protection measures is also discussed in the 4 volumes of these guidelines with references to the level of wastewater treatment, crop restriction, wastewater application method and human exposure control. The health based targets applied for the first time a reference level of acceptable risk of 10^{-6} Disability Adjusted Life Years (DALYs), a metric developed and introduced by WHO and the World Bank in 1993. The DALY is the only quantitative indicator of 'burden of disease' that reflects the total amount of healthy life lost; that is, the quality of life reduced due to a disability, or the lifetime lost due to premature mortality. A variety of measures that are adoptable and achievable, given the local socio-economic and technological conditions, are possible for health protection: (a) waste treatment, (b) crop restriction, (c) irrigation technique and application time and (d) human exposure control. In so doing, partial treatment to a less demanding standard may be sufficient if combined with other risk reduction measures to achieve the $\leq 10^{-6}$ risk.

Another principal development in the 2006 WHO Guidelines is the application of quantitative microbial risk analysis (QMRA) to wastewater use in agriculture, which provides a rational basis for microbial risk assessment and management in wastewater irrigation; it determines a numerical value of the risk of disease and/or infection as a result of a person or a community being exposed to a specified number of a particular pathogen. Thus, instead of focusing only on the quality of wastewater at its point of use, the WHO-FAO guidelines recommend defining realistic health-based targets and assessing and managing risks along the continuum – from wastewater generation to consumption of produce cultivated with wastewater – to achieve these targets. This allows a regulatory and monitoring system in line with socio-economic realities of the country or locality.

In addition to the risks from pathogen contamination, the WHO (2006) guidelines provided maximum limits in the soil for various toxic chemicals based on human exposure through the food chain. With regard to water quality for irrigation, WHO refers to the FAO guidelines, which focus on plant growth requirements and limitations (Pescod, 1992). In conclusion, regulated and well-managed irrigation under WHO-FAO guidelines can be protective of public health and the health of farm workers. More restrictive regulations are potentially prohibitively expensive in some economic contexts without necessarily improving the public health outcome. The WHO-FAO guidelines (WHO, 2006) for irrigation with reclaimed water are a science-based standard that has been successfully applied to irrigation reuse applications throughout the world.

The Australian Guidelines for Water Recycling and Reuse (2006) focus on large-scale treated sewage and grey-water reuse, including: (a) residential garden watering, car washing, toilet flushing and clothes washing, (b) irrigation for urban recreational and open space, and agriculture and horticulture, (c) fire protection and fire-fighting systems, (d) industrial uses, including cooling water; and (e) grey-water treated on-site (including in high rise apartments and office blocks) for use for garden watering, car washing, toilet flushing and clothes washing). The Australian guidelines for Water Recycling and Reuse (2009a) referring to managed aquifer recharge and to storm water harvesting and reuse were published in 2009. Also, the Australian Guidelines for Water Recycling (2009b), focusing on stormwater harvest and reuse, were published in 2009.

In Australia, effluent irrigation schemes have to comply with a range of parameters over and above pathogen contamination to protect receiving surface and groundwater resources. For example two effluent irrigation schemes operated at Picton and Dubbo (Figure 24.5) have comprehensive environmental management plans (EMP), incorporating quarterly monitoring of groundwater quality for pH, Electrical conductivity, TDS, Ca, Mg, Na, K, Sulphate, Cl, Ammonia, Nitrate and Nitrite P, TC and *E. coli*. The EMPs also include a Salt Action Plan, Occupation, Health and Safety Plan, Soil Management Plans and Contingency Plans. Essentially, the management of effluent irrigation schemes is getting more sophisticated in order to protect both public health and environmental objectives as well as to optimize crop production. It is no longer maximizing the application rate per unit area approach.

Figure 24.5 Centre pivot effluent irrigation scheme at Dubbo, Australia.

In the EU, the Water Framework Directive adopted by the European Parliament in 2000, includes a number of activities with regard to wastewater treatment and reuse. Previously, in 1991, the Urban Wastewater Treatment Directive established the principle 'Treated wastewater shall be reused however appropriate.' In 2004, the third European Union's Urban Waste Water Treatment Directive (UWWTD) implementation report concluded by stating that the challenges to implementing the Water Framework Directive (WFD) include: '...wastewater treatment, as well as wastewater reuse in order to ensure human health and protect the environment will receive further importance due to increased floods and droughts as a consequence of climate change' (World Bank, 2010). Recent developments through the Aquarec project propose seven quality categories for different types of reuses and compiled microbial and chemical limits for each category (Salgot *et al.*, 2006). This project was launched by the European Commission to investigate strategies, technologies, and management practices for local, safe, publicly acceptable, economically feasible, and sustainable use of treated wastewater for urban, peri-urban and agricultural use. The European Commission is currently looking for the most suitable EU-level instrument in order to encourage water reuse, including a regulation establishing common EU criteria. A proposal for an appropriate impact assessment, to ensure the maintenance of a high level of public health and environmental protection at the EU-level has been undertaken (Paranychianakis *et al.*, 2011).

24.4 EMERGING TRENDS

The future of land application of wastewater will be governed by the following macro drivers:

(a) *Population Growth* – It is estimated that by 2050 the world population will increase by an additional 2 billion people (e.g., a city of 145000 every day) and most of this growth will occur in developing countries that are already suffering from inadequate water resources and degraded environments. Water recycling and reuse will thus play a vital role in future urban planning (Wade Miller, 2006; Qadir *et al.*, 2010). The value of recycled water will become more apparent and there will be increased competition for this valuable resource from urban, agricultural and industrial users.

(b) *Urbanization* – The great majority of these additional people will settle in urban areas further stressing the pollution pressures and health risks in these areas. In many cases it will not be possible to simply extend existing centralized water and wastewater systems to cope with the extra water demand and waste loads. So an expected increase in decentralized self-supporting systems will emerge. In the heavily developed urban areas, direct and in-direct potable re-use schemes will become more viable (Leverenz *et al.*, 2011) as the cost of providing separate pipework for non-potable applications, in such areas, will be prohibitive.

(c) *Climate Change* – Under the global warming scenario it is predicted that the world will experience more extreme climatic conditions (bigger floods and more severe droughts). Some regions, such as Australia, are already experiencing such extremes. Future water systems will need to adapt by developing a combination of climate dependent and independent sources of water (e.g., desalination, water recycling and reuse) to improve their resilience and to avoid major social dislocation and economic loss. This will be a major challenge for developing countries with limited financial resources to cope with such a scenario.

(d) *Ageing Infrastructure Assets* – On top of the extra water demand and wastewater loads much of the existing infrastructure that support the current 6 billion people will deteriorate and will need to be renewed. This will present both a challenge and an opportunity on how to re-configure as well as to finance of water and wastewater infrastructure to meet the future challenges; and

(e) *Water Energy Nexus* – With existing sources of water reaching capacity the search for new sources of water is forcing water utilities across the globe to implement a range of higher energy sources such as long distance pumping, water recycling and reuse and desalination. As well as adding to the cost of water, supply is also contributing to the energy footprint and thus contributing to the climate change problem (Apostolidis, 2010). The question many water managers are now asking is; have we solved one problem by unintentionally creating another one? There is undoubtedly a strong correlation between energy and water use and it is important that in delivering solutions in world of climate change that proper consideration is given to the relationship between water and energy use. In the case of wastewater treatment opportunities exist to produce energy and thus reduce the overall energy footprint.

An assessment of current practice in wastewater use in agriculture all over the world by the World Bank (2010) reveals important patterns between a country's economic status (low-income, lower-middle income, upper-middle income, and high-income) with the provision of adequate water supply, sanitation, wastewater treatment and the planned use of wastewater in urban areas to be positively associated with the levels of economic development.

Australia provides a good case study of possible future trends in water recycling and reuse and how low-income countries will progress as they develop (Apostolidis *et al.*, 2011). The early land application projects in Australia were developed primarily as effluent disposal schemes to protect public health, much the same as the schemes in the US and Europe. The onset of algal blooms in the receiving waterways led to the upgrading of these schemes to reduce nutrient impacts. Increasing population, severe droughts and limited availability of new sources of water have led to recycled water being considered as a valuable resource and appropriate for use in a much wider range of applications not just land irrigation.

Today recycled water is considered a vital element of the water cycle. Effluents from wastewater treatment plants are no longer wastes but sources of water supply, nutrients and energy. All planning or water supply, wastewater and storm water is integrated and the best solution for a particular region is a case-by-case consideration. A significant number of institutional barriers for different recycled water applications have been removed and guidelines published to facilitate their implementation. Sophisticated models have been developed to help determine the optimum configurations to meet future water needs at the lowest impact on the environment. As the availability of new sources of water becomes more constrained, the competition for recycled water will increase. This will result in less water being available for agricultural use as urban and industrial users generally have a better capacity to pay for this source. This trend is supported by a recent US EPA (2012) publication. This publication identified the key drivers for further expansion of recycled water for land application and other uses across the world as being water demands and the availability of water resources and financial resources, including in parallel the global barriers to expanding planned reuse, such as institutional, organizational, economic and public perception/educational barriers.

The future use of small, medium, and large scale-commercial hydroponic systems for wastewater treatment and reuse are expected to grow more and more in both developed and developing countries since they don't need expensive wastewater treatment facilities, their operation and maintenance costs are minimal and agricultural production is higher than traditional. This trend is supported by (a) the high crop products quality as well as reasonable effluent wastewater quality, (b) nutritional added value of wastewater content, and (c) that wastewater effluent from hydroponic systems could safely be used for tree irrigation.

24.5 DISCUSSION AND CONCLUSIONS

The evolution of land application and hydroponic systems over the past 4000 years provides valuable insights into the challenges facing a world with increasing population, urbanization and weather extremes resulting from climate change.

The Hellenes and Romans recognized the importance of managing wastes from high population density communities and developed solutions that endured for many generations. As well as providing collection systems utilizing sewer and drainage networks they applied the wastes on land to grow crops. With the fall of the Hellenic and Roman Empires this knowledge was overlooked and it was not until the outbreak of widespread disease in the heavily populated growing cities of Europe that the need for wastewater management was appreciated and serious attention was given to the issue. London and Paris followed by most of the major cities around the developed world commenced building major collection systems that at least removed the problem from the urban areas. However this action only transferred the problem from location to another. The resultant pollution in the receiving waterways from large urban centers was not acceptable so authorities considered applying the effluent on land (sewage farms) as a form of treatment. The introduction of wastewater treatment plants that were able to manage large volumes of waste utilizing a small footprint saw the decline in the use of land application in the early 20th century. Further population growth and water resource constraints brought renewed focus on land application of wastewater effluent. Fast growing regions with limited water resources such as California, Australia and Israel started to recognize the benefits of utilizing treated wastewater for crop production and thus reducing pressure on scarce water resources.

Most of the industrialized countries began to establish regulatory frameworks for wastewater treatment and reuse in agriculture. Over time, these frameworks have been updated using the improved knowledge of wastewater treatment and land application system performance. There is now a better understanding of the land application systems and the management practices needed to make them sustainable over the long term. In advanced regions of the world there is now sophisticated legislation in place that imposes strict operational rules and monitoring regimes for the management of land application of wastewater effluent.

Today wastewater effluent is no longer considered a waste product. It is recognized as a valuable resource and part of the total water cycle something our ancestors also recognized.

Hydroponic systems also have a long history but mostly on a small scale. The Hanging Gardens of Babylon is perhaps the most notable application. Today there is growing interest in this concept as we reconsider the way we plan future cities to be more water sensitive and to make them more liveable.

The following conclusions can be drawn from this chapter:

(a) The application of recycled water on land for crop production has been practiced for thousands of years. It can be sustained in the future only if public health is protected, beneficial crops are produced and if there is no harm to the soils and receiving surface and groundwater resources and ecosystems.
(b) Population growth, urbanization and climate change will (i) make recycled water a more valuable resource and not for disposal, in both the developed and the developing countries, (ii) increase the competition for recycled water between urban uses and irrigation for crop production, (iii) force increasing consideration of direct and indirect potable re-use as an alternative source of water, and (iv) increase attention on energy use associated with water production and opportunities for generating energy from the wastewater.
(c) In urban planning, recycled water must be considered as a part of the total water cycle. This will require a more holistic approach in the way of water supply; wastewater and stormwater services are provided in future urban communities.
(d) Civilizations ignoring lessons from the past, spend more effort on human life and environment restoration and resources to address their problems. Thus, we can learn from the past- how to address current and future challenges.

Acknowledgement

The authors are grateful to Dr. Gareth Owens for his critical comments and contribution to editing the text.

24.6 REFERENCES

Aertgeerts R. and Angelakis A. N. (2003). State of the Art Report: Health Risks in Aquifer Recharge Using Reclaimed Water, WHO, Water, Sanitation and Health Protection and the Human Environment, WHO, Genova and WHO Regional Office for Europe, Copenhagen, Denmark, p. 212.

Andersson S. and Norström A. (2007). Potential of hydroponics for grey water treatment – two case studies. In: International Conference on Sustainable Sanitation: Food and Water Security for Latin America, November 25–28, 2007, Fortaliza, Brazil, pp. 1–6.

Angelakis A. N. (2001). Management of wastewater by natural treatment systems with emphasis on land based systems. In: Decentralized Sanitation and Reuse, P. Lens *et al.* (eds), Integrated Envir. Techn. Series, IWA Publishing, London, UK, Ch. 16, pp. 303–333.

Angelakis, A. N. and Koutsoyiannis, D. (2006). Wastewater Management in Minoan Civilization. In: 2nd Conf. on Ancient Technologies, October 17–21, 2005, Athens, Hellas. Technical Chamber of Hellas, Athens, Hellas, pp. 751–756 (in Hellenic).

Angelakis A. N., Koutsoyiannis D. and Tchobanoglous G. (2005). Urban wastewater and stormwater technologies in ancient Greece. *Water Research*, **39**, 210–220.

Apostolidis N. (2010). Australia experience in water and energy footprints. *Water Practice and Technology*, **5**(4), 76–87.

Apostolidis N., Hertle C. and Young R. (2011). Water Recycling in Australia. *Water,* **3**(3), 869–881.

Asano T. (2002). Water from (waste) water-the dependable water resource. *Water Science and Technology*, **45**(8), 24–33.

Asano T., Leong L., Rigby M. and Sakija R. (1992). Evaluation of the California wastewater reclamation criteria using enteric virus monitoring data. *Water Science and Technology*, **26**(7–8), 1513–1524.

Asano T., Burton F. L., Leverenz H. L., Tsuchihashi R. and Tchobanoglous G. (2007). Water Reuse. Issues, Technologies, and Applications. Metcalf & Eddy, Inc./AECOM. McGraw-Hill, New York.

Ash R, and Truong P. (2003). The use of Vetiver grass for sewage treatment in Australia. In: Proceedings of the Third International Conference on Vetiver and Exhibition, Guangzhou, China.

Asher C. and Loneragan J. (1995). Response of plants to phosphate concentration in solution culture: 1. Growth and phosphate content. *Soil Science*, **103**(1967), 225–233.

Austin D., Maciolek D. and Lohan E. (2003). Integrated Hydroponic and Fixed-Film Wastewater Treatment Systems and Associated Methods. http://www.wipo.int/pctdb/en/wo.jsp?IA=WO2003%2F042114&WO=2003%2F042114&DISPLAY=DESC (accessed November 2010)

Australian Guidelines for Water Recycling (2006). Managing Health and Environmental Risks (Phase 1). November 10, 2006. Natural Resource Management Ministerial Council, Environment Protection and Heritage Council/National Health and Medical Research Council. Cambera, Australia.

Australian Guidelines Water Recycling (2009a). Managed Aquifer Recharge (Phase 2). July 10, 2009. Natural Resource Management Ministerial Council, Environment Protection and Heritage Council/National Health and Medical Research Council. Cambera, Australia.

Australian Guidelines for Water Recycling (2009b). Stormwater Harvest and Reuse (Phase 2). July 31, 2009. Natural Resource Management Ministerial Council,Environment Protection and Heritage Council/National Health and Medical Research Council. Cambera, Australia.

Ayaz S. (1996). Hydroponic tertiary treatment. *Water Research*, **30**(5), 1295–1298.

Baldwin P. and Butcher D. (2007). Phytoremediation of arsenic by two hyperaccumulators in a hydroponic environment. *Microchemical Journal*, **85**, 297–300.

Becerri J. E. and Jimenez B. (2007). Potable water and sanitation in Tenochtitlan: an Aztec culture. *Water Science and Technology, Water Supply*, **7**(1), 147–154.

Berndtsson, J. C. (2001). Wastewater Fed Aquaculture, Report 3233. Department of Water Resources Engineering, Lund Institute of Technology. Lund, Sweden.

Bertoldi, F., Sant'Anna, E. and Barcelos-Oliveira, J. (2009). Chlorella vulgaris cultivated in hydroponic wastewater. In: ISHS Acta Horticulturae. International Symposium on Soilless Culture and Hydroponics, A. Rodríguez-Delfín and P. F. Martínez, (eds), October 1, 2009, Lima, Peru.

Boydena, B. and Rababah A. (1996). Recycling nutrients from municipal wastewater. *Desalination*, **106**(24), 1–246.

Brissaud F. (2013). Wastewater Reclamation and Reuse in France. http://tierra.rediris.es/hidrored/ebooks/ripda/bvirtual/articulo06.PDF

Budenheim D. (1991). Plants for water recycling, oxygen regeneration and food production. Waste *Management Research*, **9**, 435–443.

Burau G., Sheik B., Cort P., Cooper C. and Ririe D. (1987). Reclaimed water for irrigation of vegetables eaten raw. *California Agriculture*, **41**(788), 4–7.

Burian S. J. and Edwards F. G. (2002). Historical perspectives of urban drainage. *ASCE*, http://dx.doi.org/10.1061/40644(2002)284

Bouwer H. and Chaney R. L. (1974). Land treatment of wastewater. *Advance Agronomy*, **26**, 133–176.

Cahill T. (2003). Sailing the Wine Dark Sea: Why the Greeks Matter. Doubleday, New York, USA.

Chaves P., Sutherland R. and Laird L. (1999). An economic and technical evaluation of integrating hydroponics in a recirculation fish production system. *Aquaculture Economics & Management*, **3**(1), 83–91.

Cooper P. F. (2001). Historical aepects of wastewater treatment. In: Integrated Environment Technical Series, IWA Publishing, London, UK, pp. 11–38, Ch. 2, http://www.iwapublishing.com.

Crouch D. C. (1993). Water Management in Ancient Greek Sites. Oxford University Press, 200 Madison Ave., New York, USA.

Crouch D. C. (1996). Avoiding water shortages: some ancient Greek solutions. In: Diachronic Climatic Impacts on Water Resources with Emphasis on Mediterranean Region, A. N. Angelakis and A. S. Issar, (eds), Springer-Verlag, Heidelberg, Germany, pp. 129–159, Ch. 7.

De Feo, G. Antoniou, G. P., Mays, L., Dragoni, W., Fardin, H. F., El-Gohary, F., Laureano, P., Kanetaki, E. I., Zheng, X. Y. and Angelakis, A. N. (2014). Historical Development of Waste Water Management. In: Handbook of Engineering Hydrology, S. Eslamian (ed.), Taylor & Francis Group, Boca Raton, FL, USA, Vol. 2, Ch. 9, 163–217.

De Feo, Angelakis, A. N., G., Antoniou, G., Fardin, F., M., El-Gohary, F., Zheng, X. Y., Butler, D. and Yannopoulos, S. (2014). Historical development of sanitary sewers worldwide. *Sustainability*, (in press).

Diver S. (2000). Aquaponics. Integration of hydroponics with aquaculture, Horticulture system guide, National Center for Appropriate Technology (NCAT). Centre for Appropriate Technology. Appropriate Technology Transfer for rural areas, Fayetteville, Arizona, USA, p. 37.

Diver S. (2006). Aquaponics. Integration of Hydroponics with Aquaculture. Publication No. IP163 ATTRA, National Sustainable Agriculture Information Service. Fayetteville, Arizona, USA. http://www.extension.org/mediawiki/files/2/28/Hydroponics_with_Aquaculture.pdf (accessed December 2013)

Douglas S. (1975). Hydroponics. 5th edn, Bombay: Oxford University Press, Oxford, UK, pp. 1–3.

FAO (2010). The Weath of Waste: the Economics of Wastewater Use in Agriculture, FAO Water Report 35, J. Winpenny, S. Koo-Oshima. I. Heinz, M. Salgot, and J. Collado, Food and Agriculture Organization of the United Nations, Rome, ISBN 978-92-5-106578-5.

Feng R., Wei C., Tu S., Tang S. and Wu F. (2011). Detoxification of antimony by selenium and their interaction in paddy rice under hydroponic conditions. *Microchemical Journal*, **97**, 57–61.

Gericke W. F. (1940). The Complete Guide to Soilless Gardening. Published by Prentice Hall, Inc. http://www.archive.org/details/soillessgardenin031829mbp (accessed December 2013)

Guo X., Van Lersel M., Chen J., Bracket R. and Beachat L. (2002). Evidence of association of Salmonellae with tomato plants grown hydroponically in inoculated nutrient solution. *Applied and Environmental Microbiology*, **68**(7), 3639–3643.

Haddad M., Mizyed N. and Abdulah A. (2009). Performance of hydroponic system as decentralized wastewater treatment and reuse for rural communities. In: Proceedings of the 2nd International Conference on the Palestinian Environment, October 13–15, Published by An-Najah National University, pp. 91–99.

Haddad M., Mizyed N. and Masoud M. (2012). Evaluation of gradual hydroponic system for decentralized wastewater treatment and reuse in rural areas of palestine. *International Journal of Agricultural and Biological Engineering*, **5**(4), 47–53.

Hellenic Ministry of Agriculture (2007). Lavreotiki the Lavrion Museum. Archeological Receipts Fund, Publications Department. Lavrio, Hellas.

Hill R. D., Bendixen T. W. and Robeck G. G. (1964). Status of Land Treatment for Waste Functional Designs. Water Pollution Control Federation Meeting, Bal Harbour, Florida, USA.

Jiménez B. (2005). El Mezquital, Mexico: the Biggest Irrigation District that Uses Wastewater. In: Capítulo 12.1 en Water Reuse for Irrigation: Agriculture, Landscapes and Turf Grass, Lazarova y Bahri (eds), CRC Press, Florida, US, Estados Unidos, pp. 535–562, ISBN: 1-56670-649-1.

Jiménez B. and Asano, T. (2008). Water reclamation and reuse around the world. In: Water Reuse: An International Survey of Current Practice, Issues and Needs, B. Jiménez and T. Asano (eds), IWA Publishing, London, UK, Chap. 1.

Kenoyer J. M. (1998). Ancient Cities of the Indus Valley Civilization. Oxford University Press, University of Wisconsin – Madison, USA. Karachi.

Kenoyer J. M. (2006). Indus valley civilization. In: Encyclopedia of India. Stanley Wolpert (ed.), Charles Scribner's Sons, Detroit, **2**, pp. 258–267.

Lazarova V. and Bahri A. (eds) (2005). Water Reuse for Irrigation: Agriculture, Landscapes, and Turf Grass. CRC Press, Boca Raton, Florida, USA.

Leverenz H., Tchobanoglous G., Asano T. (2011). Direct potable reuse: a future imperative. *IWA Journal of Water Reuse and Desalination*, **1**(1), 2–10.

Lewis M., Yopp J., Schramm H. and Brandenburg A. (1978). The use of hydroponics to maintain quality of recirculating water in a fish culture system. *Transactions of the American Fisheries Society*, **107**, 92–99.

Matlock W. G., Davis P. R. and Roth R. L. (1972). Sewage effluent pollution of a groundwater aquifer. In: Proc. Amer. Soc. Agric. Engr., Chicago, Illinois, USA, December.

Metcalf and Eddy, Inc. (1991). Wastewater Engineering. Treatment, Disposal, Reuse. 3rd edn, McGraw-Hill Inc., New York, US.

May E., Bulter J., Ford M., Ashworth R., Williams J. and Baghat M. (1990). Comparison of chemical and microbiological processes in Gravel Bed Hydroponic (GBH) systems for sewerage treatment. In: Constructed Wetlands in Water Pollution Control, Cooper and Findlater (eds), Pergamon Press, UK.

Paranychianakis N. V., Salgot M. and Angelakis A. N. (2011). Irrigation with Recycled Water: Guidelines and Regulations. In: Treated Wastewater in Agriculture: Use and Impacts on the Soil Environments and Crops, G. Levy, P. Fine, and A. Bart-Tal, (eds), Wiley Knowledge for Generations, Hoboken, NJ. US and Oxford, UK, pp. 77–111, Ch. 3.

Pattenson G. (2010). A Brief History of Hydroponics. http://ezinearticles.com/?A-Brief-History-Of-Hydroponics&id=4673576. Listed: July 15, 2010 (accessed February 2013)

Pescod M. B. (1992). Wastewater treatment and use in agriculture, FAO Irrigation and Drainage Paper 47, Food and Agriculture Organization of the United Nations, Rome.

Pettygrove G. S. and Asano T. (eds) (1985). Irrigation with Reclaimed Municipal Wastewater: A Guidance Manual. Michigan (originally published under same title as Report No. 84-1 wr by California State Water Resources Control Board, Lewis Publishers, Sacramento, CA, USA.

Qadir M., Wichelns D. Raschid-Sally L., McCornick P. G., Drechsel P., Bahri A. and Minhas P. S. (2010). The challenges of wastewater irrigation in developing countries. *Agricultural Water Management*, **97**, 561–568.

Reed S. C. and Crites R. W. (1984). Handbook of Land Treatment Systems for Industrial and Municipal Wastes. Noyes Publications: Park Ridge, New Jersey, USA.

Reed S. C., Crites R. W. and Middlebrooks E. J. (1995). Natural Systems for Waste Management and Treatment. 2nd edn, McGraw Hill Co, New York, NY, USA.

Salgot M., Huertasa E., Weberb S. Dottb W. and Hollenderb J. (2006). Wastewater reuse and risk: definition of key objectives. *Desalination*, **187**, 29–40.

Schladweiler, J. C. (2002). Tracking down the roots of our sanitary sewers. In *Pipeline 2002: Beneath Our Feet: Challenge and Solutions*, Proceedings of the ASCE Pipeline Division of ASCE, Cleveland, OH, USA, 4–7 August 2002, G. E. Kurz (ed.), American Society of Civil Engineers, Reston, VA, USA, pp. 1–27.

Schlick G. and Bubenheim D. (1996). Quinoa: candidate crop for NASA's controlled ecological life support systems. In: Progress in New Crops, J. Janick, (ed.), ASHS Press, Arlington, VA, USA, pp. 632–640.

Schmidt K. D. (1972). Nitrate in ground water of the fresno-clovis metropolitan area, California. *Groundwater*, **10**(1), 50–64.

Scott C. A., Zarazúa J. A. and Levine G. (2000). Urban-Wastewater Reuse for Crop Production in the Water-Short Guanajuato River Basin, Mexico. Research Report International Water Management Institute, Mexico.

Shuval H. I., Adin, A., Fattal, B. Rawitz, E. and Yekutiel, P. (1986). Wastewater Irrigation in Developing Countries: Health Effects and Technical Solutions. Technical Paper No. 51. Washington, DC, World Bank.

Sparling G. P., Schipper L. A. and Russell J. M. (2001). Changes in soil properties after application of dairy factory effluent to New Zealand volcanic ash and pumice soils. *Australian Journal of Soil Research*, **39**(3), 505–518.

Stanbridge H. H. (1976). History of Sewage Treatment in Britain, Part 5 Land Treatment. The Institute of Water Pollution Control, Maidstone, Kent, England.

State of California (1978). Wastewater reclamation criteria. In: California Administrative Code Title 22, Division 4, Environmental Health, T. Asano (ed.), Dept. of Health Services, Sanitary Engineering Section (also cited in Wastewater Reclamation and Reuse Water Quality Management Library. California, USA, **10**, pp. 1477–1490.

Studymode (1999). Hydroponics Growing Without Soil. http://www.studymode.com/essays/Hydroponics-Growing-Without-Soil-10963.html. Listed Oct 1999 (accessed August 2012)

Thiyagarajan G., Umadevi R. and Ramesh K. (2007). Hydroponics. Science Tech Entrepreneur, Water Technology Centre, Tamil Nadu Agricultural University, Coimbatore, Tamil Nadu 641 003, India.

Thomas R. E. (1973). Land disposal II: an overview of treatment methods. *Journal Water Pollution Control Federation*, **45**(7), 1476–1484.

Tofflemire T. J. and Van Alstyne F. E. (1974). Literature Review. *Journal Water Pollution Control Federation*, **46**(6), 1201–1208.

Tsiknia M., Tzanakakis V. A. and Paranychianakis N. V. (2013). Insights on the role of vegetation on nitrogen cycling in effluent irrigated lands. *Applied Soil Ecology*, **64**, 104–111.

Tzanakakis V. E., Paranychianakis N. V. and Angelakis A. N. (2007). Soil as a wastewater treatment system: historical development. *Water Science and Technology Water Supply*, **7**(1), 67–76.

Tzanakakis V. A., Paranychianakis N. V., Londra P. A. and Angelakis, A. N. (2011). Effluent application to the land: changes in soil properties and treatment potential. *Ecological Engineering*, **37**, 1757–1764.

Tzanakakis V. A., Chatzakis M. K. and Angelakis A. N. (2012). Energetic environmental and economic assessment of three tree species and one herbaceous crop irrigated with primary treated sewage effluent. *Biomass and Bioenergy*, **47**, 115–124.

US EPA. (1979). A History of Land Application as a Treatment Alternative. U.S Environmental Protection Agency, EPA 430/9-79-012, Apr. 1979, Washington DC, USA.

US EPA (1980). Industrial Reuse and Recycle of Wastewaters: Literature Review. Environmental Protection Agency, EPA 600/2-80-183, September 1980, Ada OK, USA.

US EPA (1992). Guidelines for Water Reuse. Environmental Protection Agency, EPA-600/8-80-036, August 1980, Washington DC, USA.

US EPA (2004). Guidelines for Water Reuse. Environmental Protection Agency, EPA/625/R-04/108, September 2004, Washington DC and Cincinnati OH, USA.

US EPA. (2012). Guidelines for Water Reuse. Environmental Protection Agency, EPA 600/R-12/618, September 2012, Washington DC and Cincinnati OH, USA.

WHO (1973). Reuse of Effluents: Methods of Wastewater Treatment and Public Health Safeguards, Report of a WHO Meeting of Experts, Technical Report Series No. 517, WHO, Geneva.

WHO (1989). Health Guidelines for the Use of Wastewater in Agriculture and Aquaculture, Report of a WHO Scientific Group, Technical Report Series No. 778, WHO, Geneva.

WHO-FAO (2006). WHO Guidelines for the Safe Use of Wastewater, Excreta and Greywater. Vol. I: Policy and Regulatory Aspects. Vol. II: Wastewater Use in Agriculture. Vol. III: Watewater and Excreta Use in Aquaculture. Vol. IV: Excreta and Greywater Use in Agriculture, World Health Organization, Geneva.

Wade Miller G. (2006). Integrated concepts in water reuse: managing global water needs. *Desalination*, **187**, 65–75.

Walker C. and Lin H. S. (2008). Soil property changes after four decades of wastewater irrigation. *A Landscape Perspective*, **73**(1), 63–74.

Wikipedia (2009). http://en.wikipedia.org/wiki/Hydroponics (accessed on 12/05/2014).
Wolverton C., Barlow R. and McDonald R. (1976). Application of Vascular Aquatic Plants for Pollution Removal, Energy, and Food Production in a Biological System, Technical Report. N-75-22938; NASA-TM-X-72726 United States Thu Mar 24 09:13:34 EDT 2011NTISERA-02-005701; EDB-76-098781English.
World Bank (2010). Improving Wastewater Use in Agriculture. An Emerging Priority. Energy Transport and Water Department Water Anchor (ETWWA), June 30, Washington, DC, USA.

Chapter 25

Evolution and impacts of water and industrial wastewater management in Lavrion, Hellas

Kostas Komnitsas, Anthimos Xenidis, Nymphodora Papassiopi, Andreas Angelakis

25.1 INTRODUCTION

Man has always been directly depended on water as much for his drinking as for his irrigation needs and other uses (e.g., industrial and energy). The water sufficiency in a society relies on many factors such as the climatic type of a given area, the distance to the sea, the altitude, the geological structure and the hydrogeological setting of the area, the population, and so on (Mariolakos *et al.*, 2007).

One of the major problems of modern human society is the water shortage. There are many causes for this, but two of them are the most important, namely, the demographic problem and the climate change. Water shortage was also a major problem in parts of the ancient world, even though several of the previously mentioned factors such as density of population and climate change did not practically exist. It is known that ancient societies that had access to water resources and ruled the seas had developed civilization and wealth.

Lavrion is situated about 55 km SE of Athens in the hilly Lavreotiki peninsula which has an area of 200 km^2. Lavrion, Attica has been the centre of mining and processing activities for over 2700 years (Konofagos, 1980), involving the production of silver and lead from the exploitation of either carbonate or sulfide ores.

Lavreotiki peninsula was always characterized by water shortage but had several other beneficial characteristics such as short distance from the sea, altitude and geological structure that allowed the exploitation of ore reserves. All these characteristics allowed the inhabitants of the area to design brilliant technologies for the extraction of ores, the production of metals, the management of water and wastewater, the development of civilization and the generation of wealth.

The extended area of Lavrion bears carbonate [smisthonite (zinc carbonate) and cerussite (lead carbonate)] and mixed sulphide ore bodies (argentiferous galena, sphalerite and pyrite) developed along contacts of schist with marble. It is mentioned that zinc carbonate was called earlier calamine, from the Arabic word 'kalmeia' that was used for zinc ores; this term is no more used today.

Activities in the area commenced some 30 centuries BC with their peak to be during the golden age of the Athenian Republic in the 5th century BC, at the time of Pericles who was an Athenian general, politician and orator. Athenian democracy was established in 508 BC under Cleisthenes following the tyranny of Isagoras. Athens, which was a notable polis (city-state) of Attica region, is widely referred as the cradle of western civilization and the birthplace of democracy. It is underlined that from the wealth accumulated in Athens from silver production and exports, Pericles ordered the construction of a fleet consisting of 200 triremes and won the famous sea battle of Salamis.

It has been estimated that in the antiquity 13 million tons of cerussite, containing 20% lead and 400 g/ton silver, were mined and over 3500 tons of silver and 1,400,000 tons of lead were produced; 70% of this volume was produced during the *ca.* 5th and 4th centuries BC. The ratio of the mined cerussite:galena in this period was 9:1. The wastes of that period, that were used as feed in modern times, are estimated at 1 million tons of low-grade ore, 9 million tons of washery waste, 10,000,000 tons of low grade silver ores containing 7% lead and 140 g/ton silver as well as 1,500,000 million tons of slag containing 10% lead and 50 g/ton silver.

On the other hand, during the modern era mining and smelting activities, the following quantities have been produced: (a) in the period 1865–1873 from the hydro-mechanical beneficiation of ancient slags and low grade silver ores (850,000 tons containing 7% lead) 60,000 tons of lead containing silver, (b) in the period 1873–1977 the Hellenic company treated ancient slags and low grade silver ores (4,500,000 tons containing 7% Pb) and produced 310,000 tons of lead concentrate containing silver, and (c) the period 1873–1977 the French company treated 1,800,000 tons of carbonate ore containing 45% zinc and 2% lead and produced 810,000 tons of zinc and 36,000 tons of lead carbonate; it also treated 3,200,000 tons of oxide ores containing 11% lead, 2% zinc and 140 g/ton silver and produced 64,000 tons of zinc concentrate, 352,000 tons of metallic lead and 450 tons of silver; finally, it treated 3,300,000 tons of mixed sulphide ores containing 6.15% lead, 7.15% zinc and 140 g/ton silver and produced 236,000 tons of metallic zinc, 203,000 tons of metallic lead and 460 tons of silver (Konofagos, 1980; Kontopoulos *et al.*, 1995). This data show the immense mining and metallurgical activities carried out in Lavrion for almost 30 centuries. Silver mined in Lavrion contributed greatly to the prosperity of Athens. The famous silver coins (Figure 25.1) of the golden era, the Athenian Owl, were produced from silver mined in Lavrion.

Figure 25.1 The Athenian Owl (silver coin).

Mining activities in the area ceased when Romans invaded Hellas after the battle of Corinth in 146 BC. Finally, activities commenced again several centuries later, in 1865, soon after the Hellenic independence from the Turks, and ended in 1982 where in the last five years only imported ores were treated. It is worth mentioning that in 1865 about 1200 workers, a considerable number for that period, were employed in the wider Lavrion mining area.

The present chapter aims to present the evolution of water and wastewater management techniques in Lavrion during all these centuries and briefly outline the respective benefits and environmental impacts.

25.2 ANCIENT ACTIVITIES

The treatment of the ore in the ancient period aimed at the production of silver and lead. Ancient shafts and galleries had a depth of up to 120 m. Silver was extracted from underground mainly from galena and cerussite ores and was used apart from the production of silver coins also for the production of jewelries. Lead had a stable consumption and was mainly used in construction applications as well as for the production of joints for rocks and pillars, weights for looms used for weaving of fabrics, for the production of stamps and dies, for the repair of cracked jugs and in medicines. In specific periods, as for example between the 5th and the 3rd century BC, lead production and price reached peaks. It is believed though that most lead in ancient times was produced in the form of litharge. This form allowed its re-melting several centuries later (Papadimitriou, 2008). Regarding zinc, not much information is available about its potential uses at that period. It is known though that Romans had developed a technology to produce brass (alloy of copper and zinc).

Mines were owned by the city of Athens which rent them to free Athenian citizens or settlers 'μέτοικους' from other cities. Ore processing plants (laboratories) and furnaces were owned by businessmen. Mines and laboratories were usually named after a god or a hero. Both free citizens and slaves worked in the mines where conditions were usually very difficult. It is known that Athenian laws allowed basic rights to slaves as deduced from findings in archeological excavations and inscriptions dedicated to Gods which were found in cemeteries. There were several periods though when uprisings of the slaves took place, such as the one during the Peloponnese War (413 BC) when slaves took the side of Spartans when they occupied Attica as well as at the end of the 2nd century BC when slaves occupied the fortress of Lavrion.

Ancient furnaces used for the treatment of the ores had a diameter of 70–100 cm and an operating capacity of 5 tons of ore per day. It is estimated that 200 kg of solid fuel (mainly charcoal) were required per ton of ore. The cost of treatment and the selling price of the final metal are very well analyzed in a recent publication (Papadimitriou, 2008).

There is evidence of industrial water and wastewater reuse in Hellas during the Classical and Hellenistic periods, dating from the *ca.* 5th century BC. Archeological excavations in the area of Lavrion have revealed several mining installations (e.g., extraction galleries, mine shafts, washeries, and cisterns) where is precisely demonstrated the way of extraction and processing of ores as well as the operation of a washery system which constitutes a characteristic example of industrial reuse of water in ancient times (Figure 25.2) (Salliora-Oikonomakou, 2007). These washeries consisted of a network of cisterns and pipes which allowed washing of the ore and efficient reuse of water.

Figure 25.2 Ore washeries (reconstructed) at Thorikos in Lavreotiki.

Beneficiation in the washeries was taking place after ore grinding to a particle size less than 1 mm so that separation of rich and poor particles could be done with the action of flowing water. Heavy and rich in silver lead particles would settle in a short distance while lighter and poor in metal content particles were washed away. Most ancient washeries were located next to galleries so that transport of the ore was easy. Both horizontal and helicoid types of washeries were used in Lavrion. Horizontal washeries had a main water feed tank constructed of marble. It should be mentioned that Hellenes had developed brilliant quarrying techniques to extract marble blocks as well as to produce various marble objects at those times (Papadogeorgakis *et al.*, 1992). Water was flowing out of this tank through holes allowing high water flow velocity and thus transfering the light particles of the ore in the settling pond and leaving the heavy particles behind. Solid particles in wastewater were settled in four settling ponds located at the corners of rectangular washeries and finally clean wastewater was returned to the water feed tank manually (Figure 25.2).

Most cisterns found in the area had a diameter of 5–10 m and depth of 3–4 m. They were ingenious waterproof constructions using an impermeable layer of mortar and lead oxide (litharge). It has to be underlined that these constructions remain until today waterproof, indicating thus the excellent engineering properties of those layers and the level of technology developed in Lavrion several centuries BC. Furthermore, it is mentioned that several cisterns were also constructed to collect rain water used for drinking. More than 200 cisterns can be found today close to mining areas in Lavreotiki peninsula.

Along with the horizontal washeries, engineers developed the helicoid (from the word helix) ones, as shown from excavations in four different places in Lavrion (Salliora-Oikonomakou, 2007). In these washeries the water from the cistern, initially channeled into the washery tank and then through a hole on the side of the water tank met the unwashed ore in the beginning of the circular washery. As the water/ore mixture was transferred into the circular channel the heavy lead ore drop into the depressions while the low-grade ore and the gangue were settled to a primary sedimentation tank at the end of the channel. After passing through a secondary sedimentation tank, water returned back into the washery tank for reuse (Figure 25.3).

Figure 25.3 Helicoid washeries (*ca.* 4th century BC) made from marble with cavities in which the ore was placed, at Dimulaki Lavrion.

To address the lack of water in the area of Lavrion, engineers built workshops at certain locations, such as sides of hills or in valleys, in order to enable the collection of rainwater and its channeling into cisterns (Salliora-Oikonomakou, 2007). A large number of such cisterns, with a storage capacity varying from 300 to 1000 m^3, are preserved throughout Lavrion (Figure 25.4a). Usually, close to these cisterns there was a smaller one called a 'pre-cistern' (Figure 25.4b), used as a settlement or sedimentation tank and water was channeled by pipes into the large cistern. Various engineering projects were also conducted, including the building of dams on streams, waterproofing of natural depressions, or the facing of winter torrents to secure the supply of water. The above structures, shown at various places of Lavrion, provide evidence of the efforts made by the miners to discover the best method for cleaning and up-grading the ore and efficiently reusing water.

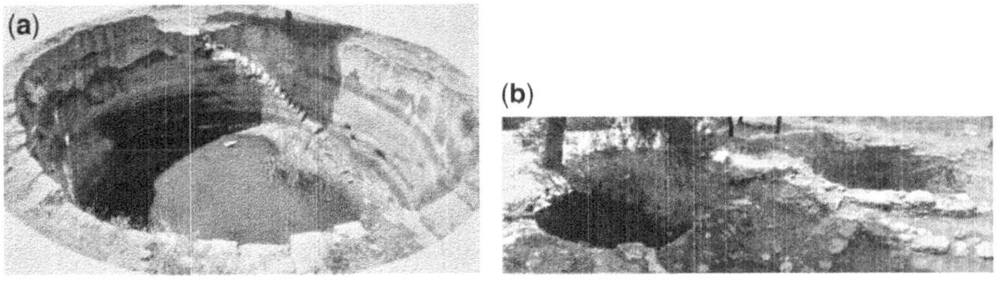

Figure 25.4 Water cisterns: (a) Cistern of collecting and storing rain water and (b) a pre-cistern located nearby the main cistern, used as settlement or sedimentation tank.

Most excavations in the wider area of Lavrion were carried out primarily by the Belgian Archeological School while some other archeological societies were also involved in certain cases. All washeries and water cisterns are made of marble and are considered unique since most other similar constructions in this period were made of wood.

Environmental impacts during the ancient times are considered negligible. Very limited production of acid mine drainage (AMD) was anticipated mainly due to the use of hydromechanical beneficiation techniques and the lack of chemicals. If some acid mine drainage was generated it is assumed that it was pretty quickly neutralized by the abundance of limestone in the area. No indication of AMD generation in the ancient times was detected in modern excavations.

In conclusion, Hellenistic societies seem to have achieved a sound ecological balance between man and environment through the development of innovative water and industrial wastewater management systems.

25.3 MODERN ACTIVITIES AND RELATED ENVIRONMENTAL IMPACTS

The metallurgical activities in the modern times commenced in 1865 and involved hydro-mechanical beneficiation of ores and ancient wastes, as well as roasting of calamine and reduction smelting of galena in kilns. It is underlined that most feed in that period consisted of ancient slags and poor ores since the technology had improved substantially over the centuries. These raw materials were scattered in the entire area of Lavreotiki peninsula and could be also found in shafts, galleries and caves. It is important to mention that in 1869 a tunnel for the first railroad in modern Hellas was constructed and used for the transfer of the ores and products to the harbor of Lavrion.

In 1863 an Italian mining engineer, Giovanni Battista Serpieri, founded the Italian-German company Roux-Serpieri-Eressynet C.A. In 1873 the Hellenic metallurgical company was founded while in 1875, Seprieri established the French company (Compagnie Francaise). Miners at the area of Plaka, Lavrion, outside gallery 65, in 1898 are shown in Figure 25.5 (Manthos, 1990).

Figure 25.5 Miners outside gallery No 65, Plaka area, Lavrion, year 1898.

The first installations in Lavrion are considered today as monuments of architectural and industrial design and National Technical University of Athens, which acquired the premises of the metallurgical plant in the '90s after a donation of the Hellenic Ministry of Culture, has carried out extensive restoration works so that many facilities were renovated and can be visited today (NTUA, 1997) (Figures 25.6.)

Figure 25.6 Industrial buildings of the modern era within the Lavrion metallurgical plant.

Most of the first activities in the modern times involved hydro-mechanical and pyrometallurgical routes. In this period most environmental problems were associated with the flue gases emitted of the kilns (containing also arsenic) and the disposal of solid wastes (mainly slags from the kilns and wastes from the hydro-mechanical beneficiation).

The situation changed after 1930 when, due to difficulties in finding new rich reserves, flotation was introduced as ore beneficiation process (Figure 25.7).

Flotation, which was invented in the beginning of the 20th century, is a water intensive process, requires chemicals and is very efficient in upgrading mixed sulphide ores to produce rich concentrates. Besides, losses of valuable components are low in the waste, which is called tailing. The main environmental concern of flotation is the generation of watery tailings which today are disposed of in well protected and isolated by impermeable geomembrane areas, called tailing dams. These constructions are build in a way that after disposal of the tailings, water flows to the center of the dam and fine particles are self-compacted. Water can then be pumped again back to the flotation unit or utilized in other industrial units. Furthermore, a trench is often constructed around the base of the dam which enables collection and treatment of leachates, often acidic if tailings are sulphidic as in the case of treatment of sulphide ores, minimizing thus environmental problems. When flotation was first introduced in Lavrion, tailings were disposed of in front of the plant and in the coastline generating thus an artificial beach, which is what we see today at Thoricos bay. After 1960 a tailings dam consisting of two ponds was constructed inside the metallurgical plant site for the disposal of the tailings, eliminating to a certain extent the environmental impacts.

Figure 25.7 The old flotation building.

Thus, modern mining and metallurgical activities in Lavrion were characterized by rather poor water and waste management practices. As a result, dozens of smaller and larger dumps of solid wastes were scattered in and around the metallurgical plant, the harbor and other parts of the city of Lavrion (Figure 25.8). These wastes consist of carbonaceous wastes, sulphide wastes and metallurgical slags.

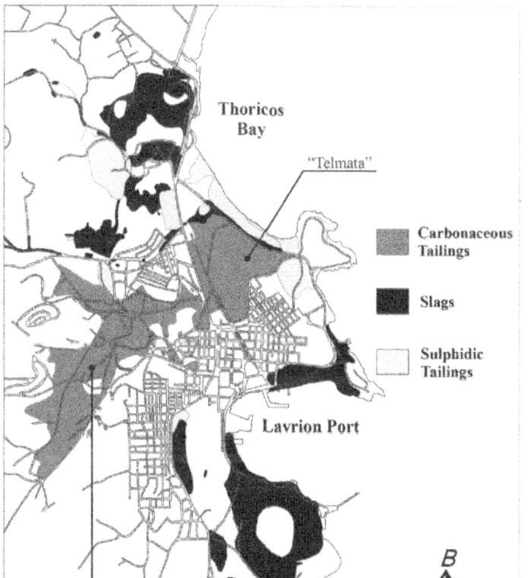

Figure 25.8 Locations in which several types of solid wastes are scattered in the wider Lavrion area. (Xenidis et al., 2003)

The Laboratory of Metallurgy of National Technical University of Athens in the frame of European Commission funded projects (Life, Contract No. 93/GR/A14/GR/4567 and Environment, Contract No. EV5V-CT93-0248) carried out a complete characterization of all tailings using extensive sampling, geotechnical, environmental and toxicity tests. It also assessed their potential to produce acidic and other hazardous leachates and the quality of ground water by sampling water from piezometers installed in several drillholes drilled in waste disposal sites.

The sulphidic tailings were classified as toxic and hazardous by considering the Toxicity Characteristics Leaching Procedure (TCLP) test, the bioavailable fraction of the toxic metals, the pore water quality and the acid generation capacity (Table 25.1) (Komnitsas *et al.*, 1995; Kontopoulos *et al.*, 1995, 1996a, b). It has been mentioned that a conceptual environmental rehabilitation strategy of the area must have as first priority to render these sources of pollution inactive through transportation of all wastes to a specific area and application of a dry vegetative cover. It is mentioned that sulphidic tailings are oxidized by the action of air and water producing acidic leachates that contaminated soils and waters.

Table 25.1 Environmental characterization of sulphidic tailings.

Quantity: 600,000 tons				Average wet bulk density: 1.62 g/cm^3			
Average Density: 2.7 g/cm^3				Average Permeability: 4×10^{-5} cm/s			
Chemical analysis							
Fe, %	S, %	Pb, %	Zn, %	Cd, mg/kg	As, %	Ca, %	Al, %
5–17	5–10	0.2–1ton	0.5–2.9	40–100	0.2–0.6	13–15	6–10
Bioavailable fraction (EDTA leachable, mg/L)				*Sequential leaching (exchangeable + carbonate), mg/L*			
Pb	Zn	Cd	As	Pb	Zn	Cd	As
124	10525	45	<20	396	12526	51	<20
Pore water analysis, mg/L (pH = 2.5, Eh = 520 mV)							
Fe	SO_4	Pb	Zn	Cd	Mg	Ca	
300	15500	1	5000	20	1000	500	
EPA TCLP test, solubility mg/L							
Pb			As			Cd	
5–25			0.1–2			0.1–5	
NNP, kg $CaCO_3$/ton							
		Surface: −200				Lower levels: 0 to +600	
Mineralogical analysis (%): FeS_2: 16, FeAsS: 0.7, PbS: 2, $CaCO_3$: 8, $FeCO_3$: 20, Insolubles: 32.5							

The complete environmental characterization of typical carbonaceous tailings is given in Table 25.2. The risk assessment study of these tailings, disposed of at an area of 50 ha, resulted in the identification of three high risk exposure routes for humans, namely direct contact, ingestion and inhalation of contaminated material. Quantification of the human health risk indicated that direct ingestion of contaminated particles is the most important exposure route for the intake of contaminants. Especially for children living in the vicinity of the tailing dumps, the intake of Pb exceeds the maximum tolerable risk (MTR) by a factor of more than 100. The maximum tolerable risk values for As and Cd are also exceeded by factors ranging between 10 and 20 for As and 2 for Cd. The direct ingestion of contaminated particles by adults was found to be the most important intake pathway, although less intense compared to children. Intake from all other routes, that is, inhalation of fugitive dusts, dermal contact and ingestion of home grown produce, were found to be lower than the corresponding MTR values (Xenidis *et al.* 2003).

Table 25.2 Environmental characterization of carbonaceous spoils

Quantity: 5,000,000 tons (depth = 8.1 m)			Particle size distribution: 17.9 % < 63 μm	
Dry bulk density: 1.8 g/cm^3			Permeability: 2.4 × 10^{-3} cm/s	
Potential for acid generation				
Ca, %	S, %		NNP, kg $CaCO_3$/ton	NPR
11.5–12.7	0.2		300 (>0)	48 (>4[a])
Total content of contaminants, mg/kg of solids				
	Pb	Zn	Cd	As
	14,700 – 24,300 (530[b])	12,000–26,000 (720[b])	100–170 (12[b])	1,800–3000 (55[b])
Distribution of contaminants in the 5-stage SEP fractions, mg/kg of solid				
Fractions	Pb	Zn	Cd	
Exchangeable	<4	12	7.2	
Carbonate	6612 (530[b])	6028 (720[b])	49.2 (12[b])	
Reducible	14,200	15,290	102.0	
Oxidisable	375	615	<0.8	
Residual	7000	5749	10.5	
EDTA-leachability of contaminants, mg/kg of solids				
	Pb	Zn	Cd	As
	6152 (530[b])	5070 (720[b])	26.4 (12[b])	30 (55[b])
EPA-TCLP test, mg/L				
	Pb	Zn	Cd	As
	87.9 (5[c])	207.3	2.48 (1[c])	<2 (5[c])
Pore water quality, mg/L				
pH	Pb	Zn	Cd	As
7.0 (6–9[d])	<0.5 (0.1[d])	0.5 (1.0[d])	0.1 (0.1[d])	<2 (0.5[d])
Phyto-accumulation of contaminants, mg/(wet kg of fruit)				
	Pb	Zn	Cd	As
Grapes	0.22 (0.1[e])	0.87 (5[e])	0.06 (0.03[e])	0.07 (0.2[e])
Olives	1.13 (0.1[e])	11.7 (5[e])	0.57 (0.03[e])	0.42 (0.2[e])

Existing limits and guidelines given in parentheses:
[a] Price *et al.* (1997).
[b] Dutch soil standards (1994).
[c] US EPA (1986).
[d] Effluents quality standards (1979).
[e] European Commission (1995).

In general, slags disposed of in the harbor of Lavrion and other nearby bays exhibit low leaching potential for lead and zinc since their average size is quite large (50–150 mm). Most leachable metals originate from the surface of the slag pieces which is later passivated (Kontopoulos *et al.*, 1996c).

Based on these studies feasible and cost effective pilot scale rehabilitation actions were proposed and implemented. These rehabilitation actions were actually the first ones implemented in Hellas for such mining and metallurgical wastes (Xenidis *et al.,* 1997; Kontopoulos *et al.,* 1998).

The tailings dam was rehabilitated by National Technical University of Athens using an innovative and low cost technique involving modification of the acid generation potential of the tailings and application of a soil cover for future development of a vegetative cover (Figure 25.9) (Komnitsas *et al.,* 2000). Several years after the construction of the cover the pore water quality of the rehabilitated tailings, as deduced from piezometer sampling, was excellent and did not pose any environmental risk. The main objective of the rehabilitation works was to prevent the generation of leachates or improve their quality and thus minimize contamination of groundwater, soils and the sea.

A similar rehabilitation scheme has been proposed for the remediation of sulphidic tailings at the coastal zone of Thoricos bay and reclamation works are currently in progress (Adam *et al.,* 2006).

Figure 25.9 View of the ongoing works for the rehabilitation of the tailings dam in Lavrion (up) and amendment of tailings with ground limestone (down).

25.4 CONCLUSIONS

It is known that Hellenes had always 'good relationships with water' and lots of characters of Hellenic Mythology are related to water. The present chapter confirms that ancient Hellenes had developed ingenious water and wastewater management techniques in the area of Lavrion for the processing of ores and the production of lead and silver.

There is evidence of industrial water and wastewater reuse in Hellas since the Classical and Hellenistic periods, dating from the *ca.* 5th century BC. Archeological excavations in the area of Lavrion have revealed several mining installations as well as the operation of a washery system which constitutes a characteristic example of industrial reuse of water in ancient times. These washeries, of horizontal and helicoid type, consisted of a network of cisterns and pipes which allowed washing of the ore and efficient reuse of water. It is important to mention, that all these washeries were made of marble and not wood, as in the rest of the world at those times, since ancient Hellenes had also developed brilliant quarrying techniques to extract marble. To combat the lack of water in Lavrion, engineers built workshops at certain locations, such as sides of hills or in valleys, in order to enable the collection of rainwater and its channeling into cisterns. Waterproofing of these cisterns was also excellent and remains intact until today.

Due to all these activities contamination in the extended area of Lavrion in ancient times is considered negligible. On the other hand processing of the ores in modern times using chemical and water intensive techniques, such as flotation, resulted in improper water management and tailings disposal and thus in much more intense environmental problems.

25.5 REFERENCES

Adam K., Drettas C., Panagopoulos, I., Voudouris N., Kourtis A., Tsakarisianos A. and Chatzigianneli C. (2006). Application of project management principles for the reclamation of old mines. The case of Thoricos Bay in Lavrion, Greece. In Proccedings of the 3rd International Conference 'Protection and Restoration of the Environment VIII', July 3–7 2006, Chania, Hellas.

Dutch Soil Standards (1994). Intervention Values and Target Values. Ministry of Housing, Spatial Planning and Environment, Athens, Hellas.

Effluents Quality Standards (1979). *Journal of the Greek Government, no 1132, 21/12/1979.* Athens, Hellas.

European Commission (1995). Compilation of Limits in Food in the Legislation of the Member States (including EEA/EFTA members). Version 3, June 1995, EC, DG III. Brussels, Belgium.

Komnitsas K., Xenidis A. and Adam K. (1995). Oxidation of pyrite and arsenopyrite in sulphidic spoils in Lavrion. *Miner. Eng.*, **8**, 1443–1454.

Komnitsas K., Xenidis A. and Tabouris S. (2000). Composite cover for the prevention of acid mine drainage. *Mining Environment Management*, **8**, 14–17.

Konofagos K. (1980). The Ancient Lavrion and the Greek Technology for Silver Production. Ekdotiki Ellados Publishing, Athens, Hellas.

Kontopoulos A., Komnitsas K., Xenidis A. and Papassiopi N. (1995). Environmental characterisation of the sulfidic tailings in Lavrion. *Miner. Eng.*, **8**, 1209–1219.

Kontopoulos A., Papassiopi N., Komnitsas K. and Xenidis A. (1996a). Environmental characterisation and remediation of the tailings and soils in Lavrion. In Conference Proceedings – Protection and Restoration of the Environment III, August 28–30, Chania, Hellas, pp. 484–493.

Kontopoulos A., Xenidis A., Komnitsas K. and Papassiopi N. (1996b). Environmental characterisation of wastes in Lavrion. In Proceedings of Fourth International Symposium on Environmental Issues and Waste Management in Energy and Minerals Production, R. Ciccu (ed.), October 7–11, Calgiari, Italy, pp. 209–216.

Kontopoulos A., Komnitsas K. and Xenidis A. (1996c). Environmental Characterisation of the lead smelter slags in Lavrion. In Proceedings of the IMM Minerals Metals and the Environment II Conference, September 3–6, Prague, pp. 405–419.

Kontopoulos A., Komnitsas K. and Xenidis A. (1998). Pollution, risk assessment and rehabilitation at the Lavrion Technological and Cultural Park, Greece. In SWEMP '98 Conference Proceedings on Environmental Issues and Waste Management in Energy and Mineral Production, A. G. Pasamehmetoglu and A. Ozgenoglu (eds), May 18–20, Ankara, pp. 105–110.

Manthos G. K. (1990). Photo Album, Mining and Metallurgical Lavrion. Municipality of Lavriotiki, Lavrion, Hellas.

Mariolakos I., Kranioti A., Markatselis E. and Papageorgiou M. (2007). Water, mythology and environmental education, *Desalination*, **213**, 141–146.

NTUA (National Technical University of Athens) (1997). Lavrion Technological and Cultural Park. NTUA Publishing, Athens, Hellas, ISBN: 960-254-526-7.

Papadimitriou G. (2008). The Production of Lead in Ancient Lavrion during Classical and Roman Times. Society of Studies of South East Attica, www.emena.gr

Papageorgakis J., Mourtzas N. and Orfanoudaki A. (1992). Bronze age quarries on the eastern coastal zone of Greece, in Ancient Stones. In: Quarrying, Trade and Provenance, M. Waelkens, N. Herz and L. Moens (eds), University Leuven Press, Leuven, Belgium, pp. 21–28.

Price W. A., Morin K. and Hutt N. (1997). Guidelines for the prediction of acid rock drainage and metal leaching for mines in British Columbia: Part II – Recommended procedures for static and kinetic testing. In Proceedings 4th International Conference on Acid Rock Drainage, Vancouver, BC, USA, pp. 15–30.

Salliora-Oikonomakou M. (2007). Lavreotiki the Lavrion Museum. Hellenic Ministry of Culture, Archeological Receipts Fund, Publications Department, Panepistimiou 57, 10564 Athens, Hellas.

U.S. EPA (1986). Toxicity characteristic leaching procedure. Appendix 1, Federal Register 51(216).

Xenidis A., Komnitsas K., Papassiopi N. and Kontopoulos A. (1997). Environmental implications of the mining activities in Lavrion. In Engineering Geology and the Environment Conference Proceedings, P. G. Marinos, G. C. Koukis, G. C. Tsiambaos and G. C. Stournaras (eds), June 23–27, Athens, Hellas, pp. 2575–2580.

Xenidis A., Papassiopi N. and Komnitsas K. (2003). Carbonate rich mine tailings in Lavrion: Risk assessment and proposed rehabilitation schemes. *Advance Environment Research*, **7**, 207–222.

Chapter 26

The Water Court of Valencia, Spain (Wastewater)

F. Hernández-Sancho and M. Molinos-Senante

26.1 ORIGEN

The world's oldest Court, the Water Court of Valencia, Spain, has overseen the fair distribution by canal of the waters of the River Turia for more than a millennium (Giner Boira, 1988). In antiquity, the canals also carried urban wastewater from the city of Valencia for reuse in the fields surrounding the city (Huerta de Valencia), while today, only river water from reservoirs is carried.

Several investigations (Almerich & Jarque, 2002; Sala Giner, 1997; Fairén Guillén, 1998) have indicated that the Court was created around 960 by the Arabs. The conquest of what would become the Kingdom of Valencia in 1238 by King Jaume I was followed by the proclamation of a law code (Fuero XXXV) that marked the official founding of the Water Court (or Tribunal de las Aguas de Valencia in Spanish). By means of the law code, King Jaume I ordered that the Arab tradition of water management (fresh and waste waters) should be continued – as indeed it is today.

There are several aspects of the Water Court that reveal its Arab origins and character. Firstly, the Court meets at the door of the cathedral as Muslims could not enter a Christian church. Another detail is that King Jaume eventually allowed the Court to meet on Thursdays, which is the equivalent of Saturday for Muslims. Finally, another reflection of the Arab character was that the president of the Court used to grant the right to speak by pointing with his foot rather than his hand (not so today) (Fairén Guillén, 1998).

26.2 THE SCOPE OF THE WATER COURT: HUERTA DE VALENCIA

The water Court responds to the circumstances of the ownership of the fields and orchards that surround the city of Valencia called *Huerta de Valencia* (Figure 26.1). These fields are divided into smallholdings, and as they all receive a great deal of sunshine they all need as much as possible of the little water that is available. Annual rainfall is just 400–500 mm and most summers are long and dry. Since ancient times, wastewater has been reused.

The Arabs were significantly responsible in the formation of the *Huerta de Valencia*, as well as the Romans since in the second century they dried and put into growing wetlands and swampy areas adjacent to the city of Valencia. The arrival of the Arabs involved a new social organization and the increase of agricultural land, improving and extending the network of ditches. New techniques of water collection and exploitation were incorporated; especially they reorganized the distribution and management of the water in a fair and proportionate way. The production of the *Huerta* in this first phase was exclusively for self-consumption and the Valencia's market.

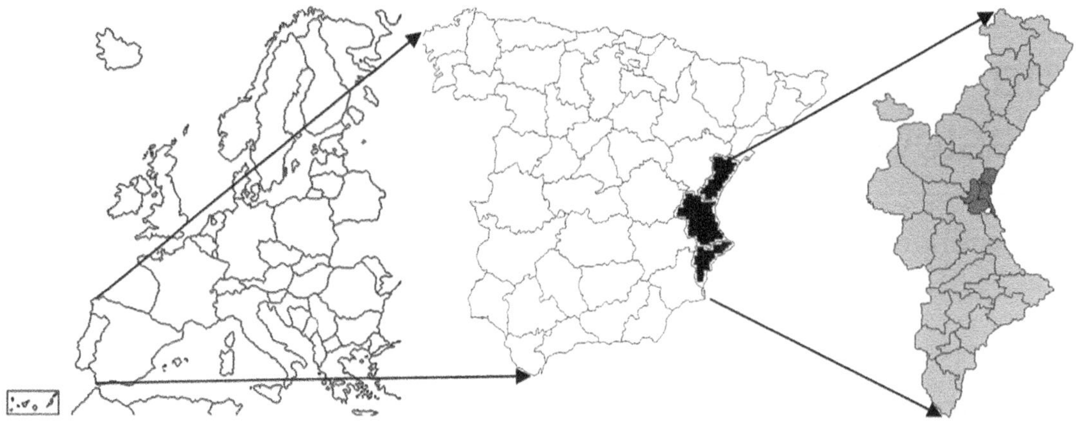

Figure 26.1 Localisation of the Huerta de Valencia.

The *Huerta de Valencia* is the result of the superposition of different cultures that have left their mark as time passed and created new needs. As well as the land cultivated was spreading due to the expansion of irrigation, the products grown also became more diversified. Thus, winter products were sprouts, artichokes, beets, beans, beans, potatoes and onions. Summer crops were tomatoes, beans, potatoes, onions, garlic, corn, lettuce and melon. Yellow nutsedge is one of the main products of the North *Huerta*, from which is extracted the 'horchata', a typical Valencian drink. In particular, between the fifteenth and seventeenth centuries the mulberry was especially important linked to the silk industry. The rise of the population of the city of Valencia involved a significant development of the *Huerta* being the wheat the main product, followed by maize and rice, beans and hemp. In the late of the nineteenth century, the improvement of the transport facilitated the entry of wheat from other regions of Spain. Its price was lower, so this traditional crop lost progressively importance until its demise. Nowadays, onion is the main crop and perhaps the most appreciated and exported from the *Huerta of Valencia* (Sala Giner *et al.*, 2002).

The river Turia is the main responsible of the creation of the *Huerta of Valencia.* The fields and orchards around the city are crossed by a series of canals that distribute water. There are eight main canals: Quart, Benager and Faitanar, Mislata, Favara and Robella take water from the right bank of the River Turia; while Tormos, Mestalla and Rascanya draw water from the left bank. The irrigated area is around 13,000 ha. The radius of this area is 15 km from the city of Valencia and from the main irrigation channels, a dense network of secondary channel depart to bring water to the last corner of each plot of land (Figure 26.2). The topography has been the determining factor: the ground, although it is a great plain, is slightly inclined towards the sea. Thus, from the mountains that delimit the *Huerta* by the West, ditches were built approximately perpendicular to the River, in direction North-South and South-North. The operation of the system is also completed with the collection, by the ditches located further down, of the excess of waters and wastewater that were again distribute. The Robella ditch through the city of Valencia was used to irrigate the gardens of the city and received some of the wastewater from Valencia (Almerich & Jarque, 2002).

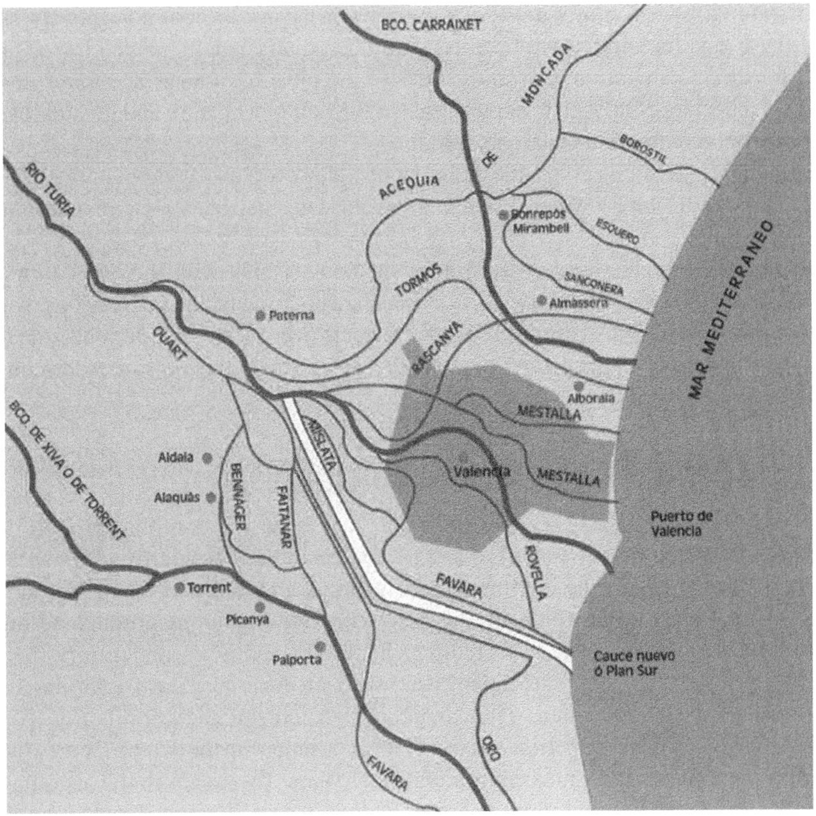

Figure 26.2 System of canals of ditches of the Huerta of Valencia.
Source: Almerich and Jarque (2002).

In the artificial range forming by the main eight canals, two ways are given in terms of its layout: (a) ditches parallel to the River in order to increase its influence area creating an alternate flow that gradually is separating from the mainstream. This is the case of the Mestalla irrigation channel (North) and Robella (South); (b) ditches leading away from the river Turia almost in perpendicular direction. This is the configuration of the following ditches: Rascanya and Tormos (North) and Favara, Faitanar, Mislata and Benager (South). Between them there is a dense network of minor channels that connect all ditches allowing water transfers and gridding the entire *Huerta*.

In relation to property and land size, in the fourteenth century farmland were expanded by the transformation of swampy areas and recovery of lost parcels, tilling new fields in low-lying areas near to the sea. It is estimated that at this time the *Huerta* was increased by 20%. Between the fifteenth and sixteenth centuries, the land belonged mainly to the Church and the nobility. In the eighteenth century, nearly half of the land remained property to the Church and the nobility, but in just half a century, the picture changed radically since the importance of the urban bourgeoisie increased very significantly, thanks in part to several confiscations. Thus, in the early nineteenth the *Huerta de Valencia* is owned of poor farmers and a new bourgeoisie that has enough purchasing power to acquire land which is worked by small tenants. In the mid-twentieth century, the Huerta of Valencia was a true mosaic of smallholdings whose fragmentation will reach minimum levels in terms of its size affecting negatively its profitability. The reasons of these smallholdings have been basically hereditary partitions and the acquisition by the tenants of plots for direct cultivation. Currently, 80% of the plots have a surface lower to 4 'hanegadas' (less than a third of a hectare). Only 7.5% of plots exceed the half of a hectare and greater than half-hectare plots are a rare exception (Almerich & Jarque, 2002).

All of the farmers are joint owners of the water rights. However, everyone's entitlement to water is in proportion to the amount of land owned. Farmland is sold with the irrigation rights and so the water is effectively tied to the land.

In formal terms, the farmers share ownership of the water rights with the other farmers who draw water from the same main canal or *acequia* – in Spanish. This organisational structure has proved very efficient as the aim is to distribute a limited supply of water over the widest possible area of land. Each acequia-community of farmers is governed by a set of ancient rules that were transmitted orally before being eventually written down.

26.3 MANAGING THE FARMER COMMUNITIES

A supervisory board is nominated every two or three years to ensure compliance with the ordinances governing each acequia-community of farmers. The chairman of this board is known as the *Síndico* –in Spanish and is elected by all the community members. Candidates for chairmen must meet three conditions: firstly, they have to work the land themselves (thereby excluding absent landlords); secondly, candidates have to be landowners (thereby excluding tenants and labourers); and thirdly, their landholding must be sufficiently large that they can make a living from farming. Each acequia-community has specific rules for electing its chairmen. The Síndico (chairman) of the community assumes various executive powers and one of his functions is to serve as a judge in the water Court. Individuals can only serve as chairmen/judges for a maximum number of years: two years in the acequia-communities of Mestalla, Rovella and Favara; and three years in Rascanya and Tormos. Re-election is possible but subject to limitations – such as having previously published the accounts and not having any lawsuits pending with the community.

The supervisory board of each acequia-community consists of a group of elected members (*vocales electos* – in Spanish). To be an elected a member it is necessary to own land with water-rights from one of a series of sections into which all the canals are divided. The board members are at the service of the community during their mandates.

Another important figure contributing to the management of the water from the canals is the warden (*guardas* – in Spanish). These are employees of the community whose main role is to ensure that water reaches all the famers at scheduled times during the week. The wardens follow orders given by the board chairman and inform the board of any infractions committed so that offenders can be subsequently called before the water Court. Each acequia-community had its own method of nominating wardens; but all wardens are required to work full-time and cannot be landowners in the community they serve.

The overseers (*veedores* – in Spanish) are employees of each acequia-community and have various procedural and administrative functions. The qualifications to be appointed overseer vary in each community: some require candidates to be landowners while others only require candidates to be farmers.

Overseers are nominated for a limited period and the number in each community reflects the amount of land irrigated by the community: the ordinances of Favara specify ten overseers; while Tormos and Mislata only specify four. The most important role of overseers is to advise the Water Court.

Various monitors (*atandadores* – in Spanish) ensure that each farmer takes a fair share of water and that enough flow remains to supply neighbouring farmers. This implies that no farmer can take water without permission from a monitor. Monitors are appointed from among the overseers and their number depends on the amount of land irrigated.

Each acequia-community also has a collector (*collector* – in Spanish) who is responsible for collecting taxes and fees.

The water available to each community now comes from rivers and public waterways and is allocated by the state in function of the area irrigated and the crops planted. In the past, the acequia-communities also had access to the wastewater generated in towns and this was reused after being mixed in the canals with water from the river.

Because of the irregularity of rainfall, farmers face significant shortages and this generates conflicts. To resolve such conflicts, a volumetric unit known as the 'fila' was created which unlike the litre or cubic meter varies in volume depending on the level of the river. The flow carried by the river at the point where the acequia communities draw their water is divided into 138 'filas' and so each 'fila' represents 1/138th part of the total river flow.

Each acequia-community is assigned a certain number of 'filas' that ensure a supply of water if the level is high (fat filas) or low (thin filas). In the past, this supply was in addition to any wastewater drawn from nearby urban areas. Conflicts were thereby avoided as each community had a number of 'filas' assigned and acequia-communities that were downstream did not find themselves disadvantaged.

26.4 CONSTITUTION AND ORGANISATION OF THE WATER COURT

The fundamental aim of the Water Court of Valencia is to distribute water among farmers in the fairest and most democratic manner possible. The Court has a dual function: judicial and administrative. However, the individuals involved in both functions are the same (Fairén Guillén, 1998).

The Water Court is composed of eight judges appointed by each of the eight acequia-communities. These judges double as the chairmen (Síndicos) of each acequia-community as both responsibilities are indissolubly united. The judges elect the president of the Court for an unlimited period. The president must be the chairmen of an acequia-community from the right side of the river, while the elected vice-president must represent one of the three communities on the left side. The explanation is that the three communities on the left side also draw water from the Moncada royal canal and so are better protected then the acequia-communities on the right side.

Both the president and the vice-president are elected by the Water Court when it is acting in either an administrative or governmental session. The president cannot vote: although he opens and closes the Court sessions; leads the discussion; and directs the questioning of the accused, experts, and witnesses. In short, the president is the public voice of the Court. At the moment of sentencing, the president calls quietly for a vote and then announces the judgement in a loud voice. To avoid any suspicion of bias, if the accused belongs to one of the acequia-communities on the right bank of the river then the activity of the Court (except for opening and closing the session) is performed by the vice-president (who comes from the other side of the river) rather than the president. When the president cannot attend a session he is replaced by the vice-chairman of his acequia-community (nominated by all eight communities) rather than the vice-president of the Court. A vice-chairman in these circumstances will temporarily acquire all the powers of the absent president.

Another important figure in the water Court is the warden (*guarda* – in Spanish) who assists in the preparation of the trial. In the performance of their duties, wardens may cross private land – and if a canal gate has been illicitly opened they may 'follow the water' to identify the farmers who are irrigating illegally. A warden also informs the judge of the names of the individuals being prosecuted before the beginning of each Court session. Wardens are also responsible for verbally ordering the accused to appear in Court in the name of the community chairman. If the accused fails to appear then a warden makes a second visit to the home of the accused and orders him to appear – this time in the name of the Court president. The warden's most important judicial role is to make accusations during Court sessions. Wardens must act when the interests of the acequia-community are threatened, and may also act when specifically requested by the community. Wardens do not need witnesses and their word is taken as evidence. For this reason, individual wardens must be carefully nominated. However, wardens can be denounced, and subsequently tried by the water Court. In addition to their Court responsibilities, wardens have a wide variety of administrative roles: maintaining reservoirs, monitoring the water flow and river level, and so on. Each acequia-community has several wardens, as well as deputy wardens who have the same powers as the wardens themselves.

The Court also has experts who are represented by overseers (*veedores* – in Spanish). This attribution mainly occurs at three points in the Court process: (a) before presenting a case to the Court; (b) during the course of the trial when they act as experts alongside the corresponding acequia-community chairmen (*Síndico* – in Spanish) to assess the damage caused by the accused; and (c) once the verbal trial has finished and sentence is pronounced the overseers further assist in valuing the damage caused by the accused. If technical expertise is required that the overseers cannot provide, the Court nominates a specialist as overseer and their testimony can be heard in Court without the need for witnesses. From the administrative point of view, overseers are responsible for ensuring that canals are cleaned and they supervise many irrigation activities.

The Water Court also appoints a sheriff (*alguacil* – in Spanish) who only reports to the Court itself. If an accused party has refused to appear before the Court after being summoned twice by the wardens, the sheriff issues a written summons. During a trial, the sheriff twice calls the names of the accused and accusers for each case. Each case is called to Court in upstream-to-downstream order and grouped as left or right of the river.

Fee-collectors are also involved in the process of the water Court and they report (directly or indirectly through a warden) individuals who fail to pay fees.

The Water Court also appoints (or dismisses) a secretary. The secretary does not attend trials, but the overseer's warden delivers Court verdicts to the secretary's office after a trial and the parties involved then collect information on sentencing from the office. Any doubts about sentences must be resolved by the president of the Court. The secretary is also responsible for recording the minutes of the Court, preparing the budgets (under the direction of the president), and issuing certificates, and so on.

26.5 WATER COURT PROCESSES

Processes always begin at the request of a party. Any member of the acequia-community may make an accusation – including the chairmen/judges (*Síndicos* – in Spanish) of each of the eight communities. However, if the chairman or judge makes an accusation then a problem arises because he cannot be both accuser and judge. The problem is resolved by the judge abandoning the position reserved for judges and standing instead in the place reserved for accusers. From this position, the judge can then make an accusation, provide evidence, and answer any of the Court's questions. After the sentence is announced, the judge returns to sit among the other judges and participate in other trials.

The other elected members of the governing boards of each of the eight acequia-communities can also make accusations. However, it is the wardens who make most of the accusations – either following complaints from officials in the community or from individual community members. As part of their job, wardens often appear in Court making accusations against farmers caught taking water illicitly. If the warden is the only accuser then he may make a claim for damages – although the beneficiary may be another farmer. If an individual makes an accusation then the warden may either: (a) join in the accusation and assume certain responsibilities if the accused is found guilty; or (b) decide to not support the accusation – in which case the Court cannot make the warden responsible for carrying out the sentence. Although a warden may substitute an individual accuser, the Court may sometimes still demand the presence of the accuser for questioning (individual who fail to appear when called may be prosecuted). In other occasions, the warden may make accusations on behalf of an overseer, a monitor, and even a community fee-collector. Wardens sometimes even make accusations on behalf of employees of the acequia-communities or the Court itself.

Officials who can formulate complaints include the overseers, care-takers, monitors, the sheriff (although rarely), and community fee-collectors. It must be remembered that any member of an acequia-community can make an accusation and several individuals often make accusations together (forming a joinder of parties).

The judges of the Water Court may themselves be denounced as they are also farmers and so liable to commit infractions punishable by the Court they represent. Members of the Court may also be denounced by the acequia-communities – although this has only happened twice (in the Acequia de Rascanya in 1961 and the Acequia de Quart in 1968) (Fairén Guillén, 1998).

Court officials and community monitors can also incur liabilities and be sentenced by the Court. In short, all community members can be accused, whether individuals or companies, as well as their tenants (this last case being the most common). The accused may band together and form – depending on the accusations – a joinder of parties.

For many centuries, the water Court was based on a doctrine, emanating from its ancient ordinances, that criminal processes required the attendance of the accused. However, as farmland started to become registered in the name of companies this situation became untenable and the options were either: (a) avoid bringing criminal cases in some cases; or (b) accept that the accused could use lawyers as accusations were not criminal. The Court is now regarded as a purely civil Court under a special jurisdiction.

26.6 SPOKEN TRIALS

The seat of the Water Court is the apostles' doorway of the Valencia cathedral. The Court opens at noon every Thursday; but the judges may decide to sit on Wednesday if Thursday is a holiday (Glick, 1970).

A low semi-circular fence is installed to separate the Court from public areas. The judges always sit in the same order (as seen from left to right when facing the cathedral): Tormos, Mestalla, Rascanya, Favara, Mislata, Robella, Benager, Faitanar and Quart.

All of the acequia-communities try avoid cases reaching the Water Court, and have internal judicial procedures that attempt to resolve conflicts at an earlier stage.

During the trial in which the judgment and sentence of an individual is discussed the judge from acequia-community of the accused remains silent. He stays seated alongside the other judges, and only speaks if asked to supply information by the president. Subsequently, the same judge will be responsible for executing the sentence of the Court with the help of the overseers.

As part of the Water Court proceedings there is a trial preparation before the judge of the acequia in which the alleged offense was committed. During the preparation, the judge may intervene to examine witnesses and the parties involved may reach an agreement. If the alleged offense affects the interests of the corresponding community and a warden has been involved from the beginning, then no pre-trial agreement can be reached. The only option is for the accused to admit guilt and promise to pay the fine and damages imposed by the Court. These pre-trial procedures are exclusively oral. Procedures may last two weeks for cases between individuals and where the community chairman/judge intervenes in an attempt at conciliation. However, the procedure lasts just one week for cases that involve the interests of the acequia-community – by the end of the week the judge reports to the Water Court and the following Thursday is scheduled for the trial.

Up to three citations are issued to the accused and the accusers (whose failure to appear has a special significance and supposes that the accusation is dropped). If the accused fails to appear after the third and written summons made by the sheriff then the trial is held in absentia as follows: the Court asks the warden or individual accuser to speak and explain the accusation. If the accusation is being made by an individual then the Court may cross examine the accuser and ask for any evidence to be shown. Subsequently, the Court will issue its verdict. The consequence of absence is not necessarily a guilty sentence as the accusation may be unfounded – however, the failure to offer a defence usually means that the accused is found guilty. Nevertheless, the Court does not systematically find accused parties guilty if they fail to appear and will always endeavour to establish the truth. In some cases, the accuser may even be found guilty.

The trial process is very simple: members of the Court enter a small enclosure bounded by a fence and stand before their respective chairs, while the sheriff remains by a small gate in the fence. The members of the Court then take a seat, and the president announces: 'The Court begins'. The judges are dressed in the classic black smock once worn by farmers around the city of Valencia. Once the Court is constituted, the sheriff calls the first case at noon. Cases are heard from each community in the order that each community's canal draws water from the river: Quart, Benager & Faitanar, Tormos, Mislata, Mestalla, Favara, Rascanya & Rovella. Calls for attendance are made twice loudly in the Valencian language. On hearing their names called, the parties enter the fenced enclosure in front of the judges. The warden announces that an accused party now stands before the Court and the warden positions himself to the right of the accused. In cases where an individual or individuals are making the accusation, the accuser(s) stands between the warden and the accused.

If the accused is from the left bank of the river, then the trial is led by president (who is always from the right bank), and if the accused is from the right bank, then the vice-president (who is always from an acequia-community on the left bank) leads the trial. The accusers begin the trial by making their accusations and providing any evidence. The presiding judge then tells the accused to defend themselves and provide any relevant evidence.

The evidence offered and admitted by the Court is as follows: (a) the expert evidence – if the parties have brought experts they give testimony and can be questioned by the other party and the Court. If no expert witness is available then the trial is suspended until the following Thursday. If an expert fails to appear on the following Thursday then the trial continues; (b) any witnesses must identify themselves to the president and answer questions from the Court and the opposing side. All evidence is submitted orally; (c) no oath is taken before declaring and testimonies may be examined by questions from the judges; and (d) if a visual inspection (known as a *visura* – in Spanish) is required then the trial is suspended and the judge from community in which the event occurred, as well a judge from the left bank and another from the right, agree a time and day to make a joint inspection with the parties involved. On the agreed day, the delegation (together with two overseers) inspects the location, listens to the explanations and observations of the parties, and questions everyone involved.

The Court passes judgment once the testimonies are completed. However, the Court may decide to make further enquiries. Therefore, if the Court is not sufficiently certain of the responsibility and guilt of the accused, the Court may order overseers and wardens to take further evidence at the place where the event occurred in order to further examine the argument of the defence. In short, the Court does not have to reach a verdict in the light of the allegations and proofs offered in Court, and may investigate further in an attempt to establish the truth of the events.

After sentencing (which we will detail later), the president (or the vice-president) of the Court will clearly order the parties to leave the Court. The warden then calls the parties for any other scheduled cases in the established order of the communities (upstream to downstream). If there are no other cases pending, the sheriff will tell the president – who then closes the session.

Sentences of the Water Court are made in public. The president counts the vote of each judge in a soft voice (the judge from the community affected does not vote) and then announces the result aloud. When the verdict is announced (either innocent or guilty) the reasons are also given. The verdicts are based on the fact that each judge has an excellent understanding of the ordinances (they were elected for this reason). The decisions as to the type of infraction committed and the relevant ordinances are reached during the public cross-examinations led by the president or vice-president of the Court.

After the session, the warden and the parties involved attend the office of the Court secretary (the secretary does not attend trials). A warden tells the secretary the events that gave rise to the trial, the offence committed, and the sentence. If the secretary has any questions, he will consult with the president (or the vice-president). The secretary fills in a form that includes the details of the parties involved and the sentence. The subsequent reading of this statement and its handing over to the parties constitutes the official notification of the sentence. Acquittals are described in a very succinct language.

26.7 ENFORCEMENT OF JUDGMENTS

The judgments of the Water Court are enforceable and final. The Court resolves every conflict presented and no appeal against the final sentence of the Court has ever been admitted.

Various ordinances state that it is the judge/chairman of the acequia-community in which the infringement occurred who is responsible for executing the sentence. When a fine is imposed the overseers play a key role in the execution. After the sentencing, the judge/chairman must ask the overseers to ratify or amend the amount of the fine. Both sides may argue that the amount is too great or too small and ask the Court to allow the overseers to modify the fine. In these cases, the Court will name a judge from the right bank and another from the left bank to evaluate the amount of the damages and the fine, and report back to the Court. The decision subsequently made by the Court is final and there is no further appeal.

It may be that expertise is needed when assessing a sentence that falls outside the normal realm of experience of the Court officials. In these cases, the Court appoints two experts to estimate the fine and report to the Court.

The most straight-forward sentence for a Water Court judge to execute is the denial of water. In practice, the judges deny water in cases where somebody has not paid a fine imposed by the Court – rather than in cases where somebody has not paid a fee that is due. The procedure is very simple: any fees already paid for water are used to pay a fine imposed by the Court. If the guilty party has not paid any fees for water that can be diverted to pay a fine, then a judge accompanied by a warden go to the place where the guilty party draws water from the canal and a series of vertical lines are painted on the water gate. The symbol is widely recognised as meaning that it is prohibited to draw water from that point. If the guilty party is subsequently found to be in contempt of Court, he will be sentenced again, and his situation will further worsen.

The Water Court can also seize and auction property to execute judgements – and in a much earlier stage, each acequia-community can issue orders to recover any payments due (although this approach is rarely necessary). The procedure is as follows: an agent of the community makes a demand for payment against the guilty party, giving the party 24 hours to make payment. If payment is made, the procedure is finished. If the debt remains unpaid after 24 hours, the agent may order the seizure of goods, for which it may require the intervention of expert appraisers. Once an embargo is placed on the goods, an official receiver of the property is announced, and the embargoed goods are officially seized by the Court. Subsequently, two auctions with a reserve price are held. If the goods are sold at auction any sums over and above the amount owed to the Court are returned to the individual whose goods were seized. If the second auction is unsuccessful then the goods are put on public display for three days and the highest offer is accepted that exceeds one-third of the reserve price set in the first auction.

26.8 OTHER ISSUES

The Water Court has a bench of eight judges – which may appear excessive or even harmful for efficient operation. However, the following considerations must be taken into account: the Court system requires that a judge operates at the level of each acequia-community and it would be unfair if only some communities had their own judge – especially as the judges are elected democratically. While the initial investigation is conducted by a single judge, the presence of eight judges in the Court offers a series of obvious advantages. Sentences are executed by a single judge – the same judge who began the proceeding. In the more difficult issues that may arise during sentence executions, a judge is bound by the opinion of the overseers and the accused may appeal to the Court and request a new expert opinion.

In some cases, the accuser may retract his accusation or accept compensation before the case reaches the Court. There are two types of cases:

(a) If the alleged offense is committed by an individual against another and community interests are unharmed, then the accuser may retract his accusation or reach a settlement before the case is heard in Court.
(b) The alleged infringement has damaged the interests of the acequia-community. In this case, the wardens, monitors, overseers, and judges automatically intervene. Two situations can be distinguished in these cases: (i) the case is dropped if during the investigation, the accuser accepts the indemnities and obligations proposed by the judge; and (ii) if the alleged offender does not accept payments or actions proposed by the judge then the next step is the first summons to appear before the Water Court.

The organisation of a legal process is determined by the president of the Court. The powers of the Water Court can be summarised as:

(a) The president, or vice-president, may question all the parties to clarify the truth without any prior oaths being given.
(b) Other judges may question the parties with the permission of the president.
(c) The Court is not subject to any prior written questioning.
(d) The witnesses and expert witnesses are questioned under the same conditions.
(e) The president calls the parties, experts, and witnesses to speak.
(f) The Court decides on the admissibility of evidence.
(g) Discussions are led by the president, who also has the power to impose disciplinary measures on those who disobey his orders, change the course of the discussion by speaking without permission, create disorder, or fail to respect the Court.
(h) if either party disagrees with the assessment of damages made by the expert appraisers the Court may request another appraisal to resolve the disagreement.

(i) The Court, according to the ordinances, appoints overseers to execute a sentence, and
(ia) The judge in charge of the execution of a sentence may stop the supply of water to the guilty party or start a process for the seizure of goods.

The principle of immediacy dominates the proceedings of the Water Court because once the trial has started the process is rapid. Accusations are made and tested before the whole Court and sentences are given immediately after the hearing. The process developed by the Water Court is widely accepted by the communities it serves – the farmers around the city of Valencia. It is also worth noting that the time required for the instruction of a process is less than eight days and processes must be finished in one session. Moreover, sentences are generally executed in two or three weeks and executions are even quicker if the decision is made to cut the water supply. The threat of cutting the water supply is so serious that matters are usually resolved quickly.

26.9 CONCLUSIONS

The Water Court of Valencia has existed for over a thousand years with the primary aim of ensuring the fair distribution of water to the fields and orchards around the city of Valencia (*Huerta de Valencia*). The waters are today taken from reservoirs near the city, while in the past the same canals were also used to transport the wastewater from Valencia to fertilise the fields.

The Court consists of eight communities of farmers and has operated without interruption over the centuries. There are several reasons why the Court is greatly respected by farmers – the first being the age of the Court. Furthermore, the fact that the judges (Síndicos) can themselves be judged by the Court gives esteem and credibility. Proof of this is the fact that most sentences are executed on a voluntary basis. However, the Court has the power to forcibly execute its sentences – although this power is rarely used – and this usually means cutting the water supply for guilty parties without the intervention of any other body.

The Court meets every Thursday at noon at the apostles' doorway of the Valencia cathedral. The Court process is oral and rapid. After a quiet deliberation, judgements are announced orally and cannot be appealed.

The duration of the process – including trial preparation, the trial (which rarely last more than an hour), and the execution of the judgment (usually voluntary) – does not exceed one month, and is often completed in ten days. The costs of the process are very low; yet the economic issues at stake can be substantial.

All of these features have enabled the survival of a Court that has a major social function in water and wastewater management.

26.10 REFERENCES

Almerich J. M. and Jarque F. (2002). La terra, l´aigua. L´home. L´Horta de Valéncia (The land, the water, the human). Conselleria d´Agricultura, Peixca i Alimentació.

Fairén Guillén V. (1988). El Tribunal de las Aguas de Valencia y su proceso. (Oralidad, concentración, rapidez, economía) (The Water Court of Valencia and its process. (Orality, concentration, rapidity, economy)). Caja de Ahorros de Valencia, Valencia.

Giner Boira V. (1988). El Tribunal de las Aguas (The Water Court of Valencia). Alzira, Graficuatre.

Glick T. F. (1970). Irrigation and Society in Medieval Valencia. The Belknap Press. Harvard University Press. Cambridge, Massachussets.

Sala Giner D. (1997). El Tribunal de las Aguas (The Water Court of Valencia). Fundación Tercer Milenio, Valencia.

Sala Giner D., Mascarell Navarro, M. J. and Tarín López, R. (2002). El Tribunal de las Aguas de Valencia (The Water Court of Valencia). In: X. Boronat, (ed). Valencia, Consellería de Agricultura, Pesca y Alimentación.

Chapter 27

An overview and synthesis of the evolution of sanitation, and wastewater technologies through the centuries: past, present, and future

J. B. Rose and A. N. Angelakis

27.1 PROLEGOMENA

Probing the past, forging the future

— Th. Kolokotronis, 1770–1843.

Humans need to have access to fresh water and historically this meant interfacing and adapting to the hydromorphology and climate of the area. This integration provided drinking water, food and transportation as well as a means to carry wastes away. Yet as more complex civilizations and settlements were formed, and the density of populations increased, humans began building infrastructure to better manage the water cycle. This included the transport of potable water and storage. Hygiene meant providing public baths. This led to the building of drainage systems with the ability to address storm and used waters. Finally this allowed the merger of stormwater and wastewater management. The importance of sanitation for life, human prosperity and cultural development was understood very early in all civilizations. In the initial stages of civilizations, odors were a strong factor promoting some collective understanding of what was 'healthy' and this was depicted in mythologies and associated with the natural forces producing the cleansing nature of the water cycle with an idea of what was fit for human purposes. Yet soon this became a more managed system by developing key principal technologies for both wastewater and stormwater.

The human race now faces even greater challenges as global waters have degraded due to population and landuse changes, exacerbated by changes in climate. We now reside in the anthropocene era and there has been a great acceleration of pollutant discharges in the last 70 years. Adaptation of the historical centralized and decentralized strategies, infrastructure and technologies for wastewater and stormwater will be crucial to our global water future.

This book is dedicated to providing a window to the past. Thus this chapter gives an overview of the historical contributions with an eye toward the future. A summary of each of the chapters is included with synthesis around key topics. An epilogue provides some commentary on the current and future issues with recommendations on the future needs. Appendix A provides a good list of achievements from

Hellenic and Roman societies adapted from DeFeo *et al.* (2012). A look into the various times and places where humans designed and built such similar waters systems, even as the technology and knowledge grew over the ages, aids in determining what path to follow into the future. Lessons on the integration of all water flows using new technologies to enhance, protect and restore quality will lead us toward a sustainable water future.

27.2 PREHISTORICAL TIMES

Modern humans (Homo sapiens) have dwelled on earth for over 200,000 years, most of that time as hunter–gatherers, and with ever increasing populations (Vuorinen *et al.,* 2007). The first human communities were scattered over wide areas and waste produced by them was returned to land and decomposed using natural cycles. Disposal problems were limited primarily because they were small communities of nomadic hunter – gatherers. A new era started when humankind established permanent settlements about 10,000 years ago, adopting an agrarian way of life. With human settlement came the ecological impacts (Lofrano & Brown, 2010). Until the birth of the first advanced civilization, the disposal of human excreta was managed through holes in the ground, covered after use as explained by the Mosaic Law of Sanitation (Deuteronomy, Chapter 23).

Because of the lack of any kind of records, it is practically impossible to evaluate the health of ancient populations. It is, however, quite safe to conclude that despite the impressive measures used to obtain pure potable water, urban centres had serious public health problems due to a lack of management of their wastewater (Vuorinen *et al.,* 2007; Larsen, 2008).

Religious laws, such as Moses' Law, writings in the Old and New Testaments and laws in the Koran, played major roles in the lives of ancient peoples. These laws mainly concentrated on the provision of personal hygiene. Dead bodies and contaminated surfaces were known to be unclean or unhygienic to touch. The importance of burying human faeces was also strongly indicated. The importance of body cleanliness before praying was a motive for maintaining the integrity of hygiene with a religious practice. Thus bathing and hygiene were linked to these practices.

The importance of hygiene and sanitation flourished at the times of Hellenic, Roman and Egyptian civilization. The use of private and public baths and latrines, cleaning of the body, shaving the head for protection from lice infestation, and the construction of water pipelines and sewage ditches were widely observed. However risks remained such as the transmission of schistosomiasis (bilharzia) which was linked to bathing and swimming in the Nile River. In these civilizations, the focus was on personal hygiene (hygiene) and human waste management (sanitation) most often at the individual level.

The evolution of urban wastewater and stormwater management in ancient Hellas that begins in Crete during the Middle Bronze and the beginning of the Late Bronze Age is considered in Chapter 1. Numerous remarkable developments occurred at various stages of Minoan civilization which flourished during these particular periods. One of its salient characteristics was the architectural and hydraulic function of its sewerage systems in the Minoan palaces and several other settlements. It might be implied, therefore, that Minoan master craftsmen in Bronze Age Crete were aware of some of the basic principles of what we call today sanitation, wastewater, and environmental technologies for developing the infrastructure. Archaeological and other evidence indicate that Minoan technicians originally, during the Bronze Age and by Hellanic philosophers, mathematicians, and engineers thereafter developed the basic principles of hydraulics. With respect to the wastewater management, Minoans developed an advanced, comfortable and hygienic lifestyle, as manifested from flushing toilets, public and private baths and very effective sewers and drains. The hydraulic and architectural function of sewer systems in palaces and cities are regarded as one of the salient characteristics of the Minoan civilization. In this Chapter the most characteristic examples of extant hydraulic works and technologies relevant to wastewater and stormwater developed during the Bronze Age in Crete, Hellas are presented and discussed.

Chapter 2

The sanitation and wastewater technologies of the Harappa and Moen-Jo-Daro civilization of the Indus Valley are considered in Chapter 2. As the name denotes, this civilization was spread along river Indus and its tributaries that is Jhelum, Chenab, Ravi, Beas and Sutlej. The work is thus about the people that inhabit the Indus and its tributaries. Geographically, the civilization was spread over an area of one million square kilometers, bounded by great Himalayas in the North, Arabian Sea in the South, Rajasthan desert in the East and rugged hills and plateaus of Balochistan in the West. The character of any civilization is determined by two major factors, the physical and the social environments. Together these two terms cover a variety of sub-factors like the climate conditions, the type of terrain, the social development of basic communities and the racial characteristics of the population. The Indus civilization represents the very modern concept of a planned urban layout for cities and towns. The entire residential areas were divided into squares or rectangles between streets and lanes. The architects who planned these layouts had very clear ideas about optimum utilization of space. The appreciation of the need for organized system of drainage and sewerage disposal was in itself unique in the ancient world. The concept of personal and environmental cleanliness was part of the character of the Indus people. The Indus people were the pioneers in these concepts of planned layouts and public hygiene. It was several centuries afterwards, that people elsewhere started using these ideas in their own planning.

Chapter 3

The Harappan civilization is the first society to have managed wastewater in South Asia, about 5000 years ago. Wastewater management was a characteristic of all the urban and semi-urban settlements of the Harappan civilization. Sewage and drainage were complex networks, especially in Mohenjo Daro and Harappa. Wastewater management consisted mainly in latrines, soak-pits, cesspools, pipes and channels. The Harappan civilization is considered to have 'disappeared' after 2000 BC, yet wastewater management spread all over the south Asian region. We find remnants of wastewater management in different non-urban settlements in the northern and the central parts of the subcontinent during the Chalcolithic period. In various sites, wastewater was canalized into cesspools, which could have been ring-wells, as in Jorwe (1375 to 1050 BC) or in Ujjain (*ca.* 500 BC). During the Early Historic period (ca. 200 BC to 300 AD), wastewater management spread all over the subcontinent, matching with the 'second urbanization phase' of South Asia. Indeed, in the 3rd century BC, earthenware drain-pipes and cesspools were already used to manage domestic wastewater in the northern part of the subcontinent (Delhi, Nasik). During the same century, similar techniques were also used in the extreme northwest of the subcontinent (Taxila, Pushkalavati), where Hellenic garrisons were settled. During the Sangam Period (*ca.* 300 BC to 300 AD), in the southeastern part of the subcontinent, industrial wastewater management was developed in Arikamedu, an Indo-Roman trading station. Thus, two phases can be considered in the development of wastewater management in ancient South Asia. During the second phase, the diffusion of wastewater management techniques seems to follow the main trade route of the region, which joined Ujjain to Taxila. Indeed, wastewater management in south-east South Asia is probably linked with the presence of Mediterranean populations, and the very same populations could even have had an influence on water engineering techniques in the northern part of the region.

27.3 HISTORICAL TIMES

Chapter 4

Although much has been written about the history of water supply systems, there is much more limited information on wastewater management. In ancient times, wastewater disposal problems were highly localized because the first human communities were small and scattered over wide areas. The waste produced by them was returned to land and decomposed using natural cycles. A new era started about 10,000 years ago, when societies moved from nomadic cultures to building permanent settlements, adopting an agrarian way of life. With human settlements, waste disposal became an important concern that has been dealt with in many different ways. In developed cities waste treatment technologies. varied by the skills the various cultures to design and build the available infrastructures. In the Egyptian city of Herakopolis (*ca.* 2100 BC), the average persons treated their wastes much like those in Ur, via individual pots where the disposal was dumped into the streets where drainage and flushing occurred during rain events. However, 'in the elite and religious quarters, there was a deliberate effort made to remove all wastes, organic and inorganic to locations outside the living and/or communal areas, which usually meant to the rivers.' Odors and hygiene were strongly linked but in the early modern period there was little understanding on the health impacts for the general population and disposal of human wastes tied closely to in-situ disposal within the city. Once linked to drainage, wastes were disposed of into rivers without knowledge of carrying capacity of the system, and water sources were being contaminated. These practices were brought to the New World. This chapter reviewed the evolution of wastewater management in Egypt from as early as ca. 3000 BC to as recently as the twentieth century. The concurrent connection between wastewater management, environmental awareness, advances in scientific knowledge and socio-economic situation evolved until not just disposal but wastewater treatment technologies were given due attention.

Chapter 5

In South America and especially in Peru, the catchment and secure provision of water had been and still remains challenging. Reasons are the particular geographic, hydrologic and climatic conditions, such as diverse and sometimes adverse environments, irregular availability of water, and immanent natural hazards, among others. Sanitation as well as returning the used waters into the natural water cycle also have been and still are main components of the adopted water resources management strategies. The present chapter gives an overview of sanitation and wastewater practices in the Central Andean Region with focus on Peru (Coastal and Andean regions) from Pre-Columbian times (3000 BC to 1532 AD) and Colonial times (1532–1821 AD) to nowadays. This long time spanning period encompasses solutions achieved by many Pre-Columbian civilizations such as Caral, Chavín, Mochica, Huari, Chimú and Inca, includes European influences brought in during Spanish Colonial times, and concludes with an insight into the actual situation, development and trends. The water systems developed for potable water, irrigation and drainage are evident. Use of the land and the drainage and irrigation canals for waste disposal, with eventual discharge to waterways with other than natural systems treatment are a continual practice.

Chapter 6

The building and use of sewerage and drainage systems have been in practice since the Minoan times (ca. 3200–1100 BC). It is evident that elaborate structures were planned, designed and built to protect the growing population centers and the agricultural land. In several Minoan palaces discovered by archaeologists in the 20th century, one of the most important elements was the provision and distribution of water and the transfer of stormwater and sewerage in drains by means of hydraulic systems. Rainwater from the flat roofs of the palace at Knossos was carried off by vertical pipes; one of these, located in the eastern wing, was emptied into a stone sewer-head from which a stone channel carried the flow of storm water. In addition, a significant integrated system appears to have been achieved during the Classical and Hellenistic periods throughout Hellas, improving and extending the similar systems yet there is no evidence suggesting the connection between the Minoan era and the Classical and Hellenistic periods regarding these water and wastewater systems. Hydraulic engineers' knowledge was not lost with the Minoans but was re-gained later in the Hellenistic and then the Roman period. During this period, the scale of baths and toilets as well as the sewerage and drainage systems in Roman cities was dramatically increased and were continued even in the Mid-evil times in Hellas based on archaeological and historical studies. Venetians' accomplishments in hydraulics are worth noting, such as the construction and operation of cisterns and drains. Many of these technologies were developed and used in the famous castles constructed during that period in several places. Water was connected to Islam, so that during the Ottoman period there was a water tap in all mosques. Hammams, which are presently also referred to as Turkish bathes were established in all the regions of the Ottoman Empire and played an important role in the Ottoman culture, serving as places of social gathering, ritual cleansing, and as architectural structures, institutes and so on. Thus again the connection between bathing and hygiene (or cleansing) was part of the society.

Chapter 7

From ancient times, the traditional cesspool (absorption well) was used as one the first means of discharging domestic wastewater, once some type of sewers or drainage had been established. They are still in use at many urban and rural areas of the country. In towns where the ground water table is high, concrete tanks (septic tanks) have been also used. In some regions of Iran (mostly in the west and northwestern parts of the country) urban centers have either been developed on rock formations (such as khorram Abad and Kermanshah) or have a high groundwater tables (such as Ahwaz and Rasht), there absorption well systems were no longer seen as effective. Therefore the traditional sewage networks consisting of rocky canals or cement pipes were used to collect and discharge wastewater from these cities. There are no maps, technical calculations or manholes for these networks to indicate that these were built according to an engineering design along with city planning. Instead, these networks are mostly combined systems that collect surface runoff during rain events and appear to have been created over time by various municipalities or communities. In addition to lacking any type of design, map or technical calculations, these networks were also devoid of any treatment and therefore it is raw wastewater that enters the farmlands from these canals for reuse. About 70 urban centers in the country possess this type of network, which are referred to as the traditional sewage networks. The historical monuments of Iran also had a certain wastewater discharge systems, which are of interest in their own merit. In the year 1991 became a milestone in the management of the national water and wastewater affairs, where by the Law for the Establishment of Provincial Water and Wastewater Companies was ratified by the Islamic Consultative Council, and thereafter and within a span of 2 to 3 years, 30 water and wastewater companies were established in provinces throughout the country. This move boosted the development of water and wastewater installations in the country and improved the operation and maintenance of existing systems. The establishment of these companies also meant the start of studies and implementation of modern wastewater collection and treatment plans, to over 200 urban centers in the country. The reuse of wastewater, through substitution of agricultural wells (which contain potable water) with treated urban effluents, is now considered as a new part of an integrated plan in Iran. Through experiences gained from the studies and implementation of wastewater plans the country is now in possession of valuable regulations and standards related to study, design and operation of wastewater reuse, which could also benefit other countries.

Chapter 8

A higher level of technology and expertise for water management appeared from ancient Macedonia, Hellas onward, as the daily requirement for drinking water and the necessity for the waste and rain water drainage increased. Special technical works, such as systems for collecting, transporting and storing groundwater, constructions for water supply, drainage and irrigation, and various other structures like dams, bridges, fountains and baths, were developed through the ages in several parts of Macedonia. The water availability was correlated with the climate conditions of the region, as well as with the water resources (surface water, groundwater and springs). Therefore the local geology and geomorphology controlled in many cases the development of the new cities since the 4th century B.C. The urban planning in ancient Macedonia is attributed mainly to Philip II and his successors who built cities according to Hippodamus system. The administrative, religious and commercial centers occupied separate areas of the city blocks among the private residences. In such cities the complex water supply and drainage systems ensured comfortable living conditions to the citizens. There were special constructions related with the water management in the most famous cities of the ancient Macedonian Kingdom, such as Pella, Dion, Aiges, Olynthos, Amphipoli and Filippoi. The collected data show that the ancient Macedonians had a deep knowledge of the hydraulic technology which was mainly developed during the Hellenistic period. In this time there was an increase in the use of baths, cisterns and fountains in Macedonia.

Chapter 9

Cyprus has developed and for thousands of years maintained, its own civilization, assimilating the various influences to which it has been subjected. These influences can be seen not only in buildings and temples but also in the drainage systems (sanitation and wastewater) of the buildings and archeological area. Ancient drainage systems have been discovered at Paphos, Kourion, Kition, Salamis, and so on, by several archeological missions. Since ancient times, as with other civilizations, Cypriots disposed of their excreta and other wastes on the site where they were generated. Probably the first disposal technologies were developed in response to a need for some kind of management scheme to minimize the aesthetic problems. Something akin to a crude pit latrine was probably first used to minimize the offensive impression created by the indiscriminate discharge of human excreta on the ground. The pit latrine or simple privy probably alleviated the nuisance and sufficed till 1950, when the installation of house –to-house water supply for all towns and villages was started. From 1950, the necessity for the disposal of liquid waste came with the development of the water-carriage waste system. The first disposal system was also a cesspool or soak-away. Septic tanks – absorption pits or ditch systems are still currently used in most of the villages in Cyprus. With this way there is decentralization which has many benefits like economies of scale and so on. In the towns of Cyprus today, the households are connected with main sewerage systems.

Chapter 10

Portuguese engineering and architecture are rich historical prestigious hydraulic and sanitary works, which still today constitute major points of attraction. These systems and the technology were deeply influenced by roman and Arab dominions. The influence of Roman masters as Vitruvio and Frontino is clearly observable in several works (e.g., spa, monumental aqueducts, pools, heating systems, water piping and sewerage systems). The Arab influence, during its 800 years of domination, is noticeable in the engineering of castles and, also, of agricultural hydraulic works. Portugal is the most ancient kingdom of Europe dating back to the Early Middle Ages, the year 1139, when the prince Afonço Henriques proclaimed himself King of Portugal. In Europe drainage sanitation in urban areas showed virtually no significant progress after the Roman Empire. During the reign of Afonço Henriques and after his death, until the fifteenth century, several prestigious works of sanitary/hydraulic structures such as bridges, hospitals, monasteries, lazarettos, fonts, aqueducts, water cisterns, spas, sewage pipes and so on, were spread all over the country. The historical elements dating back to the 15th century used the rain water drainage pipes to dispose all kinds of manures and filth. There was an excessive population concentration with particular population growth in Lisbon, plague was a common problem. The earthquake of 1755, intensified occurrence of 'fearsome floods' and problems associated with the hygiene and cleanliness of the city led to a devastating loss in lives. The season that followed the 1755 earthquake started an era of progress in the capital, marked by the rebuilding of the city and the establishment of a modern and methodical drainage system, sewer based collectors arranged in a grid, connected to the Tagus estuary. During its historical pathway Portugal was a source of competent sanitary and hydraulic engineers which spread all over the world.

Chapter 11

For their territorial expansion, the Romans were content to reap the hydraulic techniques of their time and train engineers. The exchange and development of hydraulic expertise was made within the vast territory stretching from Roman Britain via North Africa to Arabia. Water and wastewater projects in the city of Volubilis located at north of West Africa are a clear example of this expansion, comprised of aqueducts bringing water from the river to power several fountains, baths, toilets bathhouses and mills through a distribution system coupled with a system of sewage and drainage of similar consolidation. Several houses had a direct water supply with an individual sewage main and drainage system. For the most part, the Roman engineering, focused on domestic water in the Volubilis in North Africa. The water was indeed a central place in the cities and homes that were built around an atrium in the middle serving as a basin to collect rain water. The remains of cities of the Roman Empire are therefore a legacy in terms of expertise in the integrated management of water resources and wastewater and storm water as well. The end of Rome did not mean an interruption in the flow of science and hydraulic expertise. The Muslim expansion reached the Iberian Peninsula and the Great Maghreb allowed assimilating all hydraulic techniques developed in Mesopotamia and the science of collecting groundwater by completing the transition in Egypt. Inheritance and mixing of this knowledge exercised great influence around the Mediterranean, including structures like water mills, the water clocks and other hydraulics such as sewers and drains. The ancestral sewage system of the city of Fez illustrates the diversity of this knowledge on wastewater management in the Mediterranean region. The ancestral traditional latrines of Fez are located near the main and antic mosques and were equipped as traditional bathrooms to allow the faithful to their ablutions before prayer and visitors to the city of Fez to use as a public latrine and were connected to traditional pipes for sewage disposal. There is a great similarity between the wastewater collection in the both ancient Moroccan cities: Volubilis founded by Romans and Fes founded by Muslims at the end of the eighth century, but with influence of the two cultures on the evolution of wastewater systems in the Mediterranean region.

Chapter 12

The principal aspects of sanitation and wastewater technologies in ancient Roman cities with particular attention to Pompeii, Herculaneum and Ostia are of particular importance as aesthetic concerns and water infrastructure advanced as shown by contrasting these various cities. Due to the absence of knowledge about bacteria and viruses, ancient Roman cities were principally concerned with alleviating the filthy sight and obnoxious odour caused by human waste than with addressing public health issues. Pompeii was characterized by a diffuse presence of cesspits (connected to private latrines, the so called latrinae), while there was a sewerage and stormwater system present, it was not uniform. Rainwater and wastewater were mainly disposed of by means of the road slope. Since Herculaneum was smaller than Pompeii, it had fewer cesspits; whereas, its sewerage and stormwater system was much more systematic. The rainwater was disposed of through the cobbled streets of the city: the steeper slope of the streets avoided the need to raise the floor. Finally, in Ostia there were no cesspits, but a complete and uniform sewerage and drainage system.

Chapter 13

After the third Phoenician war, Tunisia was the first colony in Africa incorporated under the Roman Empire. The area witnessed prosperity and development of agriculture, urbanization, architecture and economic wealth for over 700 years. Water historically was abundant and the water supply was transported by aqueducts and underground channels, and there was access to sanitation with the building of latrines. These Roman toilets were technically sustainable and very famous. They also constituted a social place for people who paid taxes for access, and where politics and business were discussed. However, in general the sanitary system for the community was rudimentary. The public toilets consisted of a seat cut into marble benches, and the wastes were carried with water circulating underneath. Otherwise, Romans used to throw their wastes in the street, which was washed almost daily. This caused contamination of the clean water system and spread several diseases. Globally, this situation was characterized as a poor level of water and waste management and a major risk for health. Toilet hygiene was very poor and public toilets lacked privacy and no means to wash one's hand. Even in private toilets they were flushed with water from kitchens. Socially speaking, only wealthy citizens had water closets that drained into cesspools below their homes. After the dislocation of their Empire, the Romans ruins in many cities of Tunisia are testimony of the sanitary system adopted in public and private constructions, illustrated in selected cities like Carthage, Thugga, Sufetula, Bulla Regia, Thuburbo Majus, and others.

Chapter 14

Despite uneven distribution of rainfall and mountainous geographic conditions in Korea, historically, Koreans developed a unique wastewater management technology including urine separation toilet and a water system for their agricultural based society. While toilets are found in royal palaces and old temples, Korea's traditional wastewater management can be summarized as decentralized, resource recovery oriented, eco-friendly, and minimal water consumption. The practice of wastewater management in rural areas reflects ancient wisdom for cohabiting with ecological conditions. Social aspects such as proverbs and cultural tradition of wastewater management may reflect some of the history of overcoming hardships concerning water management and can also provide the basic philosophy of how to handle human waste in accordance to local environments. This implies the widespread prevalence of social responsibility among individuals and society concerning wastewater management. Some of the traditional practices have significant implications for today's engineering and science by providing an effective way of involving and educating individuals in modern wastewater management. Thus, the technological and social aspect of Korea's traditional management system may benefit wastewater management programs in future cities that are threatened by climate change and water pollution.

27.4 CASE STUDIES

Chapter 15

The public infrastructure in ancient Athens comprised a well organised system of water management, including water supply, rain water drainage and sewage disposal; it was adapted to the hydrographical net, the road system and the city walls. Eridanos River and the torrential streams pouring into Eridanos were artificially modified to serve as the backbone of the drainage and sewage disposal system of Athens. The so called Great Drain was the first stream transformed into a stone built channel early in the 5th century BC, designed mainly to drain the run off rain water in the flat area of Agora, the political centre of Athens; the channel was covered by horizontal stone plates and served as a road too. Even Eridanos itself was finally covered to function as cloacae, which is found today in excavations buried below sediments 5–10 m thick. The defensive moat trench surrounding the city walls was an auxiliary sewage collector. A dense net of surface or underground terracotta pipes along the roads collected rain water and sewage from the buildings on both sides. Ilissos River ran outside Athens and had many sanctuaries at the western part which was therefore protected from the disposal of sewage. Despite this intricate water system cross connections were possible and reuse of wastewater and storm water took place, poor sanitation conditions erupted and contributed to the terrible plague in 430 BC at the beginning of the Peloponnesian war.

Chapter 16

The characteristics of an expansive aqueduct and wastewater system of Diocletian palace in the Split Croatia is an example of the engineering feats of the times. Roman emperor Diocletian, Dalmatian by birth, built his palace in the year 305 on the eastern coast of the Adriatic Sea. During the turbulent times of the Middle Ages the Palace was inhabited by the refugees from the destroyed city of Salona, capital of roman province of Dalmatia, thereby transforming it into the city of Split. The continuity of the life inside the roman walls made Diocletian's palace the best preserved late roman palace in the world. To have water inside his Palace Diocletian had built an aqueduct 9.4 km long from the spring of the river Jadro. Diocletian's aqueduct is very well preserved as it was reconstructed by the end of the 19th century and is still in use. The aqueduct channel is 60/120 cm and originally it was covered by vault 30 cm high. Longitudinal channel slope is between 0.65 and 2.660/00 and the current capacity is approximately 470 L/sec. There are several larger buildings on its route – 4 bridges and 1 tunnel 1268 meters long. The position of castellum aquae (the ancient water cistern served by an aqueduct which then in turns served the water distribution system) has yet to be found. The Palace also had a drainage/ sewerage system for surface water and foul water. During archaeological excavations some 250 meters of total 600 meters of main sewage system of Diocletian's Palace have been explored thus far. The channels are situated in the middle of main and lateral streets and have inclination towards south-west. Longitudinal channel slope is between 1.24 and 2%. Below Cardo and western part of Decumaus its profile is 115/220 cm, and bellow lateral streets 65/160 cm. Even today there some projects to put in a functioning roman sewage as a drainage system for the city core and even as a tourist attraction.

Chapter 17

China is a country of typical dual social structures of cities and rural areas. Prior to the 1950s, the situation and evolution of sanitation had been inoculated with the ethnic cultures and local environments shaping the diversity of wastewater systems. Southwest China is a large area with most ethnic diversity and ecological diversity in China and in the case of Jinghong area of Yunnan province within the Dai nationality community, the local ethnic minority and traditional society influenced the adaptation and changes before and after 1950s. Before 1950s, the villagers built wells for drinking water and channels for daily life, made to facilitate storage. The facilities for using water for drinking, bathing, washing reflected the customs and simultaneously developed a water-rich related culture. After 1950s, water supplies began to be built and reached the villages near the city but the remote areas did not change. After 1980s, there was a rapid change focused on sanitation. The improvement to access to a water supply brought the opportunity to the villages for building toilets, bathrooms and more water facilities. This change also brought improvement to the quality of living with benefits to the local people facing tremendous water pollution. The quality of sanitation has now improved dramatically between the traditional model (use of land or rivers as toilets and rivers or wells for bathing) and the new model (water-based toilets along with tap water within each house) in many of the cities, yet in rural areas land disposal remained a staying concept for use.

Chapter 18

The evolution, which continues today, of the wastewater collection system serving the city of Rome is attributable to Tarquinius Priscus, King of Rome. The development of an important drainage system to adapt the landscape configuration of the Roman area with the needs of the city's urban growth was necessary. (Cloaca Maxima) drained the area at the foot of the Palatine Hill and discharged into the river Tiber. The construction of this infrastructure for environmental protection and health of citizens is still in use today. Similarly the technical characteristics of the sewage system were adapted to the different needs of the city. What began as a simple initial lowland drainage system evolved to a sewage and water rain catchment system and finally a unitary system in the historic part of the town which is equipped with a modern real-time control, with wastewater treatment plants conveys purified water to the Tiber downstream of the city.

Chapter 19

Since its origins, the Barcelona sewer system has evolved at the same pace as the city. The first systems, dating back to Roman times, show examples of organized networks, such as the one found in Palma de Sant Just Street, composed of stone blocks, with connections and inlets located throughout the length of the channel. After the 'fall of the Roman Empire', cities fell into a deep decline and they become more rural, simplifying the sewerage to cesspools or septic tanks. During the twelfth century, some watercourses such as Merdançar, were covered and used as sewers. Considering references from some travelers who passed through the city of Barcelona between the fifteenth and sixteenth centuries, it seems that the cleanliness of streets, achieved with the help of the sewers, was the pride of the town. In 1738, the City Council, using the Roman Law, anticipated the works for building sewers and paving of new roads at the expenses of the landowners. This included the construction of drains to convey the wastewater from the houses. The type of sewer was usually a channel with straight walls covered with slabs. In the late eighteenth century, the main sewers of Barcelona were: Rec Comtal, Riera de Sant Joan, La Rambla, Ample Street, Rondes, Passeig de Sant Joan, Gran Via and Casp Street. Nevertheless, although the sanitary and hygienic conditions were not yet adequate, it was not until the nineteenth century, that the advances in technology and science showed the relationship between hygiene and diseases. This new knowledge promoted much of the engineering planning aimed at solving the problem of unhealthiness in cities like Barcelona. This resulted in the development of the Cerdà Town Plan to organize new urban developments outside old city walls which were taken down in 1854, and the subsequent Sewer Plan from Garcia Faria was implemented.

Chapter 20

It was not only Prague which, in the second half of the 19th century, was confronted with the necessity of a modern drainage system. The improvement of hygienic conditions was an all-European trend. The condition of the Prague sewerage system – built between the end of the 18th century and the middle of the 19th century – was entirely unsatisfactory. That is why Prague invited tenders for a general project of the drainage of Prague and its surrounding villages. Five projects were presented in 1885, but none of them corresponded either to the specifications, or the needs of the Prague conurbation. That is why in 1886 a new drainage programme was prepared and the Sewerage Office was founded, which was supposed to specify data for a subsequent successful design. In January 1889 the Prague municipality asked assistance from many prominent European experts such as Dr. Hobrecht from Berlin, engineers Lindley from Frankfurt am Main, Kaumann from Wroclaw, Hallenstein from Munich and Kaftan from Prague. William H. Lindley was called on to evaluate projects and make a decision. He didn`t recommend any but offered the municipal council that he would prepare a project of his own. That is how he stood in the forefront of the design of the Prague drainage system. In 1892 Lindley`s 'Prague period' began. Despite professional, political and national disfavor he was able, as the chief clerk of the Sewerage Office, to resolve not only the drainage of Prague and its environs, but, by means of two sewerage zones, he also eliminated the problem of reverse flooding of the city from sewers and preserved the function of the city drainage system during floods. He also put through and realized the construction of a mechanical sewage treatment plan, all this according to a plan presented for approval in the middle of 1893. Six months later the project was approved. Lindley also decided that Prague sewers would be built of brick despite great pressure by powerful concrete producers. He was officially charged with supreme directorship of the Sewerage Office, so intensive design and preparatory work to build the sewerage system and the sewage treatment plant could begin. A performance contract between the City of Prague and Lindley was concluded in March 1896 and then it was renewed several times. The construction of the primary sewerage system began in the same year. In March 1901 the construction of the sewage treatment plant started in Bubeneč. It was put into operation five years later together with the sewerage system. In 1909 Lindley terminated his activities in the capital of the Czech Kingdom and turned his position over to Emanuel Heinemann, who continued the construction along Lindley`s lines. William H. Lindley left his professional mark all over Europe, from St Petersburg to Bucharest and from London to Baku. He contributed greatly to the successful solution of hygienic conditions in Prague as part of his stellar achievements.

27.5 HISTORY OF WATER BORNE DISEASES

Chapter 21

During the years between 460–377 BC the treatise *On Airs, Waters, Places* was produced and aimed at showing the crucial role of waters in the preservation and restoration of health according to Hippocrates. One of the fundamental theories of the ancient Hellenic physician was that all diseases have natural causes, which, in the treatise *On Airs, Waters, Places*, are looked for in the climatological changes of different areas. More specifically, Hippocrates gives an account of the various disease conditions that appear due to the changes of the seasons and the kinds of winds and waters in every different climate. He stresses the contribution of water towards health and sanitation, elaborating on all disease conditions that derive from unhealthy waters, describing the good and bad effects that waters have on health, covering stagnant waters, rain, river and snow waters. In general, Hippocrates presents in his treatise all kinds of waters and their effects on the inhabitants of the cities, giving many details on the disease conditions that they provoke, whether these are private or epidemic ones, stressing also the fact that, not only the waters, but the climate in general, is responsible for the shaping of the temperaments, the way of life and the ethics of the people.

Chapter 22

The relationship between sanitation and health began to be revealed by ancient authors, including Hippocratic writings, Celsus' De Medicina (10–1 ~B.C.), Naturalis Historiae of Pliny (~77–79 A.D.), several writings of Galenos (e.g., De Sanitate Tuenda & De Methodo Medendi (~150–200 A.D.)) and from the late antiquity Caelius Aurelianus' treatises (5th century) on acute and chronic diseases. The poor level of waste management, including wastewater, most probably involved a major risk for public health during antiquity. For instance, toilet hygiene must have been quite poor. Consequently water-borne infections were among the main causes of death. Dysentery and different kinds of diarrhoeas must have played havoc with the populations. Already in the Hippocratic writings (written mainly in the late 5th and early 4th century BC) different kinds of intestinal disease conditions are described. However, the retrospective diagnoses are tricky and the causative agents are difficult to identify. Summer and early autumn, when water resources were meagre in the Mediterranean world, must have been a time when drinking water was easily contaminated, and intestinal diseases were rife as presented in several passages in the Hippocratic writings. (e.g., *Airs, Waters, Places.* 7; *Aphorisms.* III, 11, 21, 22; *Internal Affections.* 26, 45) The mortality of children, especially recently weaned, must have been high, which is probably echoed in the following passage of a Hippocratic author: '*It is mostly children of five years that die from this disease* (dysenteries), *and also older ones up to ten years; other ages less.*' (*Prorrhetic II.* 22)

27.6 SANITATION IN MODERN TIMES

A number of discoveries in the 19th century were important for the understanding of communicable diseases. For example, the link between contaminated water and cholera was famously established by John Snow in 1854; the importance of hygienic handwashing before attending delivery of a baby was noted by Dr. Semmelweis in 1845. Other discoveries such as detection of microorganisms (very small organisms only visible under a microscope) was reported by Antonie Philips van Leeuwenhoek (noted as the father of microbiology) and Dr. Koch established Koch's postulates regarding the demonstration of a pathogen as a cause of disease. From the 1850s, at the same time that Louis Pasteur began addressing vaccines, following the industrial revolution in Europe in this part of the 19th century improvements in sanitation around the use of toilets, water supply and plumbing significantly reduced the occurrence of communicable diseases. The term 'environmental health' is now used to describe human health in relation to environmental factors such as these and is defined as the control of all the factors in a person's physical environment that have, or can have, a damaging effect on their physical, mental or social wellbeing. The issue of environmental health is now a global matter under the guidance of the United Nations (UN) through the World Health Organization. Traditionally hygiene and infection control have been vital factors within the environmental health paradigm, but now issues such as global warming and the links between medical conditions such as cardio-vascular disease and our environment and lifestyles are emerging. In the 21st century, our environment includes all the external influences and conditions that can affect our health, life and growth. These influences are constantly changing and the effects on our health may not be easily foreseen.

Chapter 23

From 18th century through to the 21st century sanitation has had a profound impact on health and community development. This is thanks to the scientific progress, the improvement of economical condition, the industrial revolution, the implementation of technology, and the growing attention on environmental issues. During the 18th century sanitation systems did not include any treatment process and were realized by using simple static and dynamic systems to collect and dispose the wastewater. Later, urban development, the increasing population density in specific areas and the new concept of environment protection resulted in complex sanitation system which now includes sewage collection followed by a wastewater treatment plant at a distinct location mostly far away from the city center, called nevertheless 'central wastewater treatment plant'. Yet in addition to these centralized systems several decentralized wastewater treatment systems are used or under investigation. These on-site systems often distinguish brown, yellow and grey water in order to optimize the water, nutrient and energy recovery and also fall within the eco-sanitation principle. Gained experience has demonstrated that the scientific progress is not always linear, but has developed through forward and backwards steps. This is in agreement with the Khunt's theory of scientific progress. Because scientific progress is not a linear accumulation of new knowledge, is necessary to undertake 'paradigm shifts' in order to break off from the mainstream or 'normal science' and allow for transformation. These 'paradigm shifts' are needed for sanitation systems; however it is a challenging transformation which should involve a change in society norms and not only in the technical system. The historical development of sanitation from latrine to centralized and decentralized wastewater treatment plants resulted in the treatment systems and the technologies available today. The needs and perspectives for considering 'paradigm shifts' as a starting point for the revolution of sanitary science should be discussed to bring completely different sanitation systems in urban and rural areas.

Chapter 24

Wastewater reuse has historically been practiced in the form of land application and use of hydroponic systems. In modern times the fundamental processes are well understood and the current trends, and future prospects are broad reaching. 'Land treatment' refers to the application of wastewater to the soil to achieve treatment and to meet irrigation needs of vegetation. Application of wastewater to the land was the first practice used to protect public health and control environmental pollution and has gone through different stages of development with time. Currently, its application has been expanded for the treatment of various types of wastewaters including dairy, meat, commercial effluents as well as and polluted water sources. Provided pathogenic organisms and hazardous substances are effectively removed prior to land application this method is recognized as reasonable for rural communities, clusters of homes and small industrial units due to low energy demands and low operation and maintenance costs.

Hydroponics (plant cultivation in a nutrient solution) is not a new idea; it is deep rooted in history. Egyptians, Chinese and Indians used it almost 4000 years ago. This may include the Hanging Gardens of Babylon about 660 BC. The earliest published work on growing terrestrial plants without soil was in 1627. The discoveries of German botanists, in the years 1859–65, resulted in a development of the technique of soilless cultivation. In 1929, the use of solution culture for agricultural crop production was publicly promoted. The term hydroponics was introduced in 1937 and this area developed more rapidly from 1960s onward. Hydroponics has many advantages. These systems require small space, can operate with any size of flow, greatly reduce or eliminate soil borne weeds, diseases and parasites, and do not require special drainage system. Hydroponics can support the growth of almost any plant and can be accommodated in various spaces as available around the house (various containers, channels, pipes, etc).

As human population increases, the need for water increases in domestic, agricultural, industrial and urban sectors. Wastewater reuse after treatment is gaining acceptance worldwide, as availability of fresh water sources decreases. Reuse applications of wastewater in hydroponic cultivation can be adapted to the demand of high quality, high production and low waste of water and nutrients. A number of concerns still under investigation including the transmission risk of specific types of pathogens, and chemical content of treated wastewater. It was pointed out that there are two main types of hydroponic water and wastewater systems: (a) solution culture such as nutrient film technique (NFT), passive sub-irrigation, the Wick system, continuous drip, and top irrigation systems and (b) medium culture such as hydroponic plant growth systems, aeroponic growth system, flood and drain system, static solution culture system, and deep water culture system.

Chapter 25

The evolution of industrial water and industrial wastewater management in Lavrion, Attica, Hellas is an example of the merger and transition of technology from the domestic wastewater treatment. This Chapter confirms that ancient Hellenes had developed ingenious water and wastewater management techniques in the area of Lavrion, for the processing of ores and the production of lead and silver, sice the Archaic period. In this area intensive mining and metallurgical activities, over a time span of over 2700 years, have resulted in the production of metals, including silver and lead, the development of industrial processes and the evolution of civilization. Water played a major role in industrial activities as well as in daily life in Lavrion, where precipitation is usually very limited. Due to the scarcity of water, ores were treated in ancient times using hydromechanical techniques and several man-made ingenious water reservoirs and systems have been built to store rainwater and enable reuse of wastewater. Thus, the environmental problems in the ancient times, even after considering the low production rates and the not very efficient, by today's standards processing techniques, were very few. The situation has been deteriorated though after the 2nd World War when flotation, involving the use of chemical reagents, was introduced as ore beneficiation process. In this final period of Lavrion mining history water management was rather poor. Huge volumes of watery tailings were produced and disposed of as such first at the Thoricos beach and later at a tailings dam within the complex metallurgical plant area causing thus much more intense pollution problems. Furthermore, the environmental pressure was increased by the disposal of large volumes of metallurgical slags in coastal areas and the harbor as well as in other areas around Lavrion town.

Chapter 26

The legal systems for managing water have been used historically for wastewater as well. The Court of Waters of the valley of Valencia is the oldest existing institution of justice in Europe. Their organization dates from times of the Al-Andalus (X Century). It is formed by the receivers of the eight drains that take water from the Turia River. It is presided by one chosen among them. Although nowadays the drains only transport surface water, in the past they also received the wastewater from the nearby towns. Some characteristics of the Court are as follow: (a) it has authority on all drains of the valley of Valencia; (b) its receivers have been chosen democratically among the members of the community. Hence, the most honest and right members are looked for and iii) its members are not people of legal formation but they are aware of the law that has to be applied based on decrees they dominate. The proceeding of the Court is simple. All the demanded people are mentioned for the following Thursday. The case is set out or the plaintiff is presented. There are neither lawyers nor written document. All take part in their own name. However, the Court can ask the necessary questions to better information on the case. After that and in presence of the interested ones, the Court deliberates and sentences. There is no possibility for appealing the sentence since the Receiver of the drain is in charge of it. The Court of Water is the institution that has been utilized for the Valencian agricultural community to solve effectively their water problems from the most remote times.

27.7 SANITATION IN PRESENT TIMES AND FOR CITIES OF THE FUTURE

We currently live within an era termed the 'anthropocene' representing the human initiated transformation of the earth systems away from their natural state coined by Paul Crutzen and Eugene Stoemer (2000). This has affected our water systems unlike anytime in the history of humankind. In previous eras the urban water cycle was based on a linear concept elaborated by Roman engineer, adopted and further developed by English engineers in the 19th century. During the twentieth century water works and wastewater treatment facilities were added, but the concept remained the same: water is used just once, and discharged thereafter. This concept became the symbol of the so called 'end of the pipe technology' and the 'throw-away' mentality of modern societies. The inherited concept consisted of four components: abstraction and transport of fresh water towards the city using gravity flow canals and aqueducts (a), distribution of the water within the city limits (b), collection and transport of the used water together with road and roof runoff out of the city limits (c), and discharge of the collected water into the nearest river or into the open sea (d).

In the 21st century it is a human-coupled water cycle that requires attention with new concepts to better manage the resource and protect ecosystem services. On the global scale water resources are almost all affected by overuse and pollution. Climate change has added another layer of complexity. Based on the contemporary knowledge and technology water collection, storage, treatment and quality control can be addressed in an innovative and integrated manner for a sustainable water future.

First, the importance of 'healthy waters' is now understood. Wastewater treatment is being transformed toward resource recovery facilities. Water reuse is becoming globally acceptable and finally integrated water resource management is the framework touted for decision making.

27.7.1 Healthy waters understood

The risks associated with sewage are now better understood than ever before and the technology to monitor water quality is available to aid in determining all treatment needs for managing an urban water cycle. While our ancestors did not fully understand the treats to health due to impaired water quality, there was enough knowledge to dispose of wastes and stormwater generally downstream of the water supply. Despite this knowledge, then and now, waterborne disease throughout the world remains a concern and human domestic wastewater is an important source of contaminants impacting the aquatic environment and public health. This includes more than 150 types of biological contaminants of concern associated with human, animal and zoonotic diseases, antibiotics and antibiotic resistance, as well as other pharmaceuticals and personal health care products, which are all excreted or often dumped down the drain or toilet. Nutrient enrichment of waters due to nitrogen and phosphorous is also linked to wastewater. Routine wastewater treatment does not efficiently remove many of these contaminants and data on the 35 billion gallons of sewage produced each day even in the U.S. discharged to rivers are limited. Predominantly, wastewater treatment is focused on reducing the biological oxygen demanding substances (BOD), total suspended solids and on nitrification, with some disinfection only to meet a fecal coliform standard which is inadequate to address all the multitude of viral and parasitic pathogens, antibiotics, resistance genes and other pharmaceuticals. Characterization of the hundreds of emerging contaminants in wastewater and stormwater flows has not received adequate attention or monitoring and thus the scientific data base is poor. Associated biosolids, septic tank effluent and septage all contribute to the diffuse contaminant load when applied to land. While there are regulatory approaches for the use of biosolids, little is done to address septic tank solids and effluents. In addition sewage spills and combined sewage overflows may contribute 10 times more untreated sewage to the environment as climate changes and precipitation intensity increases.

The technology to remove biological and chemical contaminants has emerged over the last decade, but ancient waterborne pathogens such as cholera and typhoid still run rampant through parts of the world.

Recommendations for the future include improvement of the water quality data base for wastewater and return flows, use of novel infrastructure/technology in a multiple barrier approach for advancing wastewater treatment; new methods for disinfection of wastewater and stormwater and coupling wastewater quality to integrated water resource management goals for the coupled human water-cycle.

27.7.2 Resource recovery facilities

In ancient times human wastes were collected in small individual containers and disposed of nearby. In some communities the reuse of night soil was part of the culture recognizing that nutrients were valuable components of the waste stream. While dry sanitation is still practiced, water-based sanitation remains, in most urban communities, the medium for removing, carrying and dissolving wastes from the urban environment for treatment and disposal. Recovering this water and restoring it to high quality should be an important aspect of all facilities in the future. Wastewater contains nutrients that are of great value, but when disposed of to the water environment, water pollution is a global threat (Liu *et al.*, 2012). The need for fertilizers in modern agricultural practices to address global food security is also greatest where there is little wastewater infrastructure such as in most parts of Sub-Sahara-Africa and where soil fertility is poor (Knudson *et al.*, 2005).

The energy recovery from wastewater, including recovery of heat, is a modern day challenge. Mo & Zhang (2013) have reviewed the potential opportunities for a resource recovery system. In regard to energy this includes for larger scale systems production and use of biogas for heat and electricity; Incineration of biosolids (with a water content of less than 30%) used primarily in the larger urban areas requiring high capital investment; effluent hydropower in areas where the flow supports such a scheme. Except in the case of hydropower, it is essential that as much water as possible is recovered from the wastewater.

Future facilities will address recovery of water, nutrients and energy. Resource recovery may be scalable and while used in centralize larger systems, de-centralized systems may be more practical in the future. The use of membrane bioreactors produces a smaller footprint and facilitate resource recovery. However, innovative technology and engineering is needed. Retrofitting existing municipal treatment plants will be expensive. Aging wastewater treatment facilities as well as stormwater infrastructure in the US alone is estimated to require $298 billion over the next twenty years for restoration (http://www.infrastructurereportcard.org/wastewater/). The ability to transform the world as engineers of the past did will require investments in science, engineering and technology. This will be driven by technological advancements in research coupled with the capacity of the big water industry players and development of good policies by governments. It is this combination of efforts that will drive the momentum of resource recovery facilities for the benefits for all humankind.

In general, by using new concepts, technologies, and process configurations, it is possible to recover and utilize the energy content present in the wastewater in order to enhance the energy sustainability of wastewater treatment facilities. Based on 450 mg of COD/L of wastewater and 13.5 MJ/kg of COD, the energy available in wastewater on a COD basis was calculated to be (Tchobanoglous *et al.*, 2014):

$$E = (450 \text{ kg COD}/1000 \text{ m}^3)(13.5 \text{ MJ/kg COD}) = 6075 \text{ MJ}/1000 \text{ m}^3.$$

On the other hand, the energy required for secondary wastewater treatment usually ranges from 1,200 to 2400 MJ/1000 m^3. This means that energy available in wastewater is 2 to 4 times the amount required for its treatment. Also the total energy content in fresh solids (removed by a 200 to 350 fabric screen filter from raw wastewater) of syngas (H_2, CH_4, CO, and CO_2) was calculated as 67,81 MJ/5 kg of fresh solids with moisture content of approximately 15% (Gikas *et al.*,, 2010). Also the electrical energy input for the above fresh solids quantity was calculated as 27MJ, making the net energy production over 2.5 times the electrical energy impact to the system.

27.7.3 Potable water reuse

Water reuse is now being used more than any time in history (Jiminez and Asano, 2008). For the urban system high quality water to serve the potable needs has been found to be acceptable due to a number of advances. While inadvertent non-planned reuse has been going on since the beginning of time, the knowledge on emerging contaminants, new technology and engineering has made it possible to produce highly purified water for potable purposes. The modern era of planned potable reuse and some of the historical systems are shown in Figure 27.1. This began as early as the 1960s recognizing that more advanced treatment was needed for wastewaters discharged and impacting potable water supplies. However the first direct potable reuse system was established in Windhoek, Namibia. This was prior to advancements in membrane technology which now is used as the 'industry standard' for producing high quality drinking water from wastewater streams. Both microfiltration (MF) and reverse osmosis are now in use for both supplementing groundwater and surface reservoirs.

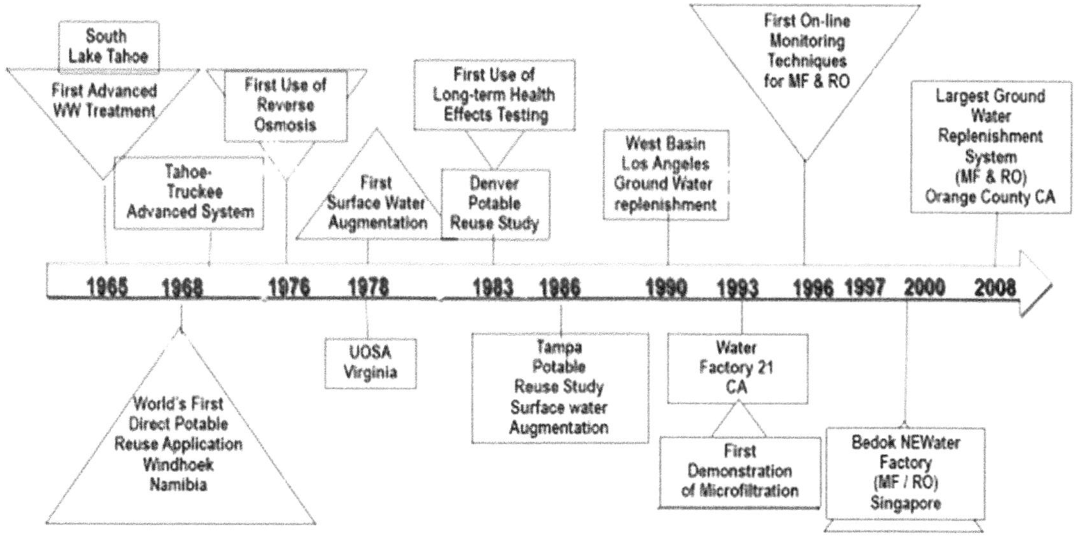

Figure 27.1 History of potable water reclamation.

The world's largest system is located in Orange County, California known as the Groundwater Replenishment System (GWRS). This is an off shoot of Factory 21 which was an advanced demonstration and research facility. Besides microfiltration, and reverse osmosis, ultraviolet radiation with hydrogen peroxide dosage is used for advanced disinfection (http://www.gwrsystem.com). The system in Singapore has had global recognition and there are now four NEWater facilities serving the city-state. (http://www.pub.gov.sg/about/historyfuture/Pages/NEWater.aspx). Research in optimizing the upstream processes, removal of emerging contaminants, advances in membrane technology, but primarily in real-time monitoring is needed.

Table 27.1 A Timeline for Historical Development of Sanitation and Wastewater Management (Adapted from De Feo *et al.*, 2014).

Period	Achievements	Comments
ca. 6500–3300 BC	First confirmed evidence of habitation and first farmers. One of the first successful efforts in urban water management was focused on stormwater drainage.	Oldest known at El Kown in present day Syria and probably in Crete and some Aegean islands
3300–2000 BC	Well-constructed vaulted storm drains and sanitary sewer systems were implemented. In Babylon clay pipes led to cesspools. The Babylonians, like other ancient civilizations, viewed uncleanliness as a taboo, not because of sanitation concerns but the moral evil it suggested.	In Mesopotamia.
ca. 2100–1100 BC	Sewerage and drainage systems became available in Crete, Hellas. Terracotta pipes were used to drain rood and road runoff to stone sewers, The first 'flushed' toilet was implemented at Knossos palace (Crete, Hellas).	Minoan setlements (e.g., Knossos, Zakros, Agia Triada, and Tylissos).
2600–1900 BC	Earliest toilets and sewerage and drainage systems were developed Mohenjo-Daro (Indus valley). Also, there is evidence of wastewater disposed to the agricultural lands.	In Mohenjo-daro city and other cities located in modern Pakistan.
ca. 1500–800 BC	The realization of the importance of clean water for people is evident from the myths of ancient cultures. Religious cleanliness and cleaning feet and hands with water became a must in various ancient cults.	In various Mycenaean cities (in south Hellas).
800–300 BC	The Etruscans build a road system with properly drained surfaces. In particular, they realized a coordinated and comprehensive plan for slopes of drainage channels on the sides of streets. They often constructed sewers and drains which were further developed and improved by the Romans.	Etruscans region (in central Italy).
After 770 BC	The first well organized drainage systems in temples and other religious structures appeared.	Early Dynasties in China and other areas of south-eastern Asia.
ca. 480–67 BC	Alcmaeon of Croton (floruit *ca.* 470 BC) was the first Hellenic doctor to state that the quality of water may influence the health of people. Also, Hippocratic treatise *Airs, Waters, Places* (*ca.* 400 BC) deals with the different sources, qualities and health effects of water. In the following, the importance of clean water for the public health was recognized and the first well organized baths, toilets, and sewerage and drainage systems appeared at different locations in the Mediterranen region. However, the first recorded epidemic of 430–426 BC, happened in Athens, causing the death of the great statesman, Pericles, which contributed significantly to the decline and fall of classical Hellas.	Hellas, Asia Minor, south Italy, and northern African states
ca. 67BC–330 AD	The importance of water for the public health was widely recognized in various parts of the world. The importance of urban sewerage and drainage became recognized. The first sewerage network system in Rome was installed and	

(*Continued*)

Table 27.1 A Timeline for Historical Development of Sanitation and Wastewater Management (Adapted from De Feo *et al.*, 2014) (*Continued*).

Period	Achievements	Comments
	connected to houses at *ca.* 100 AD. Public toilets facilities became common to all; they were cramped, however, without any privacy, and had no decent way of body cleaning. Private toilets usually lacked running water and they were commonly located near the kitchens. The rich had running water in their homes; the poor had to fetch their water from public fountains. Thus, water-borne infections were inevitable. Dysentery and different kinds of diarrheas were common.	In eastern Mediterranean, Egypt, north Africa (modern Tunisia), the Apennine peninsula (modern Italy).
ca. 330–1700 AD	With the advent of the Dark Ages in most parts of Europe water supply and sanitation lost attention. Little progress was made for 1400 years, from the late forth century on. The lack of proper sanitation increased the outbreak of epidemics in medieval towns in Europe. However, baths, toilets, sewerage and drainage systems were further improved by Byzantines and Venetians. Also at this time several Asian countries (e.g., China, India, and Vietnam) implemented various types of drainage systems in the religious temples developed under several dynasties.	The transition from the ancient world is not certain.
ca. 1700–1700 AD	The great epidemics of cholera and typhoid fever occurred in England during 1830–1850 and their association with the pollution of water sources with raw wastewater made clear the need for household sanitation, interruptingdirect contact of humans with their own feces by means of underground gravity sewers and purifying the collected wastewater prior to discharge. The need for protection of water resources motivated health agencies to set sanitation rules and environmental policies to protect public health. The first large-scale wastewater treatment systems were set at the beginning of 1800s when 'sewage farms' were developed. This practice was mainly developed from 1840 to 1890 in England, while in the 1870s the first systems appeared in the United States, France, Germany, Hellas, and other European countries.	After the beginning of the Industrial revolution.
1900–1990 AD	The basic treatment processes were invented and constructed in the developed world. For instance, in 1914 the first activated sludge treatment plant went in operation in Manchester, UK. Towards the end of the 20th century membrane separation technology started to replaced gravity sedimentation tanks opening up a new horizon of wastewater treatment and reuse. At the same time the first decentralized wastewater treatment and reuse systems appeared, particularly in high-rise buildings in Japan. Contemporary process technology allows treatment of sewage up to drinking water quality. Even when the water is not used for human consumption it can be safely used for groundwater recharge (e.g., demonstrated at Orange County, California) or as process water in industry (e.g., in Singapore).	

27.7.4 Integrated water resource management (IWRM)

IWRM has been defined over the last several decades 'as a process which promotes the coordinated development and management of water, land and related resources, in order to maximize the resultant economic and social welfare in an equitable manner without compromising the sustainability of vital ecosystems' (Global Water Partnership). The competition for water along with climate change has created a critical need for what is seen as a paradigm shift.

The United Nations and others have moved forward with a key recommendation that IWRM be a central dogma for the future management of water systems (UN Water Status report, 2012). Even though 80% of the countries have embraced the IWRM framework, it is difficult to pinpoint the scale at which implementation has taken place, yet in some cases there has been a reported development in advancing infrastructure as a result of adopting IWRM plans with appropriate financing. However revenue is often not keeping abreast of the needs.

However, IWRM has directed investment in wastewater treatment and flood control, improving water quality. This improvement has been essential to the maintenance of ecosystem services. One of the key needs in the 21st century as a result of all the technology and ability to monitor the water systems is the concurrent management of information. Climate, hydro, bio informatics systems are critical for building a knowledge management system. Establishing national water resources information systems will assist ultimately in moving toward sustainability.

27.8 EPILOGUE

This book provides solid information about the technical methods applied in the past to overcome shortcomings of water demands and sanitation in urban areas. The remaining question is, however, how ancient civilization organized planning, construction and operation of the water structures required to safeguard the wellbeing of agricultural and urban societies.

The authors of the individual chapters describe very elaborately the results of construction work such as aqueducts, pipings, toilets, sewers, just to name a few. But how were actions planned and executed leading to such structures, taking into account the limited scientifically based knowledge available at the time? In response to the ongoing and soaring progress of climate change, urbanization and globalization of consumer habits it appears tremendously necessary to revisit the inherited methods of managing water resources.

As today, in ancient times it was necessary to plan and construct reservoirs to collect often irregularly appearing precipitation in arid areas. It was necessary to build levees to protect agricultural and urban areas from occasional flooding. Canals and aqueduct had to be designed and built, as well as water distribution systems and storm water collection and transport facilities. Moreover, the manpower had to be recruited to accomplish large water structures with primitive tools and within a limited time period. During the construction the workers had to be supported with expertise, food, water and medical care. And after the structures were put into operation it was necessary to manage and control their performance.

In several ancient civilizations, storm water and wastewater public works, are characterized by simplicity, robustness of operation, and absence of complex controls. In the future, stormwater and wastewater management systems based on reapplication of old practices and philosophical approaches, using new equipment, in order to effectively meet the modern emerging challenges could be of great significance.

Finally, it is obvious that ideas, technologies, and practices developed during several periods of ancient civilizations (e.g., Minoans, Indus valley, Chinese, Hellenes, and Romans) greatly influenced our current knowledge. Today, 1.1 billion people lack access to safe drinking water, 2.6 billion people lack adequate sanitation, 1.8 million people die every year from diarrheal diseases, including 90 % of children under 5

(Bond *et al.,*, 2013). This situation is no longer bearable and there is an extraordinary need for sustainable and cost-effective water supply and sanitation facilities, particularly in cities of the developing world. Stormwater also needs to be addressed. In a list (of 182 floods) consisting of the deadliest floods worldwide with a minimum of 50 deaths there were significant disease risks and large economic damages (King, 2013). Direct flood damages annual costs from 1905 to 2012 years ranged from 39 to 29,246 million $ (http://www.nws.noaa.gov/hic/). All indicators suggest that there will be an increase of such events due to the increase in urbanization and climatic variability (http://en.wikipedia.org/wiki/List_of_deadliest_floods). Thus, applicability of selected ancient water techniques for the contemporary developing world should be seriously considered.

27.9 REFERENCES

Bond T., Roma E., Foxon K. M., Templeton M. R. and Buckley C. A. (2013). Ancient water and sanitation systems – applicability for the contemporary urban developing world. *Water Scienice and Technology*, **67**(5), 935–41.

Crutzen P. J. and Stoemer E. F. (2000). The 'Anthropocene'. *IGBP Newsletter*, **41**, 17–18.

De Feo, G. Antoniou, G. P., Mays, L., Dragoni, W., Fardin, H. F., El-Gohary, F., Laureano, P., Kanetaki, E. I., Zheng, X. Y. and Angelakis, A. N., 2014. Historical Development of Waste Water Management. In: Handbook of Engineering Hydrology, S. Eslamian (ed.), Taylor & Francis Group, Boca Raton, FL, USA, Vol. 2, Ch. 9: 163–217.

Gikas P., Noll S. A., Stedman K. and Tchobanoglous G. (2010). The charactirization and testing of wastewer fine screen fresh solids for suitability as gazification feedstock. In 2nd Intrern. Conference on Hazardous and Industrial Waste Management, October 5–8, 2010, Chania, Hellas.

Jimenez B. and Asano T. (eds) (2008). Water Reuse: an International Survey of Current Practices, Issues and Needs. Scientific and Technical Report Series No. 20, IWA Publishing, London, UK.

King R. O. (2013). The National Flood Insurance Program: Status and Remaining Issues for Congress. CRS Report for Congress, Congressional Research Service, Washington, USA.

Knudsen M. T., Halberg N., Olesen J. E., Byrne J., Iyer V. and Toly N. (2005). Global trends in agriculture and food systems. In: Global Development of Organic Agriculture: Challenges and Promises, N. Halberg, H. F. Alrøe, M. T. Knudsen and E. S. Kristensen (eds), CAB International, Wallingford, UK.

Liu C., Kroezea C, Hoekstrac A. Y. and Gerbens-Leenesc W. (2012). Past and future trends in grey water footprints of anthropogenic nitrogen and phosphorus inputs to major world rivers. *Ecological Indicators*, **18**, 42–49

Lofrano G. and Brown J. (2010). Wastewater management through the ages: a history of mankind. *Science Total Environment*, **408**(22), 5254–5264.

Mo W. and Zhang Q. (2013). Review: energy nutrients water nexus: Integrated resource recovery in municipal wastewater treatment plants. *Journal of Environmental Management*, **127**, 255–267.

Tchobanoglou G., Stensel H. D., Tsuchihashi R. and Burton F. L. (2014). Wastewater Engineering: Treatment and Resource Recovery. 5th edn, Metcalf & Eddy I AECOM, McGraw-Hill Book Company, New York, USA.

United Nations Environment Programme (2012). Status Report on The Application of Integrated Approaches to Water Resources Management. ISBN: 978-92-807-3264-1

Vuorinen H. S., Juuti P. S. and Katko T. S. (2007). History of water and Health from ancient civilizations to modern times. *Water Science and Technology*, **7**, 49–57.

IWA Publishing's authorised EU representative for General Product Safety Regulations is Diane D'Arras, 15 rue Duret, 75116 Paris, France, e-mail: safety@iwap.co.uk.

Printed and bound by CPI Group (UK) Ltd, Croydon, CR0 4YY
12/05/2026
02108870-0016